MAGILL'S ENCYCLOPEDIA OF SCIENCE

PLANT LIFE

MAGILL'S ENCYCLOPEDIA OF SCIENCE

PLANT LIFE

Volume 1

Acid Precipitation–DNA: Recombinant Technology

Editor
Bryan D. Ness, Ph.D.
Pacific Union College, Department of Biology

Project Editor
Christina J. Moose

SALEM PRESS, INC.
Pasadena, California
Hackensack, New Jersey

Some of the updated and revised essays in this work originally appeared in *Magill's Survey of Science: Life Science* (1991), *Magill's Survey of Science: Life Science, Supplement* (1998), *Natural Resources* (1998), *Encyclopedia of Genetics* (1999), *Encyclopedia of Environmental Issues* (2000), *World Geography* (2001), and *Earth Science* (2001).

∞ The paper used in these volumes conforms to the American National Standard for Permanence of Paper for Printed Library Materials, Z39.48-1992 (R1997).

Library of Congress Cataloging-in-Publication Data

Magill's encyclopedia of science : plant life / edited by Bryan D. Ness.

p. cm.
Includes bibliographical references (p.).
ISBN 1-58765-084-3 (set : alk. paper) — ISBN 1-58765-085-1 (vol. 1 : alk. paper) —
ISBN 1-58765-086-X (vol. 2 : alk. paper) — ISBN 1-58765-087-8 (vol. 3 : alk. paper) —
ISBN 1-58765-088-6 (vol. 4 : alk. paper)
1. Botany—Encyclopedias. I. Ness, Bryan D.
QK7 .M34 2002
580′.3—dc21

2002013319

First Printing

PRINTED IN THE UNITED STATES OF AMERICA

TABLE OF CONTENTS

PUBLISHER'S NOTE

Magill's Encyclopedia of Science: Plant Life is designed to meet the needs of college and high school students as well as nonspecialists seeking general information about botany and related sciences. The definition of "plant life" is quite broad, covering the range from molecular to macro topics: the basics of cell structure and function, genetic and photosynthetic processes, evolution, systematics and classification, ecology and environmental issues, and those forms of life—archaea, bacteria, algae, and fungi—that, in addition to plants, are traditionally studied in introductory botany courses. A number of practical and issue-oriented topics are covered as well, from agricultural, economic, medicinal, and cultural uses of plants to biomes, plant-related environmental issues, and the flora of major regions of the world. (Readers should note that, although cultural and medicinal uses of plants are occasionally addressed, this encyclopedia is intended for broad information and educational purposes. Those interested in the use of plants to achieve nutritive or medicinal benefits should consult a physician.)

Altogether, the four volumes of *Plant Life* survey 379 topics, alphabetically arranged from *Acid precipitation* to *Zygomycetes*. For this publication, 196 essays have been newly acquired, and 183 essays are previously published essays whose contents were reviewed and deemed important to include as core topics. The latter group originally appeared in the following Salem publications: *Magill's Survey of Science: Life Science* (1991), *Magill's Survey of Science: Life Science, Supplement* (1998), *Natural Resources* (1998), *Encyclopedia of Genetics* (1999), *Encyclopedia of Environmental Issues* (2000), *World Geography* (2001), and *Earth Science* (2001). All of these previously published essays have been thoroughly scrutinized and updated by the set's editors. In addition to updating the text, the editors have added new bibliographies at the ends of all articles.

New appendices, providing essential research tools for students, have been acquired as well:

- a "Biographical List of Botanists" with brief descriptions of the contributions of 134 famous naturalists, botanists, and other plant scientists
- a Plant Classification table
- a Plant Names appendix, alphabetized by common name with scientific equivalents
- another Plant Names appendix, alphabetized by scientific name with common equivalents
- a "Time Line" of advancements in plant science (a discursive textual history is also provided in the encyclopedia-proper)
- a Glossary of 1,160 terms
- a Bibliography, organized by category of research
- a list of authoritative Web sites with their sponsors, URLs, and descriptions

Every essay is signed by the botanist, biologist, or other expert who wrote it; where essays have been revised or updated, the name of the updater appears as well. In the tradition of Magill reference, each essay is offered in a standard format that allows readers to predict the location of core information and to skim for topics of interest: The title of each article lists the topic as it is most likely to be looked up by students; the "Category" line indicates pertinent scientific subdiscipline(s) or area(s) of research; and a capsule "Definition" of the topic follows. Numerous subheads guide the reader through the text; moreover, key concepts are italicized throughout. These features are designed to help students navigate the text and identify passages of interest in context. At the end of each essay is an annotated list of "Sources for Further Study": print resources, accessible through most libraries, for additional information. (Web sites are reserved for their own appendix at the end of volume 4.) A "See also" section closes every essay and refers readers to related essays in the set, thereby linking topics that, together, form a larger picture. For example, since all components of the plant cell are covered in detail in separate entries (from the *Cell wall* through *Vacuoles*), the "See also" sections for these dozen or so essays list all other essays covering parts of the cell as well as any other topics of interest.

Approximately 150 charts, sidebars, maps, tables, diagrams, graphs, and labeled line drawings offer the essential visual content so important to students of the sciences, illustrating such core concepts as the parts of a plant cell, the replication of DNA, the phases of mitosis and meiosis, the world's most important crops by region, the parts of a flower, major types of inflorescence, or different classifications of fruits and their characteristics. In addition, nearly 200 black-and-white photographs appear throughout the text and are captioned to offer examples of the important phyla of plants, parts of plants, biomes of plants, and processes of plants: from bromeliads to horsetails to wheat; from Arctic tundra to rain forests; from anthers to stems to roots; from carnivorous plants to tropisms.

Reference aids are carefully designed to allow easy access to the information in a variety of modes: The front matter to each of the four volumes includes the volume's contents, followed by a full "Alphabetical List of Contents" (of all the volumes). All four volumes include a "List of Illustrations, Charts, and Tables," alphabetized by key term, to allow readers to locate pages with (for example) a picture of the apparatus used in the *Miller-Urey Experiment*, a chart demonstrating the genetic offspring of *Mendel's Pea Plants*, a map showing the world's major zones of *Desertification*, a cross-section of *Flower Parts*, or a sampling of the many types of *Leaf Margins*. At the end of volume 4 is a "Categorized Index" of the essays, organized by scientific subdiscipline; a "Biographical Index," which provides both a list of famous personages and access to discussions in which they figure prominently; and a comprehensive "Subject Index" including not only the personages but also the core concepts, topics, and terms discussed throughout these volumes.

Reference works such as *Magill's Encyclopedia of Science: Plant Life* would not be possible without the help of experts in botany, ecology, environmental, cellular, biological, and other life sciences; the names of these individuals, along with their academic affiliations, appear in the front matter to volume 1. We are particularly grateful to the project's editor, Bryan Ness, Ph.D., Professor of Biology at Pacific Union College in Angwin, California. Dr. Ness was tireless in helping to ensure thorough, accurate, and up-to-date coverage of the content, which reflects the most current scientific knowledge. He guided the use of commonly accepted terminology when describing plant life processes, helping to make *Magill's Encyclopedia of Science: Plant Life* easy for readers to use for reference to complement the standard biology texts.

CONTRIBUTOR LIST

About the Editor: Bryan Ness is a full professor in the Department of Biology at Pacific Union College, a four-year liberal arts college located atop Howell Mountain in the Napa Valley, about ninety miles north of San Francisco. He received his Ph.D. in Botany from Washington State University. His doctoral work focused on molecular plant systematics and evolution. In addition to authoring or coauthoring a number of scientific papers, he has contributed to *The Jepson Manual: Higher Plants of California*, *The Flora of North America*, a multivolume guide to the higher plants of North America, and more than a dozen articles for various Salem Press publications, including *Aging*, *Encyclopedia of Genetics*, *Magill's Encyclopedia of Science: Animal Life*, *Magill's Medical Guide*, and *World Geography*. For four years he managed The Botany Site, a popular Internet site on botany. He is currently working on a book about plant myths and misunderstandings.

Stephen R. Addison
University of Central Arkansas

Richard Adler
University of Michigan, Dearborn

Steve K. Alexander
University of Mary Hardin-Baylor

Michael C. Amspoker
Westminster College

Michele Arduengo
Independent Scholar

Richard W. Arnseth
Science Applications International

J. Craig Bailey
University of North Carolina, Wilmington

Anita Baker-Blocker
Applied Meteorological Services

Iona C. Baldridge
Lubbock Christian University

Richard Beckwitt
Framingham State College

Cindy Bennington
Stetson University

Alvin K. Benson
Utah Valley State College

Margaret F. Boorstein
C. W. Post College of Long Island University

P. E. Bostick
Kennesaw State College

J. Bradshaw-Rouse
Independent Scholar

Thomas M. Brennan
Dickinson College

Alan Brown
Livingston University

Kenneth H. Brown
Northwestern Oklahoma State University

Bruce Brunton
James Madison University

Pat Calie
Eastern Kentucky University

James J. Campanella
Montclair State University

William J. Campbell
Louisiana Tech University

Steven D. Carey
University of Mobile

Roger V. Carlson
Jet Propulsion Laboratory

Robert E. Carver
University of Georgia

Richard W. Cheney, Jr.
Christopher Newport University

John C. Clausz
Carroll College

Miriam Colella
Lehman College

William B. Cook
Midwestern State University

J. A. Cooper
Independent Scholar

Alan D. Copsey
Central University of Iowa

Joyce A. Corban
Wright State University

Mark S. Coyne
University of Kentucky

Stephen S. Daggett
Avila College

William A. Dando
Indiana State University

James T. Dawson
Pittsburg State University

Albert B. Dickas
University of Wisconsin

Gordon Neal Diem
ADVANCE Education and Development Institute

David M. Diggs
Central Missouri State University

John P. DiVincenzo
Middle Tennessee State University

Gary E. Dolph
Indiana University, Kokomo

Allan P. Drew
SUNY, College of Environmental Science and Forestry

Frank N. Egerton
University of Wisconsin, Parkside

Jessica O. Ellison
Clarkson University

Cheryld L. Emmons
Alfred University

Frederick B. Essig
University of South Florida

Danilo D. Fernando
SUNY, College of Environmental Science and Forestry

Mary C. Fields
Medical University of South Carolina

Randy Firstman
College of the Sequoias

Roberto Garza
San Antonio College

Ray P. Gerber
Saint Joseph's College

Soraya Ghayourmanesh
Independent Scholar

Carol Ann Gillespie
Grove City College

Nancy M. Gordon
Independent Scholar

D. R. Gossett
Louisiana State University, Shreveport

Hans G. Graetzer
South Dakota State University

Jerry E. Green
Miami University

Joyce M. Hardin
Hendrix College

Linda Hart
University of Wisconsin, Madison

Thomas E. Hemmerly
Middle Tennessee State University

Jerald D. Hendrix
Kennesaw State College

John S. Heywood
Southwest Missouri State University

Jane F. Hill
Independent Scholar

Joseph W. Hinton
Independent Scholar

Carl W. Hoagstrom
Ohio Northern University

Virginia L. Hodges
Northeast State Technical Community College

David Wason Hollar, Jr.
Rockingham Community College

Howard L. Hosick
Washington State University

Kelly Howard
Independent Scholar

John L. Howland
Bowdoin College

M. E. S. Hudspeth
Northern Illinois University

Samuel F. Huffman
University of Wisconsin, River Falls

Diane White Husic
East Stroudsburg University

Domingo M. Jariel
Louisiana State University, Eunice

Karen N. Kähler
Independent Scholar

Sophien Kamoun
Ohio State University

Manjit S. Kang
Louisiana State University

Susan J. Karcher
Purdue University

Jon E. Keeley
Occidental College

Leigh Husband Kimmel
Independent Scholar

Samuel V. A. Kisseadoo
Hampton University

Kenneth M. Klemow
Wilkes University

Jeffrey A. Knight
Mount Holyoke College

Denise Knotwell
Independent Scholar

James Knotwell
Wayne State College

Lisa A. Lambert
Chatham College

Craig R. Landgren
Middlebury College

Contributor List

John C. Landolt
Shepherd College

David M. Lawrence
John Tyler Community College

Mary Lee S. Ledbetter
College of the Holy Cross

Donald H. Les
University of Connecticut

Larry J. Littlefield
Oklahoma State University

John F. Logue
University of South Carolina, Sumter

Alina C. Lopo
University of California, Riverside

Yiqi Luo
University of Oklahoma

Fai Ma
University of California, Berkeley

Jinshuang Ma
Arnold Arboretum of Harvard University Herbaria

Zhong Ma
Pennsylvania State University

Dana P. McDermott
Independent Scholar

Paul Madden
Hardin-Simmons University

Lois N. Magner
Purdue University

Lawrence K. Magrath
University of Science and Arts of Oklahoma

Nancy Farm Männikkö
Independent Scholar

Sergei A. Markov
Marshall University

John S. Mecham
Texas Tech University

Roger D. Meicenheimer
Miami University, Ohio

Ulrich Melcher
Oklahoma State University

Iain Miller
Wright State University

Jeannie P. Miller
Texas A&M University

Randall L. Milstein
Oregon State University

Eli C. Minkoff
Bates College

Richard F. Modlin
University of Alabama, Huntsville

Thomas J. Montagno
Simmons College

Thomas C. Moon
California University of Pennsylvania

Randy Moore
Wright State University

Christina J. Moose
Independent Scholar

J. J. Muchovej
Florida A&M University

M. Mustoe
Independent Scholar

Jennifer Leigh Myka
Brescia University

Mysore Narayanan
Miami University

Bryan Ness
Pacific Union College

Brian J. Nichelson
U.S. Air Force Academy

Margaret A. Olney
Colorado College

Oghenekome U. Onokpise
Florida A&M University

Oluwatoyin O. Osunsanya
Muskingum College

Henry R. Owen
Eastern Illinois University

Robert J. Paradowski
Rochester Institute of Technology

Bimal K. Paul
Kansas State University

Robert W. Paul
St. Mary's College of Maryland

Kenneth A. Pidcock
Wilkes College

Rex D. Pieper
New Mexico State University

George R. Plitnik
Frostburg State University

Bernard Possidente, Jr.
Skidmore College

Carol S. Radford
Maryville University, St. Louis

V. Raghavan
Ohio State University

Ronald J. Raven
State University of New York at Buffalo

Darrell L. Ray
University of Tennessee, Martin

Judith O. Rebach
University of Maryland, Eastern Shore

David D. Reed
Michigan Technological University

Mariana Louise Rhoades
St. John Fisher College

Connie Rizzo
Pace University

Harry Roy
Rensselaer Polytechnic Institute

David W. Rudge
Western Michigan University

Neil E. Salisbury
University of Oklahoma

Helen Salmon
University of Guelph

Lisa M. Sardinia
Pacific University

Elizabeth D. Schafer
Independent Scholar

David M. Schlom
California State University, Chico

Matthew M. Schmidt
SUNY, Empire State College

Harold J. Schreier
University of Maryland

John Richard Schrock
Emporia State University

Guofan Shao
Purdue University

Jon P. Shoemaker
University of Kentucky

John P. Shontz
Grand Valley State University

Nancy N. Shontz
Grand Valley State University

Beryl B. Simpson
University of Texas

Sanford S. Singer
University of Dayton

Susan R. Singer
Carleton College

Robert A. Sinnott
Arizona Agribusiness and Equine Center

Elizabeth Slocum
Independent Scholar

Dwight G. Smith
Connecticut State University

Roger Smith
Independent Scholar

Douglas E. Soltis
University of Florida

Pamela S. Soltis
Independent Scholar

F. Christopher Sowers
Wilkes Community College

Alistair Sponsel
Imperial College

Valerie M. Sponsel
University of Texas, San Antonio

Steven L. Stephenson
Fairmont State College

Dion Stewart
Adams State College

Toby R. Stewart
Independent Scholar

Ray Stross
State University of New York at Albany

Susan Moyle Studlar
West Virginia University

Ray Sumner
Long Beach City College

Marshall D. Sundberg
Emporia State University

Frederick M. Surowiec
Independent Scholar

Paris Svoronos
Queen's College, City University of New York

Charles L. Vigue
University of New Haven

James Waddell
University of Minnesota, Waseca

William J. Wasserman
Seattle Central Community College

Robert J. Wells
Society for Technical Communication

Yujia Weng
Northwest Plant Breeding Company

P. Gary White
Western Carolina University

Thomas A. Wikle
Oklahoma State University

Donald Andrew Wiley
Anne Arundel Community College

Robert R. Wise
University of Wisconsin, Oshkosh

Stephen L. Wolfe
University of California, Davis

Ming Y. Zheng
Gordon College

LIST OF ILLUSTRATIONS, CHARTS, AND TABLES

Volume-page

ALPHABETICAL LIST OF CONTENTS

Volume 1

Volume 2

Volume 3

Volume 4

MAGILL'S ENCYCLOPEDIA OF SCIENCE

PLANT LIFE

ACID PRECIPITATION

Categories: Environmental issues; pollution

Acid precipitation is rain, snow, or mist which has a pH lower than unpolluted precipitation. Increased levels of acid precipitation have significant effects on food chains and ecosystems.

Precipitation—rain, snow, hail, sleet, or mist—is naturally acidified by carbonic acid (H_2CO_3). Carbon dioxide (CO_2) in the atmosphere reacts with water molecules, lowering the pH of precipitation to 5.6. A pH scale is used to measure a solution's acidity or alkalinity; pH is defined as the negative logarithm of the concentration of hydrogen ions, H^+. A solution with a pH of 7.0 is neutral. A pH lower than 7 is acidic, and a pH greater than 7 is alkaline. Other acidic substances are also present in the atmosphere, causing "unpolluted" precipitation to have a pH approaching 5.0. Solutions with a pH of 5.0 or less have concentrations of hydroxyl ion, or OH^-, and carbonate ion, or CO_3^-, approaching zero.

Acid precipitation is the name given to rain or snow contaminated with oxides of sulfur (SO_x) and oxides of nitrogen (NO_x). These chemicals combine with water droplets to form sulfuric acid and nitric acid. SO_x is formed by combustion of materials containing sulfur, and NO_x is formed by oxidation of molecular nitrogen in the atmosphere during combustion. SO_x sometimes arises from natural sources such as volcanoes and geyser fields, and NO_x is formed by lightning. Downwind of smelting facilities, hydrochloric acid (HCl) and hydrofluoric acid (HF) may also contribute to acid precipitation. Acid precipitation may detrimentally change soil chemistry, either by stripping nutrients, especially magnesium and calcium, or mobilizing *phytotoxic trace elements* (elements toxic to plants).

Geographic Extent of Damage

Acid precipitation is a regional problem. SO_x and NO_x can travel many thousands of kilometers in the atmosphere after being emitted by large, stationary sources, especially those that have very high smokestacks. These pollutants are slowly transformed into sulfuric and nitric acid aerosols and are incorporated into precipitation, which eventually makes contact with the earth's surface. Acid precipitation in the eastern United States contains more SO_x than precipitation in the western United States, which contains more NO_x.

In North America, acid precipitation and *dry deposition* (of acid aerosol particles) are major environmental problems in New England and New York State and in Ontario and Quebec. These regions attribute much of their acid precipitation to emissions from large coal-burning plants in the American Ohio Valley. Scandinavian activists blame coal-burning power plants and factory emissions in the British Isles for that region's acid rain problems. Central Europe—including Poland, the Czech Republic, Slovakia, and eastern Germany—has many power plants and factories that burn high-sulfur coal. Acid-laden pollution plumes stretch thousands of kilometers downwind from smokestacks in that region.

Controlled Studies

Controlled experiments on individual plant species have revealed short-term damage to a limited number of those species. Experiments using *simulated acid rain* (SAR) are difficult to extrapolate to field conditions, where the specific pollutants and pH levels vary widely over time. In controlled conditions, studies showed no link between SAR and yield in Amsoy soybeans. However, field studies demonstrated that acid deposition does decrease yield in Amsoy soybeans.

Acid precipitation influences plant diseases by acting on both pathogens and host organisms. Seedlings of *Pinus rigida*, *Pinus echinata*, *Pinus taeda*, and *Pinus strobus* exposed to SAR of pH 3.0 had a 100 percent mortality rate because of fungal *damping-off*, a diseased condition of seedlings marked by wilting or rotting. Red spruce seedlings subjected to dilute sulfuric acid mist developed brown lesions on their needles, followed by needle drop. Studies showed a reduction in the growth of sugar maple seedlings following exposure to low-pH moisture, and that seedling survival decreased with increasing acidity.

Crop and Forest Decline

In field experiments, soybeans have shown reduced yields with decreasing pH (increasing acidity) of moisture applied. Yields of seed and seed protein are both reduced in soybeans exposed to high acidity. A lower number of seed pods were found in plants exposed to high acidity, compared to control plants.

Acid precipitation causes detrimental long-term effects in most ecosystems, especially forests. Root systems under acidic stress show great variability in tolerance and injury. Acidic stress on roots decreases root growth, measured by a reduction in root length, and severely damaged trees have more fine roots with opaque tip zones than do slightly damaged trees. Some scientists have suggested that the radical growth rate in yellow pines in the southeastern United States may be reduced by acid precipitation.

PhotoDisc

Activists blame coal-burning power plants and factory emissions for acid rain problems. After being emitted by large, stationary sources, especially those that have very high smokestacks, pollutants can travel thousands of kilometers in the atmosphere. Those that are transformed into sulfuric and nitric acid aerosols are incorporated into precipitation, which eventually makes contact with the earth's surface.

Since the 1960's Central European soils have been progressively acidified, altering soil buffering capacities. Acid rain containing nitrates (which are not immobilized in soil) played an important role in this soil acidification. Acidification has reduced the magnesium, calcium, and potassium available for nutrient uptake by plants and has affected root growth. One-quarter of European forests are moderately or severely damaged by acid precipitation, with dry deposition believed (by scientists and politicians) to be largely responsible. This pattern of damage, first detected in the 1980's, has been called *neuartige Waldschäden* (literally, "new-type forest decline"). It has been detected throughout Central Europe at all elevations and on all soil types. *Waldschäden* is most pronounced downwind of major air pollution sources.

Abnormally high numbers of red spruce have died in the high-elevation northern Appalachian Mountains since the 1960's. This die-off has been attributed to high rates of acid deposition (up to 4 kiloequivalents of hydronium ions per hectare per year) and exposure to acid fog droplets for up to two thousand hours per year. Very high levels of trace metals (known to be phytotoxic) have accumulated in the region.

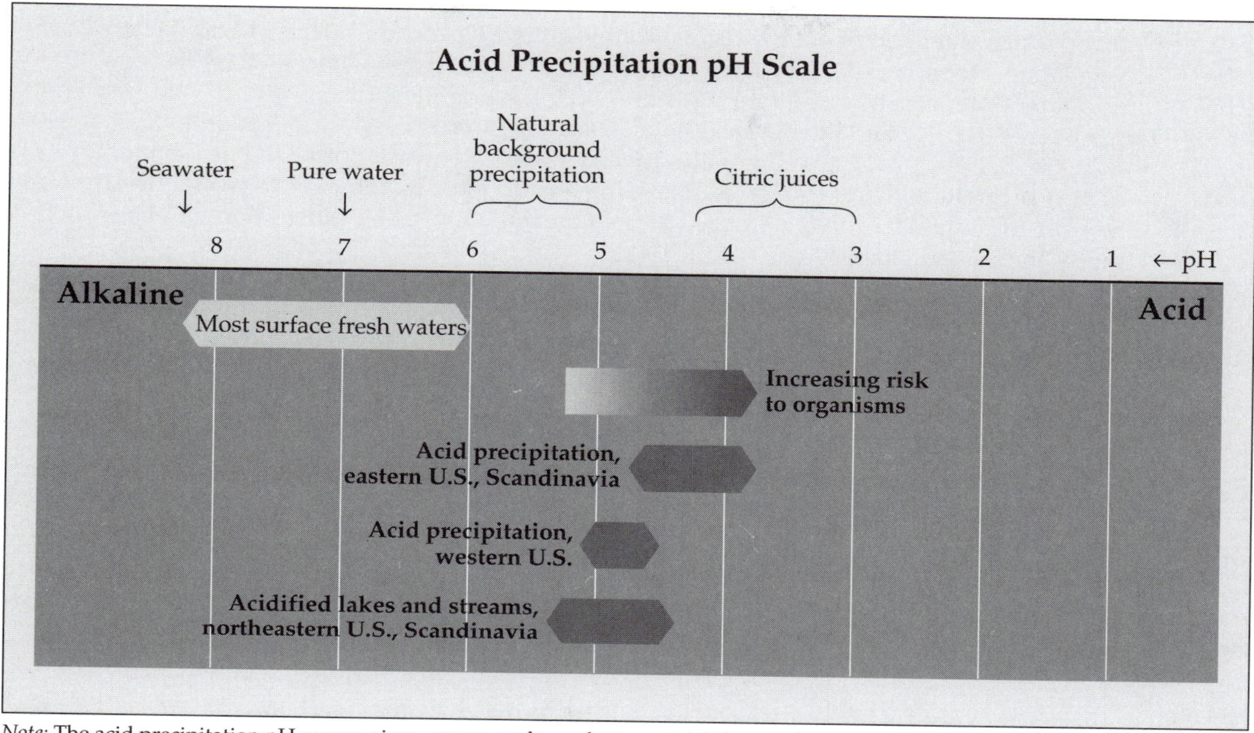

Note: The acid precipitation pH ranges given correspond to volume-weighted annual averages of weekly samples.
Source: Adapted from John Harte, "Acid Rain," in *The Energy-Environment Connection*, edited by Jack M. Hollander, 1992.

Aquatic Ecosystems

Most freshwater ecosystems range from pH 6.0 to pH 8.0. In limestone terrain, acid precipitation is neutralized by dissolution of calcium carbonate. As a freshwater environment becomes acidified, the number of species it supports declines. When conditions are more acidic than pH 5.5, dissolved inorganic carbon exists only as dissolved carbon dioxide. Planktonic algae, which can use low levels of dissolved inorganic carbon, are favored in these environments. Acid environments greatly reduce the numbers of herbivores that graze on aquatic plants; this is thought to explain why filamentous green algae are found in most acidified lakes. Scientists point out that it is difficult to separate the effects that acidification alone produces in an aquatic ecosystem.

Anita Baker-Blocker

See also: Air pollution; Old-growth forests; Root uptake systems.

Sources for Further Study

Adriano, D. C., and A. H. Johnson, eds. *Biological and Ecological Effects*. Vol. 2 in *Acidic Precipitation*. New York: Springer-Verlag, 1989. Extremely detailed volume contains individual chapters by different authors, devoted to the effects of acid precipitation on trees, crops, and freshwater biota. Reviews damage to North American and European forests. Each chapter offers an extensive bibliography of acid precipitation research relevant to the topic. Includes many charts and diagrams, fully indexed.

Ahrens, C. Donald. *Meteorology Today*. 6th ed. Pacific Grove, Calif.: Brooks/Cole, 2000. College meteorology text reviews the sources of acid precipitation and where its major effects occur. Includes illustrations, bibliography, index, and glossary.

Canter, Larry W. *Acid Rain and Dry Deposition*. Chelsea, Mich.: Lewis, 1989. About sixty pages of this book are devoted to geographic extent of acid rain and the chemical transformations and transport of pollutants in the atmosphere. Covers the effects of acid rain on vegetation, crops, and floral species. It is a guide to research done through the 1980's on acid rain. Comprehensive literature review and bibliography; index.

Lutgens, Frederick K., and Edward J. Tarbuck. *The Atmosphere.* 8th ed. Upper Saddle River, N.J.: Prentice Hall, 2001. Excellent introductory meteorology text with good description of acid precipitation and its effects, including a color photograph of trees in the Great Smokey Mountains severely damaged by acid precipitation. Color illustrations, bibliography, index, and glossary.

ACTIVE TRANSPORT

Categories: Cellular biology; physiology; transport mechanisms

Active transport is the process by which cells expend energy to move atoms or molecules across membranes, requiring the presence of a protein carrier, which is activated by ATP. Cotransport is active transport that uses a carrier that must simultaneously transport two substances in the same direction. Countertransport is active transport that employs a carrier that must transport two substances in opposite directions at the same time.

Biologists in nearly every field of study have discovered that one of the major methods by which organisms regulate their metabolisms is by controlling the movement of molecules into cells or into organelles such as the nucleus. This regulation is possible because of the semipermeable nature of *cellular membranes*. The membranes of all living cells are fluid mosaic structures composed primarily of lipids and proteins. The lipid molecules are *aliphatic*, which means that their molecular structure exhibits both a hydrophilic (water-attracted) and a hydrophobic (water-repelling) portion. These aliphatic molecules form a double layer: The hydrophilic heads are arranged opposite one another on the inner and outer surfaces, and the hydrophobic tails are aligned across from one another within the interior, sandwiched between the hydrophilic heads. The protein in the membrane is interspersed periodically throughout the lipid bilayer. Some of the protein, referred to as *peripheral protein*, penetrates only one of the lipid layers. The *integral protein*, as the remaining protein is called, extends through both layers of lipid to interface with the environment on both the internal and external surfaces of the membrane. These integral proteins can serve as transport channels and carriers.

Cellular Energy

Transport across the membrane is accomplished by three different mechanisms: *simple diffusion*, *facilitated diffusion*, and active transport. The first two mechanisms are referred to as passive processes because they do not require the direct input of cellular energy, and they involve transport down a *concentration gradient*, that is, from the side with a higher concentration to the side with a lower concentration of the substance being transported. In many instances, however, substances are transported across a membrane from the side with a low concentration to the side containing a greater concentration. This "uphill" movement across membranes is called active transport, and it requires the expenditure of cellular energy.

Cellular energy, produced by the biological oxidation of fuels such as carbohydrates, is stored as adenosine triphosphate (ATP). When this high-energy phosphate is hydrolyzed, the stored energy is released to drive cellular reactions such as active transport. The ATPase protein located in membranes belongs to one of the groups of enzymes which hydrolyze ATP. The mechanism has not been completely deciphered, but it appears as though a protein carrier molecule binds with the substance to be transported at the surface on one side of the membrane. This binding occurs at a specific activated region on the carrier protein called the *active site*. After combining with the carrier, the substance is moved across the membrane and released at the surface on the other side of the membrane. ATP is then hydrolyzed by an ATPase, and the energy released in this reaction prepares the protein carrier for attachment to another molecule to be transported by reactivating the active site. There is some question as to whether ATPase is a component of the carrier molecule or functions separately from it. Regardless of the spatial arrangement, the two molecules are intimately related in the active transport process.

Cotransport and Countertransport

There are two important modifications of the active transport process: *cotransport* and *countertransport*. Cotransport, or *symport*, involves a specialized protein molecule referred to as a symport carrier. A symport carrier has two attachment sites. One site is for the attachment of the molecule to be transported, and the other is for the attachment of a second molecule, which can be referred to as the *synergist*. Both the molecule to be transported and the synergist must be bound to the symport carrier before transport across the membrane can take place. The synergist is moved down a concentration gradient, and this downhill flow of the synergist drives the carrier to transport both molecules. In order to keep the synergist moving down a concentration gradient when attached to the symport carrier, the synergist must be pumped back across the membrane. This movement of the synergist in the opposite direction is mediated by a protein carrier activated by the energy released from the hydrolysis of ATP by an ATPase.

Countertransport, or *antiport*, also utilizes a specialized carrier with two attachment sites. This antiport carrier binds with the molecule to be transported at one of the attachment sites, and a second molecule, which can be called the *antagonist*, binds at the other. The carrier moves the molecule to be transported across the membrane while simultaneously moving the antagonist in the opposite direction. Again, both molecules must be attached to the antiport carrier before either can be transported, and the flow of the antagonist down a concentration gradient drives the transport by the carrier in both directions. The antagonist is pumped back across the membrane by a protein carrier activated by the energy released from the hydrolytic action of an ATPase on ATP. This action maintains a concentration gradient favorable for transport when the antagonist is attached to the antiport carrier.

Transport in Action

The presence of these three active transport mechanisms has been well documented. Calcium, for example, has been shown to be pumped from the cell by a carrier protein activated by the hydrolysis of ATP. Sugars for energy and carbohydrate structure must be cotransported into the cell by a symport carrier that utilizes the sodium ion as a synergist. At least two countertransport ion pumps have been identified. One pumps the potassium ion into the cell at the same time that it pumps the hydrogen ion out. The second pumps the potassium ion into the cell while the antagonist, sodium, is moved in the opposite direction. It is likely that numerous other active transport systems exist that have not yet been positively identified.

A protein carrier is one of the basic components of any active transport mechanism. Although no specific carrier molecule has yet been positively identified, there is ample indirect evidence to support the presence of such a protein. Much of this evidence comes from studies showing that active transport exhibits *saturation kinetics*. This means that the transport of a specific ion will increase as the concentration increases, up to a certain point. At this point, further increases in concentration will have no effect on transport. These results strongly suggest that the ion is binding with another molecule in the membrane, such as a carrier protein, which is limited in concentration and becomes saturated. Studies have also shown the transport of some substances to be competitively inhibited by the presence of another, very similar, substance. This indicates that both substances are competing for the same site on a membrane molecule, such as a protein carrier.

Role of Active Transport

The ability to accumulate substances against a concentration gradient is necessary for the normal function and survival of cells. There are numerous examples, however, of active transport being intimately involved in the regulation of some important biological processes. In the plant kingdom, sugar is produced by photosynthesis in the green leaves. This sugar must be transported out of the leaves and into nonphotosynthetic tissues, such as roots or fruit, through specialized transport cells in the *phloem*. The loading of sugars into the phloem is dependent on an active cotransport mechanism.

Almost every field of life science is concerned with *gene regulation*. Genes are continually being induced (activated) or repressed (deactivated) as organisms develop and change from the time of their conception until their death. Repression is usually caused by the presence of a protein molecule in the cell nucleus, but induction may very often be the result of metabolites being actively transported into the cell or nucleus. Hence, the active transport mechanisms may be a very important component of gene regulation.

D. R. Gossett

See also: Cells and diffusion; Osmosis, simple diffusion, and facilitated diffusion; Vesicle-mediated transport.

Sources for Further Study

Campbell, Neil A., and Jane B. Reece. *Biology*. 6th ed. San Francisco: Benjamin Cummings, 2002. An introductory, college-level textbook for science majors. The chapter on traffic across membranes provides a clear, concise, somewhat detailed description of the membrane transport process. The text and graphics furnish the reader with a very clear understanding of membrane structure and transport. Includes suggested readings at the end of the chapter and a glossary.

Curtis, Helena, and N. Sue Barnes. *Biology*. 5th ed. New York: Worth, 1989. An introductory, college-level textbook for science majors. The chapter "How Things Get into Cells" provides an excellent discussion of the overall process of transport mechanisms. A very readable text and well-done graphics make the active transport process understandable even to the novice. Includes suggested readings at the end of each chapter, and glossary.

Glass, A. D. M. "Regulation of Ion Transport." *Annual Review of Plant Physiology* 34 (1983): 311-326. Detailed review article outlining much of the pertinent information related to the control of ion transport in plants. A historical perspective and a list of advances are presented. A very informative review with a detailed list of literature citations.

Lee, A. G., ed. *Biomembranes: A Multi-Volume Treatise*. Greenwich, Conn.: JAI Press, 1995- . Individual volumes cover subjects such as general principles, rhodopsin and G-protein linked receptors, receptors of cell adhesion and cellular recognitions, ATPases, and transmembrane receptors and channels.

Mader, Sylvia S. *Inquiry into Life*. 9th ed. New York: McGraw-Hill, 1999. An introductory-level textbook for the nonscientist. The chapter on cell membrane and cell wall function provides a simple overview of the active transport process. Concise diagrams and a very readable text establish a general picture of the process. Key terms and suggested reading list at the end of each chapter. Includes glossary.

Nicholls, David G., and Stuart J. Ferguson. *Bioenergetics 2*. 2d ed. San Diego: Academic Press, 1992. Coverage of bioenergetics and active transport. Well illustrated, with bibliography and index.

Salisbury, Frank B., and Cleon W. Ross. *Plant Physiology*. 4th ed. Belmont, Calif.: Brooks/Cole, 1999. An intermediate, college-level textbook for science students. The chapter on absorption of mineral salts gives an in-depth view of the physiological role of active transport. An excellent explanation of the transport process is provided in text and graphics. Contains a detailed bibliography at the end of each chapter.

Stein, Wilfred D. *Channels, Carriers, and Pumps: An Introduction to Membrane Transport*. San Diego: Academic Press, 1990. Describes advances in the field and integrates them with transport kinetics, function, and regulation, linking experimental data and the construction of theoretical models. For those with a strong background in cell biology.

ADAPTATIONS

Categories: Evolution; genetics

The results of natural selection in which succeeding generations of organisms become better able to live in their environments are called adaptations. Many of the features that are most interesting and beautiful in biology are adaptations. Specialized structures, physiological processes, and behaviors are all adaptations when they allow organisms to cope successfully with the special features of their environments.

Adaptations ensure that individuals in populations will reproduce and leave well-adapted offspring, thus ensuring the survival of the species. Adaptations arise through *mutations*—inheritable changes in an organism's genetic material. These rare events are usually harmful, but occasionally they give specific survival advantages to the mutated organism and its offspring. When certain individuals in a population possess advantageous mutations, they are better able to cope with their specific environmental conditions and, as a result, will contribute more offspring to future generations than those individuals that lack the mutation. Over time, the number of individuals that have the advantageous mutation will increase in the population at the expense of those that do not have it. Individuals with an advantageous mutation are said to have a higher *fitness* than those without it, because they tend to have comparatively higher survival and reproductive rates. This is *natural selection*.

Natural Selection

Over very long periods of time, evolution by natural selection results in increasingly better adaptations to environmental circumstances. Natural selection is the primary mechanism of evolutionary change, and it is the force that either favors or selects against mutations. Although natural selection acts on individuals, a population gradually changes as those with adaptations become better represented in the total population. Most flowering plants, for example, are unable to grow in soil containing high concentrations of certain elements (for example, heavy metals) commonly found in mine tailings. Therefore, an adaptation that conferred resistance to these elements would open up a whole new habitat where competition with other plants would be minimal. Natural selection would favor the mutations, which confer specific survival advantages to those that carry them and impose limitations on individuals lacking these advantages. Thus, plants with special adaptations for resistance to the poisonous effects of heavy metals would have a competitive advantage over those that find heavy metals toxic. These attributes would be passed to their more numerous offspring and, in evolutionary time, resistance to heavy metals would increase in the population.

Types of Adaptations

Although natural selection serves as the instrument of change in shaping organisms to very specific environmental features, highly specific adaptations may ultimately be a disadvantage. Adaptations that are specialized may not allow sufficient *flexibility* (generalization) for survival in changing environmental conditions. The degree of adaptative specialization is ultimately controlled by the nature of the environment. Environments, such as the tropics, that have predictable, uniform climates and have had long, uninterrupted periods of climatic stability are biologically complex and have high species *diversity*. Scientists generally believe that this diversity results, in part, from complex competition for resources and from intense predator-prey relationships. Because of these factors, many narrowly specialized adaptations have evolved when environmental stability and predictability prevail. By contrast, harsh physical environments with unpredictable or erratic climates seem to favor organisms with general adaptations, or adaptations that allow flexibility. Regardless of the environment type, organisms with both general and specific adaptations exist because both types of adaptation enhance survival under different environmental circumstances.

Structural adaptations are parts of organisms that enhance their survival ability. Camouflage that enables organisms to hide from predators or their prey; protective spines on cacti that inhibit organisms that might feed on them; color, scent, or shape of flowers that promotes seed production—these are all structural adaptations. These adaptations enhance survival because they assist individuals in dealing with the rigors of the physical environment, obtaining nourishment, competing with others, or attracting pollinators.

Metabolism is the sum of all chemical reactions taking place in an organism, whereas physiology consists of the processes involved in an organism carrying out its function. *Physiological adaptations* are changes in the metabolism or physiology of organisms, giving them specific advantages for a given set of environmental circumstances. Because organisms must cope with the rigors of their physical environments, physiological adaptations for temperature regulation, water conservation, varying metabolic rate, and dormancy allow organisms to adjust to the physical environment or respond to changing environmental conditions.

Adaptations and Environment

Desert environments, for example, pose a special set of problems for organisms. Hot, dry environments require physiological mechanisms that enable organisms to conserve water and resist prolonged periods of high temperature. Evolution has favored a specialized form of photosynthesis in cacti and other succulents inhabiting arid regions. Crassulacean acid metabolism (CAM) photosynthesis allows plants with this physiological adaptation to absorb carbon dioxide at night, when relative humidity is comparatively high and air temperatures relatively low. Taking in carbon dioxide during the day would dehydrate plants, because opening the pores through which gas exchange takes place al-

Structural adaptations are parts of organisms that enhance their survival ability. For example, protective spines on cacti can inhibit organisms that might feed on them.

lows water to escape from the plant. CAM photosynthesis, therefore, allows these plants to exchange the atmospheric gases essential for their metabolism at night, when the danger of dehydration is minimized.

Because organisms must also respond and adapt to an environment filled with other organisms—including potential predators and competitors—adaptations that minimize the negative effects of biological interactions are favored by natural selection. Often the interaction among species is so close that each species strongly influences the others and serves as the selective force causing change. Under these circumstances, species evolve together in a process called *coevolution*. The adaptations resulting from coevolution have a common survival value to all the species involved in the interaction. The coevolution of flowers and their pollinators is a classic example of these tight associations and their resulting adaptations.

Speciation

Adaptations can be general or highly specific. General adaptations define broad groups of organisms whose lifestyles are similar. At the species level, however, adaptations are more specific and give narrow definition to those organisms that are more closely related to one another. Slight variations in a single characteristic, such as bill size in the seed-eating Galapágos finches, are adaptive in that they enhance the survival of several closely related species. An understanding of how adaptations function to make species distinct also furthers the knowledge of how species are related to one another.

Why so many species exist is one of the most intriguing questions of biology. The study of adaptations offers biologists an explanation. Because there are many ways to cope with the environment, and because natural selection has guided the course of evolutionary change for billions of years, the vast variety of species existing on the earth today is simply an extremely complicated variation on the theme of survival.

Robert W. Paul, updated by Bryan Ness

See also: Adaptive radiation; C_4 and CAM photosynthesis; Coevolution; Competition; Evolution: gradualism vs. punctuated equilibrium.

Sources for Further Study

Arms, Karen, and Pamela Camp. *Biology: Journey into Life*. 3d ed. Fort Worth, Texas: Saunders, 1994. This college-level text is well written and easily understandable because the authors use nontechnical language. Gives an excellent overview of evolution and natural selection. Several good examples of adaptations.

Brandon, Robert N. *Adaptation and Environment*. Princeton, N.J.: Princeton University Press, 1990. Explains the concept of selective environment and its influence on adaptation.

Futuyma, Douglas. *Science on Trial*. Reprint. Sunderland, Mass.: Sinauer Associates, 1995. Written for readers with little background in science; gives evidence for the theory of evolution. In presenting this evidence, Futuyma also shows how the views of creationists contrast with those of evolutionists. The illustrations and examples of adaptations are excellent.

Gould, Stephen J. *Ever Since Darwin*. Reprint. New York: W. W. Norton, 1992. Gould is well known for his ability to write humorous, slightly irreverent biological literature that the nonscientist can enjoy. No exception, this book gives an interesting view of evolution.

Ricklefs, Robert E. *Ecology*. 4th ed. New York: W. H. Freeman, 2000. The section on evolutionary ecology offers the reader an excellent but technical description of adaptations in their ecological context. Although this is an advanced college text, the material is accessible with little difficulty. The introduction gives the current views in evolutionary ecology, and the chapter summaries are excellent.

Rose, Michael R., and George V. Lauder, eds. *Adaptation*. San Diego: Academic Press, 1996. A discussion of new developments in adaptation, with new methods and new theoretical foundations and achievements. Rose provides an insightful reintroduction to the themes that Charles Darwin and his successors regarded as central to any profound understanding of biology.

Whitfield, Philip. *From So Simple a Beginning: The Book of Evolution*. New York: Maxwell Macmillan, 1993. A well-organized and well-written introduction to the concepts of evolution, written for the general reader. Illustrated.

Wickler, Wolfgang. *Mimicry in Plants and Animals*. New York: McGraw-Hill, 1968. This intriguing, small book takes up the fascinating subject of mimicry, the resemblance of one organism to another or to some object that gives the mimic a specific offensive or defensive advantage. Highly adaptive forms are beautifully illustrated in color; most examples are from the insect world.

ADAPTIVE RADIATION

Categories: Ecology; ecosystems; evolution; genetics

In adaptive radiation, numerous species evolve from a common ancestor introduced into an environment with diverse ecological niches. The progeny evolve genetically into customized variations of themselves, each adapting to survive in a particular niche.

In 1898 Henry F. Osborn identified and developed the evolutionary phenomenon known as adaptive radiation, whereby different forms of a species evolve, quickly in evolutionary terms, from a common ancestor. According to the principles of *natural selection*, organisms that are the best adapted (most fit) to compete will live to reproduce and pass their successful traits on to their offspring. The process of adaptive radiation illustrates one way in which natural selection can operate when members of one population of a species are cut off or migrate to a different environment that is isolated from the first. Such isolation can occur from one patch of plantings to another, from one mountaintop or hillside to another, from pond to pond, or from island to island. Faced with different environments, the group will diverge from the original population and in time become different enough to form a new species.

Genetic Changes

In a *divergent population*, the relative numbers of one form of *allele* (characteristic) decrease, while the relative numbers of a different allele increase. New environmental pressures will select for favorable alleles that may not have been favored in the old environment. Over successive generations, therefore, a new gene created by random *mutation* (change) may replace the original form of the gene if, for example, the trait encoded by that gene allows the divergent group to cope better with environmental factors, such as food sources, predators, or temperature. The result in the long term is that deoxyribonucleic acid (DNA) changes sufficiently through the growth of divergent populations to allow new generations to become significantly different from the original population. In time, they are unable to reproduce with members of the original species and become a new species.

Galápagos Islands Case Study

Adaptive radiation occurs dramatically when a species migrates from one landmass to another. This may occur between islands or between continents and islands. A classic example of adaptive radiation is the evolution of finches noted by Charles Darwin during his trips to the Galápagos Islands off the west coast of South America. Several species of plants and animals had migrated to these islands from the South American mainland by means of flight, wind, ocean debris, or other means of transport. Finches from the mainland—perhaps aided by winds—settled on fifteen of the islands in the Galápagos group and began to adapt to the various unoccupied *ecological niches* on those islands, which differed. Over several generations, natural selection favored a variety of finch species with beaks adapted for the different types of foods available on the different islands. As a result, several species of different finches evolved, roughly simultaneously, on these islands.

In the Hawaiian Islands, twenty-eight species of the Asteraceae *family are known together as the Hawaiian silversword alliance. The entire group of plants, including this silversword, appears to be traceable to one ancestor.*

Hawaiian Silversword Alliance

Although plants seem unable to "migrate" as birds and other animals do, adaptive radiation occurs in the plant world as well. In the Hawaiian Islands, for example, twenty-eight species of the *Asteraceae* family are known together as the Hawaiian silversword alliance. The entire group appears to be traceable to one ancestor, thought to have arrived on the island of Kauai from western North America. The silverswords—which compose three genera, *Argyroxiphium*, *Dubautia*, and *Wilkesia*—have since evolved into twenty-eight species, and this speciation came about due to major ecological shifts. These plants are therefore prime examples of adaptive radiation.

Within the silversword alliance, different species have adapted to widely varying ecosystems found throughout the islands. *Argyroxiphium sandwicense*, for example, is endemic to the island of Maui and grows at high elevations from 6,890 to 9,843 feet (2,100-3,000 meters) on the dry, alpine slopes of the volcano Haleakala. This species has succulent leaves covered with silver hairs. It is thought that the hairs lessen the pace of evaporative moisture loss and protect the leaves from the sun. In contrast, species of the genus *Dubautia* that grow in wet, shady forests have large leaves that lack hairs.

Despite their "customized" physiologies, the silverswords that have evolved in Hawaii are all closely related to one another, so much so that any two can hybridize. Studies of the silverswords have provided what geneticist Michael Purugganan called a "genetic snapshot of plant evolution." Adaptive radiation is one window into how new plant structures arise.

Jon P. Shoemaker, updated by Christina J. Moose

See also: Evolution: convergent and divergent; Hardy-Weinberg theorem; Species and speciation; Trophic levels and ecological niches.

Sources for Further Study

Givnish, Thomas J., and Kenneth J. Sytsma, eds. *Molecular Evolution and Adaptive Radiation*. New York: Cambridge University Press, 2000. International experts who study the phenomenon in a variety of organisms and geographical settings provide a survey of advances in the study of adaptive radiation. Synthesizes the recent explosion of research in this area, from orchids to monkeys.

Robichaux, Robert, et al. "'Radiating' Plants." *Endangered Species Bulletin Update*, March/April, 1999, S4-S5. Discusses the necessity of ensuring that there are opportunities for the continued adaptive radiation of Hawaii's plant and animal lineages, using the Hawaiian silversword alliance as a case in point.

Schluter, Dolph. *The Ecology of Adaptive Radiation*. Oxford, England: Oxford University Press, 2000. Reevaluates the theory, along with its most significant extensions and challenges, in the light of recent evidence. The first full-length volume to be devoted to adaptive radiation in decades.

AFRICAN AGRICULTURE

Categories: Agriculture; economic botany and plant uses; food; world regions

Soil and climatic conditions throughout Africa determine not only agricultural practices, such as which crops can be grown, but also whether plant life is capable of sustaining livestock on the land and enabling fishing of the oceans.

Rainfall—the dominant influence on agricultural output—varies greatly among Africa's fifty-six countries. Without irrigation, agriculture requires a reliable annual rainfall of more than 30 inches (75 centimeters). Portions of Africa have serious problems from lack of rainfall, such as increasing desertification and periods of drought.

Food output has declined, with per capita food production 10 percent less in the 1990's than it was in the 1980's. In most African countries, however, more than 50 percent, and often 80 percent, of the population works in agriculture, mostly subsistence agriculture. Large portions of the continent, such as Mali and the Sudan, have the potential of becoming granaries to much of the continent and producing considerable food exports.

Traditional African Agriculture

Traditionally, agriculture in Africa has been subsistence farming in small plots. It has been labor-intensive, relying upon family members. New land for farming was obtained by the *slash-and-burn* method (*shifting cultivation*). The trees in a forested area would be cut down and burned where they fell. The ashes from the burned trees fertilized the soil. Both men and women worked at such farming. Slash-and-burn agriculture is common not only in Africa but also in tropical areas around the world. In areas of heavy rainfall, the rains wash out the nutrients from soil and burned trees in a period of two to three years.

The crops grown depend upon the region. In the very dry, yet habitable, parts of Africa—such as the Sudano-Sahelian region that stretches from Senegal and Mali in the west of Africa to the Sudan in the east—a key subsistence crop is green millet, a grain. Ground into a type of flour, it can be made into a bread-like substance. In moister areas, traditional crops are root and tuber crops, such as yams and cassava. Cassava has an outer surface or skin that is poisonous, but it can be treated to remove the poison. The tuber then can be ground and used to make a bread-like substance. Other important traditional crops are rice and corn, which were introduced by Europeans when they came to Africa.

Animal husbandry, or seminomadic herding, is another form of traditional agriculture. Problems that have arisen with this type of agriculture are the availability of water and grass or hay for cattle. Regions that are very moist, such as the Gulf of Guinea, which has rain forest, are not good for cattle because of the tsetse fly, which carries diseases such as sleeping sickness.

Crops

The most widely grown crop is rice, which is grown on more than one-third of the irrigated crop area in Africa. Cultivated mostly in wetlands and valley bottoms, rice is the most common crop in the humid areas of the Gulf of Guinea and Eastern Africa. It is also grown on the plateaus of Madagascar. In the northern and southern regions, rice represents only a small portion of the total crops under water management. Wheat and corn are cultivated and irrigated, mostly in Egypt, Morocco, South Africa, Sudan, and Somalia.

Vegetables, including root and tuber crops, are present in all regions and almost every country. Vegetables are grown on about 8 percent of the cultivated areas under water management. In Algeria, Mauritania, Kenya, Burundi, and Rwanda, they are the most widespread crops under water management. *Arboriculture* (growing of fruit trees), which represents 5 percent of the total irrigated crops, is concentrated in the northern region and consists mostly of citrus fruits. Commercial crops (for cash and export) are grown mostly in the Sudan and in the countries of the southern region and consist mostly of cotton and oilseeds. Other commercial crops in Africa are sugarcane, coffee, cocoa, oil and date palm, bananas, tobacco, and cut flowers. Sugarcane is grown in all countries except in the northern region. The other commercial crops are concentrated in a few countries.

North Africa

In Morocco, Algeria, Tunisia, Libya, and Egypt, the region's agricultural resources are limited by its dry climate. Its products are those typical of the Mediterranean, steppe, and desert regions: wheat, barley, olives, grapes, citrus fruits, some vegetables, dates, sheep, and goats.

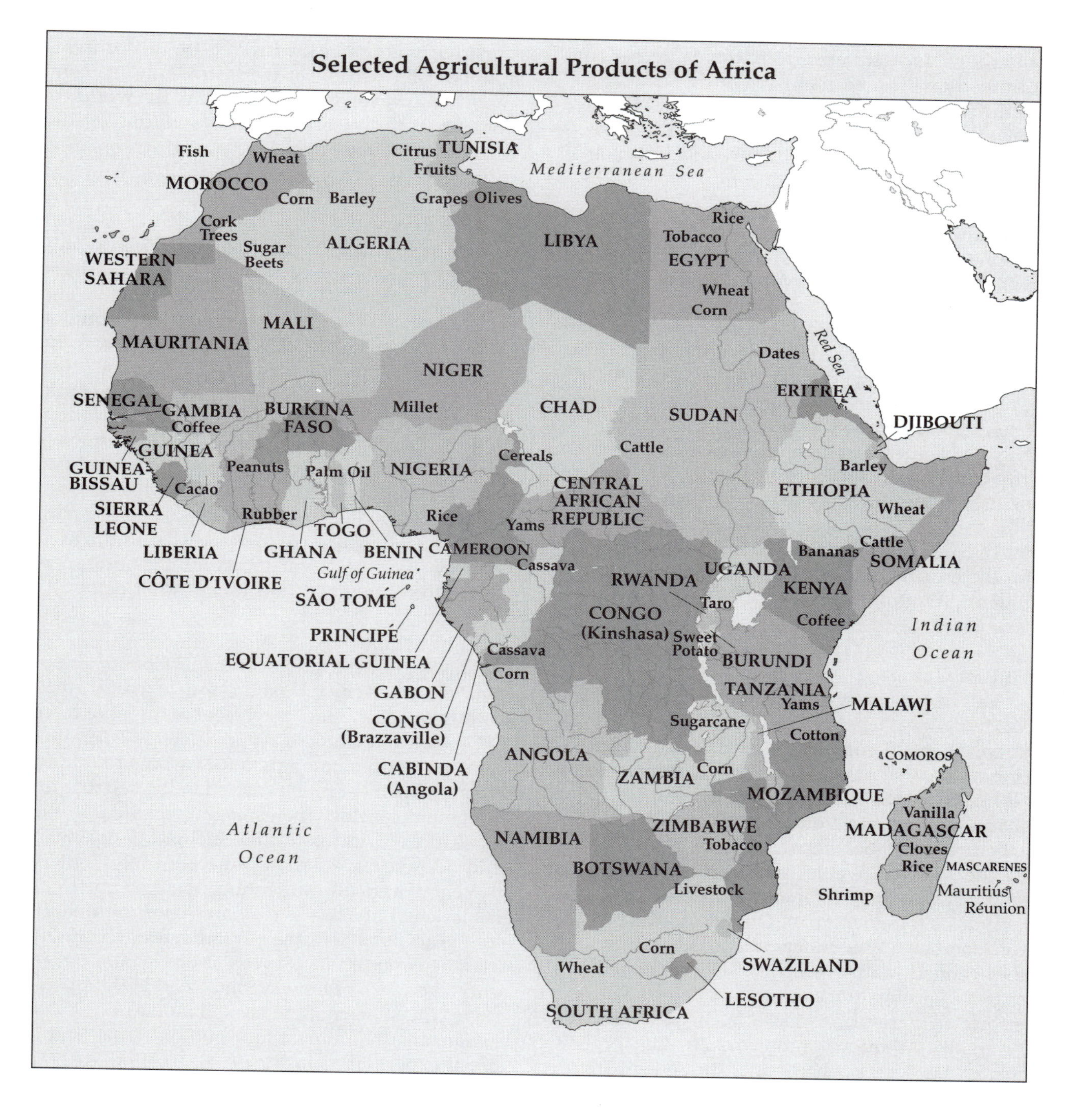

Agriculture employs less than 20 percent of the working population in Libya and as much as 55 percent in Egypt. From about the middle of the twentieth century, North Africa's production failed to keep pace with its population growth and remained susceptible to large annual fluctuations. Cropland occupies about 33 percent of Tunisia but less than 3 percent of Algeria, Egypt, and Libya. Some export crops, such as citrus fruits, tobacco, and cotton, have suffered from strong international competition. The northern region is not a major contributor to the continent's fish catch. Morocco, however, with its cool, plankton-rich Atlantic waters and access to the Mediterranean Sea, is one of the world's largest fish producers.

Sudano-Sahelian Region

This region comprises Mauritania, the western Sahara, Senegal, Gambia, Mali, Burkina-Faso, Niger, Chad, and the Sudan. Because of the region's extreme dryness, mostly subsistence farming and seminomadic herding are practiced. Millet is the primary crop. In the late twentieth century, this region was devastated by long droughts that caused famine and starvation. Mali and the Sudan have the Niger and Nile Rivers flowing through them. These great rivers provide plenty of water for irrigation of fields. During the rainy season in Mali—typically June through September—the Niger River widens into a great, extensive floodplain. This area is good for the growing of rice. Similarly, in the Sudan the Blue and White Niles meet at Khartoum to form the Nile River.

Gulf of Guinea

This region comprises Guinea-Bissau, Cape Verde, Guinea, Liberia, Sierra Leone, Côte d'Ivoire, Togo, Ghana, Benin, and Nigeria. With the exception of Nigeria, agriculture there is dominated by rice cultivation. The percentage of total land area that is under cultivation ranges from 60 percent in Liberia to just 9 percent in Sierra Leone.

The total cultivable area of Ghana is 39,000 square miles (100,000 square kilometers), or 42 percent of its total land area. Only 4.8 percent of the total land area was under cultivation at the end of the twentieth century. Much of the cultivation is subsistence farming of yams and other crops. Ghana's efforts in agriculture have been hampered by droughts. Additional problems are that organic matter has been leached out of the soils by heavy rainfall and that increasing deforestation has led to additional erosion. This is the situation in much of the Gulf of Guinea and the central regions.

About half of Nigeria's available land is under cultivation. Increasing rainfall from the semiarid north to the tropically forested south allows for great crop diversity. Principal food crops are corn, millet, yams, sorghum, cassava, rice, potatoes, and vegetables. Nigeria was the world's fourth-largest exporter of cocoa beans in 1990-1991, accounting for about 7.1 percent of world trade in this commodity. However, Nigeria's share of the world cocoa market has been substantially reduced because of aging trees, low prices, black pod disease, smuggling, and labor shortages.

Central Region

This region comprises the Central African Republic, Cameroon, Congo-Brazzaville, Congo-Kinshasa, Gabon, Equatorial Guinea, Burundi, Rwanda, and São Tomé and Príncipe. Cameroon has 14.7 million acres of arable land. In 1997, 55,000 tons of rice were produced, but the country imported 124,000 tons in 1995. In the central region, the percentage of arable land ranges from 0.4 percent for the Congo-Brazzaville to 47 percent for Rwanda. Cassava is harvested in Congo-Brazzaville, Congo-Kinshasa, Equatorial Guinea, and Gabon. Corn is harvested in Congo-Brazzaville, Congo-Kinshasa, and Burundi. In Rwanda, 17 percent of the harvested land is used to grow sweet potatoes. Agriculture is not important in the economy of São Tomé and Príncipe.

Eastern Region

This region comprises Eritrea, Djibouti, Ethiopia, Somalia, Kenya, Uganda, and Tanzania. Agriculture employs about 80 percent of the labor force in Uganda and Ethiopia. Approximately 2.5 million small farms dominate agriculture in both countries. About 84 percent of Uganda's land is suitable for agriculture—a high percentage compared to the majority of African countries, such as Ethiopia with only 12 percent. Food crops account for about 74 percent of agricultural production. Only one-third is marketed; the rest is for home consumption. In four years out of five, the minimum needed rainfall may be expected in 78 percent of Uganda but in only 15 percent of Kenya. Somalia and Ethiopia receive almost none of the needed minimum.

Tanzania has almost four million farms. Traditional export crops include coffee, cotton, cashew

Leading Agricultural Crops of African Countries with More than 15 Percent Arable Land

Country	*Products*	*Percent Arable Land*
Burundi	Coffee, cotton, tea, corn, sorghum, sweet potatoes, bananas, manioc	44
Comoros	Vanilla, cloves, perfume essences, copra, coconuts, bananas, cassava	35
Gambia	Peanuts, millet, sorghum, rice, corn, cassava, palm kernels	18
Malawi	Tobacco, sugarcane, cotton, tea, corn, potatoes, cassava, sorghum, pulses	18
Mauritius	Sugarcane, tea, corn, potatoes, bananas, pulses	49
Morocco	Barley, wheat, citrus, wine, vegetables, olives	21
Nigeria	Cocoa, peanuts, palm oil, corn, rice, sorghum, millet, cassava, yams, rubber	33
Rwanda	Coffee, tea, pyrethrum (insecticide made from chrysanthemums), bananas, beans, sorghum, potatoes	35
Togo	Coffee, cacao, cotton, yams, cassava, corn, beans, rice, millet, sorghum	38
Tunisia	Olives, dates, oranges, almonds, grain, sugar beets, grapes	19

Source: Data are from *The Time Almanac 2000*. Boston: Infoplease, 1999.

nuts, tobacco, and tea. Major staple foods (corn, rice, and wheat) are exported in times of surplus. Tanzania's climatic growing conditions are favorable for the production of a wide range of fruits, vegetables, and flowers. Drought-resistant crops (sorghum, millet, and cassava) and other substaples such as onions, Irish potatoes, sweet potatoes, bananas, and plantains are also produced.

Areas that have 20-30 inches (50-75 centimeters) of rainfall per year rely on a mixture of agriculture and livestock herding. Regions with a smaller annual rainfall or a long dry season can support only drought-resistant crops such as sorghum, millet, and cassava. Over large areas of eastern Africa, rainfall is inadequate for crop cultivation. The whole of Somalia and 70 percent of Kenya receive less than 20 inches (50 centimeters) of rain four years out of five. In these areas, the only feasible use of land is for raising livestock. Agriculture is not an important factor in the economies of Eritrea and Djibouti.

Southern Region

This region comprises Angola, Namibia, Zambia, Zimbabwe, Malawi, Mozambique, Botswana, Lesotho, Swaziland, and South Africa. The arable percentage of the total land area ranges from 14 percent in Malawi to just 1 percent in Namibia. With the exception of Mozambique, where cassava predominates, corn is the major crop in the countries in this region.

About 13 percent of South Africa's land area can be used for crop production. Rainfall varies across the country, and varied climatic zones and terrains enable the production of almost any kind of crop. The largest area of farmland is planted with corn, followed by wheat, then oats, sugarcane, and sunflowers. The nation is well known for the high quality of its fruits, such as apples and citrus.

Agriculture is the predominant economic activity in Zimbabwe, accounting for 40 percent of total export earnings—about 22 percent of the total economy—and employing more than 60 percent of the country's labor force. The main export crops are tobacco, cotton, and oilseeds. Zimbabwe is usually self-sufficient in food production. Its main food crops are corn, soybeans, oilseeds, fruits and vegetables, and sugar.

Mozambique's agriculture has been badly hindered by civil war. However, the country has considerable potential for irrigation due to the Zambezi and Limpopo Rivers. The irrigation potential is estimated to be 7.5 million acres. In the 1990's, only 110,000 acres were irrigated, growing rice, sugarcane, corn, and citrus.

Agriculture and livestock production employ about 62 percent of Botswana's labor force. Most of the country has semidesert conditions with erratic rainfall and poor soil conditions, making it more suitable to grazing than to crop production. The principal food crops are sorghum and corn. Namibia's cultivated area is only 506,000 acres—only 0.8 percent of the cultivable area. Agriculture makes up approximately 10 percent of the economy but employs more than 80 percent of the population. The major irrigated crops are corn, wheat, and cotton.

Indian Ocean Islands

This region comprises Madagascar, Mauritius, the Comoros, and the Seychelles. During the 1990's an estimated 8.7 million people lived in the rural areas, 65 percent of whom lived at the subsistence level. Only 5.2 percent of Madagascar's total land area (7.4 million acres) was under cultivation. Of the total land area, 50.7 percent supported livestock production, while 16 percent (1.2 million acres) of the land under cultivation was irrigated.

Cassava, planted almost everywhere on the island, is grown as well as corn and sweet potatoes, with smaller quantities of cotton, bananas, and cloves. The fisheries sector, especially the export of shrimp, has been the most rapidly growing area of the agricultural economy in the Indian Ocean Islands region.

Mauritius has 30,000 acres of sugarcane plantations that have had one of the highest sugarcane and sugar yields in the world. The Seychelles have a total land area of only 72 square miles (187 square kilometers), of which only 3,000 acres are cultivated. This 3 percent of the land area accounts for only 4 percent of the island nation's economy. The Comoros' agriculture is heavily weighted toward rice, the staple food of the populace.

Dana P. McDermott

See also: African flora; Agricultural revolution; Agriculture: history and overview; Agriculture: traditional; Agriculture: world food supplies; Alternative grains; Desertification; Drought.

Sources for Further Study

Gibbon, Peter, Kjieli J. Havnevik, and Kenneth Harmele. *A Blighted Harvest: The World Bank and African Agriculture in the 1980's*. Trenton, N.J.: Africa World Press, 1993. Describes the theory as well as the economic critique of agricultural adjustment. Covers reforms and agricultural production and other trends.

Gleave, M. B., ed. *Tropical African Development*. New York: John Wiley & Sons, 1992. Includes chapters on environmental constraints on development, traditional agriculture, and agricultural development.

Grove, A. T. *The Changing Geography of Africa*. 2d ed. Oxford, England: Oxford University Press, 1993. Intended for use in schools. Explores Africa at the end of the twentieth century, how it has developed, and how changes in other parts of the world have affected it. Many examples, statistics, maps, diagrams, and photos.

Mortimore, Michael. *Roots in the African Dust: Sustaining the Sub-Saharan Dry*lands. New York: Cambridge University Press, 1998. This volume, accessible to students, policymakers, and scholars, includes chapters on the extent of soil degradation, merits of agricultural intensification, and conservation of biotic resources. Offers a breadth of analysis in a holistic, wide-ranging view of rural livelihoods and landscapes. Mortimore argues that desertification is reversible.

Rached, Eglal, Eva Rathgeber, and David B. Brooks, eds. *Water Management in Africa and the Middle East: Challenges and Opportunities*. Ottawa, Canada: International Development Research Centre, 1996. Discusses management and allocation of water resources in Africa and within its various subregions. Irrigation and water-moving issues are covered.

AFRICAN FLORA

Category: World regions

With few exceptions, Africa's flora (vegetation) is tropical or subtropical. This is primarily because none of the African continent extends far from the equator, and there are only a few high-elevation regions that support more temperate plants.

Listed in order of decreasing land area, the three main biomes of Africa are subtropical desert, tropical savanna, and tropical forest. The flora in southern Africa has been most studied. The flora of central and northern Africa is less known.

The subtropical desert biome is the driest of the biomes in Africa and includes some of the driest locations on earth. The largest desert region is the Sahara in northern Africa. It extends from near the west coast of Africa to the Arabian Peninsula and is part of the largest desert system in the world, which extends into south central Asia. A smaller desert region in southern Africa includes the Namib Desert, located along the western half of southern Africa, especially near the coast, and the Kalihari Desert, which is primarily inland and east of the Namib Desert.

Where more moisture is available, grasslands predominate, and as rainfall increases, grasslands gradually become tropical savanna. The difference between a grassland and a savanna is subjective but is in part determined by tree growth, with more trees characterizing a savanna. The grassland/tropical savanna biome forms a broad swath across much of central Africa and dominates much of eastern and southern Africa.

Tropical forests make up a much smaller area of Africa than the other two biomes. They are most abundant in the portions of central Africa not dominated by the grassland/tropical savanna biome and are not far from the coast of central West Africa. Scattered tropical forest regions also occur along major river systems of West Africa, from the equator almost to southern Africa.

Subtropical Desert

The subtropical deserts of Africa seem, at first, to be nearly devoid of plants. While this is true for some parts of the Sahara and Namib Deserts that are dominated by sand dunes or bare, rocky outcrops, much of the desert has a noticeable amount of plant cover. The Sahara is characterized by widely distributed species of plants that are found in similar habitats. The deserts of southern Africa have more distinctive flora, with many species endemic to specific local areas.

Succulents of the Subtropical Desert

To survive the harsh desert climate, plants use several adaptations. *Mesembryanthemum*, whose species include ice plant and sea figs, is a widespread genus, with species occurring in all of Africa's deserts. It typically has thick, succulent leaves. Such *succulents* store water in their leaves or stems, which they retain by using a specialized type of photosynthesis. Most plants open their stomata (small openings in the leaves) during the day to get carbon dioxide from the surrounding air. This would lead to high amounts of water loss in a desert environment, so succulents open their stomata at night. Through a biochemical process, they store carbon dioxide until the next day, when it is released inside the plant so photosynthesis can occur without opening the stomata.

To prevent water loss, many succulents have no leaves at all. *Anabasis articulata*, found in the Sahara desert, is a leafless succulent with jointed stems. Cacti are found only in North and South America, but a visitor to the Sahara would probably be fooled by certain species in the spurge family that resemble cacti. For example, *Euphorbia echinus*, another Saharan plant, has succulent, ridged stems with spines. The most extreme adaptation in succulents is found in the living stones of southern Africa. Their plant body is reduced to two plump, rounded leaves that are very succulent. They hug the ground, sometimes partially buried, and have camouflaged coloration so that they blend in with the surrounding rocks and sand, thus avoiding being eaten by grazing animals. Other succulents, such as the quiver tree, attain the size and appearance of trees.

Water-Dependent Plants of the Subtropical Desert

Water-dependent plants are confined to areas near a permanent water source, such as a spring. The most familiar of these plants is the date palm, which is a common sight at desert oases. Tamarind and acacia are also common where water is available. A variety of different sedges and rushes occur wherever there is abundant permanent freshwater, the most famous of these being the papyrus, or bulrush.

Ephemerals of the Subtropical Desert

Annuals whose seeds germinate when moisture becomes available and quickly mature, set seed, and die, are called *ephemerals*. These plants account for a significant portion of the African desert flora. A majority of the ephemerals are grasses. Ephemerals are entirely dependent on seasonal or sporadic rains. A few days after a significant rain the desert turns bright green, and after several more days flowers, often in profusion, appear. Some ephemerals germinate with amazing speed, such as the pillow cushion plant, which germinates and produces actively photosynthesizing seed leaves only ten hours after being wetted. Reproductive rates for ephemerals, and even for perennial plants, are rapid. Species of morning glory can complete an entire life cycle in three to six weeks.

Tropical Savanna

Tropical savanna ranges from savanna grassland, which is dominated by tall grasses lacking trees or shrubs, to thicket and scrub communities, which are composed primarily of trees and shrubs of a fairly uniform size. The most common type of savanna in Africa is the savanna woodland, which is composed of tall, moisture-loving grasses and tall, deciduous or semideciduous trees that are unevenly distributed and generally well spaced. The type of savanna familiar to viewers of African wildlife documentaries is the savanna parkland, which is primarily tall grass with widely spaced trees.

Savanna Grasses and Herbs

Grasses represent the majority of plant cover beneath and between the trees. In some types of savanna, the grass can be more than 6 feet (1.8 meters) high. Although much debated, two factors seem to perpetuate the dominance of grasses: seasonal moisture with long intervening dry spells and periodic fires. Given excess moisture and lack of fire, savannas seem inevitably to become forests. Activities by humans, such as grazing cattle or cutting trees, also perpetuate, or possibly promote, grass dominance.

A variety of herbs exist in the savanna, but they are easily overlooked, except during flowering periods. Many of them also do best just after a fire, when they are better exposed to the sun and to potential pollinators. Plants such as hibiscus and coleus are familiar garden and house plants popular the world over. Vines related to the sweet potato are also common. Many species from the legume or pea and sunflower families are present. Wild ginger often displays its showy blossoms after a fire.

Savanna Trees and Shrubs

Trees of the African savanna often have relatively wide-spreading branches that all terminate at about the same height, giving the trees a flat-topped appearance. Many are from the legume family, most notably species of *Acacia*, *Brachystegia*, *Julbernardia*, and *Isoberlinia*. With the exception of acacias, these are not well known outside Africa. There is an especially large number of *Acacia* species ranging from shrubs to trees, many with spines. A few also have a symbiotic relationship with ants that protect them from herbivores. The hashab tree, a type of acacia that grows in more arid regions, is the source of gum arabic.

Although not as prominent, the baobab tree is renowned for its large size and odd appearance and occurs in many savanna regions. It has an extremely thick trunk with smooth, gray bark and can live for up to two thousand years. Many savanna trees also have showy flowers, such as the flame tree and the African tulip tree.

Tropical Forest

The primary characteristics of African tropical forests are their extremely lush growth, high species diversity, and complex structure. The diversity is often so great that a single tree species cannot be identified as dominant in an area. Relatively large trees, such as ironwood, iroko, and sapele, predominate. Forest trees grow so close together that their crowns overlap, forming a *canopy* that limits the amount of light that falls beneath them. A few larger trees, called emergent trees, break out above the thick canopy.

A layer of smaller trees live beneath the main canopy. A few smaller shrubs and herbs grow near the ground level, but the majority of the herbs and other perennials are *epiphytes*, that is, plants that grow on other plants. On almost every available space on the trunks and branches of the canopy trees there are epiphytes that support an entire, unique community. All this dense plant growth is supported by a monsoon climate in which 60 inches (150 centimeters) or more of rain often falls annually, most of it in the summer.

Baobab tree in Zambia.

Lianas and Epiphytes

Lianas are large, woody vines that cling to trees, many of them hanging down near to the ground. They were made famous by Tarzan movies. Many lianas belong to families with well-known temperate vine species, such as the grape family, morning glory family, and cucumber family. Other, related plants remain intimately connected to the trunks of trees. One of these, the strangler fig, is a strong climber that begins life in the canopy.

The fruits are eaten by birds or monkeys, and the seeds are deposited in their feces on branches high in the canopy. The seeds germinate and send a stem downward to the ground. Once the stem reaches the ground, it roots; additional stems then develop and grow upward along the trunk of the tree. After many years, a strangler fig can so thoroughly surround a tree that it prevents water and nutrients from flowing up the trunk. Eventually, the host tree dies and rots away, leaving a hollow tube of mostly strangler fig. Other climbers include members of the *Araceae* family, the most familiar being the ornamental philodendron.

The most common epiphytes are bryophytes, lower plants related to mosses, and lichens, a symbiotic combination of algae (or cyanobacteria) and fungus. The most abundant higher plants are ferns and orchids. As these plants colonize the branches of trees, they gradually trap dust and decaying materials, eventually leading to a thin soil layer that other plants can use. Accumulations of epiphytes can be so great in some cases that tree branches break from their weight. Epiphytes are not parasites (although there are some parasitic plants that grow on tree branches); they simply use the host tree for support.

Tropical Forest Floor Plants

Grasses are almost entirely absent from the forest floor; those that grow there have much broader leaves than usual. Some forest-floor herbs are able to grow in the deep shade beneath the canopy, occasionally being so highly adapted to the low light that they can be damaged if exposed to full sunlight. Some popular house plants have come from among these plants, because they do not need direct sunlight to survive. Still, the greatest numbers of plants occur beneath breaks in the canopy, where more light is available.

Bryan Ness

See also: African agriculture; Biomes: types; C_4 and CAM photosynthesis; Cacti and succulents; Deserts; Forests; Grasslands; Rain-forest biomes; Savannas and deciduous tropical forests.

Sources for Further Study

Alpert, Peter. "Integrated Conservation and Development Projects." *Bioscience* 46, no. 11 (December, 1996): 845-856. Examples of integrated conservation and development projects (ICDPs), which have been developed to conserve biodiversity and ecological systems in developing nations, including Rwanda, Kenya, and Namibia.

Becker, Hank. "Floral Gems." *Agricultural Research* 45, no. 9 (September, 1997): 8-13. A joint project of the governments of South Africa and the United States allowed for the development and introduction of exotic South African plants into American markets.

Boroughs, Don. "Battle for a Wild Garden." *International Wildlife* 29 (May 3, 1999). Describes devastated wild regions of South Africa and the restoration of flora there, undertaken by concerned citizens. The government has become involved as well.

Morrissey, Brian. "The Making of a Rainforest." *Natural History* 107 (June, 1998): 56-63. Account of museum scientists' and artists' trip to Africa to observe a rain forest so they could accurately recreate the forest in a diorama for the American Museum of Natural History.

Pooley, Elsa. *Flowers, Grasses, Ferns, and Fungi*. Halfway House, South Africa: Southern Book Publishers, 1998. Part of the Southern African Green Guide series, this plant identification book includes bibliographical references and an index.

Steentoft, Margaret. *Flowering Plants in West Africa*. New York: Cambridge University Press, 1988. This account of the flowering plants of West Africa south of the Sahara Desert emphasizes species of ecological or economic importance. Includes ample notes on field recognition and a bibliography.

Van Wyk, Braam, and Piet Van Wyk. *Field Guide to Trees of Southern Africa*. Cape Town, South Africa: Struik, 1997. A guide to more than one thousand species of trees found in southern Africa.

AGRICULTURAL CROPS: EXPERIMENTAL

Categories: Agriculture; economic botany and plant uses; food

Experimental crops are foodstuffs with the potential to be grown in a sustainable manner, produce large yields, and reduce people's reliance on the traditional crops wheat, rice, and corn.

Shifting from a hunter-gatherer society to an agrarian society led to increasingly larger-scale agricultural production that involved selecting local crops for domestication. In recent history there has been a reduction in the number of agricultural crops grown for human consumption. There are estimated to be at least 20,000 species of edible plants on earth, out of more than 350,000 known species of higher plants. However, only a handful of crops feed most of the world's people. These include wheat, rice, corn, potatoes, sugar beets, sugarcane, cassava, barley, soybeans, tomatoes, and sorghum. Rice, wheat, and corn together account for a majority of calories consumed. In the effort to develop experimental crops, agricultural goals include expanding the diversity of plant food in the human diet.

AP/Wide World Photos

A farmer cultivates soybeans in Mankato, Minnesota. Soybeans are thought to be native to China, but worldwide production is now the greatest of any legume.

Recent Successes

Soybeans (*Glycine max*) are a relatively new crop that gained worldwide acceptance and widespread cultivation in the second half of the twentieth century. Originally cultivated in China, soybeans gradually spread throughout Asia and became a staple food there. High in protein, soybeans were first grown in the Western world as animal feed. Concerted breeding efforts have resulted in many locally adapted varieties. Today, soybeans as both meal and oil are commonplace. Worldwide soybean production is now the greatest of any legume.

Triticale (*x Triticosecale*) is a hybrid created to combine the ruggedness and high protein content of rye (*Secale cereale*) with the high yield of wheat (*Triticum aestivum*). Triticale has not replaced wheat or rye in bread-making due to its rather low gluten content but is used to supplement bread flours. Triticale is also adaptable to marginal agricultural soils.

Kiwifruit (*Actinidia deliciosa*) is another recent success story. A previously little-known fruit originally called Chinese gooseberry, it was introduced to New Zealand at the turn of the twentieth century and renamed kiwifruit. The name change was a marketing strategy that led to worldwide popularity. Today kiwifruit cultivation and consumption are increasing worldwide. Kiwifruit grows on a deciduous vine, much like grapes. It can be harvested and then stored for several months without loss of quality.

Grains and Cereals

Quinoa (*Chenopodium quinoa*) is a grain native to the Andes Mountains of South America. It has been a staple in the diets of people living in that region for centuries. Although the leaves are edible, it is principally the tiny seed which is consumed. The seeds contain high amounts of protein, calcium, phosphorus, and the essential amino acid lysine, which is typically lacking in other cereals such as wheat, rye, and barley. Quinoa seeds must be washed or otherwise processed to remove the bitter saponins contained in the pericarp and can then be cooked and eaten much like rice. Quinoa can also be ground into flour as a supplement for bread making. Cultivation and use of quinoa have increased steadily since the 1980's.

Grain amaranths (*Amaranth*) are being rediscovered and developed as a potential new source of grain. Amaranth was a staple crop for centuries in Mexico, Central America, and South America. Amaranth is grown as an annual and yields thick, heavy seed heads containing numerous tiny seeds. The hard seed coat is removed by heating or boiling and can be prepared much like corn. Amaranth is comparable to other grains in protein, contains high amounts of lysine, and can be consumed by those allergic to typical grains. Breeding efforts over the last few decades involving *A. hypochondriacus*, *A. cruentus*, and *A. hybridus* have greatly increased seed yield as well as desirable plant growth habit. Another important characteristic is amaranth's drought resistance.

Legumes

Members of the *Leguminosae* family are particularly valuable as food sources because they contain high levels of protein. This is in part due to their ability to fix atmospheric nitrogen in *root nodules* that contain *nitrogen-fixing bacteria*. This symbiotic relationship with the bacteria means relatively little nitrogenous fertilizer is required for agricultural production of legumes. Tarwi (*Lupinus mutabilis*) is a legume native to the South American Andes that has a high protein and oil content, similar to the soybean. Tarwi is also high in the essential amino acid lysine. It grows well in poor soils and is drought-resistant. Current breeding efforts focus on reducing the bitter alkaloids, which can be removed by rinsing in water.

The winged bean (*Psophocarpus tetragonolobus*), native of tropical Asia, is entirely edible—leaves, flowers, seeds, pods, and tuberous roots. Like most legumes, the winged bean has a high protein content. This species could have tremendous potential in many tropical regions of the world, rivaling the success of the soybean.

A native of North America, the groundnut (*Apios americana*) was a major food source of many American Indian tribes. It is purported to have been offered to the Pilgrims to avert starvation. The numerous underground tubers can be prepared (cooking is necessary) like potatoes yet have a much higher protein content.

Several other legumes whose use and acceptance are likely to increase include the tepary bean (*Phaseolus acutiflius*), the pigeon pea (*Cajanus cajan*), and the bambara groundnut (*Voandzeia subterranea*).

Other Crops

There are many other potential food crops. Most have been cultivated on a small scale for years and are being rediscovered and researched for commercial production. Some of these include potato-like oca tubers (*Oxalis tuberosa*), fruits such as cherimoya (*Annona cherimola*), pepino (*Solanum muricatum*), and feijo (*Acca sellowiana*), and nuts such as egg nut (*Couepia longipendula*).

Thomas J. Montagno

See also: Agriculture: world food supplies; Alternative grains; Culturally significant plants; Fruit crops; Grains; Legumes; Nitrogen fixation; Nutrition in agriculture; Plants with potential; Vegetable crops.

Sources for Further Study

Janick, Jules, and James E. Simon, eds. *Advances in New Crops*. Portland, Oreg.: Timber Press, 1990. Proceedings of the First National Symposium NEW CROPS: Research, Development, Economics. Contains research articles on a variety of potential food, fiber, industrial, forage, and medicinal crops. Includes tables, charts, photographs, indexes.

Levins, Estelle, and Karen McMahon. *Plants and Society*. 2d ed. Boston: WCB/McGraw-Hill, 1999. This basic textbook provides historical and botanical information about the world's major and alternative food crops. Includes illustrations, photographs, and tables.

Simpson, Britnall B., and Molly C. Ogarzaly. *Economic Botany: Plants in Our World*. 3d ed. Boston: WCB/McGraw-Hill, 2001. This textbook contains a chapter on the future use of plants. Includes illustrations, photographs, bibliographical information.

AGRICULTURAL REVOLUTION

Categories: Agriculture; economic botany and plant uses; history of plant science

The agricultural revolution marked the transition by humans from hunting and gathering all their food to domesticating plants for food.

People first obtained their food by scavenging kills made by other animals, by hunting animals, and by gathering wild food plants. Between ten thousand and twelve thousand years ago, people began to use plants in new ways. Some scientists and historians call this period of time the "agricultural revolution."

Agricultural Beginnings

Before the 1960's, many scientists and historians believed that hunter-gatherers abruptly switched from foraging to farming. Those who thought that agriculture arose quickly coined the term "agricultural revolution." They suggested that this revolution spread rapidly because it was a tremendous improvement over the old foraging lifestyle, with the availability of cultivated food sources far more dependable than those of wild sources.

Since the 1960's, scientists and historians have challenged this view of agricultural beginnings. Later studies have shown that modern hunting-gathering societies have a remarkably sophisticated knowledge about native plants and plants' life cycles. Gatherers use a large number of plant species for food. Hunting and gathering cultures today do not have to plant seeds intentionally to keep from starving and most likely did not have to do so in the past.

Domesticated and Wild Plants

Domesticated plants are genetically distinct from their wild ancestors. *Domestication* involves processes by which a wild plant adapts to the needs of the farmer. The traits that make a plant desirable as a human food plant may not be ones that confer survival value on plants in their natural habitats and may actually be detrimental to the plant's survival in the wild.

People who gathered wild plants looked for traits that made gathering easier and more profitable. They would have gathered grasses, for example, that had bigger seeds or plants, had more seeds or fruits or edible parts, or had seed heads that did not shatter easily. If seeds from such plants are the ones that were planted, accidentally or on purpose, their useful traits would be reinforced in successive generations. The appearance of a domesticated plant in the archaeological record is the end result of generations of cumulative genetic transformations that might have taken hundreds or even thousands of years. Therefore, it becomes difficult to pick a single point in the past for any continent or geographic region that signals the beginning of an agricultural economy.

Geography of Agricultural Origins

The area of the Middle East called the Fertile Crescent (between the Tigris and Euphrates Rivers in what is now Iraq) seems to be the first area where formal agriculture began. The native grasses were highly productive. Wild wheat and barley grew in dense stands and were valuable food sources before cultivation began. It was this use of gathered wild grasses for food that probably led to their early domestication. Along with the grasses, complementary sources of protein were adopted, namely leguminous crops such as pea and lentil, and animals were domesticated. The plants that were domesticated in the Middle East include einkorn wheat, emmer wheat, bread wheat, barley, lentil, pea, vetch, the fava (or broad) bean, chickpea, lupine, and flax.

Agriculture originated in northeast China with the Yang Shao culture around six thousand years ago and spread quickly into Korea and Japan. Some of the plants brought under cultivation include barley, barnyard millet, common or broomcorn millet, foxtail millet (or foxtail grass), soybean, adzuki and mung beans, hemp, buckwheat, bottle gourd, Chinese cabbage, great burdock, lacquer tree, paper mulberry, and a number of fruit trees, including apricot, pear, peach, and plum.

In Southeast Asia, as well as the Pacific Islands and India, cultivated plants included sesame, the pigeon pea, eggplant, rice, sugarcane, bananas, plantains, coconuts, oranges, mango, Asian yam, betel nut, pepper, taro, bitter gourd, winter melon, snake gourd, luffa, mangosteen, durian, rambutan, breadfruit, and bamboo.

In Africa, plant domestication took place south of the Sahara Desert and north of the equator. Many crops were grown, including various kinds of millet, sorghum, okra, coffee, watermelon, several species of yam, African rice, cowpea, African oil palm, and cola nut. In Ethiopia, Ethiopian oats, coffee, enset, tef, noog, and chat were cultivated.

In Central America, archaeological evidence suggests that squash and pumpkins may have been cultivated before corn, especially in the Oaxaca region. In Oaxaca and Tamaulipas, along with squash and pumpkins, people were cultivating beans and bottle gourds, followed later by corn. In the Tehuacán Valley of Central Mexico, corn, chile peppers, avocado, beans, amaranth, and foxtail grass were among the very earliest cultivated plants. Cultivated plants either originating or cultivated early in South America include quinoa, white potato, peanut, cacao, jicama, lima bean, common bean, pineapple, chile pepper, papaya, sweet potato, yucca, and avocado. Tomatoes were cultivated in both Central and South America.

In North America, prior to the diffusion of the corn-squash-beans complex from the southwest after 1000 C.E., Indians of Eastern North America were cultivating a number of plants, including bottle gourd, erect knotweed, sumpweed, goosefoot, maygrass, little barley, and sunflower.

Carol S. Radford

See also: African agriculture; Agriculture: traditional; Asian agriculture; Central American agriculture; Fruit crops; Grains; Plant domestication and breeding; South American agriculture; Vegetable crops.

Sources for Further Study

Cowan, Wesley C., and Patty Jo Watson. *The Origins of Agriculture: An International Perspective*. Washington, D.C.: Smithsonian Institution Press, 1992. Contributions by different authors address plant domestication and the origins of agriculture around the world. Includes maps, tables.

Grigg, David B. *An Introduction to Agricultural Geography*. 2d rev. ed. London: Routledge, 1995. Provides a comprehensive introduction to agriculture in developing and developed nations, describing both human and environmental issues. Covers the physical environment, economic behavior and demands, institutional, social, and cultural influences, and the impact of farming upon the environment.

Harlan, Jack R. *The Living Fields: Our Agricultural Heritage*. New York: Cambridge University Press, 1995. Discusses both the origins of agriculture and the different regions of the world where it developed. Includes maps and photographs.

Heiser, Charles B., Jr. *Seed to Civilization: The Story of Food*. Cambridge, Mass.: Harvard University Press, 1990. Discusses many of the plants domesticated for human use.

AGRICULTURE: HISTORY AND OVERVIEW

Categories: Agriculture; disciplines; economic botany and plant uses

Agriculture is the ability to produce sufficient food and fiber to feed and shelter the population, the most important natural resource a nation can have. In modern urban societies, it is also the natural resource that is most often taken for granted.

The beginnings of agriculture predate written history. No one knows when the first crop was cultivated, but at some time in the distant past humans discovered that seeds from certain wild grasses could be collected and planted in land that could be controlled and the grasses later gathered for food. Most scholars believe this occurred at about the same time in both the Eastern and Western Hemispheres, some eight thousand to ten thousand years ago.

EPHEMERAL ARCHIVES

The earliest agricultural centers were located near large rivers that helped maintain soil fertility with the deposition of new topsoil during each annual flooding cycle.

Early Agriculture

The earliest attempts to grow crops were primarily to supplement the food supply provided by hunting and gathering. However, as the ability to produce crops increased, people began to *domesticate* plants and animals, and their reliance on hunting and gathering decreased, allowing the development of permanent settlements in which humans could live. As far back as six thousand years ago, agriculture was firmly established in Asia, India, Mesopotamia, Egypt, Mexico, Central America, and South America.

The earliest agricultural centers were located near large rivers that helped maintain soil fertility by the deposition of new *topsoil* with each annual flooding cycle. As agriculture moved into regions that lacked the annual flooding of the large rivers, people began to utilize a technique known as *slash-and-burn agriculture*. In this type of agriculture, a farmer clears a field, burns the tress and brush, and farms the field. After a few years soil nutrients become depleted, so the farmer must repeat the process at a new location. This type of agriculture is still practiced in some developing countries and is one reason tropical rain forests are disappearing at a fast rate.

Until the nineteenth century, most farms and ranches were family-owned, and most farmers practiced *sustenance agriculture*: Each farmer produced a variety of crops sufficient to feed his or her family as well as a small excess which was sold for cash or bartered for other goods or services. Agricultural tools such as plows were made of wood, and almost all agricultural activities required human or animal labor. This situation placed a premium on large families to provide the help needed in the fields.

The arrival of the Industrial Revolution changed agriculture, just as it did almost all other industries. Eli Whitney invented the cotton gin in 1793. The mechanical reaper was invented by Cyrus McCormick, and John Lane and John Deere began the commercial manufacture of the steel plow in 1833 and 1837, respectively. These inventions led the way to the development of the many different types of agricultural machinery that resulted in the *mechanization* of most farms and ranches. By the early part of the twentieth century, most agricultural enterprises in the United States were mechanized. American society was transformed from an agrarian society into an urban society. People involved in agricultural production left farms to go to cities to work in factories. At the same time, there was no longer a need for large numbers of people to produce crops. As a result, fewer people were required to produce the growing amounts of agricultural products that supplied an increasing number of consumers.

Modern Agriculture

As populations continued to grow, there was a need to select and produce crops with higher yields. The *Green Revolution* of the twentieth century helped to make these higher yields possible. Basic information supplied by biological scientists allowed agricultural scientists to develop new, higher-yielding varieties of numerous crops, particularly the *seed grains* which supply most of the calories necessary for maintenance of the world's population. These higher-yielding crop varieties, along with improved farming methods, resulted in tremendous increases in the world's food supply.

The new crop varieties also led to an increased reliance on *monoculture*. While the practice of growing only one crop over a vast number of acres has resulted in much higher yields, it has also decreased the genetic variability of many agricultural plants, increased the need for commercial fertilizers, and produced an increased susceptibility to damage from a host of biotic and abiotic factors. These latter two developments have resulted in a tremendous growth in the *agricultural chemical in-*

Major Crops and Places of Original Cultivation

Crop	*Region*	*Crop*	*Region*
Apples	Central Asia and the Middle East	Olives	Mediterranean
Apricots	China	Onions	Mediterranean
Asparagus	Mediterranean	Oranges	China, India, and Southeast Asia
Avocados	Central and South America	Papayas	Central and South America
Bananas	India and Southeast Asia	Parsnips	Mediterranean
Beans	Ethiopia	Peaches	China
Bell peppers	Central and South America	Peanuts	Central and South America
Cantaloupes	Central Asia and the Middle East	Pears	Central Asia and the Middle East
Carrots	Central Asia and the Middle East	Peas	Ethiopia
Cashews	Central and South America	Pineapples	Central and South America
Celery	Mediterranean	Potatoes	Central and South America
Cherries	China	Pumpkins	Central and South America
Coconuts	India and Southeast Asia	Radishes	India and Southeast Asia
Coffee	Ethiopia	Rhubarb	Mediterranean
Corn	Central and South America	Rice	India and Southeast Asia
Cotton	Central Asia and the Middle East	Rye	Central Asia and the Middle East
Cucumbers	China	Soybeans	China
Eggplant	China	Spinach	Central Asia and the Middle East
Figs	Central Asia and the Middle East	Sugarcane	China
Garlic	Central Asia and the Middle East	Sweet potatoes	Central and South America
Ginger	India and Southeast Asia	Tangerines	India and Southeast Asia
Grapes	Central Asia and the Middle East	Tea	China
Leeks	Central Asia and the Middle East	Tomatoes	Central and South America
Lettuce	Mediterranean	Turnips	Central Asia, the Mediterranean, and the Middle East
Lima beans	Central and South America		
Mangoes	India and Southeast Asia	Walnuts	China
Mustard	India and Southeast Asia	Wheat	Central Asia and the Middle East
Oats	Central Asia and the Middle East	Yams	India and Southeast Asia
Okra	Ethiopia		

Source: Data are from Brian Capon, *Botany for Gardeners: An Introduction and Guide* (Portland, Oreg.: Timber Press, 1990), p. 81.

dustry. Today's modern agricultural unit requires relatively few employees, is highly mechanized, devotes large amounts of land to the production of only one crop, and is highly reliant on agricultural chemicals such as fertilizers and pesticides.

Agricultural Diversity

Modern agriculture is subdivided into many different specialties. Those agricultural industries that deal with plants include *agronomy*, the production of field crops; *forestry*, the growth and production of trees; and *horticulture*. Horticulture is subdivided into *pomology*, the growth and production of fruit crops such as oranges and apples; *olericulture*, the growth and production of vegetable crops (tomatoes, lettuce); *landscape horticulture*, the growth and production of trees and plants that are used in landscape design; and *floriculture*, the growth and production of flowering plants used in the floral industry.

The various agriculture industries produce a tremendous number of agricultural products. Agricultural products that are derived from plants can be subdivided into timber products (lumber, furniture), grain products (wheat, oats), fiber products (cotton, flax), fruit products (grapes, peaches), nut crops (pecans, hazelnuts), vegetable products (lettuce, cabbage), beverage products (tea, coffee), spice and drug crops (garlic, mustard, opium, quinine), ornamental crops (carnations, chrysanthemums), forage crops (alfalfa, clover), and other cash crops such as sugarcane, tobacco, artichokes, and rubber.

Impact on Soil Resources

While there have been tremendous increases in agricultural productivity through the use of modern agricultural practices, these practices have had a significant impact on some other natural resources. *Soil* is one of the most overlooked and misunderstood resources. Most people think of soil as an inert medium from which plants grow. In reality, *topsoil*—that upper 15 to 25 centimeters (6 to 10 inches) of the earth's terrestrial surface in which nearly all plants grow—is a complex mixture of weathered mineral materials from rocks, partially decomposed organic molecules, and a large number of living organisms.

The process of soil formation is very slow. Under ideal conditions, enough topsoil can form in one year to produce a layer of about 1 millimeter (0.04 inch) thick when spread over 1 hectare (2.5 acres). With proper management, topsoil can be kept fertile and productive indefinitely. Unfortunately, many agricultural techniques lead to the removal of trees and shrubs, which provide windbreaks, or to the depletion of soil fertility, which reduces the plant cover over the field. These practices expose the soil to increased erosion from wind and moving water. As a result, as much as one-third of the world's current croplands are losing topsoil faster than it can be replaced.

Water and Irrigation

Because plants require water in order to grow, agriculture represents the largest single use of global water. About 73 percent of all fresh water withdrawn from groundwater supplies, rivers, and lakes is used in the irrigation of crops. Almost 15 percent of the world's croplands are irrigated. Water use varies among countries. Some countries have abundant water supplies and irrigate liberally, while water is very scarce in other countries and must be used very carefully. Because as much as 80 percent of the water intended for irrigation is lost to evaporation before reaching the plants, the efficiency of water use in some countries can be very low.

There is no doubt that irrigation has dramatically increased crop production in many areas, but some irrigation practices have been detrimental. Overwatering can lead to a *waterlogging* of the soil. Waterlogging cuts off the supply of oxygen to the roots, and the plants die. Irrigation of crops in dry climates can often result in *salinization* of the soil. In these climates, the irrigation water rapidly evaporates from the soil, leaving behind mineral salts that were dissolved in the water. As the salts accumulate, they become lethal to most plants. Some experts estimate that as much as one-third of the world's agricultural soil has been damaged by salinization. There is also an argument as to whether the increased usage of water for agriculture has decreased the supply of potable water fit for other uses.

Fertilizers

Plants require sunshine, water either from rainfall or irrigation, carbon dioxide from the atmosphere, and thirteen mineral nutrients from the soil. Of these, calcium, magnesium, nitrogen, phosphorus, and potassium are required in the greatest amounts. Calcium and magnesium are plentiful in soils located in dry climates, but in wetter climates these nutrients are often *leached* through the soil. In these regions, calcium and magnesium are returned to the soil in the form of lime, which is also sometimes added to soil to raise its pH (increase its alkalinity). Nitrogen, phosphorus, and potassium are the nutrients which are most often depleted from agricultural soils, and these nutrients are often referred to as the *fertilizer elements*. Because these nutrients stimulate plant growth and can greatly increase crop yields, it is necessary to apply them to the soil regularly in order to maintain fertility.

The amount of fertilizer applied to the soil increased more than 450 percent in the second half of the twentieth century. While this increase in the use of fertilizers has more than doubled worldwide crop production, it has also caused some problems. The increased production of fertilizers has required the use of energy and mineral resources that could have been used elsewhere. In many cases, farmers tend to overfertilize. Overfertilization not only wastes money but also contributes to environmental degradation. Fertilizer elements, particularly nitrogen and phosphorus, are carried away by water runoff and are eventually deposited in the rivers and lakes, where they contribute to pollution of aquatic ecosystems. In addition, nitrates can accumulate in underground water supplies. These nitrates can be harmful if ingested by newborns.

Other Resources

Modern agriculture, as it is practiced in the United States, consumes large amounts of energy. Farm machinery used in planting, cultivating, harvesting, and transporting crops to processing plants or to market consumes large supplies of liquid fossil fuels such as gasoline or diesel. The energy required to produce fertilizers, pesticides, and other agricultural chemicals is the second largest energy cost associated with agriculture. The use of fuel required by pumps to irrigate crops is also a major energy consumer. Additional energy is used in food processing, distribution, storage, and cooking after the crop leaves the farm. The energy used for these activities may be five times as much as that used to produce the crop.

About 16 percent of the total energy used in the United States is consumed by systems devoted to feeding people, and most of the foods consumed in the United States require more calories of energy to produce, process, and distribute to the market than they provide when they are eaten. The next major development in agriculture will be the *biotechnical revolution*, in which scientists will be able to use molecular biological techniques to produce new crop varieties. In the future, agricultural scientists may be able to develop crop plants that can be produced, processed, and distributed with less impact on other resources.

Commercial Impact

While fewer than 1 percent of Americans are directly involved in agricultural production, agriculture in the United States employs about seventeen million people in some phase of the industry, from production to retail sales. This includes workers hired by agricultural chemical companies which produce or sell agricultural implements and machinery, processing and canning plants, and wholesale and retail marketing firms such as grocery stores. In 1997 there were six thousand to eight thousand different agricultural products on the market. Agriculture is also big business, with assets of about one hundred billion dollars. This figure equals 88 percent of the combined capital assets of all the major U.S. corporations. American farmers now produce 76 percent more crops than the previous generation did on the same amount of land, and one-third of all American agricultural products are exported. This makes the United States the world's most agriculturally productive country and the largest exporter of agricultural products.

D. R. Gossett

See also: Agricultural crops: experimental; Agricultural revolution; Agriculture: marine; Agriculture: modern problems; Agriculture: traditional; Agriculture: world food supplies; Agronomy; Alternative grains; Asian agriculture; Australian agriculture; Biofertilizers; Biopesticides; Biotechnology; Caribbean agriculture; Central American agriculture; Composting; Corn; Drought; European agriculture; Farmland; Fertilizers; Grains; Green Revolution; Herbicides; High-yield crops; Horticulture; Hybridization; Hydroponics; Integrated pest management; North American agriculture; Organic gardening and farming; Pesticides; Plant biotechnology; Plants with potential; Rice; Slash-and-burn agriculture; Soil; South American agriculture; Sustainable agriculture; Vegetable crops.

Sources for Further Study

Acquaah, George. *Principles of Crop Production: Theory, Techniques, and Technology*. Englewood, N.J.: Prentice Hall, 2001. Comprehensive coverage of U.S. crop production issues, including plant anatomy, climate and weather, soil and land, radiation, temperature and air, and photosynthesis. Well illustrated.

Grigg, David B. *An Introduction to Agricultural Geography*. 2d rev. ed. London: Routledge, 1995. Provides a comprehensive introduction to agriculture in developing and developed nations, describing both human and environmental issues. Covers the physical environment, economic behavior and demands, institutional, social and cultural influences, and the impact of farming upon the environment.

Janick, Jules. *Horticultural Science*. 4th ed. New York: W. H. Freeman, 1986. Contains sections on the major horticultural crops.

Smith, Bruce D. *The Emergence of Agriculture*. New York: W. H. Freeman, 1994. A lucid synthesis of many disciplines, explaining the origins of agriculture and the transition from wild to domesticated crops.

Zohary, Daniel, and Maria Hopf. *Domestication of Plants in the Old World: The Origin and Spread of Cultivated Plants in West Asia, Europe, and the Nile Valley*. 3d ed. New York: Oxford University Press, 2001. A thorough review of information on the beginnings of agriculture, particularly utilizing new molecular biology findings on the genetic relations between wild and domesticated plant species.

AGRICULTURE: MARINE

Categories: Agriculture; economic botany and plant uses; food; water-related life

Marine agriculture uses techniques of artificial cultivation, such as growing, managing, and harvesting, and applies them to marine plants and animals. The products are then used for human consumption.

Marine agriculture is also known as *mariculture* or *aquaculture*, although aquaculture is a more general term referring to both freshwater and marine farming of organisms. The world's oceans cover approximately three-fourths of the globe, including vast regions of unexplored life and landforms. The potential for exploiting the oceans agriculturally is great but currently meets significant obstacles. Because of the expense of equipment and personnel involved, most marine species are not cultivated. Coastal pollution, habitat destruction, competition for land use, and economics all limit mariculture programs. Nevertheless, mariculture does offer several food, medical, and other products that are currently being marketed.

Food

Seaweeds are edible, especially the red and brown algae. The three most common types of seaweeds are known by their Japanese names: *nori* (*Porphyra*), a red seaweed high in vitamin C and digestible protein; *kombu* (*Laminaria*); and *wakame* (*Undaria*), high in calcium. They are eaten raw, cooked, or dried and have several vitamins and minerals as well as protein. Seaweeds are low in fats, and 35 to 50 percent of the dry weight of red seaweeds is protein. Seaweeds can be used to add taste and variety to foods. They are used as a hot vegetable, boiled and formed into cakes and fried, in salads, and in preparing desserts, breads, soups, casseroles, sandwiches, teas, and candy.

The world's yearly harvest of seaweeds is approximately 8.4 million tons of green seaweed, 2.8 million tons of brown seaweed, and 1.2 million tons of red seaweed. The total seaweed market in 1998 was worth more than $5 billion, with $600 million deriving from food additives alone. China is the leading harvester and the world's biggest seaweed consumer. Japan is the leading seaweed importer and, at the end of the twentieth century, employed more than thirty-five thousand people in the industry. Harvesting and marketing edible seaweed is a growing business in the United States, especially on the West Coast.

Seaweeds produce several types of *phycocolloids*, starchlike chemicals used in food processing and manufacturing. An important type called *algin*, which makes up alginic acid and alginates, is used in manufacturing dairy products such as ice cream, cheese, and toppings as well as to prevent frostings and pies from desiccation. Another extract is *agar*, used to form jellies and protect fish and meats during canning. Agar is also used in low-calorie foods and as a thickener. Red algae is a source of the agglutinant *carrageenan*, which is used in many food products as an emulsifier to give body to dairy products and other processed foods, including instant puddings. Additionally, seaweed-based food additives are common in prepared and fast foods, including hamburgers and yogurt.

Kelp farming is a major livelihood in the eastern Pacific, with approximately 140,000 tons harvested each year for the extraction of alginates used in food and food additives. Kelp is a good source of calcium, potassium, iron, iodine, bromine, and zinc. It is also low-fat, has some protein, and is a natural tenderizer. Kelp flakes are used as a low-sodium salt substitute.

Medicine

The use of marine plants in medicine is still in the early stages of exploration and faces many challenges, including identification of useful chemicals and the cultivation of significant quantities. Dinoflagellates and other microalgae are being investigated for compounds that might fight cancerous tumors. Diluted algae toxins from red tides can be used to inhibit the growth of most bacteria. Green algae has halosphaerin, a strong antibiotic. Seaweed is used in wound dressings in hospitals and as a source of iodine, A, B, D, and E vitamins, calcium, magnesium, potassium, sodium, sulfur, and trace antioxidants such as selenium and zinc. The seaweed extract agar is used in laxatives and as a medium to grow bacteria and molds.

Kelp is rich in *chlorophyll*, which can help detoxify the body, fight inflammations, and increase the formation of oxygen-carrying red blood cells. Chlorophyll is also used to fight bad breath and as an ingredient in deodorants. Kelp is used to reduce cholesterol, treat gastrointestinal, respiratory, and genitourinary disorders, and lower blood pressure. The alginic acid produced by kelp can rid the body of radioactive strontium, the most dangerous to humans of all components in the fallout from atomic explosions.

Other Uses

Marine plants are used for a variety of other purposes. Seaweed is used as a component of many fertilizers, as a food additive in animal feed, and to reduce soil acidity. Research on cattle and swine has revealed that the addition of seaweed to animal feed can enhance the immune system and makes the meat a more desirable color. It can also save cattle from the effects of fungus-infected grass.

Seaweed is used as an ingredient in cosmetics as well as to nourish, revitalize, condition, and improve the skin, hair, and body. It is used in cleansers, toners, moisturizers, scrubs, body lotions, and hair and bath products. The giant kelp (*Macrocystis*) is a major source of algin for commercial uses, as is the brown algae *Laminaria*, which is harvested in the north Atlantic. Algin is used in shampoos, shaving cream, plastics, pesticides, rubber products, paper, paints, and cosmetics. Additionally, kelp is used in emulsifiers for toothpastes and printing inks. Kelp has even been used to make fishing lines. Some research has been done on using kelp as a fuel to produce a clean-burning methane gas. Kelp can be used to ferment human waste and garbage, which can then be sold as fertilizers.

Virginia L. Hodges

See also: Agricultural crops: experimental; Agriculture: world food supplies; Algae; Brown algae; Carbohydrates; Green algae; Marine plants; Medicinal plants; Red algae.

Sources for Further Study

Castro, Peter, and Michael E. Huber. *Marine Biology*. 3d ed. Boston: McGraw-Hill, 2000. Well-organized, basic textbook with summary and bibliography with each chapter. Good section on marine agriculture. Includes illustrations, glossary, and index.

Cousteau, Jacques. *The Ocean World*. New York: Harry N. Abrams, 1985. Discusses all ma-

rine life. Includes color photographs, glossary, bibliography, and index.
Levinton, Jeffrey S. *Marine Biology: Function, Biodiversity, Ecology*. New York: Oxford University Press, 1995. Well-organized, advanced textbook. Bibliography and review questions with each chapter. Includes glossary, index.
Sumich, James L. *An Introduction to the Biology of Marine Life*. 5th ed. Dubuque, Iowa: Wm. C. Brown, 1992. Textbook format with basic explanations, review and discussion questions, and bibliography with each chapter. Includes glossary and taxonomic and subject indexes.

AGRICULTURE: MODERN PROBLEMS

Categories: Agriculture; economic botany and plant uses; environmental issues

Many current problems in agriculture are not new. Erosion and pollution, for example, have been around as long as agriculture. However, agriculture has changed drastically within its ten-thousand-year history, especially since the dawn of the Industrial Revolution in the seventeenth century. Erosion and pollution are now bigger problems than before and have been joined by a host of other issues that are equally critical—not all related to physical deterioration.

Monoculture

Modern agriculture emphasizes *crop specialization*, also known as *monoculture*. Farmers, especially in industrialized regions, often grow a single crop on much of their land. Problems associated with this practice are exacerbated when a single variety or cultivar of a species is grown. Such a strategy allows the farmer to reduce costs, but it also makes the crop, and thus the farm and community, susceptible to widespread crop failure. The corn blight of 1970 devastated more than 15 percent of the North American corn crop. The corn was particularly susceptible to the harmful organisms because 70 percent of the crop being grown was of the same high-yield variety. Chemical antidotes can fight pests, but they increase pollution. Maintaining species diversity or varietal diversity—growing several different crops instead of one or two—allows for crop failures without jeopardizing the entire economy of a farm or region that specializes in a particular monoculture, such as tobacco, coffee, or bananas.

Genetic Engineering

Growing *genetically modified* (GM) crops is one attempt to replace post-infestation chemical treatments. Recombinant technologies used to splice genes into varieties of rice or potatoes from other organisms are becoming increasingly common. The benefits of such GM crops include more pest-resistant plants and higher crop yields. However, environmentalists fear new genes could trigger unknown side effects with more serious, long-term environmental and economic consequences than the problems they were used to solve. GM plants designed to resist herbicide applications could potentially pass the resistant gene to closely related wild weed species that would then become "super weeds." Also, pests, just as they can develop resistance to pesticides, may also become resistant to defenses engineered into GM plants. The high cost of recombinant technologies calls into question the feasibility of continuing development of GM plants.

Erosion

An age-old problem, soil loss from *erosion* occurs all over the world. As soil becomes unproductive or erodes away, more land is plowed. The newly plowed lands usually are considered *marginal*, meaning they are too steep, nonporous or too sandy, or deficient in some other way. When natural vegetative cover blankets these soils, it protects them from erosive agents: water, wind, ice, or gravity. Plant cover "catches" rainwater that seeps downward into the soil rather than running off into rivers. As marginal land is plowed or cleared to grow crops, erosion increases.

Expansion of land under cultivation is not the only factor contributing to erosion. Fragile grasslands in dry areas also are being used more intensively. Grazing more livestock than these pastures can handle decreases the amount of grass in the pasture and exposes more of the soil to wind, the primary erosive agent in dry regions. *Overgrazing* can affect pastureland in tropical regions too. Thousands of acres of tropical forest have been cleared to establish cattle-grazing ranges in Latin America. Tropical soils, although thick, are not very fertile. After one or two growing seasons, crops grown in these soils will yield substantially less than before.

Tropical fields require *fallow periods* of about ten years to restore the soil after it is depleted. That is why tropical farmers using *slash-and-burn agriculture* move to new fields every few years in a cycle that returns them to the same place years later, after their particular lands have regenerated. Where there is heavy forest cover, soils are protected from exposure to the massive amounts of rainfall. Organic material for crops is present as long as the forest remains in place. When the forest is cleared, however, the resulting grassland cannot provide the adequate protection, and erosion accelerates. Lands that are heavily grazed provide even less protection from heavy rains, and erosion accelerates even more.

The use of machines also promotes erosion, and modern agriculture relies on machinery such as tractors, harvesters, trucks, balers, and ditchers. In industrialized nations, machinery is used intensely. Machinery use is on the rise in developing countries such as India, China, Mexico, and Indonesia, where traditional, nonmechanized farming methods are the norm. Farming machines, in gaining traction, loosen topsoil and inhibit vegetative cover growth, especially when farm implements designed to rid the soil of weeds are attached. The soil is then more prone to erode.

Eco-fallow farming has become more popular in the United States and Europe as a way to reduce erosion. This method of agriculture, which leaves the crop residue in place over the fallow (nongrowing) season, does not root the soil in place as well as living plants do. As a result, some erosion

AP/Wide World Photos

A Texas farmer digs an open irrigation ditch in a grapefruit orchard. With an increasing reliance on irrigation worldwide, groundwater resources are mismanaged and overtapped. In the United States, aquifer overdraft averages 25 percent over the replacement rate.

continues. Additionally, eco-fallow methods require heavy use of chemicals, such as herbicides, to "burn down" weed growth at the start of the growing season. This contributes to increased erosion and pollution.

Pollution and Silt

Besides causing resistance among harmful bacteria, insects, and weeds, pesticides inevitably wash into, and contaminate, surface and groundwater supplies. Chemicals, although problematic, are not as difficult to contend with as the increasingly heavy silt load choking the life out of streams and rivers. Accelerated erosion from water runoff carries silt particles into streams, where they remain suspended and inhibit the growth of many forms of plant and animal life.

The silt load in American streams has become so heavy that the Mississippi River Delta is growing faster than it once did. Heavy silt loads, combined with chemical residues, are creating an expanded dead zone. By taxing the capabilities of ecosystems around the Delta, sediments are filtered out slowly, plant absorption of nutrients is decreased, and salinity levels for aquatic life cannot be stabilized. Most of the world's population lives in coastal zones, and 80 percent of the world's fish catch comes from coastal waters over continental shelves that are most susceptible to this form of pollution.

Pesticide Resistance

With the onset of the *Green Revolution*, the use of herbicides, insecticides, and other *pesticides* increased dramatically all over the world. An increasing awareness of problems caused by overuse of pesticides extends even to household antibacterial cleaning agents and other products. Mutations among the genes of bacteria and plants have allowed these organisms to resist the effects of chemicals that were toxic to their ancestors. Use of pesticides leads to a cycle wherein more, or different combinations of, chemicals are used, and more pests develop resistance to these toxins. Additionally, the development of herbicide-resistant crop plants enables greater use of herbicides to kill undesirable weeds on croplands.

Increasing interest in *biopesticides* may slow the cycle of pesticide resistance. Types of biopesticides include beneficial microbes, fungi, and insects such as ladybugs that can be released in infested areas to prey upon specific pests. Biopesticides used today include naturally occurring and genetically modified organisms. Their use also avoids excessive reliance on chemical pesticides.

Fertilizers and Eutrophication

Increased use of *fertilizers* was another result of the Green Revolution. Particulate amounts of most fertilizers enter the hydrologic cycle through runoff. As a result, bodies of water become enriched in dissolved nutrients, such as nitrates and phosphates. The growth of aquatic plants in rivers and lakes is overstimulated, and this results in the depletion of dissolved oxygen. This process of *eutrophication* can harm all aquatic life in these ecosystems.

Water Depletion

With an increasing reliance on irrigation, groundwater resources are mismanaged and overtapped. The rate of groundwater recharge is slow, usually between 0.1 and 0.3 percent per year. When the amount of water pumped out of the ground exceeds the recharge rate, it is referred to as *aquifer overdraft*. An aquifer is a water-bearing stratum of permeable rock, sand, or gravel.

In Tamil Nadu, India, groundwater levels dropped 25 to 30 meters during the 1970's due to excessive pumping for irrigation. In Tianjin, China, the groundwater level declines 4.4 meters per year. In the United States, aquifer overdraft averages 25 percent over the replacement rate. The Ogallala aquifer under Kansas, Nebraska, and Texas represents an extreme example of overdraft: Depletion is 130 to 160 percent above the replacement rate annually. At this rate, this aquifer, which supplies water to countless communities and farms, has been projected to become nonproductive by 2030.

Soil Salinization

In addition, continued irrigation of arid regions can lead to soil problems. *Soil salinization* is widespread in the small-grained soils of these regions, which have a high water absorption capacity and a low infiltration rate. Some irrigation practices add large amounts of salts into the soil, increasing its natural rate of salinization. This can also occur at the base of a hill slope. Soil salinization has been recognized as a major process of land degradation.

Although surface and groundwater resources cannot be enriched by technology, conservation and improved environmental management can make the use of precious freshwater more efficient.

In agriculture, for example, drip irrigation can reduce water use by nearly 50 percent. In developing countries, though, equipment and installation costs often limit the availability of these more efficient technologies.

Urban Sprawl

As more farms become mechanized, the need for farmers and farm workers is being drastically reduced. From a peak in 1935 of about 6.8 million farmers farming 1.1 billion acres, the United States at the end of the twentieth century counted fewer than 2 million farmers farming 950 million acres.

Urban sprawl converts a tremendous amount of cropland into parking lots, malls, industrial parks, and suburban neighborhoods. If cities were located in marginal areas, then concern about the loss of farmland to commercial development would be nominal. However, the cities attracting the greatest numbers of people have too often replaced the best cropland. Taking the best cropland out of primary production imposes a severe economic penalty.

James Knotwell and Denise Knotwell, updated by Bryan Ness and Elizabeth Slocum

See also: Biopesticides; Drought; Erosion and erosion control; Fertilizers; Forest management; Pesticides.

Sources for Further Study

Hoag, Dana. *Agricultural Crisis in America: A Reference Handbook*. Santa Barbara, Calif.: ABC-Clio, 1999. Summarizes challenges facing the American agricultural industry, investigating problems and solutions. Issues covered include groundwater pollution, loss of farmland, food safety, and wildlife and animal welfare.

Jackson, Wes. *New Roots for Agriculture*. Lincoln: University of Nebraska Press, 1985. A sociological study of agricultural ecology in the United States.

Lægreid, M., O. Kaarstad, and O. C. Bøckman. *Agriculture, Fertilizers, and the Environment*. New York: Cabi, 1999. Scientific review of environmental and sustainability issues related to fertilizer use.

Paarlberg, Don, and Phillip Paarlberg. *The Agricultural Revolution of the Twentieth Century*. Ames: Iowa State University Press, 2001. Documents the development of modern agriculture. Includes new and old photos of equipment and field techniques.

Rissler, Jane, and Margaret Mellon. *The Ecological Risks of Engineered Crops*. Cambridge, Mass.: MIT Press, 1996. A careful science and policy assessment of the potential dangers of genetically engineered crops. The authors—both of whom work for the Union of Concerned Scientists—not only outline the risks but also make suggestions for ways to minimize them.

AGRICULTURE: TRADITIONAL

Categories: Agriculture; economic botany and plant uses

Two agricultural practices that are widespread among the world's traditional cultures, slash-and-burn agriculture and nomadism, share several features. Both are ancient forms of agriculture, both involve farmers not remaining in a fixed location, and both can pose serious environmental threats if practiced in a nonsustainable fashion. The most significant difference between the two is that slash-and-burn is associated with raising field crops, while nomadism usually involves herding livestock.

Slash-and-Burn Agriculture

Farmers have practiced *slash-and-burn agriculture*, which is also referred to as *shifting cultivation* or *swidden agriculture*, in almost every region of the world where farming is possible. Although at the end of the twentieth century slash-and-burn agriculture was most commonly found in tropical areas such as the Amazon River basin in South America, swidden agriculture once dominated agriculture in more temperate regions, such as northern Europe. Swidden agriculture was, in fact, common in Finland and northern Russia well into the early decades of the twentieth century.

Slash-and-burn acquired its name from the practice of farmers who cleared land for planting crops by cutting down the trees or brush on the land and then burning the fallen timber on the site. The farmers literally slash and burn. The ashes of the burnt wood add minerals to the soil, which temporarily improves its fertility. Crops the first year following clearing and burning are generally the best crops the site will provide. Each year after that, the yield diminishes slightly as the fertility of the soil is depleted.

Farmers who practice slash-and-burn do not attempt to improve fertility by adding fertilizers such as animal manure to the soil. They instead rely on the soil to replenish itself over time. When the yield from one site drops below acceptable levels, farmers then clear another piece of land, burn the brush and other vegetation, and cultivate that site while leaving their previous field to lie fallow and its natural vegetation to return. This cycle will be repeated over and over, with some sites being allowed to lie fallow indefinitely, while others may be revisited and farmed again in five, ten, or twenty years.

PhotoDisc

Farm workers in China harvest tea.

Farmers who practice slash-and-burn do not always move their dwelling places as they cultivate different fields. In some geographic regions, farmers live in a central village and farm cooperatively, with fields being alternately allowed to remain fallow and farmed, making a gradual circuit around the central village. In other cases, the village itself may move as new fields are cultivated. Anthropologists studying indigenous peoples in Amazonia, for example, discovered that village garden sites were on a hundred-year cycle. Villagers farmed cooperatively to clear a garden site. That garden would be used for about five years; then a new site was cleared. When the fields in use became an inconvenient distance from the village—about once every twenty years—the entire village would move to be closer to the new fields. Over a period of approximately one hundred years, a village would make a circle through the forest, eventually ending up close to where it had been located long before any of the present villagers had been born.

In more temperate climates, farmers often owned and lived on the land on which they practiced swidden agriculture. Farmers in Finland, for example, would clear a portion of their land, burn the covering vegetation, grow grains for several years, and then allow that land to remain fallow for five to twenty years. The individual farmer rotated cultivation around the land in a fashion similar to that practiced by whole villages in other areas but did so as an individual rather than as part of a communal society.

Although slash-and-burn is frequently denounced as a cause of environmental degradation in tropical areas, the problem with it is not the practice itself but the length of the cycle. If the cycle of shifting cultivation is long enough, forests will grow back, the soil will regain its fertility, and minimal adverse effects will occur. In some regions, a piece of land may require as little as five years to regain its maximum fertility; in others, it may take one hundred years. Problems arise when growing populations put pressure on traditional farmers to return to fallow land too soon. Crops are smaller than needed, leading to a vicious cycle in which the next strip of land is also farmed too soon, and each site yields less and less. As a result, more and more land must be cleared.

Nomadism

Nomadic peoples have no permanent homes. They earn their living by raising herd animals, such as sheep, horses, or other cattle, and they spend their lives following their herds from pasture to pasture with the seasons, going wherever there is sufficient food for their animals. Most nomadic animals tend to be hardy breeds of goats, sheep, or cattle that can withstand hardship and live on marginal lands. Traditional nomads rely on natural pasturage to support their herds and grow no grains or hay for themselves. If a drought occurs or a traditional pasturing site is unavailable, they can lose most of their herds to starvation.

In many nomadic societies, the herd animal is almost the entire basis for sustaining the people. The animals are slaughtered for food, clothing is woven from the fibers of their hair, and cheese and yogurt may be made from milk. The animals may also be used for sustenance without being slaughtered. Nomads in Mongolia, for example, occasionally drink horses' blood, removing only a cup or two at a time from the animal.

In mountainous regions, nomads often spend the summers high up on mountain meadows, returning to lower altitudes in the autumn when snow begins to fall. In desert regions, they move from oasis to oasis, going to the places where sufficient natural water exists to allow brush and grass to grow, allowing their animals to graze for a few days, weeks, or months, then moving on. In some cases, the pressure to move on comes not from the depletion of food for the animals but from the depletion of a water source, such as a spring or well. At many desert oases, a natural water seep or spring provides only enough water to support a nomadic group for a few days at a time.

In addition to true nomads—people who never live in one place permanently—a number of cultures have practiced *seminomadic farming*: The temperate months of the year, spring through fall, are spent following the herds on a long loop, sometimes hundreds of miles long, through traditional grazing areas, then the winter is spent in a permanent village.

Nomadism has been practiced for millennia, but there is strong pressure from several sources to eliminate it. Pressures generated by industrialized society are increasingly threatening the traditional cultures of nomadic societies, such as the Bedouin of the Arabian Peninsula. Traditional grazing areas are being fenced off or developed for other purposes.

Environmentalists are also concerned about the ecological damage caused by nomadism. Nomads generally measure their wealth by the number of animals they own and will try to develop their herds to as large a size as possible, well beyond the numbers required for simple sustainability. The herd animals eat increasingly large amounts of vegetation, which then has no opportunity to regenerate. Desertification may occur as a result. Nomadism based on herding goats and sheep, for example, has been blamed for the expansion of the Sahara Desert in Africa. For this reason, many environmental policymakers have been attempting to persuade nomads to give up their traditional lifestyle and become sedentary farmers.

Nancy Farm Männikkö

See also: Forest management.

Sources for Further Study

Beckwith, Carol, and Marion Van Offelen. *Nomads of Niger*. New York: Academic Press, 1987. Well-written depiction of the life of the ancient nomadic Wodaabe people, who have roamed for centuries across the sub-Saharan bushland. Photographer Beckwith spent eighteen months traveling with one particular band of Wodaabe.

Colfer, Carol J., with Nancy Peluso and Chin See Chung. *Beyond Slash and Burn: Building on Indigenous Management of Borneo's Tropical Rain Forest*. New York: New York Botanical

Garden, 1994. From the Advances in Economic Botany series; contains bibliographical resources.

Goldstein, Melvyn C., and Cynthia M. Beall. *The Changing World of Mongolia's Nomads.* Berkeley: University of California Press, 1994. This anthropological study of a Mongolian herding community presents an intimate portrait of life on the steppes.

Grigg, David B. *An Introduction to Agricultural Geography.* 2d rev. ed. London: Routledge, 1995. Provides a comprehensive introduction to agriculture in developing and developed nations, describing both human and environmental issues. Covers the physical environment, economic behavior and demands, institutional, social and cultural influences, and the impact of farming upon the environment.

Mortimore, Michael. *Roots in the African Dust: Sustaining the Sub-Saharan Drylands.* New York: Cambridge University Press, 1998. This volume, accessible to students, policymakers, and scholars, includes chapters on the extent of soil degradation, merits of agricultural intensification, and conservation of biotic resources. Offers a breadth of analysis in a holistic, wide-ranging view of rural livelihoods and landscapes. Mortimore argues that desertification is reversible.

Schmink, Marianne, and Charles H. Wood. *Contested Frontiers in Amazonia.* New York: Columbia University Press, 1992. This interdisciplinary analysis of the process of frontier change in one region of the Brazilian Amazon shows how deforestation and settlement patterns were outcomes of the competition for resources among various groups.

AGRICULTURE: WORLD FOOD SUPPLIES

Categories: Agriculture; food

Soil types, topography, climate, socioeconomics, dietary preferences, stages in agricultural development, and governmental policies combine to give a distinctive personality to regional agricultural characteristics and, hence, food supplies in various areas of the world.

All living things need food to live, grow, work, and survive. Almost all foods that humans consume come from plants and animals. Not all of earth's people eat the same foods, however. The types, combinations, and amounts of food consumed by different peoples depend upon historic, socioeconomic, and environmental factors.

History of Food Consumption

Early in human history, people ate what they could gather or scavenge. Later, people ate what they could plant and harvest and the products of animals they could domesticate. Modern people eat what they can grow, raise, or purchase. Their diets or food composition is determined by income, local customs, religion or food biases, and advertising. There is a global food market, and many people can select what they want to eat and when they eat it according to the prices they can pay and what is available.

Historically, in places where food was plentiful, accessible, and inexpensive, humans devoted less time to basic survival needs and more time to activities that led to human progress and enjoyment of leisure. Despite a modern global food system, instant telecommunications, the United Nations, and food surpluses in some places, however, the problem of providing food for everyone on earth has not been solved.

In 1996 leaders from 186 countries gathered in Rome and agreed to reduce by half the number of hungry people in the world by the year 2015. United Nations data for 1998 revealed that more than 790 million people in the developing parts of

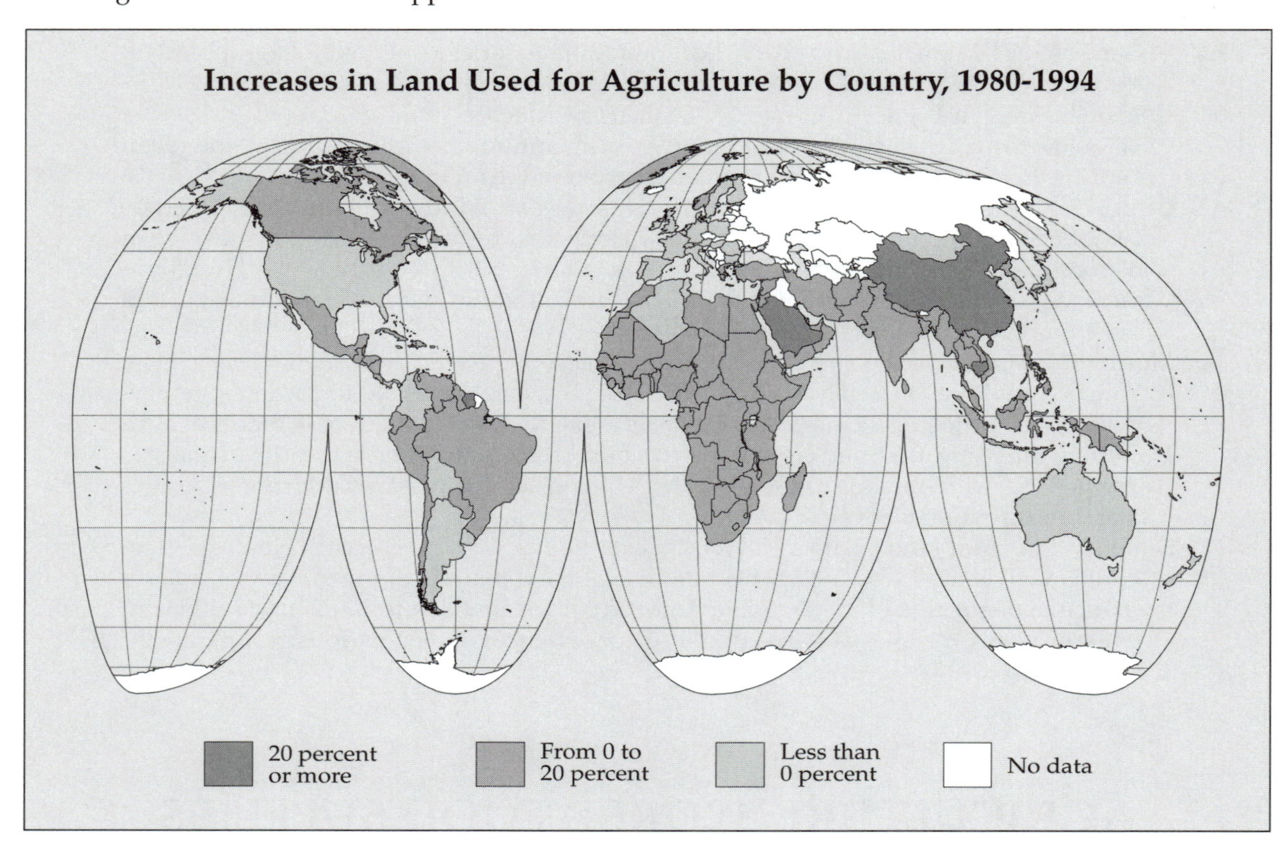

the world did not have enough food to eat. This is more people than the total population of North America and Europe at that time. The number of undernourished people has been decreasing since 1990. Still, at the current pace of hunger reduction in the world, 600 million people will suffer from "acute food insecurity" and go to sleep hungry in 2015. Despite efforts being made to feed the world, outbreaks of food deficiencies, mass starvation, and famine are a certainty in the twenty-first century.

World Food Source Regions

Agriculture and related primary food production activities, such as fishing, hunting, and gathering, continue to employ more than one-third of the world's labor force. Agriculture's relative importance in the world economic system has declined with urbanization and industrialization, but it still plays a vital role in human survival and general economic growth. Demands on agriculture in the twenty-first century include supplying food to an increasing world population of nonfood producers as well as producing food and nonfood crude materials for industry.

Soil types, topography, weather, climate, socioeconomic history, location, population pressures, dietary preferences, stages in modern agricultural development, and governmental policies combine to give a distinctive personality to regional agricultural characteristics. Two of the most productive food-producing regions of the world are North America and Europe. Countries in these regions export large amounts of food to other parts of the world.

North America is one of the primary food-producing and food-exporting continents. After 1940 food output generally increased as cultivated zacreage declined. Progress in improving the quantity and quality of food production is related to mechanization, chemicalization, improved breeding, and hybridization. Food output is limited more by market demands than by production obstacles. Western Europe, although a basic food-deficit area, is a major producer and exporter of high-quality foodstuffs. After 1946 its agriculture became more profit-driven. Europe's agricultural labor force grew smaller, its agriculture became more mechanized, its farm sizes increased, and capital investment per acre increased.

Foods from Plants

Most basic staple foods come from a small number of plants and animals. Ranked by tonnage produced, the most important food plants throughout the world are wheats, corn, rice, potatoes, cassava, barley, soybeans, sorghums and millets, beans, peas and chickpeas, and peanuts. Wheat and rice are the most important plant foods. More than one-third of the world's cultivated land is planted with these two crops. Wheat is the dominant food staple in North America, Western and Eastern Europe, northern China, the Middle East, and North Africa. Rice is the dominant food staple in southern and eastern Asia.

Corn, used primarily as animal food in developed nations, is a staple food in Latin America and southeast Africa. Potatoes are a basic food in the highlands of South America and in Central and Eastern Europe. Cassava is a tropical starch-producing root crop of special dietary importance in portions of lowland South America, the west coast countries of Africa, and sections of South Asia. Barley is an important component of diets in North African, Middle Eastern, and Eastern European countries. Soybeans are an integral part of the diets of those who live in eastern, southeastern, and southern Asia. Sorghums and millets are staple subsistence foods in the savanna regions of Africa and South Asia, while peanuts are a facet of dietary mixes in tropical Africa, Southeast Asia, and South America.

The World's Growing Population

The problem of feeding the world is compounded by the fact that population was increasing at a rate of nearly 80 million persons per year at the end of the twentieth century. That rate of increase is roughly equivalent to adding a country the size of Germany to the world every year. Compounding the problem of feeding the world are population redistribution patterns and changing food consump-

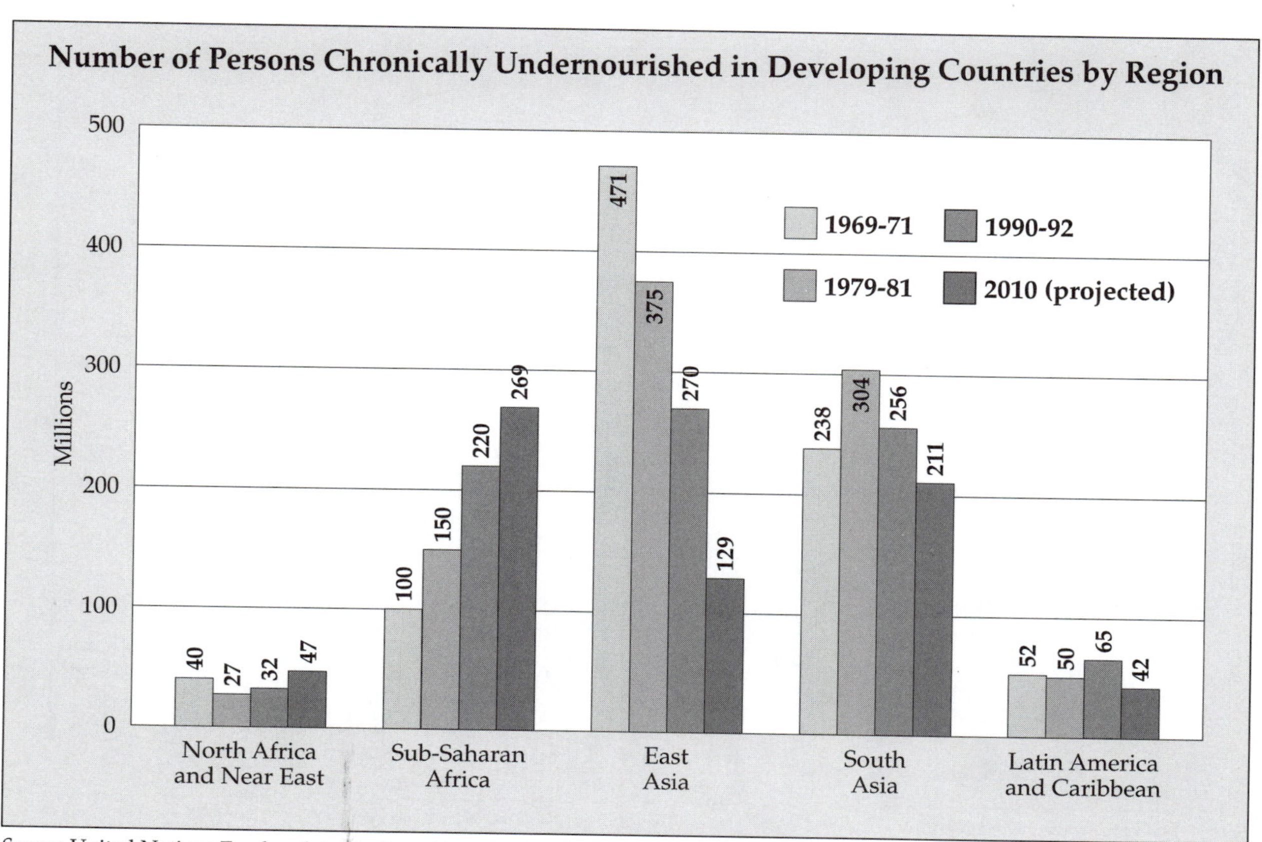

Source: United Nations Food and Agriculture Organization (FAOSTAT Database, 2000).

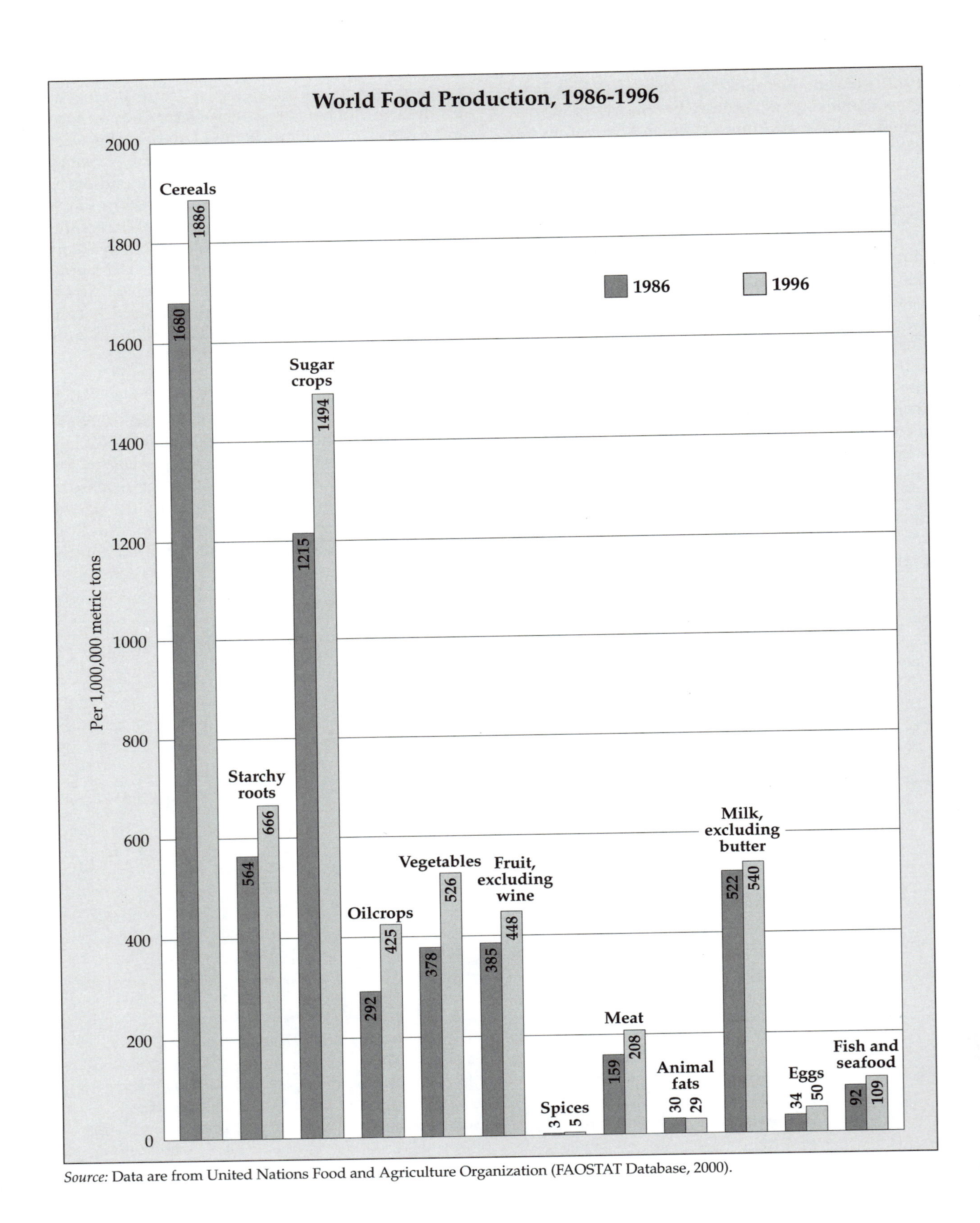

Source: Data are from United Nations Food and Agriculture Organization (FAOSTAT Database, 2000).

tion standards. By 2001, the world had exceeded the six billion mark, and the world population was projected to reach approximately ten billion people by 2050—four billion people more than were on the earth in 2000. Most of the increase in world population was expected to occur within the developing nations.

Urbanization

Along with an increase in population in developing nations is massive urbanization. City dwellers are food consumers, not food producers. The exodus of young men and women from rural areas has given rise to a new series of megacities, most of which are in developing countries. By the year 2015, twenty-six cities in the world are expected to have populations of ten million people or more.

When rural dwellers move to cities, they tend to change their dietary composition and food-consumption patterns. Qualitative changes in dietary consumption standards are positive, for the most part, and are a result of educational efforts of modern nutritional scientists working in developing countries. During the last four decades of the twentieth century, a tremendous shift took place in overall dietary habits. Dietary changes and consumption trends have contributed to a decrease in child mortality, an increase in longevity, and a greater resistance to disease. This globalization of people's diets has resulted in increased demands for higher quality, greater quantity, and more nutritious basic foods.

Perspectives

Humanity is entering a time of volatility in food production and distribution. The world will produce enough food to meet the demands of those who can afford to buy food. In many countries, however, food production is unlikely to keep pace with increases in the demand for food by growing populations. The food gap—the difference between production and demand—could more than double in the first three decades of the twenty-first century. Such a development would increase the dependence of developing countries on food imports. About 90 percent of the rate of increase in aggregate food demand in the early twenty-first century is expected to be the result of population increases. Factors that could lead to larger fluctuations in food availability include weather variations, such as those induced by El Niño and climatic change, the growing scarcity of water, civil strife and political instability, and declining food aid.

William A. Dando

See also: Agricultural crops: experimental; Agricultural revolution; Agriculture: history and overview; Agriculture: marine; Agriculture: modern problems; Agriculture: traditional; Agronomy; Biotechnology; Genetically modified foods; Green Revolution; Horticulture; Human population growth; Hydroponics; Plant biotechnology; Plants with potential.

Sources for Further Study

Conway, Gordon, and Vernon W. Ruttan. *The Doubly Green Revolution: Food for All in the Twenty-First Century*. Ithaca, N.Y.: Comstock, 1998. Discusses the need for equitable food distribution and the need to pay more attention to the environment than did the Green Revolution of the 1960's-1980's.

DeRose, Laurie Fields, Ellen Messer, and Sara Millman. *Who's Hungry? and How Do We Know? Food Shortage, Poverty, and Deprivation*. New York: United Nations University, 1998. Differentiates between food shortage (regional food scarcity), food poverty (inadequate household food supplies), and food deprivation (malnutrition) in order to identify and respond to the causes of hunger.

Lappé, Frances Moore, Joseph Collins, and Peter Rosset. *World Hunger: Twelve Myths*. 2d ed. New York: Grove Press, 1998. Three experts on food and agriculture expose the flaws within many tenaciously held beliefs, which prevent effective treatment of the problems of world hunger and global food distribution.

Shiva, Vandana. *Stolen Harvest: The Hijacking of the Global Food Supply*. Cambridge, Mass.: South End Press, 2000. Charts the impacts of globalized, corporate agriculture on small-scale farmers, the environment, and food quality. Covers genetically modified foods, seed monopoly, and the need for sound environmental thinking throughout the world.

AGRONOMY

Categories: Agriculture; disciplines; economic botany and plant uses; soil

Agronomy is a group of applied science disciplines concerned with land and soil management and crop production. Agronomists' areas of interest range from soil chemistry to soil-plant relationships to land reclamation.

The word "agronomy" derives from the ancient Greek *agros* (field) and *nemein* (manage) and therefore literally means "field management." The American Society of Agronomy defines agronomy as "the theory and practice of crop production and soil management." There are many specialties within the study of agronomy.

PhotoDisc

Agronomy is a family of disciplines, including soil management; a component of soil management is tilling and plowing techniques. This field has been plowed by tractor.

Agronomic Specialties

Agronomy is the family of disciplines investigating the production of crops supplying food, forage, and fiber for human and animal use. It studies the stewardship of the soil upon which those crops are grown. Agronomy covers all aspects of the agricultural environment, from agroclimatology to soil-plant relationships. It includes crop science, soil science, weed science, and biometry (the statistics of living things) as well as crop, soil, pasture, and range management; turfgrass; agronomic modeling; and crop, forage, and pasture production and utilization.

Within each area are subdisciplines. For example, within soil science are traditional disciplines such as soil fertility, soil chemistry, soil physics, soil microbiology, soil taxonomy and classification, and *pedogenesis*, the science of how soils form. Newer disciplines within soil science include such studies as *bioremediation*, or the study of how living organisms can be used to clean up toxic wastes in the environment, and land reclamation, the study of how to reconstruct landscapes disturbed by human activities, such as surface mining.

Scientific Goals

Chief among detrimental human activities is poor field management, which leads to reduced productivity and reduced environmental quality. Historical examples abound; one is that of the 1930's Dust Bowl in the United States. In the early 1900's much of the American Southern Plains, which had been natural grassland, was converted to wheatland. Planting and plowing methods of the time did not enable wheat to protect the ground against winds. Additionally, *overgrazing* of livestock had destroyed what grassland remained by the 1930's. The soil eroded, drought conditions which would last for most of the decade set in, and a series of wind and dust storms whipped through the region.

An estimated 50 million acres of land were destroyed before soil conservation measures, implemented under the administration of Franklin Roosevelt, began to improve the situation.

It is the role of agronomy to manage soil and crop resources as effectively as possible so that the twin goals of productivity and environmental quality are preserved. Agronomy treats the agricultural environment as humankind's greatest natural resource: It is the source of food, clothing, and building materials. The agricultural environment purifies the air humans and other animals breathe and the water they drink.

Agronomists, whatever their specific field, seek to utilize soil and plant resources to benefit society. Crop breeders, for example, use the genetic diversity of wild varieties of domesticated plants to obtain the information needed to breed plants for greater productivity or pest resistance. Soil scientists study landscapes to determine how best to manage soil resources. Integrating agricultural practices with the environment maintains soil fertility and keeps soil in place so that erosion does not reduce the quality of the surrounding environment.

Mark S. Coyne, updated by Elizabeth Slocum

See also: Agriculture: history and overview; Soil; Soil conservation; Soil management.

Sources for Further Study

Alston, Julian M., Philip G. Pardey, and Michael J. Taylor, eds. *Agricultural Science Policy: Changing Global Agendas*. Baltimore: Johns Hopkins University Press, 2001. Contains eleven studies from a 1996 conference, where scientists surveyed agricultural policy issues, such as the environment, genetic diversity, food safety, poverty, human health, animals rights, and more.

Committee on Managing Global Genetic Resources. *Agricultural Crop Issues and Policies: Managing Global Genetic Resources*. Washington, D.C.: National Academy Press, 1993. Produced by a Board of Agriculture committee; examines the structure that underlies efforts to preserve genetic material, the role of biotechnology, and other issues that affect land management and use in developing and developed countries.

Gliessman, Stephen R. *Agroecology: Ecological Processes in Sustainable Agriculture*. Boca Raton, Fla.: Lewis, 2000. This textbook provides the theoretical and conceptual framework for the study of agroecology, the application of ecological concepts to the design of sustainable agroecosystems. Only later sections assume an extensive background in ecology.

Pierce, F. J., and W. W. Frye, eds. *Advances in Soil and Water Conservation*. Chelsea, Mich.: Ann Arbor Press, 1998. An in-depth examination of the most important developments in shaping soil and water conservation in the second half of the twentieth century.

Sparks, Donald, ed. *Advances in Agronomy*. San Diego: Academic Press: 2001. Contains six reviews detailing advances in the plant and environmental soil sciences, including soil properties and quality and improvement of crops.

AIR POLLUTION

Categories: Environmental issues; pollution

Air pollution results from the contamination of air with naturally occurring or synthesized gases and particulates that reach levels that are harmful to biological systems. Air pollution damages plants in both natural and agricultural systems.

Dry *air* is a mixture of gases composed of approximately 78.08 percent nitrogen, 20.95 percent oxygen, 0.93 percent argon, and 0.036 percent carbon dioxide. The remainder is composed of trace amounts of gases, such as helium, hydrogen, neon, methane, nitrogen oxides, sulfur oxides, and others. *Environmental air* contains additional quantities of water vapor and other components related to natural geological and biological activities and human activities. The composition of environmental air is variable by time of day, season, geographic location, and altitude. Those components added by human activity are often linked to air pollution.

Types of Pollutants

Air pollutants may be classified as primary or secondary. *Primary pollutants* are classified as gaseous or particulate. *Gaseous pollutants* include compounds such as sulfur oxides, nitrogen oxides, and various types of hydrocarbons, such as methane, benzene, and chlorofluorocarbons (CFCs). Gaseous pollutants may impact plant health by directly damaging plant structure or by altering physiological processes. *Particulates* are irritants, such as soot and dust that clog plant stomata (pores). In addition, particulates may serve as carriers for harmful chemicals that adhere to the particulates' surfaces.

Secondary pollutants are produced as a result of the interactions of primary gaseous pollutants with one another or with other atmospheric components. For example, sulfur dioxide and nitrogen dioxide react with water to form sulfuric and nitric acids, respectively, which are important components of *acid precipitation*. Production of some secondary pollutants requires the presence of ultraviolet radiation. In the presence of ultraviolet light, nitrogen dioxide loses an oxygen atom, which then reacts with molecular oxygen to form the secondary pollutant, ozone. Ozone in the upper atmosphere protects the earth from harmful ultraviolet radiation but is, itself, harmful to organisms in the lower atmosphere. Peroxyacetyl nitrates (PAN) are secondary pollutants that result from the reaction of nitrogen oxides and hydrocarbons with ozone in the presence of ultraviolet light. Ozone and PAN are major components of *photochemical smog*.

Sources of Air Pollution

Anthropogenic air pollution is a result of human activities. Burning fossil fuels, such as coal and petro-

AP/Wide World Photos

Refineries in Pasadena, Texas, are shown in the foreground as downtown Houston stands in the smog.

Air Pollutant Emissions by Pollutant and Source, 1998

Source	Particulates[1]	Sulfur Dioxide	Nitrogen Oxides	Volatile Organic Compounds	Carbon Monoxide	Lead (tons)
Fuel combustion (stationary sources)						
Electric utilities	302	13,217	6,103	54	417	68
Industrial	245	2,895	2,969	161	1,114	19
Other fuel combustion	544	609	1,117	678	3,843	416
Residential	432	127	742	654	3,699	6
Subtotal	1,091	16,721	10,189	893	5,374	503
Industrial processes						
Chemical and allied product manufacturing	65	299	152	396	1,129	175
Metals processing	171	444	88	75	1,495	2,098
Petroleum and related industries	32	345	138	496	368	NA
Other	339	370	408	450	632	54
Subtotal	607	1,458	786	1,417	3,624	2,327
Solvent utilization	6	1	2	5,278	2	NA
Storage and transport	94	3	7	1,324	80	NA
Waste disposal and recycling	310	42	97	433	1,154	620
Highway vehicles						
Light-duty gas vehicles and motorcycles	56	130	2,849	2,832	27,039	12
Light-duty trucks	40	99	1,917	2,015	18,726	7
Heavy-duty gas vehicles	8	11	323	257	3,067	—
Diesels	152	86	2,676	222	1,554	NA
Subtotal	257	326	7,765	5,325	50,386	19
Off highway[2]	461	1,084	5,280	2,461	19,914	503
Miscellaneous[3]	31,916	12	328	786	8,920	NA
Total emissions	34,742	19,647	24,454	17,917	89,454	3,972

Note: In thousands of tons, except as indicated.
— Represents or rounds to zero.
NA Not available.
1. Represents both particulates of less than 10 microns and particulate dust from sources such as agricultural tilling, construction, mining and quarrying, paved roads, unpaved roads, and wind erosion.
2. Includes emissions from farm tractors and other farm machinery, construction equipment, industrial machinery, recreational marine vessels, and small general utility engines such as lawn mowers.
3. Includes emissions such as from forest fires and other kinds of burning, various agricultural activities, fugitive dust from paved and unpaved roads, and other construction and mining activities, and natural sources.
Source: Adapted from U.S. Environmental Protection Agency, *National Air Pollutant Emission Trends*, 1900-1998, EPA-454/R-00-002. From *Statistical Abstract of the United States: 2001* (Washington, D.C.: U.S. Bureau of the Census, 2001).

leum products (including natural gas, gasoline, and fuel oil), leads to the release of large quantities of carbon dioxide. Atmospheric carbon dioxide levels have increased approximately 25 percent since the beginning of the Industrial Revolution as a result of fossil fuel burning. Carbon dioxide, as a result of the *greenhouse effect*, increases the atmosphere's heat-trapping capacity and is therefore a major factor in *global warming*. Sulfur oxides are released in the burning of fossil fuels and in mineral ore processing. Nitrogen oxides are produced in side reactions during fossil fuel combustion, in which high heat induces the oxidation of atmospheric nitrogen into nitrogen oxides. Hydrocarbons are released from industrial processes such as petroleum distillation and from incomplete fuel combustion. Particulates are products of farming, industrial processes, construction, demolition, and mining. Governments of many, but not all, countries continually monitor air quality and impose regulations to control anthropogenic pollution.

Some pollutants are released from natural sources. Bacterial decomposition of organic materials in oxygen-depleted swamps leads to methane production. Methane, an important greenhouse gas, is also a product of organic decomposition in landfills and is a waste gas associated with cattle feeding operations. The haze above some forests is caused by the release of volatile organic compounds, called *terpenes*, from trees. Ultraviolet light degrades some of the terpenes and induces synthetic reactions among various molecules, leading to the production of larger secondary compounds. Sulfur oxides and particulates are released during volcanic eruptions.

Plant Damage from Air Pollution

Air pollution damages plant life in both natural and agricultural ecosystems. Pollution effects on plants depend on several factors, including pollutant concentration, duration of exposure, and life stage of the plant, along with physical factors, such as temperature, light density, humidity, and season. Resistance to pollution stress is highly variable within and among plant species. Damage usually occurs after a specific threshold is reached. Although individual compounds are often studied, pollutants do not occur independently. Thus, harmful pollutant effects may be amplified or decreased by pollutant interactions.

High atmospheric carbon dioxide levels increase biomass production in most plants and improve water-use efficiency, while sometimes reducing overall plant protein and nitrogen contents. Increased growth at the expense of nutritional quality for the ecosystem's herbivores and for livestock may be a costly exchange, leading to reduced ecosystem stability and reduced nutritional value for consumers.

Nitrogen oxides enter plants through the stomata or by diffusion through the epidermis. Low levels of nitrogen oxides may have a fertilizing effect on nitrogen-limited plants, while higher levels may lead to direct toxicity through the production of nitric acid in the cytoplasm. Reduced cytoplasmic pH (greater acidity) may influence the movement and availability of nutrients in cells. Nitrogen oxides may cause chloroplasts to swell, thus affecting photosynthesis. Visible symptoms of plant damage from nitrogen oxides include *chlorosis* (yellowing) of the leaf tips and margins, which may progress into *necrosis* (tissue death).

Ozone causes direct *oxidation damage* to cuticles and enters plants through the stomata. Inside cells, ozone acts as a strong oxidant, reacting with many cellular components. Cell walls may become thickened and pigmented, leading to a characteristic *stippling*. White to red necrotic lesions may form, as well as chlorotic flecks and general leaf chlorosis.

Sulfur oxides enter plants through the stomata. In cells, sulfur oxides enter chloroplasts and cause acidification of the stroma, which contains enzymes responsible for photosynthetic carbohydrate production. Stroma pH (alkalinity versus acidity, where a measure of 7 is considered neutral) needs to be around 9 for proper enzyme function. Acidification by only 0.5 pH units may reduce net photosynthesis by 50 percent. Long-term effects of sulfur oxide damage include decreased growth and early leaf fall. Visible signs of injury include a characteristic chlorosis of leaf margins and interveinal spaces, while tissues next to veins remain green. Leaf undersurfaces may become silver or bronze in color. Necrotic areas may be ivory, brown, black, or red, or they may fall out completely.

Darrell L. Ray

See also: Acid precipitation; Eutrophication; Greenhouse effect; Ozone layer and ozone hole debate; Rain forests and the atmosphere.

Sources for Further Study

Agrawal, Shashi B., and Madhoolika Agrawal. *Environmental Pollution and Plant Responses*. New York: Lewis, 1999. A review of plant responses to air pollution. Includes charts, graphs.

Miller, G. Tyler. *Environmental Science: Working with the Earth*. 8th ed. Albany, N.Y.: Brooks/Cole, 2000. A general introduction to environmental science. Includes coverage of air pollution.

Wellburn, Alan. *Air Pollution and Climate Change*. 2d ed. New York: John Wiley and Sons, 1994. An introduction to the biology of air pollution. Includes tables, graphs, and color plates.

ALGAE

Categories: Algae; microorganisms; *Protista*; water-related life

Algae comprise a diverse group of (with few exceptions) photosynthetic oxygen-producing organisms, ranging in size from microscopic single cells to gigantic seaweeds.

The study of algae is known as *phycology* (in Greek, *phycos* means "algae"). Currently, most authors place eukaryotic algae in the kingdom *Protista* (domain *Eukarya*) and prokaryotic algae in the domain *Bacteria*. In the past algae were considered to be lower plants because some forms look like plants. As in plants, the primary photosynthetic pigment in algae is chlorophyll *a*, and oxygen is produced during photosynthesis.

What Are Algae?

Algae can be found nearly everywhere on earth: oceans, rivers, lakes, in the snow of mountaintops, on forest and desert soils, on rocks, on plants and animals (such as within the hollow hair of the polar bear), or even on other algae. They are involved in diverse interactions with other organisms, including symbiosis, parasitism, and epiphytism. *Lichens* are symbiotic associations between algae (blue-green algae, or *cyanobacteria*) and fungi. Atmospheric nitrogen-fixing cyanobacteria occur in symbiotic associations with plants such as bryophytes, water ferns, gymnosperms (such as cycads), and the angiosperms. The aquatic fern *Azolla*, commonly used as a biofertilizer in rice fields in Asian countries, harbors the symbiotic cyanobacterium *Anabaena azollae*. *Gunnera*, the only flowering plant to house symbiotic cyanobacterium *Nostoc*, is widely distributed in the tropics.

Symbiotic dinoflagellates known as *zooxanthellae* live within the tissues of corals. Corals get their colors and obtain energy from their photosynthetic symbionts. About 15 percent of red algae occur as parasites of other red algae. Parasitic algae may even transfer nuclei into host cells and transform them. After transformation, the reproductive cells of the host algae carry the parasite's genes. Various algae live on the surfaces of plants and other algae as epiphytes. Sometimes algae can be found in strange places—the pink color of flamingos originates, for example, comes from a pigment in the algae consumed by these birds.

Algal Structure and Properties

Algal cells are bounded by a cell wall. Algal cells are either prokaryotic or eukaryotic. All prokaryotic algae belong to *Cyanophyta* (cyanobacteria) and lack both a nucleus and complex membrane-bound organelles, such as chloroplasts and mitochondria. Photosynthesis occurs in cyanobacteria in thylakoid membranes similar to those of plants. However, there is no double membrane surrounding the thylakoids of cyanobacteria.

All other algal groups are eukaryotic. Eukaryotic algae differ from cyanobacteria in that they possess chloroplasts and flagella with associated structures and in their cell wall composition. According to the *endosymbiont hypothesis*, some eukaryotic algae (red and green algae) obtained their chloroplasts by acquiring symbiotic prokaryotic cyanobacteria. This is known as primary endosymbiosis. Other eukaryotic algae probably obtained their chloroplasts by taking up eukaryotic endosymbiotic algae, a process known as secondary *endosymbiosis*. The existence of secondary endosymbiosis is indicated by the occurrence of more than two membranes around the chloroplasts of some algae, such as haptophytes, euglenophytes, dinoflagellates, and cryptomonads.

Pigments found in algae include chlorophylls, phycobilins, and carotenoids. All algae contain chlorophyll *a*. Accessory pigments vary among different algal groups.

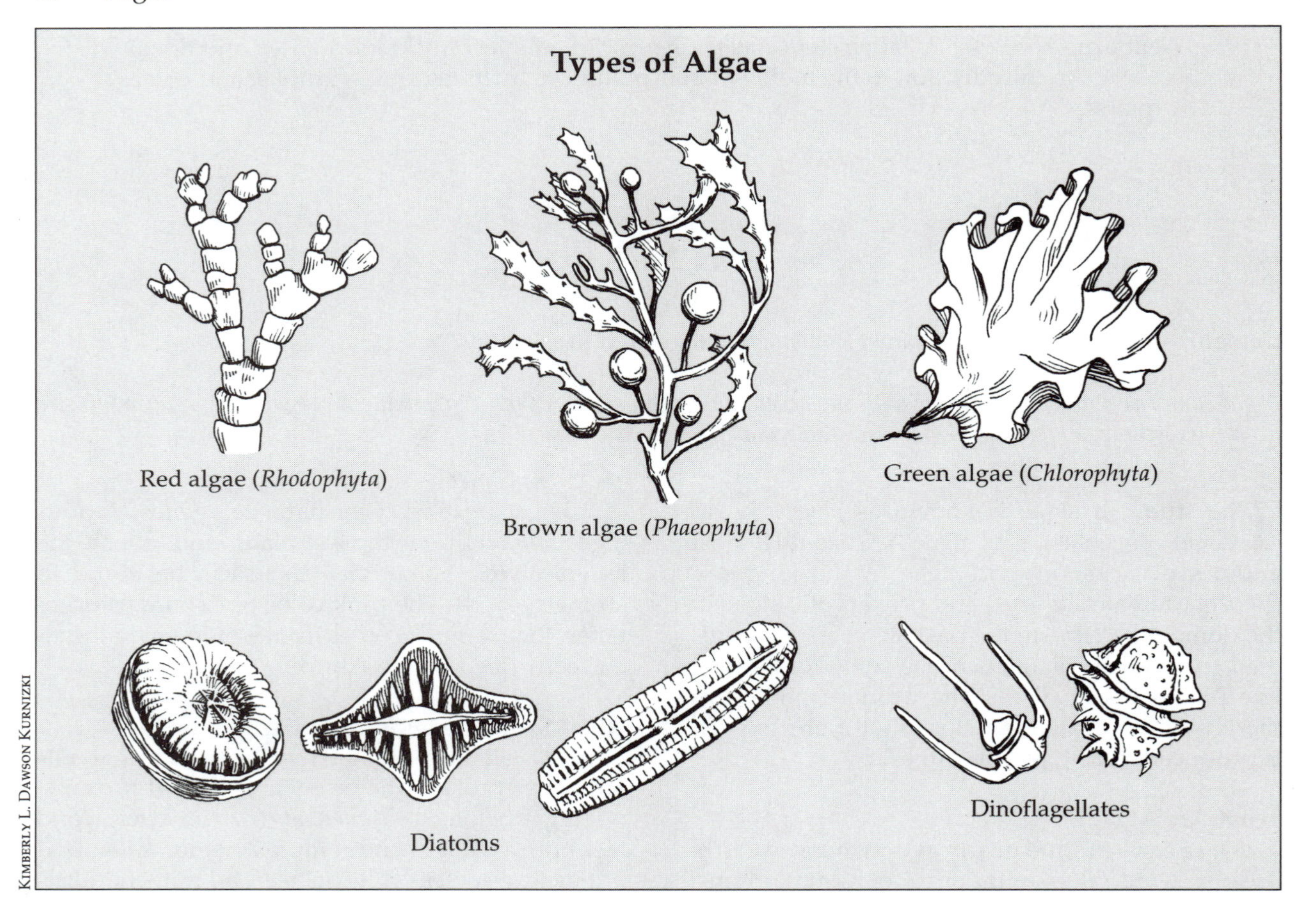

Photoautotrophy is the principal mode of nutrition in algae; in other words, they are "self-feeders," using light energy and a photosynthetic apparatus to produce their own food (organic carbon) from carbon dioxide and water. The majority of algal groups contain *heterotrophic* species, which obtain their organic food molecules by consuming other organisms. Numerous algae are *mixotrophs*; that is, they use different modes of nutrition (such as autotrophy and heterotrophy), depending on the availability of resources. The molecules used as food reserves differ among and are characteristic for algal groups. Food reserve molecules are polymers of glucose with different links between monomers.

Many algae are capable of movement. Movement is accomplished by means of flagellar action and by extrusion of mucilage. There are also peristaltic and amoeba-like algal movement. Within algal cells, movement of the cytoplasm, plastids, and nucleus also occurs. Advantages conferred by mobility include achieving optimal light conditions for photosynthesis, avoiding damage caused by excess light, and obtaining inorganic nutrients.

Algal Reproduction and Life Cycles

Algae may reproduce either asexually or sexually. Asexual reproduction among algae includes production of unicellular spores that germinate without fusing with other cells, fragmentation of filamentous forms, and cell division by splitting. In sexual reproduction, parent cells release gametes, which then fuse to form a zygote. Zygotes may either develop into new filaments or produce haploid spores by meiotic division.

Algae exhibit different types of life cycles. Some algal life cycles are characterized by an alteration of generations similar to that of plants. Two phases occur: sporophyte (usually diploid) and gametophyte (usually haploid). The sporophyte produces haploid spores through meiosis, and the haploid gametophyte produces male or female gametes by mitosis. Gametophyte and sporophyte may be structurally identical or dissimilar, depending on the algal group.

Roles of Algae

Algae have played significant roles in the earth's ecosystems since the origin of cyanobacteria (also known as blue-green algae) more than three billion years ago. Early cyanobacteria were responsible for the development of significant amounts of free oxygen in the atmosphere, which then made aerobic respiration possible. More than 70 percent of all photosynthetic activity on earth is carried out by *phytoplankton*—floating microscopic algae—rather than plants. Phytoplankton recharge the atmosphere with oxygen and simultaneously absorb carbon dioxide, helping to support the complex web of aquatic biota.

Algae are also very important in the global cycling of other elements, such as carbon, nitrogen, phosphorus, and silicon. Several algal groups—such as cyanobacteria, green algae, red algae, and the haptophyte algae—are able to generate calcium carbonate. Sedimented algae are the major contributors to deep-sea carbonate deposits (sand), which cover about half of the world's ocean floor. Calcified coralline red algae contribute to coral reefs in tropical waters. Silica sediments in oceans (sand) are based on abundant growth of another algal group, the diatoms, which contain silica in their cell walls.

Some algae (cyanobacteria) are able to fix atmospheric nitrogen and convert it to ammonia. Ammonia, in turn, can be a nitrogen source for plants and animals. On the other hand, high levels of nitrogen and phosphorus in rivers and lakes owing to pollution can cause the rapid and uncontrollable growth of algae, known as *algal blooms*. A bloom of algae is a threat to human and marine health, both directly and indirectly. It clogs fishes' gills, interferes with water filters, and ruins recreation sites. More than 50 percent of algal blooms produce toxins. Cases of human respiratory, skin, and gastrointestinal disorders associated with algal toxins have been reported. Certain blooms of algae are called *red tides*. The water appears to be red or brown because of the color of algal bodies, mainly dinoflagellates that contain the pigment xanthophyll.

Classification of Algae

Phylum	*Common Name*	*Species*
Bacillariophyta	Diatoms	100,000 or more
Chlorophyta	Green algae	17,000
Chrystophyta	Chrysophytes	1,000
Cryptophyta	Cryptomonads	200
Dinophyta	Dinoflagellates	4,000
Euglenophyta	Euglenoids	1,000
Haptophyta	Haptophytes	300
Phaeophyta	Brown algae	1,500
Rhodophyta	Red algae	6,000

Source: Data on species adapted from Peter H. Raven et al., *Biology of Plants*, 6th ed. (New York: W. H. Freeman/Worth, 1999).

Technological Applications

Algae have been used as food, medicine, and fertilizer for centuries. The earliest known reference to the use of algae as food occurs in Chinese poetic literature dated about 600 B.C.E. More recently, algae have begun to play important roles in certain biotechnological processes.

Several algae, including reds, browns, greens, and cyanobacteria, are used for food in Pacific and Asian countries, especially Japan. The annual harvest of the red alga *Porphyra* worldwide is worth several billion dollars. *Porphyra* (Japanese *nori*, Chinese *zicai*) is used as a wrapper for sushi or may be eaten alone. Another edible alga with a high iodine content is the brown alga *Laminaria* (Japanese *kombu*). The cyanobacterium *Spirulina*, with a protein level of 50 to 70 percent, was cultivated for centuries by indigenous Central Americans at Lake Texcoco near modern-day Mexico City for use as human food.

Several gelling agents are produced from red and brown algae. Agar from red algae is used as a medium for culturing microorganisms including algae, as a food gel, and in pharmaceutical capsules. Red algal carrageenan is used in toothpaste, cosmetics, and food such as ice cream and chocolate milk. Alginates from brown algae have extensive applications in the cosmetics, soap, and detergent industries. Sources of alginates are *Laminaria*, some *Fucus* species, and the giant kelp *Macrocystis*, which can grow to more than 60 meters long. Algae are also used as feed in the culture of commercially important fish and shrimp.

Mass cultivation of algae (microalgae)—in open ponds and photobioreactors for production of fuels (such as biomass) and biochemicals (such as carotenoids, amino acids, and carbohydrates) and for water purification—is a rapidly developing area based on the use of solar energy as energy source. The green alga *Dunaliella* is used in the industrial production of carotene. In wastewater treatment plants, algae are used to remove nutrients and heavy metals and to add oxygen to the water.

Algae are used worldwide as indicators (biomonitors) of water quality, helping to detect the presence of toxic compounds in water samples. Several fast-growing algae are used, including the green alga *Selenastrum capricornutum*. Many algae are widely employed as research tools because they are easy to culture and manipulate. Danish biologist Joachim Hammerling's experiments with the green alga *Acetabularia* identified the nucleus as the likely storage site of hereditary information.

Diversity

Taxonomists believe that there are between thirty-six thousand and ten million species of algae. Molecular comparisons using nucleotide sequences in ribosomal RNA (ribonucleic acid) suggest that algae do not fall within a single group linked by a common ancestor but that they evolved independently. The algae are divided into nine major phyla, which differ in their photosynthetic pigments, food reserves, cell structure, and reproduction. These groups include euglenoids, cryptomonads, dinoflagellates, haptophytes, and red algae.

Phylum *Euglenophyta* contains mostly unicellular forms with one or two flagella. Only one-third of this group possess chlorophyll-containing chloroplasts. Other euglenoids are strictly heterotrophic. The phylum contains more than nine hundred, mostly freshwater, species. The food reserve is the carbohydrate paramylon, a polymer of glucose. Euglenophytes have chlorophyll *a* and *b* as well as carotenoids as their photosynthetic pigments. There is no cell wall. Cells have several small chloroplasts; each is surrounded by three membranes. A close relative of euglenophytes is the protozoan *Trypanosoma*, which causes the human disease African sleeping sickness. Reproduction in the euglenophytes occurs by division of cells. Sexual reproduction is unknown.

Phylum *Cryptophyta* includes unicellular biflagellates. In addition to chlorophyll *a*, chloroplasts can contain chlorophyll *c*, carotenoids, and phycobilins. The carotenoid pigment alloxanthin is unique to *Cryptophyta*. Four membranes surround each chloroplast. Chloroplast endoplasmic reticulum borders the chloroplasts. The principal food reserve is starch. Instead of a typical cell wall, a periplast composed of protein plates occurs beneath the cell membrane. There are about two hundred freshwater and marine species. Reproduction is primarily asexual.

Members of the phylum *Dinophyta*, or dinoflagellates, have unicellular forms with two different flagella. There are between two thousand and four thousand marine species and about two hundred freshwater forms. Many have chlorophylls *a* and *c* as well as the unique carotenoid peridinin. Some members of *Dinophyta* have fucoxanthin. Chloroplasts have three closely associated membranes. The primary food reserve is starch, but lipids are also important storage molecules. A dinoflagellate cell is not surrounded by a cell wall but has a theca (a sort of armor) made of cellulose. Dinoflagellates can reproduce asexually and sexually.

Phylum *Haptophyta* includes primarily marine unicellular biflagellated algae. A haptophyte cell also has a flagellum-like haptonema, used to capture food. There are about three hundred species. The photosynthetic pigments include chlorophyll *a* and accessory pigments chlorophyll *c* and the carotenoid fucoxanthin. Each chloroplast has four membranes. The food reserve is chrysolaminarin, which is a polymer of glucose. Several layers of scales, or coccoliths, composed primarily of calcium carbonate may cover the haptophyte cell. Asexual and sexual reproduction is widespread.

Phylum *Rhodophyta*, or the red algae, has between four thousand and six thousand species. Red algae lack any flagellated stages. The photosynthetic pigments include chlorophyll *a* as well as accessory phycobilins and carotenoids. Two membranes surround each chloroplast. The food reserve is a floridean starch. A red algal cell is encircled by a wall composed of cellulose. Asexual and sexual reproduction, as well as alteration of generations, are widespread among *Rhodophyta*. A triphasic life cycle is unique for this group of algae.

Sergei A. Markov

See also: Bacteria; Biofertilizers; Biotechnology; Brown algae; *Charophyceae*; *Chlorophyceae*; Cryptomonads; Diatoms; Dinoflagellates; Eutrophication; Evolution of plants; Green algae; Haptophytes; Heterokonts; Lichens; Nitrogen cycle; Nitrogen fixation; Phosphorus cycle; Photosynthesis; Phytoplankton; *Protista*; Red algae; *Ulvophyceae*.

Sources for Further Study

Graham, Linda E., and Lee W. Wilcox. *Algae*. Upper Saddle River, N.J.: Prentice Hall, 2000. This comprehensive textbook focuses on diversity and relationships among the major algal types, algal roles in food webs, global biogeochemical cycling, the formation of harmful algal blooms and ways people use algae. Also provides broad coverage of freshwater, marine, and terrestrial algae.

Meinesz, Alexandre. *Killer Algae*. Chicago: University of Chicago Press, 1999. French scientist recounts the epidemic spread of a tropical green alga, *Caulerpa taxifolia*, along the Mediterranean coastline. Recommended for environmentalists and students of biology or ecology.

Sze, Philip. *A Biology of the Algae*. Boston: WCB/McGraw-Hill, 1998. This basic textbook presents an overview of different algal groups.

ALLELOPATHY

Categories: Physiology; poisonous, toxic, and invasive plants

Allelopathy refers to all the biochemical interactions, both beneficial and harmful, among all types of plants, including microorganisms.

For an allelopathic interaction to occur, chemicals must be released into the environment by one plant that will affect the growth of another. In this way allelopathy differs from *competition*, which involves removal of some factor from the environment that is shared with other plants. Allelopathy was recognized as early as Theophrastus (300 B.C.E.), who pointed out that chick pea plants destroy weeds growing around them.

GARY BASS/PHOTO AGORA

In some cases, roots secrete allelochemicals that kill seedlings of other plants. Bracken fern (Pteridium aquilinum) *is known to affect the growth of many other plants.*

Methods of Action

A variety of different allelochemicals are produced by plants, usually as secondary metabolites that do not have a specific function in the growth and development of the host plant but that do affect the growth of other plants. Originally plant physiologists thought these secondary products were simply metabolic wastes which plants had to store because they do not have an excretory system as animals do. Their various functions are now beginning to be understood.

One class of allelochemicals, *coumarins*, block or slow cell division in the affected plant, particularly in root cells. In this way growth of competing plants is inhibited, and seed germination can be prevented. Several kinds of allelochemicals, including *flavonoids*, *phenolics*, and *tannins*, suppress or alter hormone production or activity in competing plants. Other chemicals, including *terpenes* and certain antibiotics, alter membrane permeability of host cells, making them either leaky or impermeable. In some cases, membrane uptake can be enhanced, particularly for micronutrients in low concentration in the soil. Finally, a variety of allelochemicals have both positive and negative effects on metabolic activity of the affected plant.

Allelopathy in Agriculture

Most of the negative effects of weeds on crop plants have been attributed to competition; however, experiments using weed extracts have demonstrated that many weeds produce allelochemicals. Similarly, some crop plants are allelopathic to others and themselves, including wheat, corn, and rice. In these cases the residues of one year's crop can interfere with crop growth in subsequent years. This is increasingly important for farmers to consider who are incorporating low-tillage methods to reduce soil erosion. To minimize these effects, some of the traditional techniques of cover cropping, companion cropping, and crop rotation must be employed. Known allelopaths are also beginning to be used as biological control agents to manage invasive and weedy plant species.

Allelopathy in Nature

Several tree species, including black walnut, black locust, and various pines, are known to produce allelochemicals that inhibit the growth of understory species. In some cases this is a result of drip from the foliage or leachate from fallen leaves and fruit. In other cases, roots secrete allelochemicals that kill seedlings of other plants. Bracken fern (*Pteridium aquilinum*) is known to affect the growth of many other plants.

Marshall D. Sundberg

See also: Agriculture: traditional; Biochemical coevolution in angiosperms; Biofertilizers; Coevolution; Community-ecosystem interactions; Competition; Hormones; Invasive plants; Metabolites: primary vs. secondary; Organic gardening and farming; Pheromones; Soil degradation; Succession.

Sources for Further Study

Moore, Randy, W. Dennis Clark, and Darrell S. Vodopich. *Botany*. 2d ed. New York: McGraw-Hill, 1998. Introductory botany text with a better discussion of allelopathy than most.

Rice, Elroy L. *Allelopathy*. 2d ed. Orlando, Fla.: Academic Press, 1984. This is the classic reference.

ALTERNATIVE GRAINS

Categories: Agriculture; economic botany and plant uses; food

Alternative grains refers to alternatives to high-yield grain crops, the harvest of which has led to severe soil erosion and increased use of fertilizers and pesticides.

More than one-half of the calories consumed daily by humans comes from grains. Most of these grains are produced by plants of the *grass* family, *Poaceae*. Major *cereal* plants domesticated centuries ago include rice (*Oryza sativa*), wheat (*Triticum aestivum*), and corn (*Zea mays*). Other important grain crops, also plants of the grass family, include barley (originating in Asia), millet and sorghum (originating in Africa), and oats and rye (originating in Europe).

Grain Genetics

Since the early twentieth century, the scientific principles of genetics have been applied to improvements of crop plants. Some notable improvements occurred between 1940 and 1970. As a result of irrigation, improved genetic varieties, and the use of large amounts of fertilizers and pesticides, yields of major crops greatly increased. Norman Borlaug received a Nobel Prize in 1970 for his contributions to these developments, which came to be called the *Green Revolution*. However, it soon became apparent that the Green Revolution was not the boon first envisioned. For maximum yield, large-scale farming involving huge capital investment is required. Also, environmentalists became concerned over the erosion and other environmental damage caused by the use of large amounts of fertilizers and pesticides.

Buckwheat probably originated in China. It tolerates cool conditions and is adapted to short growing seasons, thus permitting it to be grown in the temperate regions of North America and Europe. In the United States, it is often associated with pancakes but is more often used for livestock feed.

Minor Cereals

Various alternatives have been proposed. For grain crops, several approaches offer promise, including more widespread use of minor cereals, especially those tolerant of unfavorable growing conditions; development of new cereal plants by hybridization or other genetic manipulation; and use of pseudocereals, nongrass crop plants that produce fruits (grains) similar to those of cereal plants.

Most *sorghum* (*Sorghum bicolor*) grown in the United States is used for silage or molasses. In Africa and India, various grain sorghums are grown in regions where rainfall is too low for most other grain crops. Well adapted to hot, dry climates, their grains are used to make a pancakelike bread. *Millet* refers to several grasses that are also useful cereal plants because they tolerate drought well. In Africa the most important are pearl millet (*Pennisetum glaucum*) and finger millet (*Eleusine coracana*). Grains of both species can be stored for long periods and are used to make bread and other foods. Other, perhaps less important, grain plants also called millet include foxtail millet (*Setaria italica*), native to India but now grown in China; proso millet (*Panicum milaeceum*), native to China but grown in Russia and central Asia; sanwa millet (*Echinochloa frumentacea*), cultivated in East Asia; and teff (*Eragrostis teff*), an important food and forage plant of Ethiopia. Such grain sorghums and millets have the potential to grow in areas with hot, dry climates far beyond the regions where they are now being utilized.

In a distinct category is *wild rice* (*Zizania aquatica*). Native to the Great Lakes region of the United States and Canada, it has been, and still is, harvested by American Indians. Like the common but unrelated rice (*Oryza sativa*), wild rice grows in flooded fields. Attempts to cultivate it since the 1950's have been somewhat successful as the result of the development of nonshattering varieties. However, it remains an expensive, gourmet item. Two cereal plants have promise because of the high-protein content of their grains. Wild oat (*Avena sterilis*) is a disease-resistant plant with large grains. Job's tears (*Coix lachryma-jobi*), native to Asia, is now planted throughout the tropics. Research on these and related species continues.

Although all important cereal plants have been improved by genetic techniques, the most notable new alternative grain plant is triticale (*Triticosecale*). The first human-made cereal, it is the result of crossing wheat with rye. The sterile hybrid from such a cross was made fertile by doubling its chromosomes. Thus, triticale varieties produce viable seeds. Triticale combines the superior traits of each of its parents: the cold tolerance of rye and the higher yield of wheat. The protein content of triticale compares favorably with that of wheat, and its quality, as measured by lysine content, is higher. However, flour made from triticale is not suitable for making bread unless mixed with wheat flour.

Pseudocereals

Pseudocereals are plants that are not of the grass family but produce nutritious, hard, grainlike fruits that can be stored, processed, and prepared for food much like grains. They belong to several plant families. Many grow under conditions not suitable for the major cereal crops. Buckwheat (*Fagopyrum esculentum*), of the *Polygonaceae* family, probably originated in China. It tolerates cool conditions and is adapted to short growing seasons, thus permitting it to be grown in the temperate regions of North America and Europe. In the United States, it is often associated with pancakes but is used in larger quantities for livestock feed. In Eastern Europe, the milled grain is used for soups.

Quinoa (*Chenopodium quinoa*) of the goosefoot family, *Chenopodiaceae*, has been cultivated by Indians of the Andes Mountains for centuries. The leafy annual produces grainlike fruits (actually achenes) with a high protein content and exceptional quality, high in lysine and other essential amino acids. After its bitter saponins have been removed, it can be cooked and eaten like rice or made into a flour. Quinoa has been cultivated in the Rocky Mountains since the 1980's and has become a gourmet food in the United States.

Most amaranths (*Amaranthus*) plants are New World weeds. They belong to the amaranth family, *Amaranthaceae*. A few species were used by Aztecs and other North American peoples, but amaranth use was banned by the Spanish. Since the late 1970's, plant breeders have targeted several species for improvement. The results are highly nutritious grains, rich in lysine, that are suitable for making flour. Research in Pennsylvania and California has resulted in improved varieties.

Thomas E. Hemmerly

See also: African agriculture; North American agriculture.

Sources for Further Study

Board on Science and Technology for International Development, National Research Council. *Lost Crops of Africa: Grains*. Washington, D.C.: National Academy Press, 1995. A report on alternative grains native to Africa and their potential for alleviating world hunger and improving economic development.

Levetin, Estelle, and Karen McMahon. *Plants and Society*. 2d ed. Boston: WCB/Mcgraw-Hill, 1999. An applied and economic botany textbook. Discusses cereal and related crop plants and their interactions with and effects on human society.

Simpson, Beryl B., and Molly C. Ogorzaly. *Economic Botany: Plants in Our World*. 3d ed. Boston: McGraw-Hill, 2001. An introductory textbook with exceptional illustrations. Discusses cereal and related crop plants.

Smith, C. Wayne, and Richard A. Frederiksen, eds. *Sorghum: Origin, History, Technology, and Production*. New York: John Wiley & Sons, 2000. A comprehensive, in-depth volume on sorghum, covering its origin, history, and technological evolution.

ANAEROBES AND HETEROTROPHS

Categories: Cellular biology; evolution; microorganisms

The first organisms to evolve on the earth are thought to have been heterotrophs and anerobes. Heterotrophs are organisms that cannot produce their own food but must fill their energy requirements by consuming organic molecules produced by other processes or organisms. Anaerobes are organisms that do not require free oxygen gas in order to survive; for some anaerobes, free oxygen may be poisonous.

Heterotrophs include many familiar organisms (such as animals) whose existence is tied to *primary producers*, those organisms that create energy-storing molecules, such as photosynthesizing plants. Anaerobes also are common, though less apparent. Typically, they are microscopic organisms restricted to living in a few surface environments where oxygen is absent. It may seem strange, then, that these organisms were perhaps the first organisms to have evolved on the earth. Yet the combination of the heterotrophic lifestyle and the anaerobic life requirement is consistent with what is known about the conditions of the early earth's surface environment.

The earliest anaerobic heterotrophs laid the biochemical foundations for the evolution of photosynthesis, free oxygen in the atmosphere, and the rise of complex organisms. All those events had the adverse impact of limiting the range of environments available to the anaerobes.

The world's first organism evolved in what has been called a "prebiotic soup" of energy-rich organic molecules. Heterotrophic organisms would exploit this environment by absorbing the molecules. A continuing supply of energy-rich molecules depended on the absence of free oxygen in the early atmosphere and the functioning of the *abiotic synthesis*.

Fermentation

The energy-rich molecules of the soup were converted to energy by a series of biochemical reactions. One of the simplest, and therefore perhaps one of the oldest, types of energy conversion reactions is *anaerobic fermentation*. During anaerobic fermentation, an energy-rich molecule, such as the simple sugar glucose, is dismantled to release energy and waste by-products. Several lines of evidence suggest that this form of energy conversion was utilized by the early heterotrophs.

One indicator that fermentation is a very ancient biochemical process is that the reaction used to release energy from the glucose molecule is very common among modern organisms. The ability to utilize the *fermentation* reaction is evident in the anaerobic reaction of yeast using sugar and releasing ethyl alcohol. Although it is not the primary energy-releasing reaction for most organisms, fermentation's widespread availability suggests that it is very old and perhaps inherited from an early, simpler ancestor.

The fermentation reaction is not very efficient. For example, it releases two units of energy for every glucose molecule, whereas oxidation of the same glucose molecule releases more than thirty energy units. Such an inefficient reaction for energy release could not be tolerated by an advanced organism with many energy demands. Alternatively, single-celled heterotrophs surrounded by, and absorbing, energy-rich molecules such as glucose (which is unlikely to decompose in an anoxic environment) do not expend much energy in gathering their food.

The Earliest Organisms

Thus, two very different types of evidence—that which points to an anoxic early atmosphere and evidence for the ancient ancestry of the glucose fermentation reaction—suggest that the earliest organism was a single-celled heterotroph that absorbed energy-rich molecules from the surrounding anaerobic environment. Modern analogues for such an organism exist. Single-celled bacteria, called *obligate anaerobes*, exist in a few anoxic environments today. It is likely that the modern obligate anaerobes have not changed significantly (especially in their morphology, or shape and size) from their Precambrian ancestors. Given this much information, paleontologists know that their search for early Precambrian fossils, the petrified remains of organisms, is not easy.

The process of fossilization—that is, the preservation of the shape of an organism in rock—is best at preserving the details of hard body parts. Hard skeletons and shells or their impressions are easier to preserve than are soft body parts. In the case of early Precambrian fossils, the most likely organisms (bacteria) not only are small but also contain no hard body parts.

Despite these barriers to preservation and despite the very poorly preserved Precambrian rock record, some early Precambrian fossil remains have been found and described. The fossils are usually found preserved in rock called chert, which probably began as a gelatinous material. Microscopic remains of organisms embedded in this gelatin were delicately preserved when the chert lost some of its water and solidified.

The oldest fossil remains identified have been found in cherts from southern Africa. These cherts, part of what is called the Fig Tree Formation, are more than three billion years old. The fossils consist of the wispy, spherical remains of what may have been a type of alga and the rod-shaped remains of a possible heterotrophic bacterium.

Richard W. Arnseth

See also: Anaerobic photosynthesis; *Archaea*; Bacteria; Cell theory; *Eukarya*; Glycolysis and fermentation; Molecular systematics; Photosynthesis; Phytoplankton; Prokaryotes; *Protista*.

Sources for Further Study

Cloud, Preston. *Oasis in Space: Earth History from the Beginning*. New York: W. W. Norton, 1988. A readable account of geologic history, concentrating the first twelve chapters on the history of the Precambrian. Data tables and illustrative diagrams supplement the text's explanation of the more technical concepts. For the college-level reader.

Gould, Stephen J., ed. *The Book of Life*. 2d ed. New York: W. W. Norton, 2001. A lucid overview of the evolution of life on earth, from bacteria to complex mammals. Chapters are written by a distinguished panel of scientists. Well illustrated with paintings, drawings, charts, and graphs.

Holland, Heinrich D. *The Chemical Evolution of the Atmosphere and Oceans*. Princeton, N.J.: Princeton University Press, 1984. A technical book that concentrates on the chemical, and especially the geochemical, aspects of atmospheric and oceanic evolution. Despite its formidable technical content, the text contains numerous references to current technical literature cited in its extensive bibliography. Accessible to those with a college chemistry background.

Schopf, J. William, ed. *Earth's Earliest Biosphere: Its Origin and Evolution*. Princeton, N.J.: Princeton University Press, 1983. A collection of works by members of the Precambrian Paleobiology Research Group, an informal group of experts in the field. One of the best sources for information on Precambrian chemical evolution, for both the biota and the environment. College-level, with chemistry and biochemistry.

Selander, Robert K., Andrew G. Clark, and Thomas S. Whittam, eds. *Evolution at the Molecular Level*. Sunderland, Mass.: Sinauer Associates, 1991. Molecular evolution is addressed in several essays.

ANAEROBIC PHOTOSYNTHESIS

Categories: Photosynthesis and respiration; physiology

Anaerobic photosynthesis, also known as anoxygenic photosynthesis, is the process by which certain bacteria use light energy to create organic compounds but do not produce oxygen. Anaerobes are those bacteria that cannot use oxygen to generate energy.

The photosynthetic process in all plants and algae, as well as in specific types of bacteria, involves the reduction of carbon dioxide to carbohydrate and the removal of electrons from water, resulting in the release of oxygen. This process is known as oxygenic or *aerobic photosynthesis*. Water is oxidized by a multi-subunit protein located in the photosynthetic membrane. This is a molecular protein feature shared among more than 500,000 species of plants on earth.

While this is a common feature among nearly every form of plant life on earth, some photosynthetic bacteria can use light energy to extract electrons from molecules other than water. These bacteria are of ancient origin and are believed to have evolved before aerobic photosynthetic organisms. These anaerobic photosynthetic organisms occur in the domain *Bacteria*. Anaerobic photosynthetic bacteria, also known as anoxygenic photosynthetic bacteria, differ from aerobic organisms in that each species of these bacteria has only one type of reaction center. In some photosynthetic bacteria the reaction center involves the oxidation of water and the reduction of the aromatic molecule plastoquinone. In

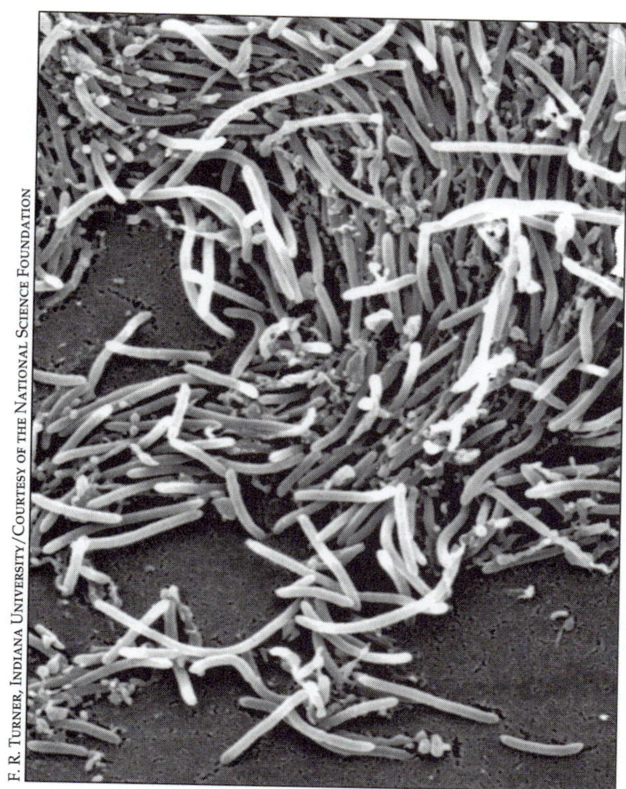

F. R. Turner, Indiana University/Courtesy of the National Science Foundation

Heliobacteria, *shown in this scanning electron micrograph, one of five groups of photosynthetic bacteria, are anaerobic photosynthetic bacteria.*

other species it involves the oxidation of plastocyanin and the reduction of ferredoxin protein.

Photosynthetic bacteria are typically aquatic microorganisms inhabiting marine and freshwater environments, including wet and muddy soils, stagnant ponds, sulfur springs, and still lakes. They are classified into five groups based on pigment composition, metabolic requirements, and membrane structure: green bacteria, purple sulfur bacteria, purple nonsulfur bacteria, heliobacteria, and halophilic archaebacteria. Some of these organisms are *strict anaerobes*; that is, they can grow only in the complete absence of oxygen. They cannot use water as a substrate, and they do not produce oxygen during photosynthesis. *Facultative anaerobes*, on the other hand, can grow either in the presence or in the absence of oxygen.

Green bacteria include two families, the *Chloroflexaceae* and the *Chlorobiaceae*. The *Chlorobiaceae* are strict anaerobes that grow by utilizing sulfide, thiosulfate, or organic hydrogen as an electron source. *Chloroflexaceae* are facultative aerobes which use reduced carbon compounds as electron donors. Purple sulfur bacteria uses an inorganic sulfur compound, such as hydrogen sulfide, as a photosynthetic electron donor. Purple nonsulfur bacteria depend on the availability of simple organic compounds such as alcohols and acids as electron donors, but they can also use hydrogen gas. Purple sulfur bacteria must fix carbon dioxide to live, whereas nonsulfur bacteria can grow aerobically in the dark by respiration on an organic carbon source.

Heliobacteria are anaerobic photosynthetic bacteria that contain a special type of bacteriochlorophyll, BChl *g*, that works as both antenna and reaction center pigment. *Halobacteria* are very unusual. They cannot grow in low salt concentrations (or their cell walls collapse). Typically, they are heterotrophs with an aerobic electron-transport chain, but they can also respire anaerobically, with nitrate or sulfur. In the absence of suitable electron acceptors they can ferment carbohydrates. Halobacteria, when exposed to light in the absence of oxygen, can synthesize a purple membrane containing a single photosensitive protein called bacteriorhodopsin which, when illuminated, begins cyclic bleaching and regeneration, extruding protons from the cell. This light-stimulated proton pump operates without electron transport. The mechanism by which halobacteria convert light is fundamentally different from that of higher organisms because there is no oxidation/reduction chemistry, and halobacteria cannot use carbon dioxide as their carbon source. As a result, some scientists do not consider halobacteria as being photosynthetic.

Process

The common features to both aerobic and anaerobic photosynthesis have been known since the mid-twentieth century:

Green plants:

$$CO_2 + 2H_2O + \text{light} \rightarrow (CH_2O) + O_2 + H_2O$$

Green sulfur bacteria:

$$CO_2 + 2S + H_2O + \text{light} \rightarrow (CH_2O) + 2S + H_2O$$

In each case, inorganic carbon (CO_2) is fixed into organic carbon (CH_2O), the source of reductant is hydrogen in either water or hydrogen sulfide, and the chemical energy required for this activity is derived from light energy. The sulfur produced anaerobically is analogous to the oxygen produced by the oxygenic photosynthesis of green plants. Photochemical processes in photosynthetic bacteria require three major components: an antenna of light-harvesting pigments, a reaction center within an intra-cytoplasmic membrane containing at least one bacteriochlorophyll, and an electron transport chain.

All photosynthetic bacteria can transform light energy into a transmembrane proton gradient used for the generation of adenosine triphosphate (ATP) and production of oxidase, but none of the anaerobic photosynthetic bacteria are capable of extracting electrons from water, so they do not evolve oxygen. Many species can only survive in low-oxygen environments. To provide the necessary electrons for carbon dioxide reduction, anoxygenic photosynthetic bacteria must oxidize inorganic or organic molecules from their immediate environment.

Despite basic differences, the principles of energy transductions are the same in anaerobic and aerobic photsynthesis. Anaerobic photosynthetic bacteria depend on bacteriochlorophyll, a group of molecules similar to chlorophyll, that absorbs in the infrared spectrum between 700 and 1,000 nanometers. The antenna systems in these bacteria consist of bacteriochlorophyll and carotenoids, serving a reaction center where primary charge separation occurs. Electron carriers include quinone and cytochrome *bc* complex. Electron transfer is coupled to the generation of electrochemical potential that drives phosphorylation by ATP synthase, and the energy required for the reduction of carbon dioxide is provided by ATP and dehydrogenase.

Randall L. Milstein

See also: *Archaea*; Bacteria; Bacterial genetics; Eutrophication; Molecular systematics; Photosynthesis; Photosynthetic light absorption; Photosynthetic light reactions; Plasma membranes; Proteins and amino acids; Respiration.

Sources for Further Study

Blankenship, R., M. T. Madigan, and C. Bauer, eds. *Anoxygenic Photosynthetic Bacteria*. Amsterdam: Kluwer Academic Press, 1995. Highly technical but complete coverage of the chemical processes of anaerobic photosynthetic bacteria.

Fenchel, Tom, and Bland J. Finlay. *Ecology and Evolution in Anoxic Worlds*. New York: Oxford University Press, 1995. University-level reading on the environmental conditions and processes of anaerobic life-forms.

Hall, D. O., and K. K. Rao. *Photosynthesis*. 5th ed. New York: Cambridge University Press, 1994. University-level overview of photosynthetic processes. Written for those with a background in chemistry.

Schlegel, Hans G. *General Microbiology*. 7th ed. New York: Cambridge University Press, 1992. Easy-to-read introduction to microbiology. Well illustrated and readily found in libraries and bookstores.

ANGIOSPERM CELLS AND TISSUES

Categories: Anatomy; angiosperms; cellular biology; physiology

Some cell types and tissues which are not found in any other groups of plants occur in angiosperms (flowering plants).

Angiosperms are a group of plants with seeds that develop within an ovary and reproductive organs in flowers. They are commonly referred to as flowering plants and represent the most successful group of plants on earth, with approximately 235,000 species. Various cell types and tissues, many of which are not found in any other groups of plants, occur in angiosperms. These cells and tissues perform varied functions, which are very efficient compared to their counterparts in other plants. These include *dermal*, *vascular* (xylem and phloem), and *ground tissues* (such as parenchyma, collenchyma, and sclerenchyma).

The growth of plants is carried on by a group of cells at their tips. These groups of cells are referred to as *apical meristems*, which are composed of *initials* and their most recent *derivatives*. The initials are the main source of body cells in plants, while the derivatives become any of the cells and tissues in the plant body. The apical meristems of both the shoot and the root show continued cell division, with cells enlarging, elongating, and differentiating in regular, distinctly organized patterns. Apical meristems bring about the increase in the length of the stems and roots and are responsible in the formation of the primary plant body. The shoot apical meristem may continually initiate the aerial components of the plant or may enter a state of periodic quiescence. In some plants, the shoot apical meristem transforms into a floral or inflorescence meristem that eventually terminates in a single flower or clusters of flowers, respectively. The *root apical meristem* is enclosed by a thimble-shaped *root cap* that hastens the penetration of roots between soil particles. Unlike the shoot apical meristem, the root apical meristem forms no appendages. In fact, the site of lateral root initiation is far removed from it.

Shoot Apex

The shoot apical meristem is typically dome-shaped but flattened, and concave outlines also exist. The outline is not constant but changes in response to *plastochron* (the time interval between the initiation of one leaf primordium and the next). At least three models describe the shoot apical meristems. Although each of these is based on one or two unique criteria, they also have a few overlapping features.

Cell Lineage Analysis

This model holds that three clonally related layers of cells characterize the shoot apical meristem. These layers can be more than one cell layer thick. L1 is the outermost layer and gives rise to the *epidermis*, L2 is the middle layer and gives rise to the *vascular tissues* and *cortex*, and L3 is the innermost layer and gives rise to the *pith*. This model was based on studies using periclinal chimeras (organs or parts of tissues of diverse genetic constitution), where one of the cell layers was genetically altered using drugs that inhibit separation of chromosomes.

Tunica-Corpus Concept

This model is based on microscopic analysis of constituent cells. It says that the shoot apical meristem is made up of two groups of cells. The *tunica*, a group of cells that form one or two stratified layers, undergoes anticlinal divisions only and gives rise to the epidermis. Partly enclosed by the tunica is the *corpus*, a group of loosely arranged cells that divide in various planes and give rise to the vascular and ground tissues. The tunica maintains its individuality by surface growth, whereas the corpus adds bulk by increase in volume.

Cytohistological Zonation

This model recognizes various definable zones in the shoot apical meristem. Three zone boundaries are distinguished by cell size: staining quality, degree of vacuolation, and frequency of cell division. The central (mother cell) zone represents a conspicuous group of enlarged and isodiametric cells that undergo infrequent cell division, possess prominent nuclei, and are often highly vacuolated. The flanking peripheral zone is derived from, and partly surrounds, the central zone. Cells of this zone are smaller, are mitotically active, and have dense cytoplasms. They give rise to the epidermis, vascular tissues, and cortex. The rib zone is located at the base of the central and peripheral zones. This zone is directly formed from the central zone, produces longitudinal files of cells by periclinal divisions, and gives rise to the pith.

Root Apex

The organization of the root apical meristem is different from that of the shoot apical meristem. Root apical meristems are commonly interpreted as having either a close or open type of organization. In a close type of organization, the dermal, vascular, and ground tissues each have their own set of initials. This organization shows a clear boundary between root cap and other tissues of the root apex. In an open type of organization, all of the root tissues share a group of initials, and therefore the boundary of the root cap is indistinguishable from the other tissues of the root apex.

Developmental Processes

The cells produced by apical meristems undergo several key developmental processes, which include growth, differentiation, and morphogenesis. Although each of these can be separated individually, they overlap in highly complex fashion. *Growth* refers to the quantitative increase in a cell's volume or mass due to enlargement and multiplication. *Differentiation* is the qualitative change in the form and function of organelles, cells, tissues, and organs, resulting in the establishment of new structures and functions. From an anatomical point of view, cell differentiation is related to changes in cell size and shape, modifications of the wall, and changes in staining characteristics of nucleus or cytoplasm, as well as the degree of vacuolation and the ultimate loss of the protoplast in some cases. *Morphogenesis* is the visible manifestation of all of the changes, brought about by growth and differentiation, as expressed in the overall morphology of the plant.

Dermal Tissues

The primary plant body is composed of three basic tissues: dermal, vascular, and ground tissues. The *dermal tissue* (or epidermis) is made up of several cell types and is involved in a variety of functions, including retention and absorption of water and minerals, protection against herbivores, and control of gas exchange. Each of these functions is attributable to one or more of the unique features of the epidermis. Most epidermal cells are flat and tightly packed, forming a single layer around stems, leaves, and other organs. The outer walls of epidermal cells are equipped with a waterproof layer made up of a fatty material called *cutin*. The tightly packed and cutinized epidermis protects the plants from desiccation by keeping moisture in.

Epidermal cells lack chloroplasts and are transparent. It is the underlying cells that give leaves and stems their green color. However, the vacuoles of some epidermal cells occasionally contain pigments and are responsible in the coloration of flowers and colored parts of variegated leaves.

Stomata are specialized structures that form part of the epidermis of leaves, stems, flowers, and fruits. They are involved in regulating the intake of carbon dioxide for photosynthesis as well as the release of oxygen. *Trichomes* are single-celled or multicellular outgrowths of epidermal cells that are involved in deterring herbivores and restricting transpiration. *Root hairs* are also outgrowths of epidermal cells that are specialized for absorbing water and minerals from soil. They occur near the tip of the root and function to increase its absorptive surface area several-thousandfold.

Vascular Tissues

Vascular tissues are of two types: xylem and phloem. *Xylem* occurs throughout the plant body, and the type that differentiates directly from the apical meristem is called primary xylem. (Secondary xylem is formed from the vascular cambium.) Primary xylem is formed as stems and roots elongate. The two kinds of conducting cells in xylem are tracheids and vessels, or vessel elements. Both are dead at maturity and have thick, lignified secondary cell walls. *Tracheids* are long, slender cells with tapered, overlapping ends. They are the only water-conducting cells in most gymnosperms (an evolutionary line of plants that includes conifers). Water moves upward in roots and stems from tracheid to tracheid through thin areas in their cell walls called pits. With only a few exceptions, all angiosperms contain vessel elements and tracheids. *Vessel elements* are short, wide in diameter, and connected end to end. Their end walls are partly or wholly dissolved, forming long hollow vessels through which water moves. All these features of vessels enable them to transport water more rapidly than tracheids.

Phloem transports dissolved organic materials throughout the plant. The conducting cells of the phloem are called *sieve elements*, which are devoid of nuclei but otherwise have intact cytoplasm. They also have thin areas along their cell walls called sieve areas that are perforated. Solutes move from sieve element to sieve element through these pores.

Ground Tissues

The three types of *ground tissue* are parenchyma, collenchyma, and sclerenchyma. *Parenchyma cells* are the most abundant and versatile cells in plants. These cells are isodiametric, are alive at maturity, are highly vacuolated, and have a primary cell wall. Parenchyma functions as food- and water-storage tissue as well as sites of metabolism in plants. Chlorenchyma cells are chloroplast-containing parenchyma specialized for photosynthesis.

Collenchyma cells are relatively long, with unevenly thickened primary walls. They support growing regions of shoots and are common in petioles, elongating stems and expanding leaves. Collenchyma cells are well adapted for support because their cell walls are able to stretch. They often form in strands or a cylinder just beneath the epidermis; such location maximizes support, as would a rod located in the center of a stem or petiole (leaf base).

Sclerenchyma cells are rigid; produce thick, nonstretchable secondary walls; and are usually dead at maturity. They occur in, support, and strengthen mature regions of plants, including stems, roots, and leaves. There are two types of sclerenchyma cells: sclereids and fibers. *Sclereids* are relatively short and variable in shape and usually occur in small groups. *Fibers* are long and slender and occur in strands or bundles. Sclereids are found in the roots, leaves, and stems. They produce the gritty texture of pears and mostly make up the tough core of apples as well as the seed coats of peanuts and walnuts. Fibers are often associated with vascular tissues and, compared to sclereids, are typically elongated cells that vary in length from a few millimeters to more than half a meter long.

Danilo D. Fernando

See also: Angiosperm plant formation; Angiosperms; Cell wall; Flower structure; Leaf anatomy; Plant fibers; Plant tissues; Root uptake systems; Roots; Shoots; Stems; Water and solute movement in plants; Wood.

Sources for Further Study

Dickison, William C. *Integrative Plant Anatomy*. San Diego: Academic Press, 2000. Presents a comparative study of plant cells and tissues. Includes photographs.

Raven, Peter H., Ray F. Evert, and Susan E. Eichhorn. *Biology of Plants*. 6th ed. New York: W. H. Freeman/Worth, 1999. Several chapters cover the structure and development of angiosperm plant bodies. Includes photographs.

Rudall, Paula. *Anatomy of Flowering Plants: An Introduction to Structure and Development*. 2d ed. New York: Cambridge University Press, 1993. Presents details of angiosperm anatomy and physiology. Includes photographs.

Uno, Gordon, Richard Storey, and Randy Moore. *Principles of Botany*. New York: McGraw-Hill, 2001. This basic text gives simple descriptions of plant cells and tissues. Includes diagrams, photographs.

ANGIOSPERM EVOLUTION

Categories: Angiosperms; evolution

Angiosperms (flowering plants) appeared about 130 million years ago and today dominate the plant world, with approximately 235,000 species.

In early Devonian-age rocks, approximately 363-409 million years old, fossils of simple vascular and nonvascular plants can be seen. Ferns, lycopods, horsetails, and early *gymnosperms* became prominent during the Carboniferous period (approximately 290-363 million years ago). The gymnosperms were the dominant flora during the Age of Dinosaurs, the Mesozoic era (65-245 million years ago). More than 130 million years ago, from the Jurassic period to early in the Cretaceous period, the first flowering plants, or angiosperms (phylum *Anthophyta*), arose. Over the following 40 million years, angiosperms became the world's dominant plants.

The angiosperms show high species diversity, and they occupy almost every habitat on earth, from deserts to high mountain peaks and from freshwater ecosystems to marine estuaries. Angiosperms range in size from eucalyptus trees well over 100 meters (328 feet) tall with trunks nearly 20 meters (66 feet) in circumference to duckweed, simple floating plants barely 1 millimeter (0.003 inch) long.

Special Characteristics

Some of the defining characteristics of angiosperms involve their physical appearance or morphology and internal anatomy: the presence of flowers and fruits containing seeds, stamens with two pairs of pollen sacs, a microgametophyte (the male, haploid stage of the life cycle contained in the pollen) with three nuclei, a megagametophyte (the female, haploid stage of the life cycle enclosed in the ovary) with eight nuclei, companion cells, and sieve tubes in the phloem (vascular tissue important in the transport of organic molecules). Some of these characteristics involve life-cycle features, such as *double fertilization*, that are distinct from almost all other members of the plant kingdom. (Double fertilization is also known in the genera *Ephedra* and *Gnetum*, members of the gymnosperms.)

Because angiosperms possess so many unique features, plant taxonomists have long believed that angiosperms originated from a single common ancestor. Because the first flowers and pollen grains appear in fossils from the early Cretaceous period, up to about 130 million years ago, it is probable that angiosperms actually arose more than 130 million years ago. As the findings of *paleobotanists* (botanists who study plants in the fossil record) have been combined with more recent knowledge from evolutionary genetics and biochemistry, a clearer picture of angiosperm evolution has emerged.

AP/Wide World Photos

In 1998 Ge Sun, David Dilcher, and their colleagues published their discovery of the oldest angiosperm fossil to date, estimated to be between 122 million and 145 million years old. The plant was given the name Archaefructus liaoningensis. *The fossil's age implies that angiosperms may have arisen as early as the Jurassic period, more than 145 million years ago.*

Proposed Ancestors

Because gymnosperms (the other large group of seed plants) have long been considered ancestral to the angiosperms, researchers have attempted to develop models for the evolution of the ovule-bearing structures of flowering plants from the similar, naked ovule-bearing structures of gymnosperms. Some lines of evidence indicate that groups of extinct cycad-like gymnosperms known as the *Bennettitales* and the *gnetophytes*, a modern division of the gymnosperms which show up in the fossil record about 225 million years ago, are the seed plants most closely related to angiosperms. All three groups, the *Bennettitales*, the gnetophytes, and the angiosperms, share, or shared, superficially similar flowerlike reproductive structures. The strobili, or cones, of some gnetophytes closely resemble flowers, and the xylem (vascular tissue specialized for transporting water) of some gnetophytes is similar to the xylem found in angiosperms.

Seed Ferns

Other lines of evidence suggest that a group of plants called the seed ferns, or *pteridosperms*, might represent the ancestors of the angiosperms. The seed ferns, which predate the angiosperms by many millions of years, had seed-bearing cupules and specialized organs that produced pollen. Many plant taxonomists believe that the seed-bearing cupules in some groups of seed ferns could have evolved into the carpels of flowers.

Earliest Flowers

Most paleobotanists assume that the first flowers were small, simple, and green in color and by modern standards were rather unattractive. Their petals and sepals were probably not clearly differentiated. In November of

1998, Ge Sun and David Dilcher and their colleagues published their discovery of the oldest angiosperm fossil to date, estimated to be at least 122 million years old and possibly as old as 145 million years. Either age qualifies it as the oldest. The fossils were discovered in China, and the fruits show the characteristic enclosed ovule (a *carpel*) that is distinctive to angiosperms. It was given the scientific name *Archaefructus liaoningensis*. Given its great age, this find implies that angiosperms may have arisen as early as the Jurassic period, more than 145 million years ago.

Other early flowers produced pollen with a single *aperture*, or opening, a trait that the monocot branch of the angiosperms shares with cycads and ginkgos. Plant taxonomists believe that pollen with a single opening is an ancestral feature that some plants have kept as they evolved. The pollen of eudicots, with its three apertures, is thought to be a derived feature (that is, a later evolutionary development).

Recent studies of angiosperm evolution, using data from deoxyribonucleic acid (DNA) sequences, have led to the proposal that an obscure shrub from the South Pacific island of New Caledonia, called *Amborella trichopoda*, represents what is left of the ancestral *sister group* (a related organism that branched off before the evolution of another group of organisms) to all the angiosperms. As a sister group to all the angiosperms, it is considered to be the most primitive (in an evolutionary sense) of the angiosperms and therefore should resemble what the ancestor to the angiosperms was like. It does possess some of the expected primitive traits for a primitive angiosperm, such as small, greenish-yellow flowers and a lack of vessels for conducting water from the ground to the leaves.

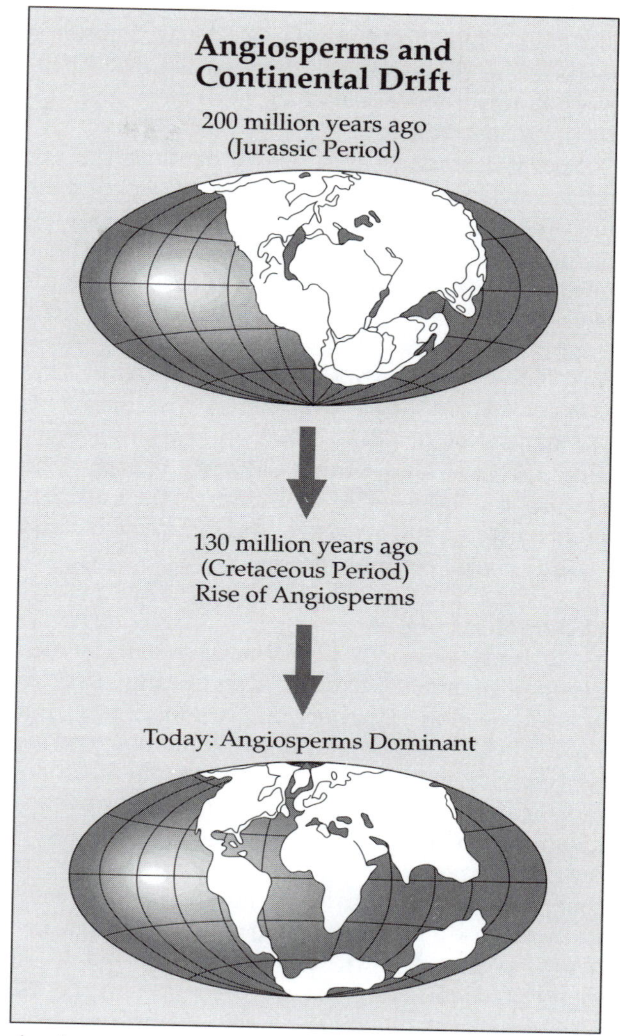

Angiosperms are proposed to have evolved approximately 130 million years ago, sometime between the Late Jurassic period (208-144 million years ago) and the Early Cretaceous period (144-65 million years ago), when South America, Africa, and India were much closer together.

Angiosperm Classification

Approximately 97 percent of angiosperm species are classified as either *Monocotyledones* (monocots), with approximately 65,000 species, or *Eudicotyledones* (eudicots), with about 165,000 species. The monocots include such familiar plants as the grasses, lilies, irises, orchids, cattails, and palms. The more diverse eudicots include most of the familiar trees and shrubs (other than the conifers) and many of the herbaceous plants. The remaining 3 percent of angiosperms are called the *magnoliids*, a group of plants considered to have primitive features, among them pollen with a single aperture. Many magnoliids also feature oil cells with ether-containing oils providing the characteristic scents of laurel and pepper, for example. The magnoliids are typically divided into the woody magnoliids and paleoherbs.

Woody magnoliids have large, often showy, bisexual flowers with multiple free parts. Magnolia trees and tulip trees (both in *Magnoliaceae*, or the magnolia family) are examples of this group. The *paleoherbs* have small, often unisexual flowers and usually just a few flower parts. Modern paleoherbs include the pepper family (*Piperaceae*), the birthwort family (*Aristolochiaceae*), and the water lily family (*Nymphaeaceae*).

Recent studies of angiosperm evolution, using data from DNA sequences, have also sharpened the understanding of some of the relationships among monocots, eudicots, and magnoliids. If these groups are displayed as an evolutionary tree (or *phylogenetic* tree), the magnoliids are *polyphyletic* (that is, they do not share a single common ancestor). The magnoliids branch off near the base of the tree on several different branches. The monocots are monophyletic (that is, they share a single common ancestor) and form a separate branch from among the magnoliid branches. The eudicots branch off last and represent the most diverse and evolutionarily complex group.

Geographic Origins

As hotly debated, perhaps, as exactly which group of plants were ancestral to the angiosperms is the question of where the angiosperms first evolved. Some botanists believe that angiosperms first developed in the Northern Hemisphere; others look at the Southern Hemisphere. At the time angiosperms are proposed to have evolved, the continents were not arranged the way they are now. At that time, all of the world's major landmasses were grouped into a supercontinent called Pangaea. The southern part of this continent is referred to as Gondwanaland, and the northern part is called Laurasia. Based on what is known about late Cretaceous angiosperms and their habitats, some scientists suggest that the westernmost, semiarid regions of Gondwanaland may be the place where angiosperms first evolved.

As Pangaea broke up, the separate continents moved in different positions, resulting in new configurations. India collided with Asia, raising the Himalaya Mountains and the Tibetan Plateau. Antarctica slipped over the South Pole, and Australia became isolated. These plate movements created new climatic regimes, opening up new niches that were rapidly exploited by the angiosperms.

Diversification and Spread

Regardless of their geographic origins, by about ninety million years ago the flowering plants were well on their way to dominating the world's flora. The early angiosperms were well adapted to drought and cold. Adaptations that conferred resistance to these conditions included strong leaves, efficient water-conducting cells, and tough, resistant seed coats. Some woody flowering plants evolved the ability to lose their leaves, called the *deciduous habit*. This characteristic allows the shutdown of metabolism during adverse environmental conditions, such as during seasonal droughts or winter weather. Because of greater climate instability during the past fifty million years or so compared to earlier times, the above-mentioned traits were important in allowing the flowering plants to adapt to different and often harsher climatic conditions.

Pollination

A major innovation that likely led to some of the great diversity seen in angiosperms was pollination by insects or other animals. As plants adapted to the various available pollinators, the pollinators also adapted to the plants, sometimes in very specific ways. Many pollination systems include specialized colors or markings, flower shapes, and flower scents. This process of evolving "together" is called *coevolution*. Coevolution has also occurred between plants and their predators. Evolution of chemical compounds to deter herbivory have, in turn, led to adaptations in many animal groups to circumvent the toxicity of the chemical compounds.

Carol S. Radford

See also: Algae; Angiosperm life cycle; Bacteria; Biochemical coevolution in angiosperms; Cell wall; Coevolution; Eukaryotic cells; Flower structure; Green algae; Photosynthesis; Plant tissues.

Sources for Further Study

Coulter, Merle C. *The Story of the Plant Kingdom*. 3d ed. Revised by Howard J. Dittmer. Chicago: University of Chicago Press, 1964. A discussion of the evolution and biology of plants.

Margulis, Lynn, and Karlene V. Schwartz. *Five Kingdoms: An Illustrated Guide to the Phyla of Life on Earth*. 3d ed. New York: W. H. Freeman, 1998. All of the major divisions of life on earth are discussed in this book.

Raven, Peter H., Ray F. Evert, and Susan E. Eichhorn. *Biology of Plants*. 6th ed. New York: W. H. Freeman/Worth, 1999. This standard college textbook discusses the systematics and morphology of the different divisions of the plant kingdom.

Stern, Kingsley R. *Introductory Plant Biology*. 8th ed. Boston: McGraw-Hill, 2000. Introductory botany textbook includes discussions of the different divisions of the plant kingdom.

ANGIOSPERM LIFE CYCLE

Categories: Anatomy; angiosperms; physiology; reproduction and life cycles

The word "angiosperm" comes from the Greek words for "vessel" and "seed" and translates roughly as "enclosed seed." In part, angiosperms (the flowering plants, phylum Anthophyta*) are defined by the fact that their seeds are enclosed by an ovule. The life cycle of an angiosperm is defined by the formation of the seed and its development to a full-grown plant which, in turn, produces seeds.*

Angiosperms are vascular plants with flowers that produce seeds enclosed in an *ovule*—a fact that is recognized as the *angiospermy condition*.

Reproductive Flower Parts

In general, angiosperms have a floral axis with four floral parts, two of which are fertile. At the receptacle, or tip, of the axis there is an ovule-bearing leaf structure known as the carpel. The ovule or ovules can be found inside the *pistil*. Three portions compose the pistil: the ovary, the style, and the stigma, where the pollen usually germinates. The mature ovule consists of the placenta, the integuments that are modified leaves that cover the entrance to the embryo sac, the micropyle, and the chalaza. These latter two parts of the ovule complement each other in their positions and functions. While the micropyle receives and guides the pollen tube, the chalaza relates to the vascular supply of the ovule, nutrition, and support. The *stamens*, which are often composed of the filament and sporangia sacs that make up the anther, surround the pistil. Stamens carry the male gametes, and the pistil carries the female gamete needed for sexual reproduction.

It is believed that the great diversity and adaptability of the angiosperms is related to the presence of a unique reproductive cycle. This cycle consists of an *alternation of generations* and the production of a pair of spores on two types of sporophylls: microspores (which become male gametophytes) and megaspores (which become female gametophytes).

Male Gamete Development

The angiosperm reproductive cycle begins with the process of *microsporogenesis*, or microspore formation. The stamen consists of a filament and the anther, also known as the *microsporangium*. In most of the cases, the anther consists of four pollen sacs, or locules. Within each locule, the archesporial cell develops through mitosis and extends as a row of cells throughout the entire length of the young anther. These mitotic cell divisions generate the anther wall, which is made up of several cell layers, the outermost of which transforms itself into the epidermis. The layer of cells below the epidermis is known as the endothecium. During anther development, the endothecial cells acquire thickenings whose function is related to anther opening and pollen release. The innermost layer of the anther wall is the tapetum, whose primary function correlates with the nourishment of the young pollen and the deposition of the exine, a coating of the pollen grain.

As development proceeds, the sporogenous cells located below the tapetum transform into microsporocytes. These microsporocytes will undergo meiosis, and tetrads (units of four) of microspores will form. Shortly after their formation, the tetrads separate into individual microspores. Each microspore is haploid, and often it will enlarge and separate from the tetrad, becoming sculptured by the deposition of sporopollenin and other substances that will turn into the ornamented surface of the pollen grain.

The second phase of pollen development is known as *microgametogenesis*. The microspore is the first cell of the gametophytic generation, the cell that generates the mature pollen. The single-nucleus microspore develops into the male gametophyte before the pollen is released. This developmental process occurs through two or three unequal mitotic divisions of the nucleus and subsequent cytokinesis (cell separation). The two daughter nuclei and cells differ in size and in form. The larger cell represents

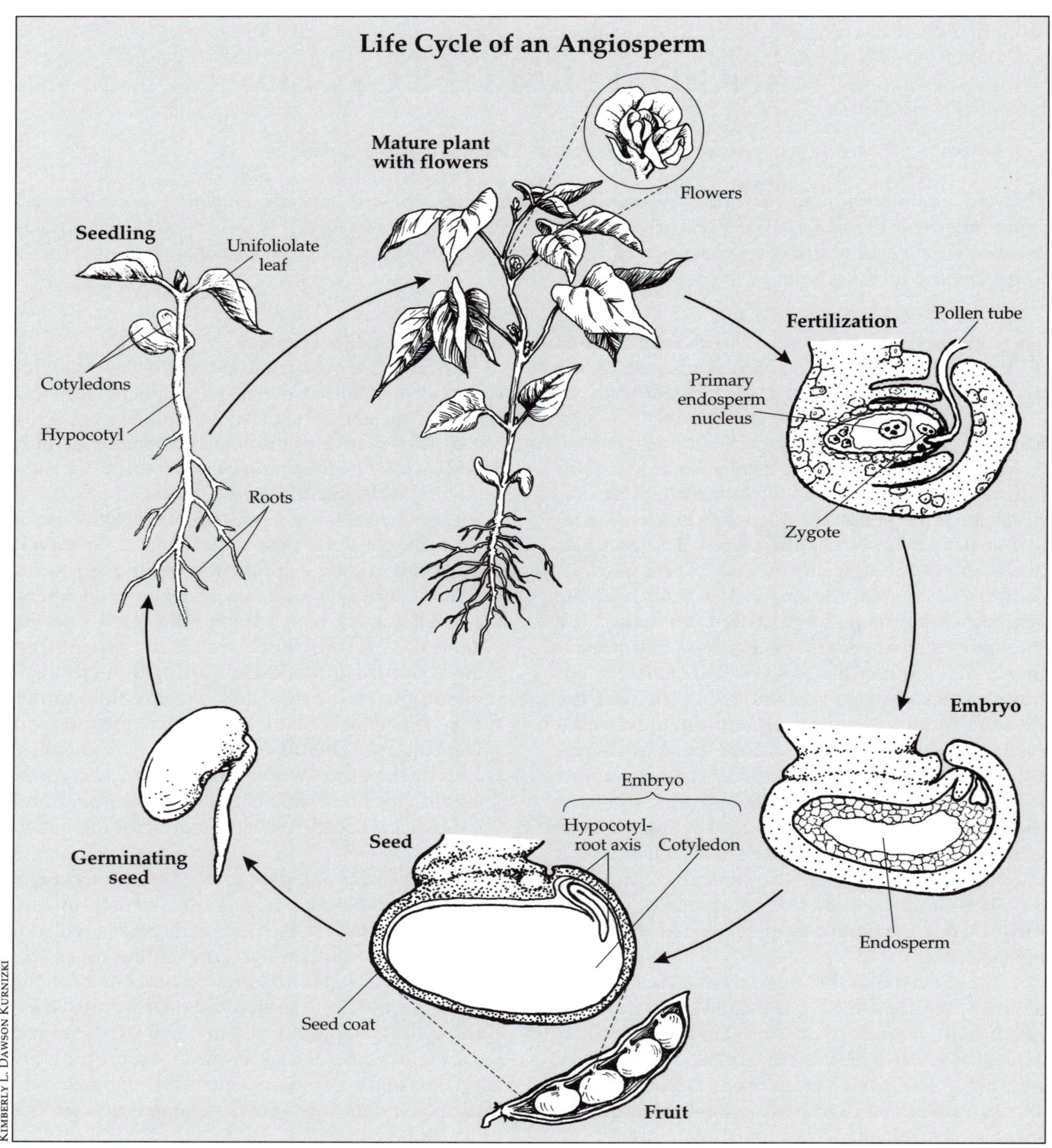

the tube cell and nucleus, while the smaller cell represents the generative cell and nucleus. At maturity, the grain can be shred in two or three nucleate conditions. When the anther opens, the mature male gametophytes or pollen grains will be disseminated and ready for germination.

Female Gamete Development

The ovule (female sex organ) consists of two opposite ends: the micropyle, where the integuments come together, and a more distant end, where the ovular tissue is more massive. This part is also known as the chalaza, and it lies directly opposed to the micropyle. The mature ovule is composed of three layers: the outer integument; the inner integument; and, underneath the integuments, the nucellus. During ovular development, one cell lying below the nucellar epidermis changes into a primary archesporial; this will divide to form the primary parietal cell and primary sporogenous cell. The primary sporogenous cell functions as the megaspore mother cell, which divides meiotically, originating four haploid megaspores. In the majority of angiosperms, three of the megaspores will degenerate, and only the chalazal one will develop into the megagametophyte (embryo sac).

After the completion of the embryo sac stage, a series of cellular events occurs, ending with the formation of the mature embryo sac, ready for fertilization by the male gametes. The chalazal megaspore enlarges and undergoes three mitoses, giving rise to eight haploid cells. The mature megagametophyte consists of two groups of four cells located at both ends of the embryo sac. The result is three antipodals at the chalazal end: the egg apparatus (consisting of the egg and two synergids at the micropylar end) and the polar nuclei. These two cells, present at both ends, usually fuse before pollination, and during fertilization they form the primary endosperm nucleus.

Pollination

The plant reproductive structures are now ready for the union of male and female gametes or fertilization, which eventually will produce a seed with a viable embryo and cotyledons. Before that step takes place, however, the pollen must be transferred from the anther to the stigma. Biotic agents (such as birds, insects, or mammals) or abiotic agents (such as wind or water) can accomplish this transfer process, known as *pollination*.

After landing on the stigma, pollen tubes will emerge through the grain apertures if the environment is high in humidity. Successful germination of the pollen in the stigma requires nutrients. In most plants, growth of the pollen tube lasts between twelve and forty-eight hours, from pollen germination to fertilization. Pollen germination starts with pollen-tube initiation, elongation, and penetration of the stigmatic tissue. During this period intense metabolic activity takes place, for the tube must synthesize membrane material and cell wall for growth and expansion. Simultaneously, at its tip the tube carries the vegetative cell nucleus, followed by the germinative cell.

Angiosperms have evolved complex breeding systems that ensure they will be pollinated by their own species. Today it is recognized that two pollination syndromes exist: *self-pollination* and *cross-pollination*. In self-breeding species, the pollen comes from the anther of the same flower. In cross-pollination (or outcrossing) species, the pollen comes from the anthers of a different flower or even a different plant of the same species. In these plants, incompatibility in the stigma guarantees that only pollen from other flowers will germinate.

Fertilization

The union of one sperm with the egg is known as *fertilization*. However, several developmental processes in the vegetative and germinative cells prepare the two sperms for a process known as *double fertilization*. A mitotic division of the germinative cell generates the sperm cells. This process that can take place on the growing pollen tube or inside the pollen grain. In a growing pollen tube, the vegetative nucleus disintegrates and the sperm cells will take the lead and enter the embryo sac for successful fertilization. Usually, the interactions between the pollen grain and the pistil ensure that the sperm cells will often reach the micropyle of the ovule.

Once the sperm reach the micropyle, the growth of other tubes stops. In the embryo sac (female gametophyte), four cells are located at the micropylar side. Of those four, the first pair that the sperm cells will encounter are the synergids. One of these is always bigger than the other and carries the filiform apparatus, a structure resembling hairs that degenerates after pollination and before fertilization. The synergids act as chemical attractants to the pollen tube, which penetrates the synergids via the filiform apparatus and then releases the two sperm cells. One of the sperm cells will fuse with the egg, producing the zygote; the other sperm cell will fuse with the primary endosperm nucleus, generating the endosperm. The remaining cells of the female gametophyte are the antipodals; they usually degenerate after fertilization has taken place.

Seed and Fruit Formation

Once fertilization has occurred, the ovule will go through a series of metabolic steps ending with the formation of the *seed* and the *fruit*. The recently created zygote transforms into a multicellular and complex *embryo* that has two well-defined polar ends: the radicle, or primary root, and the embryonic apical meristem with the first leaves. After successive mitosis, the mature endosperm usually grows close to the embryo and may provide nutrients needed for germination. The integuments will undergo further transformation, replication, and elongation and will become the seed coat—of variable texture, consistency, and colors, depending on the type of plant.

In general, after pollination or during fertilization, the ovary undergoes a series of physiological changes regulated by synchronized hormonal and genetic alterations that will modify the size of the parenchyma cells and its sugar and organic acids contents. This process turns the ovary into fruit—in many cases familiar as the edible fruits familiar in human diets. The fruit provides nourishment for the seed until it ripens and drops to the ground, where the next stage in the life cycle begins.

Germination, Seedling Development, and Maturation

Seeds are released from the fruit in a large variety of ways that have evolved to ensure the survival of species. Whether ingested by mammals and passed through their feces to the ground, borne by wind on feathery "wings," or simply falling from rotting fruit that has abscissed and dropped from the plant, the seed must next undergo a process called *germination*, in which the embryo enclosed in the seed begins its growth. The embryo develops a *hypocotyl* (root axis) and a fleshy part known as the *cotyledon*; in monocots there is one cotyledon, in dicots, two.

Germination requires certain conditions, such as the softening of the seed coat, moisture, and adequate warmth, to occur. During germination, the hypocotyl begins growing downward to become the root; the cotyledon(s) will develop into the shoot, stems, and leaves. The process of germination results in the sprouting through the ground's surface of the *seedling*, which will develop into the mature plant with flowers. The cycle then begins again.

Miriam Colella

See also: Angiosperm cells and tissues; Angiosperm evolution; Angiosperm plant formation; Angiosperms; Dormancy; Flower structure; Flower types; Flowering regulation; Fruit: structure and types; Germination and seedling development; Plant life spans; Reproduction in plants; Seeds.

Sources for Further Study

Cresti, Mauro, S. Blackmore, and J. L. van Went. *Atlas of Sexual Reproduction in Flowering Plants*. New York: Springer-Verlag, 1996. Ninety-three plates, including images from scanning, transmission electron microscopes, and light microscopes, document sexual reproduction in higher plants in three sections, "Anther Development," "Pistil Development," and "Progamic Phase and Fertilization." Of interest to students of genetics, plant breeding, and cellular biology.

Dowden, Anne Ophelia Todd. *From Flower to Fruit*. New York: Crowell, 1984. Text and drawings (some in color) that explain how flowers mature into seed-bearing fruit. For younger audiences.

Jensen, William A. *The Embryo Sac and Fertilization in Angiosperms*. Honolulu: Harold L. Lyon Arboretum, University of Hawaii, 1972. Brief (thirty-two-page) overview with illustrations.

Johri, B. M., K. B. Ambegaokar, and P. S. Srivastava. *Comparative Embryology of Angiosperms*. New York: Springer-Verlag, 1992. A review of the developmental processes of sexual reproduction in flowering plants, with tables, figures, and numerous literature citations. Aimed at students, researchers, taxonomists, breeders, and horticulturists.

Williams, Elizabeth G., Adrienne E. Clarke, and R. Bruce Knox. *Genetic Control of Self-Compatibility and Reproductive Development in Flowering Plants*. Boston: Kluwer Academic,

1994. Advances in molecular biology and genetic engineering of plant reproductive processes are narrated by the scientists in whose laboratories the advances were made, with emphasis on self-incompatibility and pollen development as it relates to male sterility. For advanced students and researchers.

ANGIOSPERM PLANT FORMATION

Categories: Anatomy; angiosperms; physiology; reproduction and life cycles

Angiosperms are flowering plants. Their formation entails development from embryo to seed, through germination to seedling, and finally to mature plant.

The life cycle of angiosperms (flowering plants) involves an alternation of generations between a dominant sporophytic (spore-producing) phase and a reduced gametophytic (gamete-producing) phase. The first cell of the sporophyte is the fertilized egg, or *zygote*, which undergoes repeated divisions, growth, and differentiation to form an embryo enclosed in the ovule. After fertilization, the ovule is transformed into the seed, which germinates into a seedling. The seedling becomes the adult plant; the plant produces flowers in which the sperm and egg—representing, respectively, the male and female gametophytic generations—are formed. Fertilization occurs, and seeds are produced to continue the life cycle.

Dicot Embryo Formation

In most angiosperms, embryo development, or *embryogenesis*, is initiated with a division of the fertilized egg into a small apical cell and a large basal cell, forming a two-celled proembryo. The apical cell generates the embryo proper, and the basal cell forms a filamentous *suspensor* that anchors the embryo. Two weeds, *Capsella bursa-pastoris* (shepherd's purse) and *Arabidopsis thaliana* (mouse ear cress or wall cress), both belonging to the *Brassicaceae* family, have attained prominence as textbook examples of embryogenesis in typical *dicots* (plants with two *cotyledons*, or seed leaves; a *monocot* has one seed leaf).

In these plants, the apical cell of the proembryo divides by two successive longitudinal walls to form a quadrant that is immediately partitioned by transverse walls into an octant, composed of an upper and lower tier of four cells each. The fates of the two tiers are already fixed in the octant embryo, as the upper tier forms the *shoot apex* and much of the cotyledons. The lower tier, in addition to providing derivatives to the remaining part of cotyledons, generates the *hypocotyl*, the *radicle*, and the *root apex*. However, the central region of the root cap, known as the columella, and the quiescent center of the root apical meristem are derived from the terminal cell of the *suspensor* closest to the embryo, known as the *hypophysis*. The apicobasal pattern of the future seedling plant is established in the octant embryo.

A series of divisions separating eight peripheral cells from a core of eight inner cells heralds histogenesis in the embryo. The result is the formation of a sixteen-celled, globular embryo, in which the peripheral cells form the protoderm (precursor cells of the embryonic epidermis), and the inner cells differentiate into the procambium and ground meristem (precursors of the vascular tissues and ground tissues, respectively) of the mature embryo. This initiates the formation of radial-pattern elements made up of concentric tissue layers in the basal part of the embryo.

The globular stage of the embryo is completed by approximately three additional rounds of divisions, mostly in the inner core of cells. The suspensor attains its genetically permissible number of six to nine cells by this stage. Gradually the cells begin to lose connection from one another and from the embryo and disintegrate.

Emerging from the globular stage, the embryo expands laterally by cell divisions to form the cotyledons and becomes heart-shaped. The heart-shaped stage is followed by the torpedo-shaped

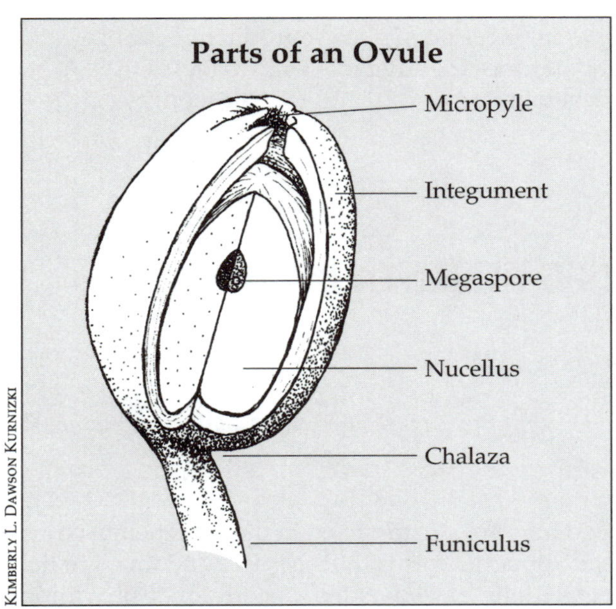

Kimberly L. Dawson Kurnizki

Angiosperms are vascular plants with flowers that produce seeds enclosed in an ovule (female sex organ), a fact that is recognized as the "angiospermy condition." The ovule consists of two opposite ends: the micropyle, where the integuments come together, and the chalaza, where the ovular tissue is more massive. The mature ovule is composed of three layers: the outer integument; the inner integument; and, underneath the integuments, the nucellus (megasporangium), in which the megaspore, which will become the female gametophyte, is embedded. The stalk to which the ovule is attached is the funiculus.

stage, in which elongation of the cotyledons and hypocotyl, as well as extension of the vascular tissues, occurs. The basic body plan of a shoot-root axis becomes unmistakably clear at this stage, with the establishment of the shoot apical meristem in the depression between the cotyledons and the organization of a root apex by incorporation of derivatives of the hypophysis at the opposite end of the embryo.

During further growth, the cotyledons bend toward the hypocotyl (bent cotyledon or walking-stick shaped stage), and the embryo is phased into the mature stage. A mature embryo of *Arabidopsis* has fifteen thousand to twenty thousand cells and, under favorable conditions of growth, develops in about nine days from the time of fertilization to the mature embryo stage. Sensitive genetic screens have led to the isolation of *Arabidopsis* mutants defective in apicobasal and radial patterning of embryos. Characterization of the mutant genes and their protein products has unraveled to some extent the molecular components of the embryonic pattern-forming system in this plant.

Monocot Embryo Formation

The early divisions of the zygote in monocots follow the same pattern as in dicots. However, in the *Poaceae* (grasses) family, which includes wheat and the other cereals, the sequence and orientation of later divisions in the proembryo are irregular, resulting in highly complex mature embryos. The main feature of the cereal embryo is the development of an absorptive organ known as the *scutellum* (considered equivalent to the single cotyledon). Other organs of the embryo for which there are no counterparts in the dicot embryo are a sheathlike tissue covering the root (*coleorhiza*), a tissue that covers the shoot (*coleoptile*), and an internode known as the *mesocotyl*. On one side of the coleorhiza there is also a small, flaplike outgrowth called the *epiblast*.

Embryo Maturation to Seed

As the embryo matures, the ovule progressively desiccates to become the seed enclosed within the ovary. Concomitantly, the integuments of the ovule harden to form the protective seed coat. Within the ovule itself, the primary endosperm nucleus formed after double fertilization begins to divide, ahead of the zygote, to produce the endosperm charged with nutrient substances. In seeds of many plants, including *Arabidopsis*, *Capsella*, bean, and pea, the endosperm is utilized by the developing embryo.

In other plants, especially the cereals, the bulk of the seed (grain) is made up of the endosperm surrounding the small embryo. The mature embryo enclosed in the seed consists of an axis bearing the *radicle* (embryonic root) at one end and the *plumule* (the embryonic shoot consisting of the shoot apex and one or two leaves) at the other end, and one (in monocots) or two (in dicots) cotyledons. The part of the embryo axis above the point of attachment of the cotyledon(s) is known as the *epicotyl*, whereas the part below the attachment point connecting to the radicle is called the *hypocotyl*.

Seed Germination

The dry seed enclosing the mature embryo may not germinate immediately; if it does not, it enters a period of *quiescence* or dormancy. Quiescent seeds germinate when provided with the appropriate conditions for growth, such as water, a favorable temperature, and the normal composition of the atmosphere. Dormant seeds germinate only when some additional hormonal, environmental, metabolic, or physical conditions are met. In almost all seeds, the first part of the embryo to emerge during germination is the radicle. It forces it way through the soil and forms the primary root of the seedling. However, the manner in which the shoot emerges and develops varies considerably in different seeds.

In the *epigeous* type of germination (for example, in beans), emergence of the radicle is followed by the elongation of the hypocotyl, which arches above the soil surface as a hook. As the hook straightens, it pulls out the cotyledons and plumule above the soil surface. In the *hypogeous* type of germination (in peas, for example) the cotyledons enclosed within the seed coat remain in the soil during germination. It is the epicotyl that arches above the soil surface, and as the hook straightens out, it carries the plumule along with it to the surface of the soil. In the monocot, such as the onion, after emergence of the radicle the single cotyledon arches aboveground and subsequently straightens.

Members of the *Poaceae* display a type of germination in which, following the outgrowth of the radicle, the coleoptile enclosing the plumule grows out of the grain and appears above the soil. The seedling leaves force their way, breaking the coleoptile, and appear outside as the first photosynthetic organs. The growth of the coleoptile during germination of grains is facilitated by the elongation of the mesocotyl. These various types of germination ensure an efficient use of food materials stored in the embryo or in the endosperm for the growth of the seedling until it becomes autotrophic.

Embryo to Adult Plant

Although the question as to whether the seedling will become a gigantic tree or a small, herbaceous plant is determined by its genetic blueprint, certain common postgermination growth and developmental episodes mark the development of the seedling into an adult plant. In dicots, continued growth of the primary root produces an extensively branched root system consisting of secondary roots or lateral roots. In monocots, the primary root disintegrates shortly after it is formed, and so the root system is constituted of numerous adventitious roots which arise from the base of the stem.

Although the cotyledons retain their photosynthetic capacity for some time after germination of the seed, the seedling becomes completely autotrophic as the shoot apex produces new leaves and branches arise in the axils of leaves. These activities are coordinated by the division of cells in the root and shoot apical meristems and the differentiation of cells into specialized tissues and organs. The shoot and root apical meristems, considered analogous to the stem cells of animals, remain active throughout the life of the plant and, hence, are known as indeterminate meristems.

V. Raghavan

See also: Angiosperm cells and tissues; Angiosperm evolution; Angiosperm life cycle; Angiosperms;

Digital Stock

As the angiosperm embryo matures, the ovule progressively desiccates to become the seed enclosed within the ovary. These milkweed seeds will germinate when particular hormonal, environmental, metabolic, or physical conditions are met.

Dormancy; Flower structure; Flower types; Flowering regulation; Germination and seedling development; Plant life spans; Pollination; Reproduction in plants; Seeds.

Sources for Further Study

Raghavan, V. *Developmental Biology of Flowering Plants*. New York: Springer-Verlag, 2000. Textbook that describes the main events of the angiosperm life cycle and organ systems of the plant from a developmental perspective. With 157 illustrations.

_______. *Molecular Embryology of Flowering Plants*. New York: Cambridge University Press, 1997. Reference book that describes the modern genetic and molecular aspects of reproductive processes in angiosperms.

Raven, Peter H., Ray F. Evert, and Susan E. Eichhorn. *Biology of Plants*. 6th ed. New York: W. H. Freeman/Worth, 1999. College-level textbook that describes the basic stages of the angiosperm life cycle.

ANGIOSPERMS

Categories: Angiosperms; economic botany and plant uses; food; medicine and health; *Plantae*; taxonomic groups

The name "angiosperms" has long been used by botanists to refer to the flowering plants, a group of approximately 235,000 species. All angiosperms are members of the phylum Anthophyta.

The name "angiosperm" is actually derived from two Greek words, *angeion*, meaning "vessel" or "container," and *sperma*, meaning "seed." The name was given in reference to the fact that the seeds of all flowering plants develop from *ovules* that are enclosed in a structure called a *carpel*. This characteristic sets the angiosperms apart from all other plants, which either do not have seeds or have seeds that are not developed in structures resembling a carpel. Although the name angiosperm is used widely, plant taxonomists and many botanists typically refer to them by the more formal name *Anthophyta*, the phylum that contains the flowering plants.

Unique Features of Angiosperms

In addition to possessing enclosed seeds, *Anthophyta* differs from other plant phyla in a number of ways. The most obvious distinguishing feature is the *flower*, a complex structure containing the reproductive parts of the plant. The reproductive structures in other plants are much less complex and showy. The angiosperm life cycle differs from that of almost all other plants. The *sporophyte* is the dominant, *diploid* stage and is the more visible form of the plant, with the leaves, stems, roots, and flowers. The *haploid gametophyte* is confined to life inside the *ovary* or *anther* of the flower, unlike the typically free-living gametophytes of most other plants.

Fertilization is also unique in angiosperms. Many angiosperms rely on insects or other animals to transfer pollen from one flower to another. *Pollen grains* produce two haploid *sperm* that travel through a *pollen tube* from the *stigma* into the ovary of the flower and into one of the *embryo sacs*. Within the embryo sac one of the sperm fertilizes the egg, which will lead to formation of the diploid embryo, and the other sperm fuses with two or more polar nuclei to form the *endosperm*, which will nourish the embryo and young seedling. This process is often referred to as *double fertilization*. Other, less obvious features set *Anthophyta* apart as well, including a unique *vascular anatomy*, pollen structure, and various biochemical characteristics.

Size and Geographic Diversity

There are approximately 235,000 species of flowering plants, and they are found in almost all terrestrial habitats, with the exception of extremely high elevations and some polar regions. A small proportion of flowering plants are aquatic (that is, found in

Common Monocot Families

There are at least four major taxonomic systems for classifying flowering plants, as well as less formal systems. While names of phyla (divisions), subdivisions, classes, subclasses, and orders vary, along with the placement of families within those larger groups, the names of families, genera, and species remain fairly constant, with fewer alterations and controversies (although subject to changes as well). Families found in the United States are followed by their common names in parentheses.

Acoraceae (calamus)
Agavaceae (century plant)
Alismataceae (water plantain)
Aloeaceae (aloe)
Aponogetonaceae (cape pondweed)
Araceae (arum)
Arecaceae (palm)
Bromeliaceae (bromeliad)
Burmanniaceae (burmannia)
Butomaceae (flowering rush)
Cannaceae (canna)
Centrolepidaceae
Commelinaceae (spiderwort)
Corsiaceae
Costaceae (costus)
Cyanastraceae
Cyclanthaceae (Panama hat)
Cymodoceaceae (manatee grass)
Cyperaceae (sedge)
Dioscoreaceae (yam)
Eriocaulaceae (pipewort)
Flagellariaceae
Geosiridaceae
Haemodoraceae (bloodwort)
Hanguanaceae (hanguana)
Heliconiaceae (heliconia)
Hydatellaceae
Hydrocharitaceae (tape grass)
Iridaceae (iris)
Joinvilleaceae (joinvillea)
Juncaceae (rush)
Juncaginaceae (arrow grass)
Lemnaceae (duckweed)
Liliaceae (lily)
Limnocharitaceae (water poppy)
Lowiaceae
Marantaceae (prayer plant)
Mayacaceae (mayaca)
Musaceae (banana)
Najadaceae (water nymph)
Orchidaceae (orchid)
Pandanaceae (screw pine)
Petrosaviaceae
Philydraceae (philydraceae)
Poaceae (grass)
Pontederiaceae (water hyacinth)
Posidoniaceae (posidonia)
Potamogetonaceae (pondweed)
Rapateaceae
Restionaceae
Ruppiaceae (ditch grass)
Scheuchzeriaceae (scheuchzeria)
Smilacaceae (catbrier)
Sparganiaceae (bur reed)
Stemonaceae (stemona)
Strelitziaceae
Taccaceae (tacca)
Thurniaceae
Triuridaceae
Typhaceae (cattail)
Velloziaceae
Xanthorrhoeaceae
Xyridaceae (yellow-eyed grass)
Zannichelliaceae (horned pondweed)
Zingiberaceae (ginger)
Zosteraceae (eelgrass)

Source: Data are from U.S. Department of Agriculture, National Plant Data Center, *The PLANTS Database*, Version 3.1, http://plants.usda.gov. National Plant Data Center, Baton Rouge, LA 70874-4490 USA, and the Texas A&M Bioinformatics Working Group, Texas A&M University, http://www.csdl.tamu.edu/FLORA/newgate.

freshwater habitats), and an even smaller number are marine (found in saltwater habitats). The greatest species richness is in tropical regions, especially tropical rain forests, and species richness steadily decreases at increasing latitudes north and south of the equator.

Angiosperms have been so successful in terrestrial ecosystems that they represent the majority of the herbs and shrubs and many of the trees as well. The diversity of growth forms is tremendous, represented by such diverse families as *Poaceae* (grasses and bamboos), which have greatly reduced and modified flowers; *Cactaceae* (cactuses), which have spines instead of leaves and very showy flowers; and *Lemnaceae* (duckweed), which has a highly reduced plant body sometimes comprising a single leaf with no true roots or stem and the smallest flowers of any angiosperm. Other families include *Asteraceae* (sunflower or aster family), with reduced disc and ray flowers crowded together into *inflorescences* called heads; *Salicaceae* (willow family), a widespread, water-loving family of trees and shrubs with reduced flowers arranged in catkins; and *Orchidaceae* (orchid family), with some of the showiest and most intricate flowers of all, which have extremely numerous and minute seeds.

Economic Importance

Economically, angiosperms have made a profound impact. Essentially all of the world's food crops, from rice, wheat, and corn to other fruits and vegetables, are derived from flowering plants. In fact, it is almost impossible to think of more than a handful of foods or food ingredients from plants that are not flowering plants. The same is true of ornamental plants. Although a few gymnosperms (such as conifers) and ferns are common as ornamentals, most of the remaining plants, even many valued for their foliage rather than their blooms, are flowering plants. The only area where angiosperms do not dominate economically is in forest products, where conifers account for a significantly larger proportion of the harvest, but even there, hardwoods predominate for certain applications.

Medicine has also reaped many benefits from angiosperms. In fact, it was primarily the herbalists, from the Middle Ages to the Scientific Revolution, who expanded humankind's understanding of flowering plants. Knowledge of flowering plants for food and medicine among many indigenous peoples has always been widespread. Modern medicine has capitalized on much of this knowledge and has

Common Dicot (Eudicot) Families

There are at least four major taxonomic systems for classifying flowering plants, as well as less formal systems. While names of phyla (divisions), subdivisions, classes, subclasses, and orders vary, along with the placement of families within those larger groups, the names of families, genera, and species remain fairly constant, with fewer alterations and controversies (although subject to changes as well). Families found in the United States are followed by their common names in parentheses.

Acanthaceae (acanthus)
Aceraceae (maple)
Achariaceae
Achatocarpaceae (achatocarpus)
Actinidiaceae (Chinese gooseberry)
Adoxaceae (moschatel)
Aextoxicaceae
Aizoaceae (fig marigold)
Akaniaceae
Alangiaceae
Alseuosmiaceae
Alzateaceae
Amaranthaceae (amaranth)
Amborellaceae
Anacardiaceae (sumac)
Ancistrocladaceae
Anisophylleaceae
Annonaceae (custard apple)
Apiaceae (carrot)
Apocynaceae (dogbane)
Aquifoliaceae (holly)
Araliaceae (ginseng)
Aristolochiaceae (birthwort)
Asclepiadaceae (milkweed)
Asteraceae (aster, also *Compositae*)
Austrobaileyaceae
Balanopaceae
Balanophoraceae (balanophora)
Balsaminaceae (touch-me-not)
Barbeyaceae
Barclayaceae
Basellaceae (basella)
Bataceae (saltwort)
Begoniaceae (begonia)
Berberidaceae (barberry)
Betulaceae (birch)
Bignoniaceae (trumpet creeper)
Bixaceae (lipstick tree)
Bombacaceae (kapok tree)
Boraginaceae (borage)
Brassicaceae (mustard, also *Cruciferae*)
Bretschneideraceae
Brunelliaceae (brunellia)
Bruniaceae
Brunoniaceae
Buddlejaceae (butterfly bush)
Burseraceae (frankincense)
Buxaceae (boxwood)
Byblidaceae
Cabombaceae (water shield)
Cactaceae (cactus)
Caesalpiniaceae
Callitrichaceae (water starwort)
Calycanthaceae (strawberry shrub)*
Calyceraceae (calycera)
Campanulaceae (bellflower)
Canellaceae (canella)
Cannabaceae (hemp)
Capparaceae (caper)
Caprifoliaceae (honeysuckle)
Cardiopteridaceae
Caricaceae (papaya)
Caryocaraceae (souari)
Caryophyllaceae (pink)
Casuarinaceae (she-oak)
Cecropiaceae (cecropia)
Celastraceae (bittersweet)
Cephalotaceae
Ceratophyllaceae (hornwort)
Cercidiphyllaceae (katsura tree)
Chenopodiaceae (goosefoot)
Chloranthaceae (chloranthus)
Chrysobalanaceae (cocoa plum)
Circaeasteraceae
Cistaceae (rockrose)
Clethraceae (clethra)
Clusiaceae (mangosteen, also *Guttiferae*)
Cneoraceae
Columelliaceae
Combretaceae (Indian almond)
Compositae (aster, also *Asteraceae*)
Connaraceae (cannarus)
Convolvulaceae (morning glory)
Coriariaceae
Cornaceae (dogwood)
Corynocarpaceae (karaka)
Crassulaceae (stonecrop)
Crossosomataceae (crossosoma)
Crypteroniaceae
Cucurbitaceae (cucumber)
Cunoniaceae (cunonia)
Cuscutaceae (dodder)
Cyrillaceae (cyrilla)
Daphniphyllaceae

(continued)

*Some systems classify magnoliids (about 3 percent of flowering plants) separately from monocots and dicots, including these families.

even expanded the search for new medicines. Flowering plants have been the original source of many precursors to modern medicines, including aspirin (willows, *Salix*), quinine (*Cinchona* species), and digitalin and digoxin (*Digitalis* species).

Lifestyle Diversity

Along with the diversity in structure comes a diversity in lifestyles. Most angiosperms are free-living, that is, receiving their primary energy and carbon from photosynthesis and their nutrients from the soil. A few groups of plants receive their energy or nutrients in other ways. Some are *saprophytes*, which receive their energy and carbon from decaying organic material in the soil and their nutrients from other soil components, much like other plants. Some of the best-known saprophytes are in *Ericaceae* (heath family). Their most distinc-

Datiscaceae (datisca)
Davidsoniaceae
Degeneriaceae
Dialypetalanthaceae
Diapensiaceae (diapensia)
Dichapetalaceae
Didiereaceae
Didymelaceae
Dilleniaceae (dillenia)
Dioncophyllaceae
Dipentodontaceae
Dipsacaceae (teasel)
Dipterocarpaceae (meranti)
Donatiaceae
Droseraceae (sundew)
Duckeodendraceae
Ebenaceae (ebony)
Elaeagnaceae (oleaster)
Elaeocarpaceae (elaeocarpus)
Elatinaceae (waterwort)
Empetraceae (crowberry)
Epacridaceae (epacris)
Eremolepidaceae (catkin-mistletoe)
Ericaceae (heath)
Erythroxylaceae (coca)
Eucommiaceae
Eucryphiaceae
Euphorbiaceae (spurge)
Eupomatiaceae
Eupteleaceae
Fabaceae (pea or legume, also *Papilionaceae*)
Fagaceae (beech)
Flacourtiaceae (flacourtia)
Fouquieriaceae (ocotillo)
Frankeniaceae (frankenia)
Fumariaceae (fumitory)
Garryaceae (silk tassel)
Geissolomataceae
Gentianaceae (gentian)
Geraniaceae (geranium)
Gesneriaceae (gesneriad)
Globulariaceae
Gomortegaceae
Goodeniaceae (goodenia)
Greyiaceae
Grossulariaceae (currant)
Grubbiaceae
Gunneraceae (gunnera)
Gyrostemonaceae
Haloragaceae (water milfoil)
Hamamelidaceae (witch hazel)
Hernandiaceae (hernandia)
Himantandraceae
Hippocastanaceae (horse chestnut)
Hippocrateaceae (hippocratea)
Hippuridaceae (mare's tail)
Hoplestigmataceae
Huaceae
Hugoniaceae
Humiriaceae
Hydnoraceae
Hydrangeaceae (hydrangea)
Hydrophyllaceae (waterleaf)
Hydrostachyaceae
Icacinaceae (icacina)
Idiospermaceae
Illiciaceae (star anise)
Ixonanthaceae
Juglandaceae (walnut)
Julianiaceae
Krameriaceae (krameria)
Lacistemataceae
Lamiaceae (mint, also *Labiatae*)
Lardizabalaceae (lardizabala)
Lauraceae (laurel)*
Lecythidaceae (brazil nut)
Leeaceae
Leitneriaceae (corkwood)
Lennoaceae (lennoa)
Lentibulariaceae (bladderwort)
Limnanthaceae (meadow foam)
Linaceae (flax)
Lissocarpaceae
Loasaceae (loasa)
Loganiaceae (logania)
Loranthaceae (showy mistletoe)
Lythraceae (loosestrife)
Magnoliaceae (magnolia)*
Malesherbiaceae
Malpighiaceae (barbados cherry)
Malvaceae (mallow)
Marcgraviaceae (shingle plant)
Medusagynaceae
Medusandraceae
Melastomataceae (melastome)
Meliaceae (mahogany)
Melianthaceae
Mendonciaceae
Menispermaceae (moonseed)
Menyanthaceae (buckbean)
Mimosaceae
Misodendraceae
Mitrastemonaceae
Molluginaceae (carpetweed)
Monimiaceae (monimia)
Monotropaceae (Indian pipe)
Moraceae (mulberry)
Moringaceae (horseradish tree)
Myoporaceae (myoporum)
Myricaceae (bayberry)
Myristicaceae (nutmeg)
Myrothamnaceae
Myrsinaceae (myrsine)
Myrtaceae (myrtle)
Nelumbonaceae (lotus lily)
Nepenthaceae (East Indian pitcher plant)
Neuradaceae
Nolanaceae

(continued)

*Some systems classify magnoliids (about 3 percent of flowering plants) separately from monocots and dicots, including these families.

tive feature is that they are either white or some shade of pink or red and are never green. *Monotropa uniflora* (Indian pipes), for example, is a ghostly white and has no chlorophyll.

Parasitism is an alternative for some angiosperms. One well-known parasite is the mistletoe (*Loranthaceae*), popular as a Christmas decoration, which is a branch parasite on trees. Many types of mistletoe have green foliage and therefore receive some of their energy from photosynthesis, but their primary nourishment comes from the host tree. Some species have foliage that is brown or yellow and do not photosynthesize much at all. The seeds of mistletoe are spread from tree to tree when birds eat their berries and defecate the seeds on the branch of another tree. Probably the most unusual

Common Dicot (Eudicot) Families *(continued)*

Nothofagaceae
Nyctaginaceae (four-o'clock)
Nymphaeaceae (water lily)
Nyssaceae (sour gum)
Ochnaceae (ochna)
Olacaceae (olax)
Oleaceae (olive)
Oliniaceae
Onagraceae (evening primrose)
Oncothecaceae
Opiliaceae
Orobanchaceae (broom rape)
Oxalidaceae (wood sorrel)
Paeoniaceae (peony)
Pandaceae
Papaveraceae (poppy)
Paracryphiaceae
Passifloraceae (passionflower)
Pedaliaceae (sesame)
Pellicieraceae
Penaeaceae
Pentaphragmataceae
Pentaphylacaceae
Peridiscaceae
Physenaceae
Phytolaccaceae (pokeweed)
Piperaceae (pepper)
Pittosporaceae (pittosporum)
Plantaginaceae (plantain)
Platanaceae (plane tree)
Plumbaginaceae (leadwort)
Podostemaceae (river weed)
Polemoniaceae (phlox)
Polygalaceae (milkwort)
Polygonaceae (buckwheat)
Portulacaceae (purslane)
Primulaceae (primrose)
Proteaceae (protea)
Punicaceae (pomegranate)
Pyrolaceae (shinleaf)
Quiinaceae
Rafflesiaceae (rafflesia)
Ranunculaceae (buttercup or ranunculus)
Resedaceae (mignonette)
Retziaceae
Rhabdodendraceae
Rhamnaceae (buckthorn)
Rhizophoraceae (red mangrove)
Rhoipteleaceae
Rhynchocalycaceae
Rosaceae (rose)
Rubiaceae (madder)
Rutaceae (rue)
Sabiaceae (sabia)
Saccifoliaceae
Salicaceae (willow)
Salvadoraceae
Santalaceae (sandalwood)
Sapindaceae (soapberry)
Sapotaceae (sapodilla)
Sarcolaenaceae
Sargentodoxaceae
Sarraceniaceae (pitcher plant)
Saururaceae (lizard's tail)
Saxifragaceae (saxifrage)
Schisandraceae (schisandra)
Scrophulariaceae (figwort)
Scyphostegiaceae
Scytopetalaceae
Simaroubaceae (quassia)
Simmondsiaceae (jojoba)
Solanaceae (potato)
Sonneratiaceae (sonneratia)
Sphaerosepalaceae
Sphenocleaceae (spenoclea)
Stachyuraceae
Stackhousiaceae (stackhousia)
Staphyleaceae (bladdernut)
Sterculiaceae (cacao)
Stylidiaceae
Styracaceae (storax)
Surianaceae (suriana)
Symplocaceae (sweetleaf)
Tamaricaceae (tamarix)
Tepuianthaceae
Tetracentraceae
Tetrameristaceae
Theaceae (tea)
Theligonaceae
Theophrastaceae (theophrasta)
Thymelaeaceae (mezereum)
Ticodendraceae
Tiliaceae (linden)
Tovariaceae
Trapaceae (water chestnut)
Tremandraceae
Trigoniaceae
Trimeniaceae
Trochodendraceae
Tropaeolaceae (nasturtium)
Turneraceae (turnera)
Ulmaceae (elm)
Urticaceae (nettle)
Valerianaceae (valerian)
Verbenaceae (verbena)
Violaceae (violet)
Viscaceae (Christmas mistletoe)
Vitaceae (grape)
Vochysiaceae
Winteraceae (wintera)
Xanthophyllaceae
Zygophyllaceae (creosote bush)

Source: Data are from U.S. Department of Agriculture, National Plant Data Center, *The PLANTS Database*, Version 3.1, http://plants.usda.gov. National Plant Data Center, Baton Rouge, LA 70874-4490 USA, and the Texas A&M Bioinformatics Working Group, Texas A&M University, http://www.csdl.tamu.edu/FLORA/newgate.

parasite is *Rafflesia*, from Malaysia and Sumatra. It parasitizes species of *Tetrastigma*, a vine that grows on the forest floor and has no stems or leaves of its own. When it blooms it has the largest flowers in the world, and it is often called the corpse flower because it has a very strong odor, like that of rotting flesh.

Other parasites receive varying proportions of their energy and nutrients from their host and conventional means, and when the contributions are nearly equal they are referred to as *hemiparasites*. Hemiparasites are common in *Castilleja* (paintbrushes), and many species invade the roots of other plants to obtain part of their nutritional needs.

A unique approach to obtaining nutrients is represented by *insectivorous plants*, commonly known as *carnivorous plants* These plants use a variety of adaptations for trapping and absorbing nutrients from insects. Sundews (*Droseraceae*) have special glands on their leaves that excrete a sticky fluid that traps insects like flypaper. Pitcher plants (*Nepenthaceae* and *Sarraceniaceae*) have special tubular leaves that resemble cups or pitchers. The inside of the leaves fill with water near the base, and the lip and inside surface of the pitcher are slippery. Once an insect gets inside, it slips into the water at the bottom. Venus's flytrap (*Dionaea*, also in *Droseraceae*) is even more intricate, with leaves specially modified with traps that spring shut when an insect lands or walks on them. There is even an aquatic carnivore, the bladderwort (*Utricularia*), which has saclike leaves with small openings that can close after a small aquatic insect or crustacean is sucked in. Although insectivorous plants do obtain some of their nutrients from insects, they also obtain nutrients from the soil or, in the case of bladderworts, surrounding water.

Angiosperm Classification

Traditionally *Anthophyta* has either been considered as a single class *Angiospermae* or *Magnoliopsida*, with two subclasses, or has been divided into two classes, *Eudicotyledones*, or *Magnoliopsida*, and *Monocotyledones*, or *Liliopsida*. The second of these two options is more commonly accepted by contemporary plant taxonomists, and the two classes are often referred to by the common names dicotyledons or dicots and monocotyledons or monocots, respectively.

The monocot/dicot dichotomy has long been considered a major evolutionary split in the angiosperms. The two classes a differ from each other in a number of ways. Monocots generally have bladelike leaves with parallel venation, whereas dicots more typically have pinnate or palmate venation. Monocots have fibrous root systems without taproots; dicots typically have taproots. The flower parts in monocots occur typically in threes, whereas they occur most often in fours and fives in dicots. Monocots lack cambial secondary growth, which is common in dicots. Monocots have scattered vascular bundles in their stems, as opposed to the more orderly arrangement seen in dicot stems.

It has long been proposed that the monocots branched off from the dicots very early in the evolution of the angiosperms, but until recently it was difficult to sort out the probable events and the resulting classification system that would be needed to reflect them. With the advent of molecular tools, such as deoxyribonucleic acid (DNA) sequencing, the study of early angiosperm evolution is getting much more attention. It has now become clear that, if the classification system is to reflect evolutionary history, *Anthophyta* must be divided into more than just two classes. Currently there is no agreement on how many other classes there should be, but *Monocotyledones* and *Eudicotyledones* will retain most of the taxa. This new approach to the classification of *Anthophyta* has also resulted in changing the common name of the "dicots" to "eudicots," meaning "true dicots."

Many of the remaining taxa not included in the monocots or eudicots are now often referred to as *magnoliids* and are considered to represent taxonomic groups that have branched off from the early angiosperms before the monocot/eudicot split. Some of these groups include the orders *Magnoliales* (which includes *Magnoliaceae*, long considered as having many primitive characteristics), *Winterales*, and *Laurales*. The placement of a few taxa, such as *Ceratophyllaceae* and *Chloranthaceae*, is particularly controversial. With continued analyses of DNA sequences it is hoped that a clearer picture of the relationships among the magnoliids and related taxa will be obtained and a more phylogenetically based classification system can be devised.

Bryan Ness

See also: Angiosperm evolution; Angiosperm life cycle; Angiosperm plant formation; Biochemical coevolution in angiosperms; Cacti and succulents; Carnivorous plants; Eudicots; Flower structure;

Flower types; Fruit: structure and types; Garden plants: flowering; Germination and seedling development; Grasses and bamboos; Growth habits; Inflorescences; Legumes; Medicinal plants; Monocots vs. dicots; *Monocotyledones*; Orchids; Parasitic plants; Pollination; Reproduction in plants.

Sources for Further Study

Heywood, V. H., ed. *Flowering Plants of the World*. Englewood Cliffs, N.J.: Prentice-Hall, 1993. A profusely illustrated guide to flowering plants. Contains information on most of the generally recognized families of the world, including descriptions, taxonomy, geographic distribution, and economic uses.

Hickey, Michael, and Clive King. *Common Families of Flowering Plants*. New York: Cambridge University Press. 1998. Contains information, detailed descriptions, and extensive, detailed line drawings on twenty-five of the most common angiosperm families. Within the section for each family there is detailed information and line drawings on a common species.

Raven, Peter H., Ray F. Evert, and Susan E. Eichhorn. *Biology of Plants*. 6th ed. New York: W. H. Freeman/Worth, 1999. This is a standard college-level textbook with two full chapters on angiosperms and other chapters that deal, in part, with anatomy and physiology of angiosperms.

Takhtajan, Armen, and A. L. Takhtadzhian. *Diversity and Classification of Flowering Plants*. New York: Columbia University Press, 1997. An updated system of classification of the angiosperms by one of the foremost living plant taxonomists. Many new data have been used to improve the classification system first proposed by Takhtajan in 1967. Includes a key to families, with descriptions.

ANIMAL-PLANT INTERACTIONS

Categories: Animal-plant interactions; ecology; ecosystems; evolution

The ways in which certain animals and plants interact have evolved in some cases to make them interdependent for nutrition, respiration, reproduction, or other aspects of survival.

Ecology represents the organized body of knowledge that deals with the relationships between living organisms and their nonliving environments. Increasingly, the realm of ecology involves a systematic analysis of plant-animal interactions through the considerations of nutrient flow in *food chains* and *food webs*, exchange of such important gases as oxygen and carbon dioxide between plants and animals, and strategies of mutual survival between plant and animal species through the processes of *pollination* and *seed dispersal*.

A major example of animal-plant interactions involve the continual processes of photosynthesis and cellular respiration. Green plants are classified as ecological *producers*, having the unique ability, by photosynthesis, to take carbon dioxide and incorporate it into organic molecules. Animals are classified as *consumers*, taking the products of photosynthesis and chemically breaking them down at the cellular level to produce energy for life activities. Carbon dioxide is a waste product of this process.

Mutualism

Mutualism is an ecological interaction in which two different species of organisms beneficially reside together in close association, usually revolving around nutritional needs. One such example is a small aquatic flatworm that absorbs microscopic green algae into its tissues. The benefit to the animal is one of added food supply. The mutual adaptation is so complete that the flatworm does not actively feed as an adult. The algae, in turn, receive adequate supplies of nitrogen and carbon dioxide and are literally transported throughout tidal flats in marine habitats as the flatworm migrates, thus exposing the algae to increased sunlight. This type of mutualism, which verges on parasitism, is called *symbiosis*.

CORBIS

Nonsymbiotic mutualism, one kind of animal-plant interaction, can be demonstrated in the often unusual shapes and colorations that flowering plants have developed to attract birds, such as this hummingbird, as well as insects and mammals, for pollination and seed dispersal purposes.

Coevolution

Coevolution is an evolutionary process wherein two organisms interact so closely that they evolve together in response to shared or antagonistic selection pressure. A classic example of coevolution involves the yucca plant and a species of small, white moth (*Tegitecula*). The female moth collects pollen grains from the stamen of one flower on the plant and transports these pollen loads to the pistil of another flower, thereby ensuring cross-pollination and fertilization. During this process, the moth will lay her own fertilized eggs in the flowers' undeveloped seed pods. The developing moth larvae have a secure residence for growth and a steady food supply. These larvae will rarely consume all the developing seeds; thus, both species (plant and animal) benefit.

Although this example represents a mutually positive relationship between plants and animals, other interactions are more antagonistic. *Predator-prey relationships* between plants and animals are common. Insects and larger herbivores consume large amounts of plant material. In response to this selection pressure, many plants have evolved *secondary metabolites* that make their tissues unpalatable, distasteful, or even poisonous. In response, herbivores have evolved ways to neutralize these plant defenses.

Mimicry and Nonsymbiotic Mutualism

In *mimicry*, an animal or plant has evolved structures or behavior patterns that allow it to mimic either its surroundings or another organism as a defensive or offensive strategy. Certain types of insects, such as the leafhopper, walking stick, praying mantis, and katydid (a type of grasshopper), often duplicate plant structures in environments ranging from tropical rain forests to northern coniferous forests. Mimicry of their plant hosts affords these insects protection from their own predators as well as camouflage that enables them to capture their own prey readily. Certain species of ambush bugs and crab spiders have evolved coloration patterns that allow them to hide within flower heads of such common plants as goldenrod, enabling them to ambush the insects that visit these flowers.

In *nonsymbiotic mutualism*, plants and animals coevolve morphological structures and behavior patterns by which they benefit each other but without living physically together. This type of mutualism can be demonstrated in the often unusual shapes, patterns, and colorations that more advanced flowering plants have developed to attract various insects, birds, and mammals for pollination and seed dispersal purposes. Accessory structures, called fruits, form around seeds and are usually tasty and brightly marked to attract animals for seed dispersal. Although the fruits themselves become biological bribes for animals to consume, often the seeds within these fruits are not easily digested and thus pass through the animals' digestive tracts unharmed, sometimes great distances from the parent plant. Some seeds must pass through the digestive plant of an animal to stimulate germination. Other types of seed dispersal mechanisms involve the evolution of hooks, barbs, and sticky substances on seeds that enable them to be easily transported by animal fur, feet, feathers, or beaks. Such strategies of dispersal reduce competition between the parent plant and its offspring.

Pollinators

Because structural specialization increases the possibility that a flower's pollen will be transferred to a plant of the same species, many plants have evolved a vast array of scents, colors, and nutritional products to attract *pollinators*. Not only does pollen include the plant's sperm cells; it also represents a food reward. Another source of animal nutrition is a substance called nectar, a sugar-rich fluid produced in specialized structures called nectaries within the flower or on adjacent stems and leaves. Assorted waxes and oils are also produced by plants to ensure plant-animal interactions. As species of bees, flies, wasps, butterflies, and hawkmoths are attracted to flower heads for these nutritional rewards, they unwittingly become agents of pollination by transferring pollen from stamens to pistils.

Some flowers have evolved distinctive, unpleasant odors reminiscent of rotting flesh or feces, thereby attracting carrion beetles and flesh flies in search of places to reproduce and deposit their own fertilized eggs. As these animals copulate, they often become agents of pollination for the plant itself. Some tropical plants, such as orchids, even mimic a female bee, wasp, or beetle, so that the insect's male counterpart will attempt to mate with them, thereby encouraging precise pollination.

Among birds, hummingbirds are the best examples of plant pollinators. Various types of flowers with bright, red colors, tubular shapes, and strong, sweet odors have evolved in tropical and temperate regions to take advantage of hummingbirds' long beaks and tongues as an aid to pollination. Because most mammals, such as small rodents and bats, do not detect colors as well as bees and butterflies do, some flowers instead focus upon the production of strong, fermenting, or fruitlike odors and abundant pollen rich in protein. In certain environments, bats and mice that are primarily nocturnal have replaced day-flying insects and birds as pollinators.

Thomas C. Moon, updated by Bryan Ness

See also: Parasitic plants; Trophic levels and ecological niches.

Sources for Further Study

Abrahamson, Warren G., and Arthur E. Weis. *Evolutionary Ecology Across Three Trophic Levels: Goldenrods, Gallmakers, and Natural Enemies*. Princeton, N.J.: Princeton University Press, 1997. A case study in plant-insect mutualism, offering a practical look at plant-animal interactions and a consideration of their theoretical implications.

Barth, Friedrich G. *Insects and Flowers: The Biology of a Partnership*. Princeton, N.J.: Princeton University Press, 1991. This beautifully illustrated book, rich in both color photographs and electron micrographs, describes all aspects of flower-insect interactions. Includes bibliography and index. The writing style is appropriate for both general readers and introductory-level biology students.

Buchmann, Stephen L., and Gary Paul Nabhan. *The Forgotten Pollinators*. Washington, D.C.: Island Press/Shearwater Books, 1997. Introduces a variety of pollinators with an emphasis on conservation of species diversity.

Howe, Henry F., and Lynne C. Westley. *Ecological Relationships of Plants and Animals*. New York: Oxford University Press, 1990. Offers introductory students in botany, zoology, and ecology a comprehensive summary of both field and experimental studies on the ecology and evolution of plant and animal interactions. Excellent black-and-white photographs, illustrations, charts, and tables.

Lanner, Ronald M. *Made for Each Other: A Symbiosis of Birds and Pines*. New York: Oxford University Press, 1996. A short and well-argued analysis of the relationship of the Clark's nutcracker and the whitebark pine and their adaptations to fill each other's needs.

Price, Peter W., G. Wilson Fernandes, Thomas H. Lewinsohn, and Woodruff W. Benson, eds. *Plant-Animal Interactions: Evolutionary Ecology in Tropical and Temperate Regions*. New York:

John Wiley & Sons, 1991. A collection of essays from a symposium, covering many aspects of plant-animal coevolution, especially comparing the ecosystems of temperate and tropical environments.

ANTARCTIC FLORA

Categories: Ecosystems; world regions

The harsh climate of Antarctica makes it one of the most inhospitable places on the earth, allowing only a relatively small number of organisms to live there. Permanent terrestrial (land) animals and plants are few and small. There are no trees, shrubs, or vertebrate land animals. Native organisms are hardy, yet the ecosystem is fragile and easily disturbed by human activity, pollution, global warming, and ozone layer depletion.

The Antarctic continent has never had a native or permanent population of humans. In 1998 the United States, Russia, Belgium, Australia, and several other countries signed one of an ongoing series of treaties to preserve Antarctica. The continent is used for peaceful international endeavors such as scientific research and ecotourism.

Terrestrial Flora

There are only two types of flowering plants in Antarctica, a grass and a small pearlwort (*Deschampsia antarctica*). These are restricted to the more temperate Antarctic Peninsula. Antarctic hairgrass (*Colobanthus quitensis*) forms dense mats and grows fairly rapidly in the austral summer (December, January, and February). At the end of summer, the hairgrass's nutrients move underground, and the leaves die. Pearlwort forms cushion-shaped clusters and grows only 0.08 to 0.25 inch (2 to 6 millimeters) per year.

Numerous species of primitive plants, such as lichens, mosses, fungi, algae, and diatoms, live in Antarctica. *Lichens* are made up of an alga and a fungus in a symbiotic (interdependent) relationship. They can use water in the form of vapor, liquid, snow, or ice. Lichens grow as little as 0.04 inch (1 millimeter) every one hundred years, and some patches may be more than five thousand years old. Mosses are not as hardy as lichens and also grow slowly; a boot print in a moss carpet may be visible for years. Fungi are found in the more temperate peninsula, and most are microscopic.

Algae grow in Antarctic lakes, runoff near bird colonies, moist soil, and snow fields. During the summer, algae form spectacular red, yellow, or green patches on the snow. Bacteria are found in lakes, meltwater, and soils. As elsewhere on the earth, bacteria play a role in decomposition. Because of the extreme conditions, they are not always as efficient in Antarctica as they are in warmer climates, and carcasses may lie preserved for hundreds of years.

Kelly Howard

See also: Arctic tundra.

Sources for Further Study

Moss, Sanford. *Natural History of the Antarctic Peninsula*. New York: Columbia University Press, 1988. A study of the Antarctic ecology and environment. Includes bibliography, glossary, and excellent illustrations.

Øvstedal, D. O., and R. I. Lewis Smith. *Lichens of Antarctica and South Georgia: A Guide to Their Identification and Ecology*. New York: Columbia University Press, 2001. Part of the series Studies in Polar Research. Includes plates, maps, and bibliography.

Pyne, Steven J. *The Ice: A Journey to Antarctica*. Seattle: University of Washington Press, 1998. A scientific study of life in Antarctica, covering the minimal plant life in the context of biology, geography, geology, and geophysics.

AQUATIC PLANTS

Categories: Economic botany and plant uses; *Plantae*; water-related life

Aquatic plants are any "true" plants, members of the kingdom Plantae, *that are able to thrive and complete their life cycle while in water, on the surface of water, or on hydric soils.*

Hydric soils develop when the ground is flooded or ponded long enough during the growing season to become anaerobic (depleted of oxygen) in the rooting zone. These soils include organic (peats and mucks) and inorganic (mineral) sediments. Aquatic plants grow in fresh, brackish, and salt water but are most common in fresh water. Their habitats include flowing waters (rivers, streams, brooks), standing waters (lakes, ponds), and *wetlands* (bogs, fens, marshes, swamps), which are categorized as *riverine*, *lacustrine*, and *palustrine* communities, respectively. Wetland plants are sometimes referred to as *helophytes*. Marshes are dominated (that is, more than half covered) by herbaceous species and swamps by woody species. Bog plants are aquatics that grow in acidic organic soils. Fen plants occur in alkaline organic soils.

Aquatic plants (also known as *hydrophytes*, *macrophytes*, and *water plants*) occur throughout the plant kingdom. The term "macrophyte" distinguishes them from microscopic aquatic algae, which are not true plants. Aquatic plants have evolved repeatedly, having more than 250 independent origins by some estimates. They occur occasionally in spore-producing plants such as ferns, liverworts, lycopods, and mosses but are relatively rare among nonflowering seed plants (gymnosperms), with bald cypress (*Taxodium*) a notable exception. Flowering plants (angiosperms) contain the greatest hydrophyte diversity, with more species proportionally in monocotyledons than in dicotyledons. Nevertheless, fewer than 2 percent of flowering plant species are aquatic.

Life-Forms

Regardless of their taxonomic affinities, aquatic plants are often classified ecologically by their life-forms. Categories include the following:

- Floating (*acropleustophytes*), with stems and leaves floating completely on the water surface and stems not rooted in the bottom, such as duckweed (*Lemna*) and water hyacinths (*Eichhornia*).
- Emergent (*hyperhydrates*), with stems and leaves extending mainly above the water surface and stems rooted in the bottom, such as cattails (*Typha*) and reeds.
- Phragmites (*planmergents* or *ephydrates*), floating-leaved, with some or all leaves floating on the water surface and stems rooted in the bottom, such as floating-leaved pondweed (*Potamogeton natans*), water chestnut (*Trapa natans*), and water lily (*Nymphaea*).
- Submersed (*hyphydrates*), with stems and leaves completely under water and stems rooted in the bottom, such as Eurasian water milfoil (*Myriophyllum spicatum*) and wild celery (*Vallisneria*).
- Suspended (*mesopleustophytes*), with stems and leaves completely under water and stems not rooted in the bottom, such as the bladderwort (*Utricularia vulgaris*) and the coontail (*Ceratophyllum*).

Benthophyte and *pleustophyte* are used respectively to differentiate between forms that are either rooted in the substrate or unrooted. Species with elongate, leafy stems are termed *vittate* or *caulescent* (such as coontail). Those with leaves clustered in a basal rosette are *rosulate* (such as wild celery), and those not clearly differentiated into stems and leaves are *thalloid* (such as duckweed). Species whose floating or emergent leaves differ morphologically from their submersed leaves are *heterophyllous* (such as floating-leaved pondweed).

Adaptations

Water plants are anatomically and structurally reduced. Watermeal (*Wolffia*), the world's smallest angiosperm, contains plants only 0.4 millimeter long. Submersed species often lack water-conducting tissue (xylem), mechanical tissue (sclerenchyma), and cuticle. Some lack roots entirely. Support and floatation of underwater stems are accommodated by buoyant tissue (aerenchyma) and extensive air spaces (lacunae) which also transport oxygen throughout the plant. Submersed plants usually possess either highly dissected (compound) or thin, ribbonlike leaves. Some leaves become fenestrate, that is, lacking tissue between the veins. Such leaf shapes increase surface area-to-volume ratios for more efficient nutrient uptake and to reduce damage from water currents.

Floating leaves are normally flat and circular, with stomata on their upper surfaces. They may reach 2.5 meters in diameter (such as *Victoria*). For stability, the stalks (petioles) of most floating leaves are positioned centrally by emargination of the base, as in the water lily, or are peltate by complete fusion of leaf lobes, as in the water shield (*Brasenia*) and *Victoria*. Physiological adaptations enable aquatic plants to tolerate deleterious effects of anaerobic hydric soils.

Reproduction

Most aquatic plants are perennials that reproduce vegetatively (asexually). Species survive winters or other unfavorable periods as intact plants, by dying back to dormant stem apices, by means of modified stems (rhizomes, stolons, tubers), or by use of specialized dormant structures (*hibernacula*) in the sediment. "Winter buds" are a kind of hibernaculum; buds are insulated by normal foliage leaves on shortened internodes. They usually remain attached to the plant. *Turions* are specialized hibernacula that produce modified, morphologically distinct leaves to protect the enclosed buds. Turions always detach from the plant and function as propagules for dispersal. Water plants also disperse vegetatively by fragmentation of stems, which are characteristically brittle, due to the lack of mechanical tissue. Detached stems can establish themselves by production of adventitious roots.

The few aquatic plants that are annuals produce seeds as their dormant stage. Some aquatic annuals also multiply vegetatively by fragmentation during the growing season. Generally, sexual reproduction is rare in submersed species, more common in floating-leaved species, and quite common in emergent species (and annuals).

Pollination in water plants is facilitated by insects (entomophily), wind (anemophily), and water (hydrophily). Most aquatics are insect-pollinated; about one-third of them are wind-pollinated. Less than 5 percent of aquatic species are hydrophilous, with pollen transported on the water surface (ephydrophily) or under the water surface (hyphydrophily). Most marine angiosperms (seagrasses) are hydrophilous.

Seeds, fruits, and vegetative propagules are dispersed locally by water currents and more widely by waterfowl. Waterfowl transport propagules in plumage, in mud adhering to their feet, and by excretion of seeds consumed as food. Many water plants are distributed broadly, with some species achieving worldwide distributions.

NOAA/National Estuarine Research Reserve Archive

A freshwater pond with large water lilies near Charleston, South Carolina.

Uses

Aquatic plants are important economically. Foods include rice (*Oryza sativa*), which sustains more human life than any other plant on earth. Aquatic plants are important horticulturally as aquarium and water-garden ornamentals. Some aquatic plants, such as the water hyacinth, are invasive weeds that interfere with shipping, irrigation, or recreation and cost millions of dollars to eradicate. The beauty of many water plants, especially water lilies, has inspired art and religion since ancient times.

Donald H. Les

See also: Adaptations; Angiosperm evolution; Angiosperms; Eutrophication; Invasive plants; Marine plants; Peat; Pollination; Rice; Wetlands.

Sources for Further Study

Arber, Agnes. *Water Plants: A Study of Aquatic Angiosperms*. London: Cambridge University Press, 1920. First comprehensive treatment of aquatic plants written in English.

Cook, Christopher D. K. *Aquatic Plant Book*. Rev ed. Amsterdam: SPB Academic Publishing, 1996. A comprehensive taxonomic key for identifying most major aquatic plant groups (families and genera) worldwide.

Kasselmann, Christel. *Aquarium Plants*. Melbourne, Fla.: Krieger, 2002. A thorough reference on the cultivation of ornamental aquatic plants.

Sculthorpe, C. Duncan. *The Biology of Aquatic Vascular Plants*. London: Edward Arnold, 1967. Excellent overview of all aspects of water plant biology.

ARCHAEA

Categories: Bacteria; evolution; paleobotany; taxonomic groups

The domain Archaea *represents a diverse group of prokaryotes originally found in environments once considered to be hostile to life, now known to be widely distributed in nature.*

The cycling of plant nutrients, such as carbon, nitrogen, and sulfur, requires the activity of microorganisms that convert these elements to forms readily available to plants. These microorganisms, which are generally found in both soil and water, include both prokaryotic organisms of the domain *Bacteria* and the domain of prokaryotes called *Archaea*, which play significant roles in nutrient cycling. Along with *Eukarya*, to which protists, fungi, plants, and animals belong, the *Archaea* form one of the three domains of life. The *Archaea* are related to both *Bacteria* and *Eukarya* and, in some respects, appear to be more closely related to *Eukarya*. Biochemical and genetic studies, including information obtained from whole genome sequencing, suggest that *Archaea* may be closely related to an ancestor that gave rise to both *Bacteria* and *Eukarya*. Thus, *Archaea* may provide some insight into the processes that resulted in the evolution of higher life-forms, including plants and animals.

A Third Domain

For more than fifty years, biologists categorized living organisms into two groups based on their cellular organization and complexity: prokaryotes (originally all classified in kingdom *Monera*), the single-celled organisms whose chromosomes are not compartmentalized inside a nucleus (which include the domain *Bacteria*), and eukaryotes, consisting of all other organisms, whose cells contain a nucleus. In the late 1970's studies on a unique group of microorganisms led investigators to question the accepted classification of prokaryotes. Originally called *Archaebacteria* by molecular biologist Carl Woese and his colleagues in 1977, these microorganisms were isolated from environments characterized by extremes in heat, acidity, pressure, or salinity, and many were found to be able to utilize sulfur and molecular hydrogen as part of their growth process.

Like all prokaryotes, *Archaea* do not have a nucleus. However, in their biochemistry and the structure and composition of their molecular machinery, they are as different from bacteria as they are from eukaryotes. Woese and his colleagues analyzed and compared specific molecules of ribonucleic acid (RNA) present within the ribosome in all organisms, called *ribosomal RNA* (rRNA). Their findings suggested that all extant life is composed of three distinct groups of organisms: the eukaryotes, or domain *Eukarya*, which includes plants and animals, and two different prokaryotes, domains *Bacteria* and *Archaea*. In 1990 Woese and others recommended the replacement of the simple prokaryote/eukaryote view of life with a new tripartite scheme based on three domains: the *Bacteria*, *Archaea*, and *Eukarya*. Since 1990 the three-domain classification has been the subject of considerable debate, and as a consequence, both old and new terminology are used in scientific and popular literature.

Characteristics

Generally, the size and shape of *Archaea* are similar to those of *Bacteria*. They are single-celled microscopic organisms that, in some cases, are motile (capable of self-movement) and may be found in chains or clusters. *Archaea* multiply in the same manner as bacteria: via *binary fission*, budding, or fragmentation. Like *Bacteria*, archaeal chromosomes are circular, indicating the absence of breaks or discontinuities, and many genes are organized in the same fashion as those found in *Bacteria*. On the other hand, the specific chemical composition of *Archaea* plasma membranes and cell walls is unique to the *Archaea* and is quite different from the composition of these structures typically found in either *Bacteria* or *Eukarya*. In fact, the distinctive ether-linked isoprenoid lipids that compose the external membranes of *Archaea* are a hallmark of these microorganisms.

Another unique characteristic of *Archaea* is the composition of the molecular genetic machinery, which is a mosaic of the components found in *Bacteria* and *Eukarya*. For example, the ribosomes (which are responsible for protein synthesis) of *Archaea* resemble the ribosomes of *Bacteria* in shape and composition and are distinct from the ribosomes of *Eukarya*. On the other hand, the enzyme utilized by *Archaea* in the production of RNA, namely RNA polymerase, is quite different from the enzyme found in *Bacteria*. In *Bacteria*, RNA polymerase molecules are composed of four major proteins, while in the *Archaea*, RNA polymerase molecules consist of more than ten proteins and are surprisingly similar to the enzyme found in *Eukarya*. In fact, archaeal RNA polymerase is so similar to the eukaryotic enzyme that combining certain proteins from both archaeal and eukaryotic sources results in a functional enzyme, a manipulation that is not possible with any bacterial RNA polymerases.

Among species of the *Archaea*, there is a variety of metabolic processes that differ greatly from the better-known metabolic routes of *Bacteria* and *Eukarya*. Many of the archaeal pathways used to convert food sources to energy and building blocks for growth involve enzymes having biological activities not found in any other biological systems. In some cases, the enzymes require the involvement of rare metals, such as tungsten. While a requirement for metals in the activity of many bacterial and eukaryotic enzymes is ubiquitous, the use of tungsten appears to be unique to *Archaea*.

Diversity

A fascinating feature of *Archaea* is that they are found in niches that support the growth of few other organisms. These include highly reduced (oxygen-free) environments or very high-temperature environments found near hot springs or undersea hydrothermal vents as well as sites that are sulfur-rich and highly acidic. *Archaea* are also found in highly saline marine environments and hypersaline lakes where the salinity is as much as ten times that in seawater.

Based on the comparison of ribosomal RNA sequences as well as physiological and metabolic characteristics, the *Archaea* have been divided into three subdomains: *Euryarchaeota*, *Crenarchaeota*, and *Korarchaeota*. The *Euryarchaeota* includes members of the *methanogenic* (methane-producing) and halophilic (salt-requiring) *Archaea* as well as many that grow at very high temperature, the thermophilic and extremely thermophilic, or hyperthermophilic, *Archaea*. Representatives of hyperthermophilic *Archaea* are found in the *Crenarchaeota*, which also includes cold-dwelling *Archaea* that have been isolated in association with certain marine sponges. The *Korarchaeota* also includes hyperthermophilic

Archaea, although these were not isolated or characterized as of 2001, but whose presence in hot spring and deep-sea samples has been identified by molecular biological techniques.

Methanogenic *Archaea*

Methane-producing *Archaea* are found in strictly anaerobic environments. They have no tolerance for oxygen: Trace amounts are inhibitory for growth, and too much is lethal. These *Archaea* obtain energy for growth by a process called *methanogenesis*, which results in the conversion of carbon dioxide to methane gas.

Methane production requires several enzymes that use coenzymes unique to methanogenic *Archaea*. The production of methane is of great importance to carbon cycling in many anaerobic environments, and microorganisms that produce this gas have been known for centuries. In 1776 the scientist Alessandro Volta demonstrated that air generated from sediments rich in decaying vegetation, such as those present in bogs, streams, and lakes, could be ignited. It is now known that methanogenic *Archaea* are responsible for generating this "marsh gas."

Because methanogens require an oxygen-free environment for growth, they are found only where carbon dioxide and hydrogen are available and oxygen has been excluded. Thus, methanogens thrive in stagnant water, natural wetlands, paddy fields, and in the rumen of cattle and other ruminants as well as in the intestinal tracts of animals and the hindguts of cellulose-digesting insects, such as termites. Methanogens are also found in hot springs and the deep ocean and are major components of the anaerobic process in waste treatment facilities. It has been estimated that production of methane by the methanogenic *Archaea* may account for almost 90 percent of the total methane released into the atmosphere each year. In addition to playing a role in carbon cycling, several methanogenic *Archaea* are also involved in nitrogen cycling, as they are able to convert molecular nitrogen into organic nitrogen via nitrogen fixation, a process that is shared by only a few prokaryotes.

Thermophilic *Archaea*

Thermophilic *Archaea* live in environments ranging in temperature from 55 degrees Celsius (131 degrees Fahrenheit) to 80 degrees Celsius (176 degrees Fahrenheit). Hyperthermophilic *Archaea* grow at temperatures near or greater than the boiling point of water and as high as 113 degrees Celsius (235 degrees Fahrenheit). These *Archaea* have been isolated from hot sulfur springs, sulfur-laden mud at the base of volcanoes, and near very hot deep-sea hydrothermal vents where superheated water is emitted at very high temperatures under pressure. Species that can use oxygen, as well as those that have no tolerance for oxygen, are known. Many of the anaerobic representatives obtain energy for growth by the metabolism of elemental sulfur.

In addition, many are found in environments that are extremely acidic, including those that are members of *Thermoplasmatales*. This group is noted for its ability to grow at a pH of 2.0 and below (on a scale where pH 7.0 is neutral), which is equivalent to the acid in car batteries. A representative is *Thermoplasma*, which does not possess a cell wall but has

PhotoDisc

Hyperthermophilic Archaea *grow at temperatures as high as 113 degrees Celsius.* Archaea *have been isolated from hot sulfur springs, such as those at Yellowstone National Park.*

a chemically unique structure composed of a lipid-polysaccharide (tetraether lipid with mannose and glucose units) that is distinctly different from the unusual ether-linked lipids found in the membrane components of typical *Archaea*.

Halophilic *Archaea*

The salt-dependent halophilic *Archaea* require extremely high concentrations of salt for survival, and some grow readily in saturated brine, where the salt concentration reaches 32 percent (in seawater it is approximately 3.5 percent) and where very alkaline conditions are not uncommon. Halophilic *Archaea* are found in salty habitats along ocean borders and inland waters such as the Dead Sea and the Great Salt Lake.

The reddish-purple color observed in salt evaporation ponds is due to production of red- and orange-colored carotenoids and other pigments associated with the massive growth of halophilic *Archaea*.

Some halophilic *Archaea* are capable of harvesting light to provide energy for growth by a mechanism that does not involve chlorophyll pigments. Light harvesting by these halophilic *Archaea* is done by a membrane-bound protein called *bacteriorhodopsin* that is equivalent to the mammalian eye pigment *rhodopsin* in both function and structure. Bacteriorhodopsin contains retinal, a purple carotenoidlike molecule used for light trapping. Interestingly, retinal is produced via a pathway that contains many of the same enzymes used for the production of lycopene by tomatoes during ripening.

Window to the Past

The extreme conditions in which *Archaea* are found suggests that these organisms have adapted to environments thought to exist during early life on earth, three billion to four billion years ago. Thus, the *Archaea* might be considered as a window into the past, and they may shed light on the processes involved in evolution as well as their relationships with *Bacteria* and *Eukarya*. In order to survive in their unique environments, *Archaea* possess molecules that withstand heat or cold, acids, salt, and in some cases, pressure—characteristics that are tailor-made for specific applications in molecular biology and biotechnology.

Uses

A number of important applications have been developed as a consequence of studying the *Archaea*. These include the identification of heat-stable enzymes for analyses used in genetic fingerprinting and cancer detection (certain polymerase chain reaction enzymes), the use of halophilic pigments for holographic applications, optical signal processing and photoelectric devices, and methanogenesis as an alternative fuel source.

Harold J. Schreier

See also: Bacteria; Biotechnology; Carbon cycle; DNA in plants; Environmental biotechnology; *Eukarya*; Eukaryotic cells; Molecular systematics; Nitrogen cycle; Nitrogen fixation; Nucleus; Photorespiration; Prokaryotes; Ribosomes; RNA.

Sources for Further Study

Howland, John L. *The Surprising Archaea: Discovering Another Domain of Life*. New York: Oxford University Press, 2000. General book describing the *Archaea*, their environment, and their unique place in evolution and in the world ecosystem as well as their commercial importance. Includes illustrations, figures.

Madigan, Michael T., John M. Martinko, and Jack Parker. *Brock Biology of Microorganisms*. 9th ed. Upper Saddle River, N.J.: Prentice Hall, 2000. This basic microbiology textbook includes several excellent chapters on the ecology, biology, and molecular biology of the *Archaea*. Includes illustrations, figures, and index.

Reysenbach, A.-L., et al., eds. *Thermophiles: Biodiversity, Ecology, and Evolution*. Boston: Kluwer Academic/Plenum, 2000. A synthesis of recent advances in the biology, biotechnology, and management of thermophilic organisms, with specific examples drawn largely from thermal springs in Yellowstone National Park.

Seckbach, J., ed. *Enigmatic Microorganisms and Life in Extreme Environments*. Boston: Kluwer Academic/Plenum, 1999. A comprehensive reference source containing research data on extremophiles, reviewing the three kingdoms of life, and discussing dry habitats and thermophilic, psychrophilic, and halophilic niches.

ARCTIC TUNDRA

Category: Biomes

The Arctic tundra is a biome representing the northernmost limit of plant growth on earth. Arctic tundra has a circumpolar distribution in the Northern Hemisphere, extending from the ice cap southward to the forested taiga of North America, Europe, and Asia. Tundra is also found on islands within the Arctic Ocean and along coastal Greenland.

The term "tundra" was derived from the Finnish word for a treeless or barren landscape. The Arctic tundra biome is located within one of the harshest climates on earth for plant growth, with winter temperatures averaging –34 degrees Celsius (–30 degrees Fahrenheit). The climate is comparatively dry, with annual precipitation of 150 to 250 millimeters (6 to 10 inches). Locked in snow or frozen within soil, the majority of moisture is not available for plant use.

In addition to surviving extreme temperatures and dry conditions, plants must adapt to seasonal variation in available sunlight; winter nights, for example, last twenty-four hours. The tundra's growing season is very short, extending over only about sixty days. Continuous sunlight during warmer months, July and August, contributes to the productivity of tundra plant communities that can yield 227-454 kilograms (500-1,000 pounds) of vegetation per acre. This *biomass* serves as an important food source for caribou, musk ox, and migratory waterfowl. Tundra vegetation is made up of *herbaceous* plants (grasses, forbs, and sedges), mosses, lichens, and shrubs that grow close to the ground, where temperatures are highest. By providing an insulating layer, snowfall is advantageous for tundra plants during cold winter months.

Herbaceous Plants

Rushlike tundra sedges belong to the flowering plant family *Cyperacaeae*. Common to the tundra, cottongrass is really a sedge within the genus *Eriophorum*. Perennial forbs are broadleaf plants that survive winter months as bulbs that are protected below the ground level. During warm months the plants begin to grow rapidly and will develop flowers and seeds when temperatures climb above 10 degrees Celsius (50 degrees Fahrenheit).

Lichens and Low-Growing Shrubs

Acting as a single organism, pioneering lichens growing on rock surfaces represent a symbiotic relationship between fungi and algae. The fungi anchor to the rock, absorbing water directly into their cells, while the algae occupy this moist area, creating food through photosynthesis that is shared with the fungi. Tundra lichens are found in fruiticose (stalklike), crustose (crustlike), or foliose (leaflike) forms.

The heath (*Ericaceae*) family includes several species of shrub, many of which have tough, evergreen leaves. Examples include rhododendron, cranberry, blueberry, and Labrador tea. Another heath, the alpine azalea (*Loiseleuria procumbens*), forms a mat or cushion where several plants clump tightly together.

Adaptations

In many ways tundra vegetation must adapt to many of the same environmental conditions as *grasslands* or *deserts*, such as little precipitation, strong winds, and extreme temperature variations. As a result of the brief growing season, plant reproduction in the tundra must take place rapidly. Other adaptations include compact plant size that protects from cold temperatures, hairy stems that help retain heat, and dark-colored leaves that absorb sunlight. Some plants have hollow stems that require fewer nutrients to grow. A unique adaptation made by the Arctic poppy (*Papaver radicatum*) and mountain aven (*Dryas integrifolia*) allows them to orient their flowers to track the sun's movement across the sky, maximizing solar radiation received.

Although sunlight is usually beneficial to plant growth, some plants such as Arctic algae must implement protective measures to avoid damage from ultraviolet radiation. The green alga *Ulva rigida*, also called sea lettuce, produces amino acids and carotenoid pigments that absorb harmful radiation. Cushion plants grow in tight but low-profile clumps, forming windbreaks that protect them from the cold, and may trap airborne dust and soil used as a source of nutrients. Many tundra plants are capable of carrying out photosynthesis under relatively low light intensities. With short growing seasons, some plants reproduce by budding and division instead of by the creation of seeds. Plants may also store nutrients in *rhizomes*, underground stems that survive after root systems die.

Edaphic Influences

Soils of the tundra are principally thin soils (inceptisols). Contributing to the lack of soil development are cold temperatures that inhibit the growth of soil-producing organisms such as bacteria. The tundra's treeless plain may be interrupted by patterned ground made up of stone polygons, soil circles, or soil stripes. These unusual features are formed by the thrusting action of repeated freezing and thawing in soil that overlies rock or permanently frozen ground called *permafrost*. Impenetrable permafrost that inhibits root system development, rather than cold temperatures, is thought to be responsible for the lack of tree growth in the tundra. Warmer summer temperatures lead to a thaw in permafrost that extends only about a meter below the surface. Ponds and boggy areas form in places where soil above the permafrost melts and cannot move downward, creating a source of moisture for plants.

Environmental Concerns

As a result of growing under harsh conditions, tundra plants are slow to recover from disturbances. Vehicles can destroy tundra plants. Other concerns include oil spillage, damage caused by pipeline construction, and other impacts tied to petroleum production.

Thomas A. Wikle

See also: Adaptations; Antarctic flora; Asian flora; Biomes: definitions and determinants; Biomes: types; European flora; North American flora; Tundra and high-altitude biomes.

AP/Wide World Photos

In the Opingivik area of Baffin Island, Canada, continuous sunlight during the warmer months contributes to the productivity of tundra plant communities that are a food source for animals such as these caribou. Tundra vegetation is made up of herbaceous plants, mosses, lichens, and shrubs that grow close to the ground, where temperatures are highest.

Sources for Further Study

Barbour, Michael G., and W. D. Billings. *North American Terrestrial Vegetation*. New York: Cambridge University Press, 1989. Provides a comprehensive look at major vegetation types of North America as well as an examination of ecosystems and management problems.

Reynold, James F., and John D. Tenhenen. *Landscape Function and Disturbance in Arctic Tundra*. Berlin: Springer, 1996. Additional information about environmental problems within Arctic tundra can be found within this book, especially with regard to a United States Department of Energy study that examined energy-related disturbances caused by petroleum exploration in Northern Alaska.

Stonehouse, B. *Polar Ecology*. New York: Blackie and Son, 1988. This book offers a general overview of the ecology of polar regions.

Wielgolaski, F. E. *Polar and Alpine Tundra*. New York: Elsevier, 1997. Supported by photographs, maps, graphs, and other illustrations, this book serves as a comprehensive guide to polar vegetation.

ASCOMYCETES

Categories: Economic botany and plant uses; fungi; microorganisms; taxonomic groups

The ascomycetes are fungi (phylum Ascomycota *or* Ascomycotina*) that produce sexual spores in a specialized cell called an ascus. These diverse fungi, with more than thirty thousand species, can be found in almost every ecosystem worldwide. One of the most famous members of the ascomycetes is the truffle.*

Ascomycetes, one of the four phyla of the fungus kingdom, by definition possess an *ascus*, a single cell inside of which sexual spores are produced. The reproductive process has been well documented and occurs when the *dikaryotic mycelium* (the mass of hyphae forming the body) undergoes changes that precede the formation of the ascus. Dikaryotic is the genetic state in which two haploid nuclei are present in the cell. One nucleus is donated by each parent.

The first change occurs when the end cell of a hyphal strand begins to form a small bend. The cell divides into three cells; the outer two cells are haploid, and the middle cell is dikaryotic. The middle cell then elongates, and the nuclei migrate into its center. The two haploid nuclei then fuse to form a single diploid nucleus, which undergoes mitosis and meiosis to form eight haploid nuclei. Cell walls form around the nuclei producing eight haploid *ascospores*. The ascospores are then liberated from the ascus.

The ascus wall determines the kind of dispersal of the spores. Some asci have a thin, single-layer wall, which breaks down to liberate spores. *Unitunicate* asci have a multilayer cell wall with a pore at the end of the ascus. Spore release is active through the pore. *Bitunicate* asci have multilayer cell walls, and release of spores is by the separation of the layers of the cell wall, with the inner layer inflating to several times its normal size and then lifting off of the ascus, allowing the spores to be released. The spores are released into the environment, where they germinate and produce haploid hyphae. The haploid hyphae fuse with compatible haploid hyphae, forming dikaryotic hyphae, and the process begins to repeat itself.

There are five different ways in which asci are formed in nature. First, asci can be produced by exposure to the environment, as are the asci of yeasts or the ascus of the peach leaf-curl pathogen *Taphrina deformans*. With these fungi, the ascospores are released by the breakdown of the ascus wall.

The other four ways of production of asci all take place inside structures made from mycelium, called *ascocarps*. These structures range from totally closed to open, like a cup. The totally closed ascocarps are called *cleistothecia*. Within these, the asci are scattered, and the spores are released by breakdown and decomposition of the fungal tissue.

Ascomycetes as Pathogens

Some members of the ascomycetes are very important plant and animal pathogens, causing serious diseases. One of the more impressive plant pathogens is the ergot fungus (*Claviceps purpurea*). This fungus colonizes the ovaries of grains, such as rye (*Secale cereale*). It produces a mass of mycelium, called a *sclerotium*, which is hard and has a density similar to that of a seed. Because of this, the sclerotia are often found in the threshed grain. Sclerotia contain an accumulation of alkaloids and other secondary metabolites. When the sclerotia are ground into flower and baked into bread, many of these secondary metabolites are passed into the bread. During the Middle Ages this fungus was responsible for a human disease called St. Anthony's Fire. Today, this fungus is used for the natural production of a coagulant which is used in medicine.

Rob and Ann Simpson/Photo Agora

*Morels (*Morchella*) are a choice edible ascomycete. They have an appealing aroma and are a favorite of humans and many wild animals.*

Another group of plant pathogens are the powdery mildews. There are several hundred species of these fungi, which produce a powdery spore mass on the outer surfaces of plant leaves. If a leaf is infected before it has expanded, it will remain small and puckered and may drop from the plant. The powdery mildews are superficial and send hyphae through the leaf cuticle into the epidermis. The fungus then grows over the surface of leaf, giving it a powdery appearance. During the winter, the fungus produces cleistothecia on the surface of the leaf. Powdery mildew can occur on most plant species and can be very damaging to crop and ornamental plants.

Economic Uses

Truffle is a generic term for fungi that form mycorrhizae (a symbiotic association) with the roots of various trees. The fungus grows into the roots and helps the plant tolerate stress, providing the plant with increased absorption of phosphorus from the soil. In return, the plant gives the fungus metabolites that it needs for growth. These fungi then produce fruiting bodies, either in the soil or upon the surface of the soil. The truffles are produced in the soil up to a depth of 1 foot.

Truffles can be located in the soil using a trained sow or dog that is able to sniff out the volatile chemicals that are produced. However, it is important to note that edible fungi such as truffles can often be mistaken for highly toxic fungi and should never be gathered or eaten without expert identification. Truffles are used as condiments and are able to impart unique aromas and sensations to food. They are shaped like small balls, varying in size from that of a pea to that of a golf ball. Truffles are only produced in nature and therefore fetch high prices. The price of truffles depends on the species, size, and freshness. Prices of the famous French Black Périgord truffle (*Tuber melanosporum*) can reach into thousands of dollars per kilogram (2.2 pounds). In the United States, the Oregon white truffle is quite appealing and can be found in the Pacific Northwest.

Morels (*Morchella*) are another choice edible ascomycete. Shaped like a little hat sitting upon a stalk, they are brown in color and have an appealing aroma. They add flavor to any food and are a favorite of many wild animals.

Scientific Uses

Some members of the ascomycetes are used for genetic studies. Such is the case for *Neurospora*, a common fungus found growing on soil and organic matter and one of the first organisms to be found in an area after a fire. As with all ascomycetes, there are two compatible mating types, which makes for easy genetic study. Using spore characteristics of shape, color, and texture, it is possible to see how mitosis and meiosis occur in the ascus by determining the placement of the spores. The fungus is readily mutated, which further enhances genetic study.

J. J. Muchovej

See also: *Basidiomycetes*; Basidiosporic fungi; Chytrids; Deuteromycetes; Fungi; Lichens; Mitosporic fungi; Mushrooms; Mycorrhizae; Yeasts; Zygomycetes.

Sources for Further Study

Alexopoulos, Constantine J., and Charles E. Mims. *Introductory Mycology*. New York: John Wiley, 1995. Covers the biology, anatomy, and physiology of fungi. Includes illustrations, bibliographic references, and index.

Manseth, James D. *Botany: An Introduction to Plant Biology*. 2d ed. Sudbury, Mass.: Jones and Bartlett, 1998. General botany text with section on fungi, their importance, and their biology. Includes illustrations, bibliographic references, and index.

Raven, Peter H., Roy F. Evert, and Susan E. Eichhorn. *Biology of Plants*. 6th ed. New York: W. H. Freeman/Worth, 1999. General botany text with section on fungi, their importance in the environment, and their biology. Includes illustrations, bibliographic references, and index.

Smith, Alexander H., and Nancy Smith Weber. *The Mushroom Hunter's Field Guide*. Ann Arbor: University of Michigan Press, 1996. This illustrated field guide gives an overview of fleshy fungi found in the United States. Includes illustrations, photographs, bibliographic references, and index.

ASIAN AGRICULTURE

Categories: Agriculture; economic botany and plant uses; food; world regions

Land constraints and growing population and urbanization throughout Asia underscore the need for environmentally sound technologies to sustain agricultural growth.

The first agricultural revolution occurred in Asia and involved the domestication of plants and animals. It is believed that *vegeculture* first developed in Southeast Asia more than eleven thousand years ago. In vegeculture, a part of a plant—other than the seed—is planted for reproduction. The first plants domesticated in Southeast Asia were taro, yam, banana, and palm. *Seed agriculture*, now the most common type of agriculture, uses seeds for plant reproduction. It originated in the Middle East about nine thousand years ago, in the basins of the two major rivers of present-day Iraq, the Tigris and the Euphrates. Wheat and barley were probably the first crops cultivated there. Although many plants were domesticated simultaneously in different parts of the world, rice, oats, millet, sugarcane, cabbage, beans, eggplant, and onions were domesticated originally in Asia.

Asia supports about 60 percent of the global population on only about 23 percent of the world's agricultural land. As a result, Asian agriculture is far more intensive than on any other continent. Despite the population pressure on arable land, Asia has made remarkable progress in agricultural productivity. Between 1966 and 1995, wheat production grew 5.5 percent annually, and rice production 2.2 percent. In Asia as a whole, food production has outpaced the growth of population. In most Asian countries, particularly in the low-income countries of South Asia, per-capita food availability has risen.

Agrarian Structure

Most people in Asia are farmers, owning an average of about 2.5 acres (1 hectare) of land per family. Topographic and climatic conditions, to a large extent, determine farm size. Agricultural potential is limited in Nepal, for example, because of the Himalaya Mountains, and in Saudi Arabia because of the Arabian Desert. In these countries, average farm size is larger relative to countries like Bangla-

Leading Agricultural Crops of Asian Countries with More than 15 Percent Arable Land

Country	*Products*	*Percent Arable Land*
Armenia	Fruit, vegetables	17
Azerbaijan	Cotton, grain, rice, grapes, fruit, vegetables, tea, tobacco	18
Bangladesh	Rice, jute, tea, wheat, sugarcane, potatoes	73
Burma (Myanmar)	Paddy rice, corn, oilseed, sugarcane, pulses, hardwood	15
India	Rice, wheat, oilseed, cotton, jute, tea, sugarcane, potatoes	56
Israel	Citrus and other fruits, vegetables, cotton	17
Korea, South	Rice, root crops, barley, vegetables, fruit	19
Lebanon	Citrus, vegetables, potatoes, olives, tobacco, hemp (hashish)	21
Nepal	Rice, corn, wheat, sugarcane, root crops	17
Pakistan	Cotton, wheat, rice, sugarcane, fruits, vegetables	27
Syria	Cotton, wheat, barley, lentils, chickpeas	28
Thailand	Rice, rubber, corn, tapioca, sugarcane, soybeans, coconuts	34
Turkey	Cotton, tobacco, cereals, sugar beets, fruits, olives, pulses, citrus	32
Vietnam	Rice, corn, potatoes, rubber, soybeans, coffee, tea, bananas	17

Source: Data are from *The Time Almanac 2000*. Boston: Infoplease, 1999.

desh, which contains a vast, fertile floodplain and receives abundant rainfall.

Another feature of Asian agrarian structure is the inequitable distribution of farmland. For example, in India more than 25 percent of cultivated land is owned by less than 5 percent of farming families. Farm holdings in most Asian countries are highly fragmented, and tenancy is widespread. Fragmentation of farms inhibits agricultural mechanization, and land consolidation efforts have had limited success in most Asian countries.

Most Asian farmers are subsistence farmers, cultivating crops for family consumption. Almost all farm operations are done manually or with the help of draft animals. Exceptions are found in Japan, South Korea, and Taiwan, where small-scale equipment similar to garden tractors is widely used. Only recently have Asian farmers started to use chemical fertilizer; for water, they largely depend on rain. As a result, yields are low, which compels farmers to cultivate the land intensively. Double-cropping is the norm; some farmers grow three crops a year. Therefore, only a small fraction of the arable land in humid regions of Asia remains fallow. Farming is labor-intensive, and the extended family is the main source of labor. This helps to explain why family size is generally large in agrarian countries of Asia.

Rice and Wheat

The coastal areas and inland river valleys of East, Southeast, and South Asia are the agricultural cores of the continent. More than half of the crop area of these regions is used to cultivate food crops such as rice and wheat. Rice is the principal food crop of all Asian countries located east and north of India and for the people of southern and eastern India.

Rice is the staple food of more than half of the world's population, and 90 percent of it is grown in coastal and deltaic plains and in the river valleys of Monsoon Asia. This region encompasses a broad geographic area characterized by a distinctive climate, stretching from Japan in the east, through Indonesia in the south, and west to Pakistan. Rice farming there is practiced mostly at the subsistence level, using traditional methods.

Wheat is the primary food crop of northern and western India and all Asian countries located west of India. People of the wheat-producing region consume rice as a secondary staple. Eighteen of the twenty-five top rice-producing countries of the world are in Asia. India has the largest area devoted to rice cultivation of the world's countries, but China is first in total production. Indonesia and Bangladesh rank third and fourth in world rice production.

With the exceptions of Japan, South Korea, China, and Taiwan, yields of all crops—particularly rice and wheat—are low in Asia compared to world standards. Although crop yields increased significantly after 1970, typical yields in Asia remained low for several reasons: Fertilizer use and the area under irrigation are among the lowest in the world. Also, most Asian farmers practice traditional farming methods, where high yields are atypical. Another major obstacle to increasing crop yields is the preponderance of small farms. Because small farms do not have access to assured irrigation and cannot afford modern agricultural inputs, their average yield is generally much lower than that of medium and large farms.

Slash-and-Burn Agriculture

In the tropical rain forests of Southeast Asia, the mountainous and hilly parts of South Asia, and in southern China, a type of primitive agriculture known as *shifting cultivation* or *slash-and-burn agriculture* is practiced. Shifting cultivators plant different crops, such as rice, corn, millet, yams, sugarcane, oilseeds, potatoes, taro, vegetables, and cotton, on one site. These farmers must abandon their fields and establish new ones every few years. As a result, a large area of land is required to support a small population. The land devoted to shifting cultivation is declining at a rapid rate worldwide because of the demand for forest resources for other uses.

Dry Agriculture

Farmers in the colder, drier parts of Asia (northeastern China, northern Japan, southeastern East Asia, northeastern Southeast Asia, and the western half of South Asia) and the river valleys of the Middle East practice a system of intensive subsistence agriculture called peasant grain-and-livestock farming, or *dry agriculture*. The dominant grain crops are wheat, barley, sorghum, millet, oats, and corn, while cotton, tobacco, and sugarcane are grown as cash crops. In arid areas, such as the Middle Eastern river valleys, irrigation helps support dry farming. Traditional water-lifting devices, such as the shaduf (a counterweighted, lever-mounted bucket), and the naria (waterwheel), permit limited double-cropping in the dry season near the rivers of the Middle East.

In the arid and semiarid parts of South Asia and the Middle East, and in the dry and cold western two-thirds of East Asia, nomadic herders graze cattle, sheep, goats, and camels. Nomadic herders move from place to place with their livestock in search of forage. As in other places, nomadic herding is declining in Asia.

Mediterranean Agriculture

A distinctive type of subsistence agriculture, called *Mediterranean agriculture*, is practiced along the Mediterranean coast of the Middle East and in the northern part of Turkey that borders the Black Sea. Traditional Mediterranean agriculture is based on wheat and barley cultivation in the rainy winter season. Farmers of this region also cultivate vine and tree crops, such as grapes, olives, and figs, and raise small livestock.

Export Crops

Plantation crops such as tea, rubber, coconuts, and coffee are grown in Asia. Tea is indigenous to China, which is the world's largest producer, fol-

PhotoDisc

An elephant works a rice field in Thailand.

lowed by India, Sri Lanka, and Bangladesh. Tea is Sri Lanka's largest export crop, accounting for about one-third of annual exports by value. Tea is grown in the central highlands of Sri Lanka and in the hilly regions of northeastern India and Bangladesh. Rubber is grown in Malaysia and Indonesia—which account for about 75 percent of total world production—and Cambodia, India, and Sri Lanka. Malaysia, Indonesia, and Sri Lanka are world-leading exporters of coconuts and coconut products.

Green Revolution

A dramatic growth in food production in Asia began with the *Green Revolution* in the late 1960's, particularly for wheat and rice. Cultivation of the new varieties of rice and wheat caused an impressive increase in the use of fertilizer and the expansion of irrigation, particularly the exploitation of groundwater through tube wells. With proper and timely application of fertilizers and water, yields of wheat can be tripled, and yields of rice can be doubled.

Critics of the Green Revolution have concentrated on the negative impacts of increased use of fertilizer and pesticides, which causes surface water pollution. With high-yield seeds, three crops a year can be cultivated. Adopting this practice has two consequences: It causes overuse of land, a major source of land degradation, and it leads to increasing monocultures of rice and wheat, reducing the genetic diversity of food crops.

Without the Green Revolution, feeding current Asian populations at prevailing nutritional standards would have been impossible. New agricultural practices enabled Asia to avoid the famine that was widely predicted in the 1970's. The new rice and wheat varieties also have stimulated agricultural employment, because more people are needed to cultivate, harvest, and handle the increased production.

Throughout Asia, agricultural growth and the increase in food production were somewhat slower in the 1990's than in the 1980's. The opportunity for bringing more land under cultivation has largely been exhausted. Therefore, any increase in crop output will have to come largely from an increase in yields. Rice and wheat yields are still relatively low in many Asian countries, primarily because of low use of modern agricultural inputs. For example, the use of chemical fertilizer in South Asian countries has not reached the levels of neighboring regions.

Demand for fruits, vegetables, meat, fish, milk, and eggs is likely to grow with the increased urbanization and industrialization of Asia. This will reduce the demand for cereal crops. In Japan, South Korea, and Taiwan, consumption of rice has already begun to decline. Increased crop production is required to feed the growing populations of most Asian countries. While increasing agricultural production, Asian policymakers must also promote environmentally sound technologies and implement effective land reforms to address the problems of inequality and poverty caused by landlessness. Better crop management and better management of irrigation water are also needed to sustain agricultural growth in Asian countries.

Forestry

Forests of significant economic importance are found primarily in northeastern East Asia and Southeast Asia. The softwood forests of northeastern East Asia cover most of Japan and parts of North Korea. Trees grown there are used for construction lumber and to produce pulp for paper. Tropical hardwood forests cover all Southeast Asian countries and the south central part of China, several places in India, and the northern part of Iran.

Trees grown in those forests are used primarily for fuel wood and charcoal, although an increasing quantity of special-quality woods are cut for export as lumber. Nearly 80 percent of the world's hardwood log exports in the early 1990's came from Malaysia. Cambodia, Malaysia, Indonesia, the Philippines, Thailand, and Myanmar export large quantities of forestry products.

Overexploitation of hardwoods and conversion of forest lands for other uses have become serious concerns. Rates of forest conversion are most rapid in continental Southeast Asia, averaging about 1.5 percent a year. *Deforestation* has important local, regional, and global consequences, ranging from increased soil and land degradation to greater food insecurity, escalating carbon emissions, and loss of biodiversity. Small-scale, poor farmers clearing land for agriculture to meet food needs and the gathering of wood to be used for cooking account for roughly two-thirds of the deforestation in Southeast Asia. Commercial logging and urban expansion account for most of the remaining deforestation.

Bimal K. Paul

See also: Asian flora; Deforestation; Green Revolution.

Sources for Further Study

Airriess, Christopher A. "Monsoon Asia." In *World Regional Geography: A Development Approach*, edited by David L. Clawson and J. S. Fisher. 7th ed. Upper Saddle River, N.J.: Prentice-Hall, 2001. The book's four introductory chapters set the stage for discussions of economic development, followed by the chapter on monsoon Asia, among others.

Babu, Suresh Chandra, and Alisher Tashmatov, eds. *Food Policy Reforms in Central Asia: Setting the Research Priorities*. Washington, D.C.: International Food Policy Research Institute, 2000. In examining food policy issues in central Asia, the book also identifies relevant information gaps and sets priorities for policy research into the region's food, agriculture and natural resource sectors.

Das, Raju J. "Geographical Unevenness of India's Green Revolution." *Journal of Contemporary Asia* 29, no. 2 (1999): 167-186. Discusses new agricultural technology's applications based on ecological and historical conditions versus the developmental role of the state as it is influenced by class structure and the political power of classes.

Hafner, James A. "A Perspective on Agriculture and Rural Development." In *Southeast Asia: Diversity and Development*, edited by Thomas R. Leinbach and Richard Ulack. Upper Saddle River, N.J.: Prentice-Hall, 1999. The book is a comprehensive survey of Southeast Asia, addressing problems of the region's uneven economic development compared to the rest of the world, within the contexts of geography, history, culture, and society.

Held, Colbert C., and Mildred McDonald Held. *Middle East Patterns: Places, Peoples, and Politics*. 3d ed. Boulder, Colo.: Westview Press, 2000. Colbert Held worked as a Middle East geographer for the U.S. State Department. This analysis shows the importance of natural resources and human culture and covers soils, water resources, climate, and topography, as well as human settlement.

Jarring, Gunnar. *Agriculture and Horticulture in Central Asia in the Early Years of the Twentieth Century with an Excursus on Fishing*. Stockholm: Almqvist & Wiksell International, 1998. The Eastern Turki texts translated and presented in this volume depict the situation of agriculture along the southern Silk Road at the turn of the twentieth century.

Yang, Yongzheng, and Weiming Tian. *China's Agriculture at the Crossroads*. Palgrave, 2000. Examines key policy points of China's agricultural reforms and issues of food self-sufficiency versus integration into the world's economy.

ASIAN FLORA

Category: World regions

Asia has the richest flora of the earth's seven continents. Because Asia is the largest continent, it is not surprising that 100,000 different kinds of plants grow within its various climate zones, which range from tropical to Arctic.

Asian plants, which include ferns, gymnosperms, and flowering vascular plants, make up 40 percent of the earth's plant species. The endemic plant species come from more than forty plant families and fifteen hundred genera. Asia is divided into five major vegetation regions based on the richness and types of each region's flora: tropical *rain forests* in Southeast Asia, temperate *mixed forests* in East Asia, tropical rain/dry forests in South Asia, *desert* and *steppe* in Central and West Asia, and *taiga* and *tundra* in North Asia.

Tropical Rain Forests

The Asian regions richest in flora, tropical rain forests, are found in the island nations of Southeast Asia, which extend from Kinabalu in the north to Java in the south, and from New Guinea in the east to Sumatra in the west. In this vast archipelago, the longest island chain between Asia and Australia, are thirty-five thousand to forty thousand vascular plant species. Tropical rain forests grow there year-round because of the region's warm temperatures and plentiful rainfall. The forests contain great varieties of tall trees, some towering 148 feet (45 meters) high. Within any 1-square-mile area, one can see as many as one hundred tree species with no single species dominant.

The rain forests have mostly broad-leafed *evergreens*, with some palm trees and tree ferns. The uppermost branches of the trees form *canopies* that cover and protect the earth below. Because little sunlight penetrates the dense canopies, few shrubs or herbs grow in the rain forests. Instead, many vines, *lianas*, *epiphytes*, and *parasites* are twined on tree branches and trunks. Mangroves fringe the tropical rain forests along the coasts.

Temperate Mixed Forests

Second in floral richness, East Asia's temperate mixed forests contain thirty thousand to thirty-five thousand plant species. This region ranges from Japan in the east to the Himalayan nations (Bhutan, Sikkim, and Nepal) in the west, and from Russia's Amur River Valley in the north to China's Hainan Island in the south. East Asia's temperate weather is similar to the climate of eastern North America, with hot summers and cool winters. From south to north and from the east coasts to lower elevations in mountainous areas in the west, the vegetation changes from evergreen to *deciduous* broad-leafed forests, with dense shrubs, bamboo, and herbs in different layers beneath the forest canopy. The major tree species are of the magnolia, oak, tea, laurel, spurge, azalea, and maple families. Herbs include members of the primrose, gentian, pea, carrot, foxglove, composite, buttercup, and rose families.

The Himalayan range is the point where the regions of South Asia, East Asia, and Central and West Asia join. From the Qinghai-Tibet Plateau in southwest China to the lower areas of the Himalayas, elevation usually is between about 5,000 and 13,000 feet (1,500 and 4,000 meters). Mountains with deep valleys showcase complex, multiple vegetation types—from mixed forests and dense shrubs to alpine meadows in mountain plains. Many *primary seed plants* (gymnosperms and flowering plants), grow there.

Untouched native vegetation in East Asia is usually found only in mountainous or remote areas. On mountains at high elevations, the points where the temperatures are so cold that trees cannot grow form what is called the *tree line*. Near the tree line, only plants related to *coniferous* and *alpine* species grow. Above about 13,000 feet (4,000 meters) in high mountain areas, no vegetation grows. Instead, snowcaps or icebergs exist year-round.

Tropical Rain/Dry Mixed Forests

The third-richest region, tropical rain/dry forests, is found in South Asia, which reaches from the Philippines in the east to Pakistan in the west, and from the Himalayas in the north to Thailand in the south. Twenty-five thousand to thirty thousand species of plants grow there. This region has both tropical rain forests and tropical seasonal dry forests. The tropical rain forest is mainly found in the region's lowlands and the seasonal dry forests in the highlands or mountainous areas. More often, these two types of forests are combined.

The tropical seasonal dry forests usually grow in a climate with wet and dry seasons or under a somewhat cooler climate than the tropical rain forests. The canopy, formed from primarily deciduous broad-leafed species, is much thinner than the canopy in the tropical rain forest, so more sunlight reaches plants below. Many different plant species live together, forming tropical jungles. Tall, thick-trunked trees, colorful orchids, ferns, dense mosses, and twined vines and lianas dominate this vast region. The major components of these kinds of forests are members of the dipterocarpas, sweetsop, laurel, piper, fig, dissotis, akee, gardenia, periwinkle, milkweed, African violet, palm, and aroid families. In central and southern India and in some areas of Pakistan there are tropical *grasslands*, called the *savanna*. Because of the savanna's hot, dry weather, mainly coarse grasses grow there.

Desert and Steppe

The desert and steppe region in Central and West Asia has twenty to twenty-five thousand species of plants. This region stretches from north and northwest China and Mongolia in the east to Turkey in the west, and from Kazakhstan in the north to the Arabian Peninsula in the south. This region's vegetation changes from semidesert or desert to the temperate grassland called the steppe.

Central and West Asia contains the largest desert-steppe landscape in the Northern Hemisphere. Few plant species grow in the steppe and nearly none in the desert. The herbs and few woody plants that grow in these dry areas are members of the grass, pink, mustard, pea, saxifrage, stonecrops, lignum vitae, forget-me-not, and lily families. Because the desert environment is so dry, plant species must be able to survive in the arid weather for long periods of time. Central and West Asia—with its steppe between the desert in the south and coniferous forests in the north—forms one of the world's largest foraging areas, providing food resources for both wild and domestic animals, such as camels, sheep, goats, cows, and horses.

Taiga and Tundra

The poorest region in floral richness, with only about five thousand vascular plant species, is North Asia. This region is primarily Siberia, the eastern part of Russia, reaching from the Ural Mountains in the west to the Bering Strait in the east and from the Arctic Circle in the north to Mongolia and Kazakhstan in the south. The region's weather is temperate, with short, mild summers and long, cold winters. The predominant vegetation in North Asia is *coniferous* (boreal) forest. This region, called the taiga, contains mainly pine, spruce, fir, larch, and some species in the birch, aspen, and willow families. Because the trees there are straight and tall, the taiga provides timber for Russia's forestry industry. Small, perennial herbs and a few types of shrubs grow in the taiga's swamps or marshes.

Farther north is the cooler Arctic area called the tundra. Plants that grow in tundra are resistant to the cold climate. During the summer they complete their life cycle quickly, before winter comes. Tundra plant species are members of such common families as composites, peas, grasses, and reeds. Far beyond the tundra is Arctic ice.

Asia's native plant species provide shelter and food for animals. For example, arrow bamboo and umbrella bamboo, found in the forests of central to southwest China, are the main food of the giant panda. Many plants in Asia also provide food, ornaments, or medicine for humans.

Food Crops

Rice is the main food for humans in Asia, especially in the tropics. In temperate Asia, wheat—one of the world's main food sources—joins rice as a primary food source. Various beans and peas provide plant protein in the human diet and are eaten with vegetables and grains. Asia has many tropical fruit plants, such as the mango, banana, litchi, citrus fruits, and breadfruit. Pears, apples, grapes, peaches, and strawberries are temperate fruits. The kiwi, one of the most nutrient-rich fruits, is cultivated in New Zealand but originally came from central China. The Chinese not only eat kiwi but also make kiwi wine. Palm dates are another important fruit in West and Southwest Asia (the Arabian Peninsula). Vegetables grown in Asia include various cabbages, lettuce, onions, garlic, celery, carrots, soybeans, cucumbers, and squash. Ginger originally came from Asia.

Soybean oil is the major cooking oil in Asia. Although soybeans are native to Asia, they have become the biggest crop grown in the United States. According to the U.S. Department of Agriculture, in 1999 U.S. farmers harvested 73.3 million acres of soybeans, 2.3 million acres more than corn and 18.6 million acres more than wheat. Another oil plant, the sunflower, is grown in temperate Asia. In tropical Asia, people use mustard oil, palm oil, cotton oil, and peanut oil. In Central and West Asia, the most popular oil is olive oil. Many other foods people enjoy throughout the world are native Asian plants, for example, tea and coconuts. Black pepper and sugarcane also are grown in tropical Asia.

Ornamental and Medicinal Plants

Many of Asia's plant species have great ornamental value. Azaleas, dogwood, primroses, camellias, peonies, roses, lotus, daisies, cherries, and begonias are frequently planted in gardens. Ornamental conifers from Asia include pines, spruces, cedars, junipers, umbrella pines, and yews. Thousands of wildflowers originating in Asia include poppies, snapdragons, slippers, columbine, trillium, marigolds, buttercups, gentian, lilies, bluebells, and violets. Europeans who explored Asia centuries ago brought ornamental plants back to their home countries. As a result, many of these plants are now grown throughout the world. The world's largest flower, *rafflesia*, grows in the tropical rain forests of Sumatra. In full bloom, the flower's diameter is about 3 feet (1 meter).

Plants make up a large part of traditional Chinese medicine, which has been practiced for thousands of years. Today, some of these plants are used in alternative medicine in the West. They include ephedra, eucommia, cinnamon, ginseng, sanqi, and ginkgo.

Scientific Value

Botanists view the region ranging from central China to the Himalayas to the northern part of South Asia as a key area for research into the origin of flowering plants. Native plant species in Asia are numerous; botanists also study Asian plants that are relics of ancient times, from millions of years ago, as well as fossils.

Ancient species include such gymnosperms as the dawn redwood of central China. Dawn redwood is similar to California's redwood and giant sequoia. Another fossil-like tree is East Asia's ginkgo. This species not only has great ornamental value but also has great commercial value as an alternative medicine. Ginkgo trees are also dust-resistant, which makes them a favorite in urban landscaping and the ornamental industry. Other Asian plants such as the magnolia and its allied families may represent the most primitive flowering plants.

Introduced Plants

Asian flora today also includes introduced plant species from other parts of the world that play important roles in people's lives. For example, the rubber tree of South America is cultivated in tropical Asia. This tree produces raw material for the natural rubber industry, in which Asia is the largest producer in the world. Cacao, a tree species that provides the basis of chocolate, was introduced from tropical America. Corn, one of the most common crops in Asia, was introduced from America several thousand years ago. Several vegetables, including the tomato, potato, eggplant, green pepper, hot pepper, and chili (all from the nightshade family), were introduced to Asia long ago. Peanuts, originally from Brazil, are also cultivated in Asia.

Coffee, an increasingly popular beverage in Asia, came originally from Africa. An introduced fruit is the pineapple, which came from tropical America but now is popular in tropical Asia. The sweet potato, from Central America, is also cultivated in Asia. Tobacco is another crop introduced to and cultivated in Asia. It originally came from tropical America, but its yield in China has made that nation a leading producer.

Impact of Human Activity

Asia's highly diversified flora have contributed positively to the daily lives of people around the world, but the demands of a rapidly growing population are a constant threat. *Deforestation*, *overgrazing*, and *urbanization* have become major reasons for heavy losses of Asian flora, especially in South and East Asia. In China alone, eight key plant species were added to the first-class protection list in the Red Book of 1992 (equivalent to the U.S. endangered species list). Among them are the Chinese silver fir, dawn redwood, and ginseng. These plants only grow in several isolated locations and are rare in their original range. As natural vegetation is cut for farming, grazing, or simply for cooking and heating fuel, fewer plants remain. Although scientists from around the world have worked on this problem for decades, the situation has not improved.

Guofan Shao and Jinshuang Ma

See also: Asian agriculture.

Sources for Further Study

Chan, Elizabeth, and Luca I. Tettoni. *Tropical Plants of Southeast Asia*. North Clarendon, Vt.: Periplus Editions, distributed in the United States by Charles E. Tuttle, 1999. A beginner-level plant identification guide.

Grubov, V. I., ed. *Plants of Central Asia: Plant Collections from China and Mongolia*. New York: Science Publishers, 1999. Part of a multivolume series offering in-depth studies of the flora of central Asia.

Musgrave, Toby, et al. *The Plant Hunters: Two Hundred Years of Adventure and Discovery Around the World*. London: Seven Dials, 1999. Biographical studies of ten British botanists who explored the world in search of exotic plants. Covers, among others, Sir Joseph Dalton Hooker, who introduced the rhododendron to Britain from the Himalayas,

and George Forrest, who discovered the star gentian in Yunan Province, China.
Quattrocchi, Umberto. *CRC World Dictionary of Plant Names: Common Names, Scientific Names, Eponyms, Synonyms, and Etymology*. 4 vols. Boca Raton, Fla.: CRC Press, 1999. Covers the origins and meanings of names of about 22,500 genera of plants and many thousand of species, biographies of botanists, and all varieties of common and botanical names in many languages for more than 200,000 species.

ATP AND OTHER ENERGETIC MOLECULES

Categories: Cellular biology; physiology

All cells contain a small collection of compounds called common intermediates. These compounds transfer energy between processes that produce energy and those that require it. The best-known example is adenosine triphosphate (ATP), but several other esters play a similar role. Because of their function in energy transfer, cells are absolutely dependent on such compounds, and a cell deprived of ATP will quickly die.

ATP (adenosine triphosphate) is the principal energy carrier for most life processes, including those carried out within plant cells, including the chemical reactions of photosynthesis. The central importance of *common intermediates* (energy-transferring compounds) is evident in the effects of various chemicals that prevent ATP synthesis. For example, in cells that require oxygen, ATP synthesis is coupled to the oxidation of food-related molecules. Cyanide and carbon monoxide interfere with such cellular oxidations and, therefore, completely prevent the ATP synthesis that is coupled to them. Thus, these compounds are extremely toxic, and cells are often killed by them in a matter of seconds.

Free Energy

Many biological processes require energy in order to occur; that is, they are not by themselves spontaneous. Examples of energy-requiring processes include contraction of muscle, beating of cilia, emission of light by fireflies, heat production by birds and mammals, and establishment of a voltage difference across a cellular membrane. The synthesis of proteins from their constituent amino acids, the formation of complex membrane fats, and, indeed, the synthesis of many other chemical compounds are not themselves spontaneous and, hence, require energy. Energy-dependent processes and reactions are referred to as *endergonic*.

On the other hand, many processes in organisms are spontaneous and do occur without any other energy source. Most of these are chemical reactions, and all such reactions, being spontaneous, can serve as energy producers. For this reason, they are often called *exergonic*. They can, in principle, provide energy for events, such as the secretion of nectar or aromatic oils by flower cells, that require it. The form of that energy is *free energy*.

There is a change in free energy, measured in units of calories or joules, associated with any chemical reaction or physical event. If the change in free energy is more than zero, the process requires energy to proceed; if less than zero, it yields energy and is spontaneous. Thus, the numerical value for the change in free energy of a reaction is very useful for predicting whether it can occur.

The other important aspect of free energy is that its numerical value can be altered by changing the concentrations of the chemicals reacting or the product formed when the chemicals react. For example, consider a hypothetical biochemical reaction in which A is the reactant and B is the product. Imagine that this reaction is not spontaneous; therefore, its free energy change is greater than zero, and it does not "go." Yet, though A becoming B is not spontaneous, its reverse, B becoming A, is. In other words, if a reaction in one direction has a change in free energy greater than zero, the reverse reaction will have one less than zero. Most important, if a reaction in a particular direction is not spontaneous, it can be made to "go" anyway, simply by increasing the concentration of the reactant until the change in free energy falls below zero. This is called the *law of mass action*: Adding more reactant (or subtracting some of the product) will often force a reaction. Stubborn reactions can be pushed or pulled.

Spontaneous Reactions

A second law explains completely how spontaneous reactions can provide energy for those that are not spontaneous. It is called the *law of the common intermediate* and states that reactions can be linked, as far as energy is concerned, by the products of one reaction serving as the reactants of another. Here are two biochemical reactions:

$$A + B \leftrightarrow C + D$$

$$X + D \leftrightarrow Y + Z$$

If the first reaction is spontaneous, with a change in free energy less than zero, it produces a substantial amount of C and D. The second reaction would not normally be spontaneous (with its positive change in free energy), but notice that D, the product of the first reaction, is a reactant in the second. When the two reactions are together in the same cell, the buildup of D tends to push the second reaction, reducing the difference in free energy between reactants and products. In other words, the energy produced by the first reaction is being transferred to the second reaction, allowing it to occur. In this case, D is the common intermediate that connects the two.

ATP

In organisms, the sum of all chemical reactions is called *metabolism*. All the systems of reactions that require energy are called *anabolism* and those producing it, *catabolism*. Anabolism and catabolism are tightly linked by common intermediates. A good example of a common intermediate is ATP. Using ATP as an example, it can be seen that the linkage between anabolism and catabolism is a two-way street. ATP is synthesized from adenosine diphosphate (ADP) and phosphate, with the required energy often coming from light (photosynthesis) or from the oxidation of foods (respiration). When ATP is used to provide energy for muscle, for example, it is broken down by hydrolysis to ADP plus phosphate. Thus, the product of catabolism (ATP) is utilized in anabolism, and the products of anabolism (ADP plus phosphate) are utilized in catabolism to regenerate ATP. Thus, the role of ATP is cyclic.

Most common intermediates are *phosphate esters* such as ATP, although a few are *thioesters*. It was once thought that such ester bonds were somehow unusual, perhaps containing an abnormally large amount of available free energy. They were often called high-energy bonds. In fact, these ester bonds are not particularly abnormal in their energy; there are many esters with much more energy—compounds that are not important common intermediates. Apparently, the reason that so few compounds have evolved as links between catabolism and anabolism is that their chemical structures are unique in enabling them to participate readily in both.

It appears likely that a very early common intermediate in ancient cells was not ATP but pyrophosphate, which is simply two phosphates hooked together by an ester bond. Apparently, cells evolved the central role for ATP at a later stage. In all these cases, when the energy of a common intermediate is utilized in an anabolic reaction, it is by means of a *hydrolysis reaction*, the breaking of an ester bond by the addition of water. In many instances, the hydrolysis occurs in more than one step, but it is a hydrolysis, nevertheless, and the free energy that matters in such cases is the free energy of the hydrolysis.

Other Intermediary Compounds

ATP usually gets the most attention, which is proper, because evolution, over the long run, appears to have exhibited the same preference. There are, however, a few other compounds that serve as common intermediates. Several are quite similar to ATP, with a ring-shaped organic part and three phosphates attached in series at one end. The organic parts are a little different, however, and the compounds have different names, with abbreviations such as GTP, UTP, and ITP. These compounds, which are, with ATP, termed *nucleoside triphosphates*, are readily interchangeable with ATP. For example, ATP can be manufactured from ADP by a phosphate transfer reaction:

$$ADP + GTP \leftrightarrow ATP + GDP$$

In addition, pyrophosphate, a common intermediate in early cells, transfers energy today in a variety of plant, animal, and bacterial species. In some organisms, pyrophosphate is made (and used) under conditions in which it is, for whatever reason, difficult to make ATP. A few reactions use energy from the hydrolysis of acetyl phosphate, which can be chemically described as a derivative of acetic acid (the active ingredient of vinegar) with a phosphate attached. Also, a number of thioesters, esters with sulfur in place of oxygen, are important common intermediates; they are often large, complicated molecules.

Finally, one may wonder why ATP and similar compounds have as large negative free energies as they do. A complete answer would involve advanced chemistry, but one factor is both important and readily understood. When ATP is hydrolyzed,

$$\text{ATP} + \text{water} \leftrightarrow \text{ADP} + \text{phosphate}$$

both of the products happen to be negatively charged. It is well known that unlike charges attract each other and like charges repel. Therefore, the two products are driven apart from each other, and they are unlikely to react together to produce a back reaction. Thus, the reaction tends to go well in the direction of ADP and phosphate but not the reverse. Another way of saying it is that the reaction is highly spontaneous, or that the reaction exhibits a solidly negative change in free energy, making such compounds good candidates for energy transfer.

John L. Howland

See also: Active transport; Calvin cycle; Cell-to-cell communication; Cells and diffusion; Chloroplasts and other plastids; Energy flow in plant cells; Exergonic and endergonic reactions; Glycolysis and fermentation; Krebs cycle; Lipids; Mitochondria; Nitrogen fixation; Oxidative phosphorylation.

Sources for Further Study

Cramer, W. A., and D. B. Knaff. *Energy Transduction in Biological Membranes: A Textbook of Bioenergetics*. New York: Springer-Verlag, 1990. A complete course, starting with basic principles, emphasizes concepts and information that are central to modern membrane biochemistry, biophysics, and molecular biology. Includes examples, problems with solutions, figures and tables, and extensive references. Intended for advanced study in biochemistry, plant physiology, and biology.

Garby, Lars, and Paul S. Larsen. *Bioenergetics: Its Thermodynamic Foundations*. New York: Cambridge University Press, 1995. Outlines the biophysical foundation of bioengergetic mechanisms, focusing on the laws of conservation of energy and increase of entropy, thermodynamic equilibrium and nonequilibrium, and energy balance.

Harris, David A. *Bioenergetics at a Glance*. Cambridge, Mass.: Blackwell Science, 1995. A clear and concise introduction to the study of energy use and conversion in living organisms, with an emphasis on the biochemical aspects of plant science and physiology and on cell biology.

Lehninger, Albert, David L. Nelson, and Michael M. Cox. *Principles of Biochemistry*. 3d ed. New York: Worth, 2000. A university-level textbook of biochemistry, known for its uncommonly clear presentation. Accessible to a general reader. Chapters specifically address the role of ATP and the ATP-ADP cycle as well as ATP synthesis and utilization. The author was a leader in the field of biological energetics and a long-time student of the place of ATP in the biological world. Particularly useful references.

Mukohata, Yasuo, ed. *New Era of Bioenergetics*. San Diego: Academic Press, 1991. The results of a Japanese research project on the bioenergetics of cation pumps, redox chains, ATP synthesis, and extremophiles.

Peusner, Leonardo. *Concepts in Bioenergetics*. Englewood Cliffs, N.J.: Prentice-Hall, 1974. Written by a leading theoretical biologist, covers matters of biological energy and its regulation. The treatment is quite mathematical in places, but clear diagrams and good analogies make it accessible to general readers. Discusses high-energy compounds and coupling between different chemical reactions.

AUSTRALIAN AGRICULTURE

Categories: Agriculture; economic botany and plant uses; food; world regions

Agriculture is an important part of Australia's economy. Australia's exports were overwhelmingly agricultural products until the 1960's, when mining and manufacturing grew in importance.

Agriculture occupies 60 percent of the land area of Australia, but much of this is used for open-range cattle grazing, especially in huge areas of the states of Queensland and Western Australia. Only 5 percent of Australia's agricultural land is used for growing crops. Western Australia and New South Wales have the largest areas of cropland. The limited area suitable for growing commercial crops is limited mainly by climate, because Australia is the world's driest continent.

Annual rainfall of about 20 inches (500 millimeters) is necessary to grow crops successfully without irrigation; less than half of Australia receives this amount, and the rainfall is often variable or unreliable. Years of drought may be followed by severe flooding. High temperatures throughout most of Australia also mean high evaporation rates, so rainfall figures alone are not a good guide to the feasibility of agriculture.

Australian soils usually require the application of fertilizer to grow crops successfully. The east coast of Australia is suitable for growing sugarcane,

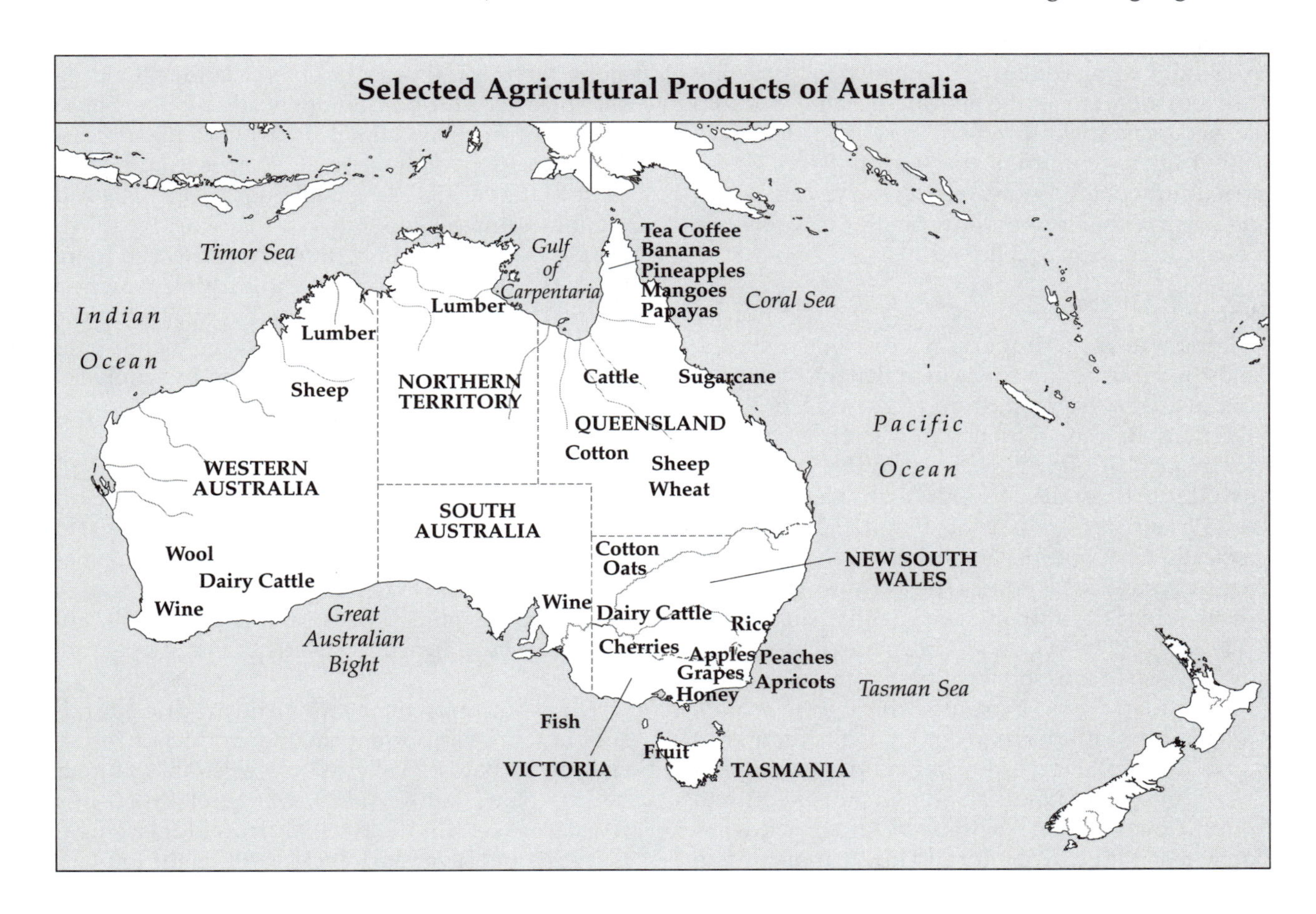

CORBIS

Australia has 242,060 acres of vineyards and more than one thousand wineries.

while the cooler southern parts are suited to growing wheat. Irrigation has opened up large areas of drier land to agriculture, especially for growing fruit, but *salinization* (the buildup of salt in the soil) has become a major problem in some areas, especially near the mouth of the Murray River. Major agricultural plant exports from Australia are wheat and sugar. Other important agricultural exports are fruits, cotton, rice, and flowers.

Wheat

Long the most important crop of Australia, wheat is produced in the Wheat Belt, a crescent of land just west of the Eastern Highlands, or Great Dividing Range, which extends from central Queensland through New South Wales to Victoria, as well as in the south of South Australia and southwest Western Australia. More than 120,000 farms in Australia grow grains, and wheat is the principal crop on some 25,000 farms. The average Australian wheat farm is family-owned and has an area of 3,700 acres (1,500 hectares). Crops are rotated, usually because of low soil fertility.

Australian wheat is planted during the winter, which is much milder than winter on the prairies of North America. Harvesting begins in September in the warm state of Queensland and moves south to Victoria and Western Australia by January. Australian wheat is high in quality and low in moisture, so it is easy to mill. Wheat crops are frequently affected by drought; another problem is Australia's markets, because the nation competes with the United States in wheat export.

When the British first came to Australia, convicts planted wheat on a government farm in what is now inner Sydney. They had difficulty growing wheat because of poor soils, unfamiliar climate, and inexperience, causing fear of widespread hunger. As settlement spread beyond the coastal plain and into the interior, wheat production rose dramatically. The rapid increase in population after the gold discoveries of the 1850's also led to increased demand for wheat. Australia began exporting wheat in 1845 and is now the world's fourth-largest exporter of wheat.

Sugar

Sugarcane is grown in a series of small regions along the tropical coast of Queensland, extending slightly across the border into northern New South Wales. A warm, wet climate is required for successful cultivation of sugarcane, so it is confined to parts of the coastal plain with good, deep soils and reliable rainfall. Australia is the world's third-largest exporter of sugar. Sugar is grown on more than six thousand small, individually owned farms. Until the 1960's, cane was cut by hand. Now it is harvested mechanically and taken by light rail to a nearby mill. There are twenty-five mills in Queensland and three in New South Wales.

Fruit

Fruit growing has a long history in Australia and is strongly influenced by climatic considerations. In Queensland, tropical fruits such as bananas, pineapples, mangoes, and papaya (called pawpaw in Australia) are cultivated. In the cooler south, apples, peaches, apricots, cherries, and grapes are grown.

Grapes are grown for eating and are dried as raisins, but more important is wine production, especially in the Barossa Valley (South Australia), Hunter Valley (New South Wales), Margaret River area (Western Australia), and the Murrumbidgee Irrigation Area and Riverina. John Macarthur, the founder of the Australian wool industry, established the first commercial vineyard, in New South Wales. Later European settlers planted vineyards in Victoria and South Australia. In the 1960's modern plantings and production methods were introduced. Australia has 242,060 acres (98,000 hectares) of vineyards and more than one thousand wineries. Wine is an important export for Australia, with the European Union purchasing 60 percent of wine exported.

Other Agricultural Products

Cotton is grown in drier interior parts of northern New South Wales and in part of central Queensland. Cotton is usually grown in conjunction with sheep farming, on family farms. Indonesia is the major customer for Australian cotton. Rice has been grown commercially in Australia since 1924, using irrigation. New South Wales is the main producer, where the Murrumbidgee Irrigation Area dominates rice production. Australia exports most of its rice crop and in 1999 was the world's eighth-largest exporter of rice. Oats are grown where the climate is too cool and too moist for wheat. In Australia, this is in the interior southeast with a small area in Western Australia. This state and New South Wales are the biggest producers of oats, which is used mainly for livestock fodder.

Other agricultural products of Australia include barley; grain sorghum; corn, called maize in Australia; vegetables, including potatoes, peas, tomatoes, and beans; oil seeds such as sunflower; soybeans; and tea and coffee in northern Queensland. Australia is a major producer of honey, with more than eight hundred commercial apiarists. Blossoms of the eucalyptus tree produce distinctive-tasting honey, which is sold mainly to European Union countries.

Ray Sumner

See also: Australian flora.

Sources for Further Study

Dyster, Barrie, and David Meredith. *Australia in the International Economy in the Twentieth Century*. New York: Cambridge University Press, 1990. Discusses agriculture in the context of Australia's contribution to world economy.

Lines, William J. *Taming the Great South Land: A History of the Conquest of Nature in Australia*. 2d ed. Athens: University of Georgia Press, 1999. A strong critique of the effects of European-derived agriculture and environmental attitudes on the natural resources of Australia.

Malcolm, Bill, Peter Sale, and Adrian R. Egan. *Agriculture in Australia: An Introduction*. New York: Oxford University Press, 1996. An introductory textbook for Australian agriculture students, taking a whole-farm approach.

Smith, David. *Natural Gain: A Case Study of the Grazing Lands of Southern Australia*. Seattle: University of Washington Press, 2001. A somewhat controversial study arguing that agriculture has fostered dynamic and sustainable ecosystems in Australia's temperate grassland regions.

AUSTRALIAN FLORA

Category: World regions

Australia broke off from the supercontinent Pangaea more than fifty million years ago, and the species of plants living at that time continued to change and adapt to conditions on the isolated island. This led to distinctive plants, differing from those of the interconnected Eurasian-African-American landmass, where new immigrant species changed the ecology.

Many species of plants in Australia are found nowhere else on earth, except where they have been introduced by humans. Such species are known as endemic species. This distinctiveness is the result of the long isolation of the Australian continent from other landmasses. Australian vegetation is dominated by two types of plants—the eucalyptus and the acacia. There are 569 known species of eucalypts and 772 species of acacia. Nevertheless, a great deal of botanical diversity exists throughout this large continent.

Climate and Ecology

Climate is a major influence on Australian flora, and the most striking feature of the Australian environment as a whole is its aridity. Nutrient-poor soils affect the nature of Australia's vegetation, especially in arid areas. Half of the continent receives less than 11.8 inches (300 millimeters) of rainfall per year; small parts of Australia receive annual rainfall of 75 inches (800 millimeters). Therefore, forests cover only a small percentage of Australia. A close correlation exists between rainfall and vegetation type throughout Australia. Small regions of tropical *rain forest* grow in mountainous areas of the northeast, in Queensland. In the cooler mountains of New South Wales, Victoria, and Tasmania, extensive temperate rain forests thrive.

More extensive than rain forest, however, is a more open forest known as *sclerophyllous forest*, which grows in the southern part of the Eastern Highlands in New South Wales and Victoria, in most of Tasmania, and in southwest Western Australia. A huge crescent-shaped region of woodland vegetation, an open forest of trees of varying height, with an open canopy, extends throughout northern Australia, the eastern half of Queensland and the inland plains of New South Wales, and to the north of the Western Australian sclerophyllous forest.

Beyond this region, the climate is arid, and shrubs, forbs (smaller herbaceous plants), and grasses predominate. The tropical north of inland Queensland, Northern Territory, and a smaller part of Western Australia have extensive areas of *grassland*. Much of Western Australia and South Australia, as well as interior parts of New South Wales and Queensland, are *shrubland*, where grasses and small trees grow sparsely. In the center of the continent is the desert, which has little vegetation, except along watercourses.

Eucalypts

Many people familiar with the song "Kookaburra Sits in an Old Gum Tree," may not realize that *gum tree* is the common Australian term for a eucalyptus tree. When the bark of a eucalyptus tree is cut, sticky drops of a transparent, reddish substance called *kino* ooze out. In 1688 the explorer William Dampier noticed kino coming from trees in Western Australia and called it "gum dragon," as he thought it was the same as commercial resin. Kino is technically not gum, as it is not water-soluble.

The scientific name *Eucalyptus* was chosen by the first botanist to study the dried leaves and flowers of a tree collected in Tasmania during Captain James Cook's third voyage in 1777. The French botanist chose the Greek name because he thought that the bud with its cap (*operculum*) made the flower "well" (*eu*) "covered" (*kalyptos*). The hard cases are commonly called gum nuts.

The more than five hundred species of eucalypts in Australia range from tropical species in the north to alpine species in the southern mountains. Rainfall, temperature, and soil type determine which particular eucalypt will be found in any area. Eucalyptus trees dominate the Australian forests of the east and south, while smaller species of eucalyptus grow in the drier woodland or shrubland areas. It is easier to mention parts of Australia where eucalypts do not grow: the icy peaks of the Australian Alps, the interior deserts, the Nullarbor Plain, and the tropical and temperate rain forests of the Eastern Highlands.

The scientific classification of the eucalypts proved difficult to European botanists. Various experts used flowers, leaves, or other criteria in their attempts to arrange the different species into a meaningful and useful taxonomy, or classification scheme. George Bentham eventually chose the shape of the anthers—the part of the *stamen* that holds the pollen—together with fruit, flowers, and nuts.

A simpler classification of the eucalypts, commonly used by foresters, gardeners, and naturalists, arranges them into six groups based on their bark: gums have smooth bark, which is sometimes shed; bloodwoods have rough, flaky bark; ironbarks have very hard bark with deep furrows between large pieces; stringybarks have fibrous bark that can be peeled off in long strips; peppermints have mixed but loose bark; boxes have furrowed bark, firmly attached. This system was devised in 1859 by Ferdinand von Müller, the first government botanist of the Colony of Victoria and the father of Australian botany.

A eucalyptus forest in Victoria, Australia.

Many of the native plants of Australia, along with eucalypts, show typical adaptations to the arid climate, such as deep taproots that can reach down to the water table. Another common feature is small, shiny leaves, which reduce transpiration. Eucalyptus leaves are tough or leathery and are described as sclerophyllous. Sclerophyllous forests of eucalypts cover the wetter parts of Australia, the Eastern Highlands, or Great Dividing Range, and the southwest of Western Australia. The hardwood from these forests is generally not of a quality suitable for building, so areas are cleared and the trees made into wood chips that are exported for manufacture of newsprint paper. This has been a controversial use of Australian forests, especially where the native forest has been cleared and replaced with pine plantations.

The southwest corner of Western Australia has magnificent forests featuring two exceptional species with Aboriginal names, *karri* and *jarrah*. Karri is one of the world's tallest trees, growing to 295 feet (90 meters) tall. This excellent hardwood tree is widely used for construction. The long, straight trunks are covered in smooth bark that is shed each year, making a colorful display of pink and gray. These forests are now protected. Jarrah grows to 120 feet (40 meters) in height and is a heavy, durable timber. It was used for road construction in the nineteenth century, but now the deep red timber is prized for furniture, flooring, and paneling.

During the nineteenth century, Australia could also claim to be home to the world's tallest trees, the mountain ash. The tallest tree, which observers claimed was 433 feet (132 meters) high and with the top broken away, was felled in 1872. The tallest accurately measured tree was 374 feet (114 meters) high.

The most widely distributed of all Australian eucalypts is the beautiful river red gum. These trees grow along riverbanks and watercourses throughout Australia, especially in inland areas; their spreading branches provide wide shade and habitat for many animals. Koalas eat leaves from this tree. In the song "Waltzing Matilda," Australia's unofficial national anthem, a man camps "under the shade of a coolabah tree." This word might apply to any eucalypt, but it is most likely a river red gum.

In drier interior areas and in some mountain areas, there are more than one hundred smaller species of eucalypts that are known by the Aboriginal name *mallee*. These many-trunked shrubs have underground lignotubers, roots that store water. Much of this marginal country was cleared for farming, creating a situation similar to that of the 1930's Dust Bowl in the United States. In the dry Australian summers bushfires are a great danger in the mallee and in any eucalyptus forest. The volatile oils of the eucalyptus trees can lead to rapid spread in the tree crowns, jumping across human-made firebreaks. On the other hand, several Australian trees not only can survive fires but actually require fire for their seeds to germinate. Eucalypts have been introduced to many countries, including Italy, Egypt, Ethiopia, India, China, and Brazil, and they are common in California, where they have been growing as introduced trees for 150 years.

Acacias

These plants are usually called *wattles* in Australia, because the early European convicts and settlers used the flexible twigs of the plant for wattling, in which twigs are woven together, making a firm foundation for a thatched roof or for walls. The walls are then covered inside and out with mud. This style of building, known as wattle-and-daub, was common throughout Australia in pioneering days.

Wattles frequently have masses of colorful flowers, usually bright yellow. One species is the national flower of Australia. Other interesting acacias include the mulga, which has an attractive wood.

Rain Forests

Although rain-forest vegetation covers only a small area of Australia, it is exceptionally varied and of great scientific interest. Neither of the two general types of rain forest found in Australia has eucalypts. Rain forests are located along the Eastern Highlands, or Great Dividing Range, where rainfall is heavy. In small areas of tropical Queensland, where rainfall is also heavy, true tropical rain forest is found. The flora are similar to those in Indonesian and Malaysian rain forests. The tropical rain forest contains thousands of species of trees, as well as lianas, lawyer vine, and the fierce *stinging tree*, whose touch could kill an unwary explorer. Toward the north of New South Wales, a kind of subtropical to temperate rain forest grows. Cool, wet Victoria and Tasmania have extensive areas of temperate rain forest, where only a few species dominate the forests. Arctic beech trees are found, as well as sassafras and tall tree ferns.

Sclerophyllous Forest

This is the typical Australian *bush*, which grows close to the coast of New South Wales, Victoria, and Tasmania. The bright Australian sun streams down through the sparse crowns and narrow leaves of the eucalypts. As the climate becomes drier, farther inland from the coast, the open forest slowly changes to a shrubbier woodland vegetation.

Grasslands

Moving farther inland, to still drier regions, woodlands give way to grasslands, where cattle are raised for beef in the tropics, and sheep are raised for wool in the temperate areas. Before Europeans came to Australia, there were native grasslands in the interior—tropical grasslands in the monsoonal north, and temperate grasslands in the south and southwest. Kangaroo grass and wallaby grass once grew in the temperate interior of New South Wales, but much of this has been cleared for agriculture, especially for wheat farming. Mitchell grass is another *tussock grass*, which grows in western Queensland and into the Northern Territory. Cattle and sheep graze extensively on this excellent native pasture.

The most common grassland type in Australia is dominated by *spinifex*, a spiky grass that grows in clumps in the arid interior and west. Even cattle cannot feed on spinifex grass, so this ecosystem is less threatened than most other grasslands. The northern grasslands are dotted with tall red termite mounds; those that are aligned north-south for protection from the hot sun are built by so-called magnetic termites.

Other Trees and Plants

Many people think that *macadamia* nuts are native to Hawaii, which produces 90 percent of the world's crop, but in fact, the tree is native to Australia. It was discovered by Ferdinand von Müller in 1857 on an expedition in northern Australia. Müller named the tree after his Scottish friend, John Macadam. The trees were introduced to Hawaii in 1882.

Cycads are plants from an ancient species which still thrive in Australia. The *Macrozamia* of North Queensland is a giant fernlike plant. Similarly old are the *Xanthorrhea* grass-trees, which used to be called "blackboys." A single spearlike stem rises from a delicate green skirt on this fire-resistant species.

In northwest Australia, the *baobab*, or bottle tree, can be found. This fat-trunked tree collects water in its tissue. One is said to have served as a temporary prison. The only other baobabs are found in Africa, a reminder that these continents were once joined. *Bottlebrush* is an Australian shrub with colorful flowers that has become popular with gardeners in many parts of the world.

Extinct and Endangered Plants

Human activities in Australia have led to the extinction of more than eighty species of plants, and the list of endangered plants contains more than two hundred species. Many nonnative species have been introduced to Australia by Europeans. Some have become pests, such as the blackberry in Victoria, the lantana in north Queensland, and water hyacinth, found throughout the continent. There are 462 national parks in Australia, as well as other conservation areas, where native flora are protected.

Aboriginal Plant Use

The Australian Aborigines used plants as sources of food and for medicinal purposes. Food plants included nuts, seeds, berries, roots, and tubers. Nectar from flowering plants, the pithy center of tree ferns, and stems and roots of reeds were eaten. Fibrous plants were made into string for weaving nets or making baskets. Weapons such as spears, clubs, and shields, as well as boomerangs, were made from hardwoods such as eucalyptus. The bunya pine forests of southeast Queensland were a place of great feasting when the rich bunya nuts fell.

Ray Sumner

See also: Australian agriculture.

Sources for Further Study

Berra, Tim M. *A Natural History of Australia*. San Diego: Academic Press, 1998. A wide-ranging overview of Australia's natural history, covering the continent's geology, flora, and fauna. Useful for travelers.

Dallman, Peter R. *Plant Life in the World's Mediterranean Climates: California, Chile, South Africa, Australia, and the Mediterranean Basin*. Berkeley: University of California Press, 1998. Compares the climate and flora of southwestern Australia with similar environments elsewhere in the world.

Kirkpatrick, J. B. *A Continent Transformed: Human Impact on the Natural Vegetation of Australia*. New York: Oxford University Press, 1994. An accessible introduction to Australia's biogeography, focusing on changes created by human habitation and suggestions for preserving the continent's natural resources.

Pyne, Stephen J. *Burning Bush: A Fire History of Australia*. Seattle: University of Washington Press, 1998. Traces the effects of fire on Australia's ecology, from naturally occurring fires to its use by Aboriginal societies and Europeans.

Watkins, T. H. "Greening of the Empire: Sir Joseph Banks." *National Geographic* (November, 1996): 28-53. A biographical article on one of the most important botanists of the eighteenth century, who made his first discoveries on a voyage to Australia and the South Pacific in 1768.

Young, Ann R. M., and Ann Yound. *Environmental Change in Australia Since 1788*. 2d ed. New York: Oxford University Press, 2000. Discusses the effects of humans on the natural environment of Australia since European colonization and assesses public policy as it relates to environmental affairs.

AUTORADIOGRAPHY

Category: Methods and techniques

Autoradiography produces an image formed by a substance's own radioactivity when exposed to a photographic film. This technique is often used for investigation of biological processes.

In 1896 Antoine-Henri Becquerel was working with rocks containing uranium ore. By chance, he put one rock sample into a dark drawer on top of a box of unexposed photographic film. When the film later was developed, it showed a clear outline of the uranium rock. Evidently, some radiation had been emitted from the rock, penetrated through the wrapping paper, and exposed the film inside. An autoradiograph, that is, an image produced by radioactivity, was visible on the film. Autoradiography, much refined, is now a valuable technique for investigating biological processes.

Hungarian chemist Georg von Hevesy pioneered the use of radioactive tracers in biological research in the 1920's, and two developments in the 1930's greatly expanded their use. First was the discovery of *induced radiation* by Frédéric Joliot-Curie and Irène Joliot-Curie, which raised the exciting possibility that artificially created radioactivity could be induced in almost any element found in nature. The second development was the invention of the *cyclotron*. The cyclotron beam was used to bombard various elements to produce new *radioactive isotopes*, including radioactive sodium, potassium, sulfur, and iron. After 1950 radioactive hydrogen and carbon also became available as *tracers*, allowing organic molecules such as carbohydrates and proteins to be labeled.

Macroautoradiography

Whole-body autoradiography has been widely used to trace the routes of molecules in *metabolism*. First, a radioactive tracer is administered to an organism by ingestion or injection. After a period of time, individual samples of tissue are removed and pressed directly against X-ray film for several days, to expose the film wherever the radioactivity has become concentrated. The film is then developed and viewed, frequently with the aid of a microscope. This process has been used to trace the uptake of nutrients by plants from the soil into the leaves or buds. Experimentation with whole organisms is called *macroautoradiography*.

Microautoradiography

A refinement of this methodology, called *microautoradiography*, has been developed for studying subcellular structures, even those as small as individual strands of deoxyribonucleic acid (DNA). Much interesting information has been learned about the mechanisms of cell division and other processes in cell biology. The cells being studied are given a nutrient solution containing molecules that have been labeled, usually with radioactive tritium, carbon, or phosphorus.

After a period of incubation, some cells are transferred to a glass slide. The slide is dipped into a liquid photographic emulsion containing light-sensitive silver bromide, which clings to the slide in a thin layer. The slide with the cells covered by emulsion is then placed into a light-tight box for several days to allow time for radioactive decay. The beta particles from tritium cause the photographic emulsion to become exposed.

The emulsion is then developed and fixed as any photographic negative would be. The developer washes the soluble silver bromide away and leaves behind the insoluble grains of silver, which show up as small black dots. A stain may be applied to show the outlines and structures within. Finally, the cell is examined with a microscope. Autoradiographs typically show the black dots of exposed silver grains against a faint background of the surrounding cell structure. When higher magnification and resolution are desired, an electron microscope can be used.

In some studies, the radioactive nutrient is supplied to the cell for a short time interval, perhaps only a few minutes. This procedure is called *pulse-labeling*. Only those molecules that are being freshly synthesized in the cell during the "pulse" will incorporate radioactive atoms. Autoradiography will then show which cells were active.

When autoradiography is applied to chromosomes or other subcellular structures, the matter of resolution becomes very important. For high resolution to be obtained, the radioactive particles should have a short range within the photographic emulsion; the black dots of silver in the developed film should pinpoint the source of radioactive decay as precisely as possible. Tritium works very well because it emits low-energy beta particles, which travel only a few millimeters in the emulsion, producing a well-localized image on the film. Radioactive carbon 14 emits higher-energy beta particles, so the silver grains in the film are more diffuse and the resolution is not as high.

Autoradiography and Electrophoresis

Autoradiography also has been very useful in biochemistry research when combined with the methodology of *electrophoresis*. Living cells are exposed to radioactively labeled amino acids, which are gradually absorbed into the proteins. For electrophoresis, the cells are transferred to a gel to which a voltage has been applied. The protein molecules will diffuse along the gel and be sorted out by their relative molecular weights. A photographic emulsion then is placed over the gel. Radioactivity from the proteins exposes the film, producing an image with black spots that show the distances that the different molecules drifted in the gel. The relative molecular weight of complex molecules that contain many thousands of atoms can be determined in this way.

One commercial catalog of radioactive materials lists many hundreds of organic chemicals that have been labeled with radioactive tritium. The other most common radioactive isotopes used for autoradiography are carbon 14, phosphorus 32, and sulfur 35. A large inventory of labeled chemicals has become available for the continuing use of radioactive tracers in biological research.

Applications

Autoradiography has been used in biology on the macroscopic level to study the uptake of radioactive tracers by both plant leaves and animal organs. Since the 1960's the technique has been applied to successively smaller structures, such as individual cells, chromosomes and organelles within a cell, strands of DNA, and protein molecules. It is easier to understand the microscopic applications after first looking at a large-scale example.

In one experiment, bean plants were grown in a nutrient solution containing radioactive phosphorus. The phosphorus moved from the roots to the leaves as expected, shown by an autoradiograph of a leaf pressed against photographic film. When the bean plant is allowed to continue growing in a nonradioactive solution, autoradiography shows that radioactive phosphorus is withdrawn from older leaves and translocated to new leaves and buds. Evidently, nutrients not only travel up from the roots but also move around the plant. In another experiment, a solution containing phosphorus was sprayed directly onto the leaf surface and was shown to migrate away from it. Redistribution of nutrients on an even larger scale takes place in deciduous trees, where as much as 90 percent of some minerals are withdrawn from leaves before they fall.

Practical Applications

In agricultural research, the effectiveness of herbicides, insecticides, and fertilizers is studied to determine which ones can increase productivity without causing serious environmental problems. Radioactive phosphorus can be used in this regard to study plant metabolism. The uptake of iron or zinc from the soil and their circulation in a plant can be studied to ascertain the effect of soil acidity and chemical form. Sometimes the presence or absence of other elements can inhibit translocation of an essential nutrient. New plant growth regulators may move from one plant through the soil to a nearby untreated plant. Autoradiography is an important analytical technique for observing the route of micronutrients and discovering what factors can change their mobility in a plant.

The sequence of bases in DNA molecules can be decoded by using electrophoresis combined with autoradiography, and the study of DNA sequences is crucial to research in many diverse areas of biology. Although alternatives to using autoradiography in DNA sequencing are now common, autoradiography is still a standard technique used in many other aspects of molecular biology.

Hans G. Graetzer, updated by Bryan Ness

See also: Electrophoresis.

Sources for Further Study

Becker, Wayne M. *The World of the Cell*. 4th ed. San Francisco: Benjamin/Cummings, 2000. A college-level textbook gives an introductory overview of cell biology, followed by sections on energy, membranes, information, and specialization in cells. Maintains a good balance between description of experimental methods and discussion of results.

Prescott, David M. *Cells*. Boston: Jones and Bartlett, 1988. A college-level textbook written by a past president of the American Society for Cell Biology. The technique of combining autoradiography with high-resolution electron microscopy is described, with diagrams included. Replication of the DNA molecule in a chromosome and RNA synthesis in the cell nucleus are shown in a sequence of photographs.

Wharton, John, and Julia M. Polak, eds. *Receptor Autoradiography: Principles and Practice*. New York: Oxford University Press, 1993. A practical guide to the methodology and application of autoradiography. Numerous examples illustrate both the benefits and the potential pitfalls of the technique. Although it concentrates on uses of autoradiography in animal and human subjects, the introductory chapters cover general topics that should be of interest to botanists.

BACTERIA

Categories: Bacteria; medicine and health; microorganisms; taxonomic groups

The earliest organisms to appear on earth, bacteria form one of the three domains of life in the three-domain system. They are structurally the smallest, physiologically the simplest, and ecologically the most widespread and abundant of all living organisms.

Although only a few thousand species are known, Bacteria are often extremely abundant; a handful of soil or a spoonful of the organic muck at the bottom of a pond may hold several million bacteria. So tiny are these smallest of living organisms that a quarter of a million of them can be squeezed into the period at the end of this sentence. Bacteria are undoubtedly the most widely distributed of all organisms in nature. They occur everywhere: in soil, water, and in the air, as well as within the bodies of virtually all plants and animals.

Bacteria represent the earth's earliest, and in many ways the earth's simplest, living organisms. Fossils of bacteria have been found in rocks estimated at 3.5 billion to 3.7 billion years old. Thus, bacteria predate eukaryotic cells by at least a billion years.

Although small and simple, bacteria are ecologically and economically among the most important of all living organisms. They are the major organisms of decay and decomposition, fermentation, and nitrogen fixation. Activities such as spoiling of food, rotting of flesh, decomposition of plants, and causing of disease are all evidence of bacterial activity; although the smallest, bacteria are hardly the safest of organisms. While the vast majority of bacteria are harmless or helpful, a small percentage causes serious and often fatal diseases in humans, other animals, and plants.

Algae and Bacteria

Bacteria are traditionally included in botany studies because they show a number of structural and physiological similarities to algae. One group, the blue-green bacteria (formerly called the blue-green algae), contain chlorophyll and manufacture food via photosynthesis. The bacterium *Prochloron* exemplifies the prochlorophytes, which are a group of photosynthetic bacteria that have chlorophylls *a* and *b* and carotenoids, which are precisely the same pigments that are found in green algae and plants. Other similarities to plants include flagella, found in some bacteria, that are similar to the paired flagella found in some algae but unlike the flagella characteristic of some of the simplest animals, the protozoa.

Characteristics and Classification

All bacteria are single-celled prokaryotic organisms (a prokaryote is a cell that lacks a nucleus). Traditionally, there are two groups of prokaryotic bacteria, the *Archaea* or *Archaebacteria* (ancient bacteria), and the true bacteria, which are classed as *Bacteria* or *Eubacteria*. Both groups comprise minute organisms that vary in size from about 0.2 to 10 micrometers in diameter. In comparison, eukaryotic cells range from about 10 to 100 micrometers in diameter.

Bacteria number some four thousand to five thousand species, but evidence based on recent deoxyribonucleic acid (DNA) studies indicates that many more species probably await identification. Part of the problem in discovering new species of bacteria is their lack of sexual reproduction, which is often an important criterion used in identifying species in other organism groups.

Known bacteria are classified on the basis of shape, DNA or ribonucleic acid (RNA) arrangements, locomotion, pigments, and staining properties. Taxonomists increasingly rely on DNA and RNA comparisons to identify new species and determine relationships among bacteria.

In the five-kingdom classification scheme, all bacteria are placed in the kingdom *Monera*, which consists of all of the prokaryotic animals. In the three-domain classification of life, the bacteria are placed in the domain *Bacteria*, while the other group of prokaryotes, the *Archaebacteria* (ancient, simple bacteria thought to represent earliest forms of life but now limited in distribution) are placed in the domain *Archaea*. All other living organisms, including the protistans, fungi, plants, and animals, are classed in the domain *Eukarya* (because they are formed by eukaryotic cells, that is, cells that have a nucleus).

Structure

Bacteria are single-celled organisms encased within a cell wall and lacking a central nucleus. They have a relatively homogenous cytoplasm devoid of membrane-bound organelles, such as Golgi bodies, endoplasmic reticulum, mitochondria, or plastids, but sites along the cell membrane perform some of these organelle functions. Bacteria do have ribosomes for protein synthesis, but these are about half the size of ribosomes that are found in the eukaryotic cells of higher plants and animals.

The nucleic acid of bacterial DNA occurs as a single naked, ringlike form called a nucleoid, which is attached to the cell wall. Bacteria also have small rings of DNA called plasmids, which are dispersed within the cytoplasm. The plasmids replicate separately from the nucleoid, suggesting that they were originally separate organisms that were somehow incorporated into the bacterial system.

The cell walls of bacteria help maintain shape and provide rigid protection. Unlike cell walls of plants, which are composed mostly of cellulose, bacterial cell walls contains *peptidoglycans*, which consist of large amino acid molecules cross-linked to molecules of polysaccharides. Antibiotics, such as penicillin, work by inhibiting cell wall formation and repair, thereby destroying the bacterial cell.

Reproduction

Most bacterial reproduction is asexual, with individual cells undergoing a form of mitosis called binary fission, during which the nucleoid and plasmids replicate themselves and then migrate to opposite ends of the dividing cell. The cell wall pinches inward to form an interior wall that separates the dividing cell into the two new daughter cells. Under ideal conditions, binary fission can take place every ten to twenty minutes.

Some species can also exchange genetic material by a process called *conjugation*. Conjugation occurs when two compatible bacteria come into close contact with each other. The cell wall of one of the two bacteria evaginates (grows outward) toward its partner and merges with it to form a hollow, connecting tube called a pilus. When the tube is complete, the DNA segment migrates through the tube to the recipient cell, where it merges with and becomes part of the recipient cell's nucleoid.

Another process of genetic transfer occurs when a bacterium picks up fragments of DNA released by fragmented dead bacterial cells and incorporates the fragments into its own nucleoid. The absorbed DNA increases the genetic variability of the bacterial cell, and new characteristics may result from the interactions of the original genetic material with the newly acquired genetic material.

Shapes

Bacteria occur in three basic body forms, which also serve as a simple method for recognizing and classifying them. *Cocci bacteria* are round, elliptical, or spheroid in shape. Rod-shaped bacteria are called *bacilli*. The third bacterial form, called *spirilla*, are greatly elongated coils and are often corkscrew-shaped. Spiral bacteria are often twisted in the form of a helix or spiral. Cocci bacterial cells may occur in several different forms; in irregular clusters they are called staphylococci, which cause the well-known staph infections. Those in filimentlike chains or beadlike chains are called streptococci, which cause strep throat, while cocci that occur in pairs are called diplococcus. In all of these forms, each cell is completely independent.

Sheaths and Surfaces

Bacteria can develop slimy or gummy capsule-like sheaths around their cells, which, in addition to body shapes and pigments, helps taxonomists to classify them. Outside the cell walls of some bacteria are sticky capsules, or slime layers, that are made up of polysaccharide or protein. These capsules help certain bacteria that cause disease avoid being detected by an animal's immune system.

Some bacteria have a mass of hairlike projections called *pili* covering their surface. They are made of protein, and they generally function to attach to the bacteria of other cells. The tubelike pili found in some bacteria serve as attachment structures that enable the bacteria to remain fastened in place to a suitable substrate. Infectious bacteria, such as the bacteria involved in the sexually transmitted disease gonorrhea, use the pili to attach to the cell membranes of the host, causing infection.

Motility

Although many bacteria are nonmotile, some filamentous bacteria have slender flagella that slowly rotate, propelling them through the medium in a spiraling glide. Flagella help bacteria to move into new habitats, to follow nutrients, and to leave environments that are nonbeneficial. Flagellated bacteria may move toward or away from stimuli, a behavior which is known as a *taxis*. Bacteria that move in response to chemicals in their environment are called chemotactic; those that move toward or away from light are *phototactic*, while *magnetotatic* bacteria respond to the earth's magnetic field. Occurring in aquatic habitats, the magnetotatic bacteria are able to detect the earth's magnetic field using iron crystals within their cytoplasm that act as tiny magnets.

If environmental conditions become unfavorable, some bacteria form structures called endospores, which consist of genetic material along with a few enzymes enclosed inside a thick protective layer. Endospores can survive for long periods of time in extremely unfavorable conditions and are also important dispersal mechanisms because they can travel for long distances in the air or water, then produce new bacteria quickly as soon as they find conditions that are favorable.

Nutrition

A major reason for the success of the bacteria is in their various forms of nutrition. All bacteria are either autotrophic (able to manufacture their own food) or heterotrophic (obtaining food by feeding on plants or animals or their remains). The majority of heterotrophic bacteria are saprobes that obtain food directly from the environment by absorbing it across the cell wall, but some are important parasites that cause disease. Other types of bacteria are *chemosynthetic*, gaining energy through reactions that combine oxygen with inorganic molecules, such as sulfur, ammonia, or nitrite. During the process, they release sulfates and nitrates, crucial plant nutrients, into the soil.

Certain types of bacteria have the ability to break down cellulose, the primary component of plant cell walls. Some of these types of bacteria have formed a symbiotic relationship with mammals called ruminants, which include deer, sheep, and cows. The cellulytic bacteria manufacture and release cellulose-digesting enzymes. They are considered symbiotic because they provide enzymes that enable the animal to digest food it would otherwise be unable to process. In turn, they inhabit an optimum environment deep within the animal's stomach and have nutrients supplied directly to them. Other symbiotic bacteria live in the intestines of humans and other animals. They feed on undigested food passing through the gut and synthesize vitamins K and B_{12}, which are absorbed into the human body.

Another extremely important group of symbiotic bacteria are the *nitrogen-fixing bacteria*, which are one of the very few groups of organisms able to extract molecular nitrogen from the atmosphere and incorporate it in organic compounds. This ecologically and economically important group grows within root nodules—small, rounded clumps that cluster along the roots of certain plants, such as alfalfa, soybeans, lupines, and clover.

Archaea

Classed as either the *Archaebacteria* or *Archaea* and considered to be the most primitive prokaryotes, the *Archaea* differ from true bacteria (in some classification schemes) in the unique structure of their RNA molecules, in lacking muramic acid in their cell walls, and in the production of distinctive lipids. The *Archaea* are usually divided today into three groups: methane bacteria, salt bacteria, and thermophilic bacteria.

Methane bacteria are the most diverse *Archaea*. They are anaerobic organisms that live in the mud of swamps and marshes, in the murky debris of ocean floors, hot springs, lake sediments, animal intestines, sewage treatment plants, and other areas in which free oxygen is not available. Methane bacteria are so-named because they metabolically generate methane gas from carbon dioxide and hydrogen during their energy-producing process.

The salt bacteria, or *halophiles*, are another ancient group that today are mostly confined to life in shallow, saltwater evaporation ponds common in parts of the western United States. When abundant, salt bacteria can give these ponds a distinctive red color. Their metabolism enables them to thrive under conditions of extreme salinity. Salt bacteria carry on simple photosynthesis using a red pigment called bacterial rhodopsin.

The *thermophiles*, or heat-loving bacteria, consist mostly of the sulfur-metabolizing bacteria that

Some Bacterial Diseases of Plants

Bacterium	Diseases
Argobacterium	Cane gall, crown gall, hairy root, twig gall
Clavibacter, Rhodococcus	Fasciation, spots, ring rots, tomato cankers, wilts
Erwinia	Blights, wilts, soft rots
Pseudomonas	Banana wilts, bud blasts, cankers, leaf spots, lilac blights, olive galls
Xanthomonas	Black venation, bulb rots, citrus cankers, cutting rots, walnut blights
Rhizobium	Root nodules (legumes)
Streptomyces	Potato scabs

Source: Data are from Peter H. Raven et al., *Biology of Plants*, 6th ed. (New York: W. H. Freeman/Worth, 1999).

thrive in sulfur hot springs, thriving at temperatures ranging from 80 degrees Celsius to 100 degrees Celsius. Some of the thermophilic bacteria are heterotrophic, but most are chemosynthetic and obtain their metabolic energy by forming hydrogen sulfide from elemental sulfur and water.

Cyanobacteria

The true bacteria have muramic acid in their cell walls. The majority are heterotrophic or parasitic, but some very important groups are autotrophic. Two groups of true bacteria are recognized, eubacteria and the cyanobacteria or blue-green bacteria.

The *cyanobacteria*, or blue-green bacteria, formally known as the blue-green algae, are different from true algae as they are prokaryotic, while all of the green algae are eukaryotes. Blue-green bacteria possess chlorophyll *a*, which is also found in green plants. Many blue-green bacteria also have a blue pigment called phycobilin, a red pigment called phycoerythrin, and carotenoids, which help gather light energy for photosynthesis. Their distinctive blue-green color is caused by the combination of chlorophyll and phycocyanin pigments. Blue-green bacteria are the only organisms that can fix nitrogen and produce oxygen at the same time, producing a nitrogenous food reserve called cyanophycin. They are also able to produce and store carbohydrates and lipids.

Blue-green bacteria occur in chains or hairlike filaments. Some form irregular, spherical, or platelike colonies held together by gelatinous sheaths. These sheaths may be colorless or pigmented with shades of yellow, red, brown, green, blue, violet, or blue-black. Blue-green bacteria lack flagella but move by rotating on their longitudinal axis, which gives them a forward-gliding motion.

Blue-green bacteria occur in soil, in water, on moist surfaces, and in root nodules of plants. They are common in temporary pools and ditches and often very abundant in freshwater habitats. Blue-green bacteria are among the first invaders of newly formed habitats, such as the ash fields around volcanoes and newly opened fissures along deep-sea volcanic ridges and mounts. They are even found in tiny fissures in desert rocks. Some species occur in jungle soils and on the shells of turtles and snails. Others live as symbionts within amoebae, protozoans, diatoms, sea anemones, fungi, and the roots of tropical cycads.

Over billions of years, the photosynthetic activity of blue-green bacteria transformed the oxygen-free early atmosphere into the modern atmosphere, in which oxygen plays such an important role for all higher plants and animals. The accumulation of oxygen in the upper atmosphere produced the high-altitude ozone layer, which shields animals and terrestrial plants from the damaging effects of ultraviolet radiation.

Ecologically, blue-green bacteria form the base of many food chains, especially in freshwater and marine habitats. During the warm months of the year, blue-green bacteria can temporarily become abundant and form floating mats of pond scum that often cover quiet waters of ponds and wetlands in late summer. The sudden increase of populations produce algal blooms (eutrophication) that cause massive die-offs of plant and animal populations because the bacteria populations consume all of the available oxygen in the water, thereby asphyxiating other organisms. In reservoirs and other human water supplies, large populations of blue-green bacteria clog filters, corrode steel and concrete structures, and cause a natural softening of water.

They produce odors and discoloration which sometimes make water unpalatable.

Eubacteria

The *eubacteria* include the unpigmented bacteria, purple bacteria, green sulfur bacteria, and the prochlorophytes. The eubacteria also include two groups of uncertain relationships: the mycoplasmas and the prochlorophytes. Most eubacteria are heterotrophic saprobes that absorb food directly from their environment. Many of these bacteria are extremely important decomposing organisms which, along with soil fungi, are responsible for the decay of all types of organic matter and the release of minerals back into the soil.

A few eubacteria are autotrophic. The purple sulfur bacteria, the purple nonsulfur bacteria, and the green sulfur bacteria all are photosynthetic bacteria that use sunlight energy to fragment hydrogen sulfide into sugar, water, and sulfur. Chemoautotrophic bacteria obtain energy by oxidizing compounds of iron, hydrogen, and sulfur. Some of the true bacteria are parasitic, living in or on living organisms and depending on them for food. Many of these parasitic forms are responsible for serious diseases in humans and other animals.

Beneficial True Bacteria

The number of true bacteria that are beneficial organisms for humans is much greater than the number of disease-causing bacteria. The use of bacteria as biological control agents is typified by *Bacillus thuringiensis* (*B.t.*), bacteria available in garden shops as a spray or powder. Placed on plants, it kills caterpillars and worms. It is sold as a mass-produced, stable, moist dust containing millions of spores of the bacteria. It is harmless to humans, birds, earthworms, and other creatures, except moth or butterfly larvae. When the caterpillar ingests the bacterial spores, they develop into bacilli, which multiply in the digestive tract and paralyze the gut of the worm. The crop pest usually dies as a result of ingestion in two to four days. Other varieties, such as *Bacillus thuringiensis* (variety *israelensis*, or *B.t.i.*),is used to control mosquitoes, while Japanese beetles can be controlled by application of a bacterial power containing *Bacillus popilliae*, which is specific to this beetle.

Other bacteria have important uses in the service of humans. Some are used in bioremediation projects, such as the cleanup of oil spills, sewage treatment plants, and toxic waste dumps. *Pseudomonas capacia*, for example, is used to decompose oil spills. More benefits from bacteria are being developed using genetic engineering techniques.

Bacteria also play a major role in the dairy industry as cultures in the production of buttermilk, acidophilus milk, yogurt, sour cream, kefir, and cheese. Whey, the watery part of milk left in cheese production, is used in the manufacture of lactic acid. Lactic acid from lactate bacteria is also used in the textile industry, in the preparation of laundry products, in leather tanning, and in the treatment of calcium and iron deficiencies.

Bacteria are cultured in vats and used to manufacture chemicals, such as acetone, butyl alcohol, dextran, sorbose, citric acid, some vitamins, and medicinal preparations. Bacteria are also used to cure vanilla pods, cocoa beans, coffee, and black tea. They are used in the production of vinegar, sauerkraut, and dill pickles. Fibers from linen cloth are separated from flax stems by bacterial action. Green plant material is fermented in silos, providing food for livestock through the action of bacteria. Lastly, bacteria producing the amino acid glutamic acid are used to produce monosodium glutamate (MSG), a common food ingredient.

Plant-Pathogenic True Bacteria

Almost all plants are susceptible to a host of bacterial diseases which cause infections, rotting, and death. Economically, the plant-pathogenic bacteria result in enormous losses of crops and other plant products. It has been estimated that about one-eighth of crops are annual losses to plant diseases.

Plant diseases caused by bacteria include galls, rots, wilts, spots, fascination, blights, and soft rots. *Blights* are caused when bacteria invade plant tissues of stems, leaves, and flowers, producing dead and discolored areas of infection. Fruits and vegetables not completely destroyed are sufficiently discolored to be unmarketable. *Soft rots* typically occur in fleshy storage organs, such as potatoes, eggplants, squashes, and tomatoes. Bacterial infections that cause drooping or wilting of plant tissues are called *wilts*. Wilts occur when bacteria colonize xylem vessels, where they block or interfere with water transport, eventually leading to dysfunction and destruction of the plant. *Galls* are plant swellings produced when bacteria and other organisms invade leaves and stems and lodge in parenchyma tissue. The infected tissue swells to produce a gall that encases the bacterial colony.

Human-Pathogenic Bacteria

Although bacteria do provide some benefits with their feeding habits, certain bacteria have feeding habits that threaten the health of humans. The largest threats stem from bacterial infection. These disease-producing bacteria, which are scientifically referred to as pathogens, synthesize toxic substances that cause disease symptoms. For example, bacteria such as *Clostridium tetani* and *Clostridium botulinium*, which cause, respectively, tetanus and botulism, the latter a deadly form of food poisoning. These bacteria produce toxins that attack the nervous system. Such bacteria are known as anaerobes, and they thrive as spores until they are introduced into a desirable environment. Tetanus enters the body through a deep puncture wound, protecting the bacteria from having contact with oxygen. As the bacteria multiply, they release their poison into the bloodstream.

Bacteria can enter the human body from the air. Every time someone coughs, sneezes, or speaks loudly, they produce an invisible spray of saliva droplets containing bacteria. The fluid around these bacteria quickly evaporates, but the bacteria cling to protein flakes that were also expectorated and enter the lungs during breathing. Once inside the lungs, bacteria access the circulatory system, via which they are transported to tissues and organ systems. Legionnaire's disease is caused by a bacterium that lives in small amounts of water in air conditioning systems. It can be transmitted throughout an entire building by airborne particles that are blown through the air conditioning ducts. In a particularly deadly form of infection, anthrax bacilli can be inhaled, lodge in the lungs, and quickly kill the host.

Bacteria also gain access to the body through the ingestion of contaminated food and water. Bacterial infections caused by eating contaminated food are widespread, especially in the less developed countries of the world. Open sewers and unsanitary toilet conditions increase the risk of waterborne bacterial diseases, such as cholera, dysentery, and food poisoning, such as salmonella. Bacteria may also be ingested via improperly stored foods, such as raw chicken, shellfish, and eggs. Once ingested, the bacteria multiply in the intestinal tract and can be passed with urine.

Some bacteria gain access to the body through direct contact. Most of the sexually transmitted diseases are caused by bacteria. Examples include syphilis and gonorrhea. Other diseases, such as anthrax and brucellosis, enter the body through the skin or mucous membranes. Contact anthrax (less deadly than inhalational anthrax, if treated quickly) is a disease of cattle and other farm animals which occasionally infects humans. It is transmitted to workers in the tanning industry who handle hides and wool. Brucellosis is also a disease from farm animals and is transmitted by contaminated milk. It is sometimes called undulant fever and is characterized by a daily rise and fall of temperature associated with the cyclic release of the toxins by the bacteria.

Other bacteria that live in the soil enter the body through wounds. The tetanus or "lockjaw" bacterium, mentioned above, is a common soil organism. Puncture wounds caused by stepping on dirty nails and other sharp devices introduce the tetanus bacteria deep into the body. Once inside, it produces toxins that are extremely powerful. Tetanus is easily controlled by immunizations that are developed from an attenuated horse serum. Some soil bacteria can cause a disease called gas gangrene. These bacteria respire anaerobically in the body and basically destroy tissues while emitting damaging gases (hence their name). If untreated, a serious infection by gas gangrene bacteria can result in the loss of a limb.

Many bacteria gain access to the human body through the bites of insects and other organisms. If an organism transmits a disease-causing organism to a different host, the first organism is called a *vector*. Many of the most feared and deadly of human diseases are caused by bacteria that are transmitted by vectors. Bubonic plague (black death) and tularemia are transmitted by fleas, deer flies, ticks, or lice. Of these, the most famous—and historically the most deadly—is bubonic plague, which is transmitted by fleas of rats and other rodents. Throughout recorded history, periodic visitation of plagues resulted in catastrophic die-off of thousands and even millions of people. The populations of Thebes, Athens, Rome, Vienna, and other early cities suffered and survived many plague years. The most potent plague years were in the late Middle Ages, when nearly a third of the European population succumbed to the bacterial disease. Bubonic plague is still found today in the United

States in ground squirrel populations and other rodents as well.

Mycoplasmas and Other "Small" Bacteria

The *mycoplasmas*, *rickettsiae*, and *chlamydiae* are small and simple prokaryotes that are sometimes collectively labeled "small bacteria." Of these, the mycoplasmas are the smallest of all bacteria and therefore the smallest of all living organisms, since viruses do not qualify as life. Mycoplasmas are distinctive in that, unlike other bacteria, they lack a cell wall. The relationships and diversity of mycoplasmas are still poorly understood, but the mycoplasmas have the distinction of being the smallest known organisms to cause human disease. One form causes a sexually transmitted disease and another, *Mycoplasma pneumonia*, causes a form of pneumonia commonly called walking pneumonia.

The *rickettsiae* are tiny bacteria that were first described by Howard Ricketts in 1909. *Rickettsiae* species are mostly transmitted by animal vectors, such as ticks and lice, and cause typhus and Rocky Mountain spotted fever. The *chlamydiae* are usually considered to be a subgroup of rickettsiae. Some *chlamydiae* are airborne diseases, among them *Chlamydia psittaci*, which causes parrot fever, or psittacosis. Humans are exposed to parrot fever by inhaling the dried droppings or dust from infected birds. Another airborne chlamydial disease of humans is transmitted by respiratory droplets containing *Chlamydial pneumonia*. This causes a mild form of walking pneumonia in humans. Other chlamydia cause sexual diseases which resemble gonorrhea and can result in serious infections of the urogenital tract if undetected or left untreated.

Prochlorophytes

Also designated as the *prochlorobacteria*, these bacteria were discovered in 1976 by Ralph Lewin of the Scripps Institute of Oceanography. The prochlorophytes are of considerable scientific interest because they have photosynthetic pigments (chlorophyll *a* and *b*) that are found in higher plants, thereby hinting at possible relationships between this group of bacteria and higher plants. In structure and characteristics, the prochlorophytes are closely related to true bacteria, however.

Genetic Engineering with Bacteria

Since they are relatively simple organisms that can easily be cultivated in large numbers, bacteria are one of the most important of all groups of organisms used in scientific experiments. As a result, they have been the subject of much genetic engineering, which is the artificial introduction of DNA into bacteria to change their characteristics. Genetic engineering is also called gene splicing and begins with the isolation of plasmid DNA from a bacterium. The DNA molecule is broken up into separate genes or groups of genes by special enzymes called restriction enzymes. The resulting DNA fragments are then mixed with repair enzymes and injected into another bacterium, which absorbs the new genes and functions in a new way under the influence of the new genes or combination of genes.

Dwight G. Smith

See also: Algae; Anaerobic photosynthesis; *Archaea*; Biotechnology; Chemotaxis; DNA: recombinant technology; Eutrophication; Food chain; Genetically modified bacteria; Halophytes; Nitrogen fixation; Photosynthesis; Prokaryotes.

Sources for Further Study

Alcamo, I. E. *Fundamentals of Medical Microbiology*. 6th ed. Sudbury, Mass.: Jones and Bartlett, 2001. Excellent introduction to the classification and biology of prokaryotes in general and the bacteria in particular.

Balows, A., et al., eds. *The Prokaryotes*. 2d ed. 4 vols. New York: Springer-Verlag, 1992. This four-volume set is the ultimate reference about prokaryotes. Topics include taxonomy, ecology, physiology, isolation, classification, diseases, and applications.

Black, Jacquelyn G. *Microbiology: Principles and Explorations*. 5th ed. New York: John Wiley & Sons, 2002. This is but one of many college-level textbooks. Provides good coverage of bacteria along with an excellent treatment of the techniques needed to study bacteria.

Brooks, G. F., J. S. Butel, and S. A. Morse. *Jawetz, Melnick, and Adelberg's Medical Microbiology*. Stamford, Conn.: Appleton and Lange, 1998. The first section of this book deals primarily with bacterial aspects of medical microbiology.

Salyers, A. A., and D. D. Whitt. *Bacterial Pathogenesis*. Washington, D.C.: ASM Press, 1994. All aspects regarding the onset, cure, and prevention of bacterial diseases are thoroughly discussed in this textbook.

BACTERIAL GENETICS

Categories: Bacteria; disciplines; genetics; microorganisms; reproduction and life cycles

Bacterial genetics is the study of the genetic material of bacterial DNA, which can provide valuable insights into the process of mutation because of bacteria's rapid rate of reproduction.

Plants were the original candidates for genetic studies, which began in the late 1800's. Studies with animals soon followed; bacteria did not become candidates for such study until the mid-1940's, when adequate technology for handling bacteria developed. Bacteria have become extremely useful organisms for genetic studies since the early 1950's. Two major features of bacteria make them desirable subjects. First, bacterial cells typically divide every twenty minutes. Their rapid rate of reproduction allows a very large number of bacteria to be produced in a short time. This, in turn, provides the researcher with more opportunity to detect the "rare genetic events" of *mutation* or *recombination*. Even more important, unlike all other organisms, bacteria have a single *chromosome* with a single set of *genes*. Thus, genetic modifications are more likely to result in immediately observable changes. In organisms that have multiple chromosomes, a change in a single gene may go undetected because its effect is masked by genes on other chromosomes.

Bacterial DNA

All bacteria have a single circular chromosome, composed of deoxyribonucleic acid (DNA). The DNA is subdivided into specific message areas known as genes, and the chromosome carries from four thousand to five thousand individual genes. For many bacteria, this constitutes the entirety of its genetic information. A number of bacteria, however, have additional DNA in the form of *plasmids*. A plasmid is a small additional circular piece of DNA, independent of the chromosome, which can hold an additional twenty to one hundred genes. Plasmid-containing cells often have several plasmids.

Many researchers have described the plasmid genes as nonessential to the normal activities of bacteria. Under certain circumstances, however, those genes might provide a survival advantage to the possessor. For example, genes for antibiotic resistance are often carried on a plasmid. Normally, antibiotics are not present in the bacteria's environment; such resistance genes would therefore be unnecessary. If the bacteria later were to come into contact with antibiotics, however, having antibiotic-resistant genes would be to their distinct advantage.

Two major types of plasmids exist: F plasmids, or *fertility plasmids*, and R plasmids, or *resistance plasmids*. Both types can carry resistance genes. Only the F plasmids, however, are able to control the formation of a special cytoplasmic tube known as the sex pilus. Cells with the F plasmids are known as F+, or donor cells. Cells without the F plasmids are called F-, or recipient cells.

Conjugation

The plasmid is a prerequisite to one type of genetic exchange, conjugation. During conjugation, the donor cell copies its plasmids and transfers them to a recipient cell to which it has attached itself by means of a sex pilus. The recipient cell can now take advantage of whatever additional genes it has received. If, in the process, it received an F plasmid, it has also become a potential donor cell. Whenever bacterial cells undergo cell division, any plasmids they possess are typically passed on to their progeny. Originally it was thought that conjugation could occur only between members of the same species, but that is not always true. For example, it is now known that some strains of the bacteria responsible for causing gonorrhea, *Neisseria gonorrhoeae*, have received antibiotic-resistant genes from unrelated species of bacteria.

There is one other type of donor cell, the Hfr+, or high-frequency recombinant, cell. Instead of the plasmid remaining independent of the cell's chromosome, it inserts itself into the chromosome. When that plasmid gets ready to copy itself, the chromosomal genes are the first to be copied. Unless the donor and recipient cells are able to maintain direct contact for a fairly long period of time, which almost never occurs, the recipient cell will not receive the plasmid. It will, however, receive numerous chromosomal genes from the donor. Those genes may later be incorporated into the chromosome of the recipient, causing gene replacement.

Not all species of bacteria participate in conjugation. Some rely on transduction as a means of receiving new genetic information. This is how *Staphylococcus aureus* has developed resistance to many antibiotics.

Transduction

There are two types of transduction: generalized and specialized. In both cases, a donor cell becomes infected with a *bacteriophage*, a virus that attacks bacteria. Upon the death of that donor cell, fragments of donor DNA are transferred as the escaping bacteriophage infects another bacterium.

In generalized transduction, a bacteriophage infects a bacterial cell. Shortly after infection, the bacterial chromosome becomes fragmented, and viral components are produced. Later the viral components are assembled to form a complete virus particle. Occasionally during this assembly process, a particle becomes contaminated with fragments of the bacterial chromosome or plasmids. After assembly is completed, the bacterial cell ruptures, allowing the escape of all virus particles. Eventually these virus particles will invade other bacterial cells. Any cells that are invaded by contaminated bacteriophage particles are said to be transduced, because they have received DNA from another bacterium. The DNA received in this manner is strictly random.

Specialized transduction involves what is known as a *latent bacteriophage*. After the initial invasion of a bacterium, the bacteriophage inserts itself into a specific region of that cell's chromosome. At some later time, the bacteriophage removes itself from the chromosome and accidentally takes a few bacterial genes located near its original insertion point. When the bacterial cell finally begins making new bacteriophage components, it behaves as if those particular genes are part of the bacteriophage and replicates them as such. Therefore, all the newly formed bacteriophage particles will contain those bacterial genes. Transduction then occurs when these bacteriophage particles invade other bacterial cells.

Transformation

The final method of genetic transfer is transformation. An extensively utilized organism for such investigation has been *Streptococcus pneumoniae*. The most famous studies involved converting nondisease-causing strains of *Streptococcus pneumoniae* into disease-causing strains. Transformation also occurs in a wide variety of other bacteria. The process of transformation requires that a population of actively reproducing bacteria come into contact with DNA fragments, often from closely related dead bacteria. These DNA fragments are referred to as either naked or cell-free DNA.

Genetic Modification

A small portion of that DNA can be absorbed and utilized by the growing bacteria. These recipients can then take advantage of any usable genes that the fragments might contain, incorporating them into their chromosome in place of their own copies of these genes by the process of recombination.

Conjugation, transduction, and transformation are all mechanisms of genetic change within a bacterial population. These mechanisms allow a specific characteristic to be spread throughout the population within a few hours. A wide number of bacterial genes have been found to be transferred by these methods, including genes that control a bacterium's ability to cause disease, to produce toxins, and to develop resistance to antibiotics and other drugs as well as genes that control a number of other characteristics. The purpose of these mechanisms, as far as the bacteria are concerned, is to enable the bacteria to adapt to changing environmental conditions so that their survival is ensured. Scientists, however, have found ways to adapt some of these mechanisms for human benefit.

Scientists have used the mechanisms of genetic transfer along with new technology from DNA research to perform *genetic engineering* on bacteria. They can use genes and specially engineered plasmids, called plasmid vectors, to make recombinant DNA in the laboratory. Recombinant plasmids can then be used to transform bacteria such as *Escherichia coli* (*E. coli*). The bacteria will treat these recombinant plasmids just like ordinary plasmids, replicating them and, for expression vectors, expressing any genes included in them. In this manner, bacteria can be used to produce a wide variety of products for medicine, agriculture, and industry. Genetic engineering and the products that result from it would not be possible without the knowledge of genetic transfer gained from studies of bacterial conjugation, transduction, and transformation.

Randy Firstman, updated by Bryan Ness

See also: Bacteria; Bacterial resistance and super bacteria; Bacteriophages.

Sources for Further Study

Brock, Thomas D., and Michael T. Madigan. *Brock Biology of Microorganisms*. 9th ed. London: Prentice Hall, 2000. The chapter on microbial genetics covers kinds of genetic recombination and gene transfer, plasmids, genetic mapping, and general genetics of eukaryotic microorganisms. There are many illustrations and diagrams. Explanations are in-depth and tend to be technical. Aimed toward an intermediate college-level audience. Supplementary references are listed.

Lim, D. V. *Microbiology*. 2d ed. Boston, Mass.: McGraw-Hill, 1998. The chapter "Information Processing: Mutation and Recombination," discusses such topics as mutation, recombination, and genetic engineering. Includes good illustrations and diagrams. Explanations can easily be followed by anyone with an interest in science. Supplementary references are listed.

Tortora, Gerard J., Berdell Funke, and Christine Case. *Microbiology*. 7th ed. San Francisco: Benjamin Cummings, 2001. Includes discussions of DNA replication in general, general bacterial genetics, gene expression, genetic transfer and recombination, and genetic engineering. The chapter on microbial genetics is especially valuable. Well written and not overly technical, it has illustrations and diagrams. References are listed.

BACTERIAL RESISTANCE AND SUPER BACTERIA

Categories: Bacteria; environmental issues; medicine and health; microorganisms

Inappropriate use of antibiotics has caused bacteria to develop resistance to the most common antibiotics. Bacterial pathogens that have developed resistance to multiple antibiotics that previously could be used to control them are referred to as super bacteria.

Bacteria are the most adaptable living organisms on earth, found in virtually all environments—from the lowest ocean depths to the highest mountains. Bacteria can resist extremes of heat, cold, acidity, alkalinity, heavy metals, and radiation that would kill most other organisms. *Deinococcus radiodurans*, for example, grows within nuclear power reactors, and *Thiobacillus thiooxidans* can grow in toxic acid mine drainage. The term "super bacteria" generally refers to bacteria that have either intrinsic (naturally occurring) or acquired resistance to multiple antibiotics. Because many of the bacteria that acquire resistance are pathogens that previously could be controlled by antibiotics, development of antibiotic resistance is regarded as a serious public health crisis, particularly for those individuals who have compromised immune systems.

History of Antibiotic Use

In the early twentieth century German chemist Paul Ehrlich received worldwide fame for discovering Salvarsan, the first relatively specific prophylactic agent against the microorganisms that caused syphilis. Salvarsan had undesirable side effects because it contained arsenic. In addition, secondary infections resulting from hospitalization were still a leading cause of death in the early twentieth century. Scottish bacteriologist Alexander Fleming discovered a soluble antimicrobial compound produced by the fungus *Penicillium*. Two English scientists, Howard Florey and Ernst Chain, took Fleming's fungus and produced purified penicillin just in time for use during World War II.

Antibiotics such as penicillin are low-molecular-weight compounds excreted by bacteria and fungi. Antibiotic-producing microorganisms most often belong to a group of soil bacteria called *actinomycetes*. *Streptomyces* are good examples of antibiotic-producing actinomycetes, and most of the commercially important antibiotics are isolated from *Streptomyces*. It is not entirely clear what ecological role the antibiotics play in natural environments.

The success of penicillin as a therapeutic agent with almost miraculous effects on infection prompted other microbiologists to look for naturally occurring antimicrobial compounds. In 1943 Selman Waksman, an American biochemist born in Ukraine, discovered the antibiotic streptomycin, the first truly effective agent to control *Mycobacterium tuberculosis*, the bacterium that causes tuberculosis. Widespread antibiotic use began shortly after World War II and was regarded as one of the great medical advances in the fight against infectious disease. By the late 1950's and early 1960's, pharmaceutical companies had extensive research and development programs devoted to isolating and producing new antibiotics.

Antibiotics were so effective, and their production ultimately so efficient, they came to be routinely prescribed for all types of infections, particularly to treat upper respiratory tract infections. When it was discovered that low levels of antibiotics also promoted increased growth in domesticated animals, antibiotics began to appear routinely as feed supplements.

Development of Antibiotic Resistance

The widespread use and, ultimately, misuse of antibiotics inevitably caused antibiotic-resistant bacteria to appear, as microorganisms adapted to new selective pressure. There are now many strains of pathogenic organisms on which antibiotics have little or no effect.

Streptococcal infections are the leading bacterial cause of morbidity and mortality in the United States. In the mid-1970's *Streptococcus pneumonia* was uniformly susceptible to penicillin. However, penicillin-resistant strains were being isolated as early as 1967. A study in Denver, Colorado, showed that penicillin-resistant *S. pneumonia* strains increased from 1 percent of the isolates in 1980 to 13 percent of the isolates in 1995. One-half of the resistant strains were also resistant to another antibiotic, cephalosporin.

Tuberculosis was once the leading cause of death in young adults in industrialized countries, so common and feared that it was known as the White Plague. Before 1990 multidrug-resistant tuberculosis was uncommon. However, by the mid-1990's there were increasing outbreaks in hospitals and prisons, in which the death rate ranged from 50 to 80 percent. Likewise, multiple drug resistance in *Streptococcus pyogenes*, the so-called flesh-eating streptococci, was once rare. There are now erythromycin- and clindimycin-resistant strains.

Many old pathogens have become major clinical problems because of increased antibiotic resistance. The number of resistant isolates in England rose from 1.5 percent in 1989 to more than

The Growth of Antibiotic Resistance

Antibiotic	*Enterococcal Species*	*% of Resistant Bacteria*		
		1995	*1996*	*1997*
Ampicillin	*Enterococcus faecium*	69.0	77.0	83.0
	Enterococcus faecalis	0.9	1.6	1.8
Vancomycin	*Enterococcus faecium*	28.0	50.0	52.0
	Enterococcus faecalis	1.3	2.3	1.9

Note: Measurements taken over a three-year period indicate a general rise in enterococcal resistance to two common antibiotics.
Source: U.S. Centers for Disease Control.

34 percent in 1995. Gonorrhea, caused by *Neisseria gonorrhoeae*, is the most common sexually transmitted disease. Physicians began using a class of broad-spectrum cephalosporin antibiotics called fluoroquinolones because *N. gonorrhoeae* had become resistant to penicillin, tetracycline, and streptomycin. There is now real concern in the medical community that exclusive use of fluoroquinolones, the only antibiotics to which *N. gonorrhoeae* are routinely susceptible, has led to rapidly developing resistance. Even newly discovered pathogens such as *Helicobacter pylori*, which is associated with peptic ulcers, are rapidly developing resistance to the antibiotics used to treat them.

The development of resistance to some antibiotics appears to be linked to antimicrobial use in farm animals. Shortly after antibiotics appeared, it was discovered that subtherapeutic levels could promote growth in animals. One such antimicrobial drug, avoparcin, is a glycopeptide (a compound containing sugars and proteins) that is used as a feed additive. Vancomycin-resistant enterococci such as *Enterococcus faecium* were first isolated in 1988 and appeared to be linked to drug use in animals. Antibiotic resistance in enterococci has been more prevalent in farm animals exposed to antimicrobial drugs. Prolonged exposure to oral glycoproteins in tests led to vancomycin-resistant enterococci in 64 percent of the subjects.

Soap and Water

Even the use of household antibacterial agents contributes to bacterial resistance. Following are some everyday ways to deprive harmful bacteria of the chance to build up resistance and become even stronger, from the October, 1998, article "Antibacterial Overkill" in the *Tufts University Health and Nutrition Letter.*

- Use regular soap for hand-washing and regular dishwashing liquid in the kitchen, rather than cleaners containing an antibacterial agent. Soap loosens bacteria from dishes; running water then rinses them away. For the same reason, frequent hand-washing is the best way to get rid of unwanted bacteria.
- Cleaning products that contain bleach and chlorine disinfect (get rid of bacteria on) surfaces in the kitchen and bathroom. Because these cleaners are not antibacterial agents, however, they will not encourage the growth of resistant bacteria.
- Do not insist on antibiotics for a viral infection, such as the flu; they will not work. Antibiotics attack bacteria but not viruses. At a seminar conducted by Dr. Stuart B. Levy of Tufts University, more than 80 percent of the physicians present admitted to having written antibiotic prescriptions against their better judgment because of patient demands.
- Do not share prescription antibiotics with family or friends. Not knowing the dose they may need, or even whether they need any, could encourage the bacteria in their systems to develop resistance they might otherwise not have had.

How Antibiotic Resistance Occurs

Antibiotic resistance occurs because the antibiotics exert a selective pressure on the bacterial pathogens. This pressure eliminates all but a few bacteria that can persist through evasion or mutation. One reason antibiotic treatments may be prescribed for several weeks is to ensure that bacteria that have evaded the initial exposure are killed. Terminating antibiotic treatment early, once symptoms disappear, has the unfortunate effect of stimulating antibiotic resistance without completely eliminating the original cause of infection.

Mutations that promote resistance occur with different frequencies. For example, spontaneous resistance of *Mycobacterium tuberculosis* to cycloserine and viomycin may occur in 1 in 1,000 cells; resistance to kanamycin may occur in only 1 in 1 million cells; and resistance to rifampicin may occur in only 1 in 100 million cells. Consequently, one billion bacterial cells will contain several individuals resistant to at least one antibiotic. Using multiple antibiotics further reduces the likelihood that an individual cell will be resistant to all antibiotics used. However, it can cause multiple antibiotic resistance to develop in bacteria that already have resistance to some of the antibiotics.

Bacterial pathogens may not need to mutate spontaneously to acquire antibiotic resistance. There are several mechanisms by which bacteria can acquire the genes for antibiotic resistance from microorganisms that are already antibiotic-resistant. These mechanisms include *conjugation* (the exchange of genetic information through direct cell-to-cell contact), *transduction* (the exchange of genetic information from one cell to another by means of a virus), *transformation* (acquiring genetic information by taking up deoxyribonucleic acid, or DNA, directly from the environment), and transfer of *plasmids*, small, circular pieces of the genetic material DNA that frequently carry genes for antibiotic resistance.

Genes for antibiotic resistance take many forms. They may make the bacteria impermeable to the antibiotic. They may subtly alter the target of the antibiotic within the cell so that it is no longer affected. The genes may code for production of an enzyme in the bacteria that specifically destroys the antibiotic. For example, fluoroquinolone antibiotics inhibit DNA replication in pathogens by binding to the enzyme required for replication. Resistant bacteria have mutations in the amino acid sequences of this enzyme that prevent the antibiotic from binding to this region.

New Strategies

The increased use of antibiotics has led to increases in morbidity, mortality caused by previously controlled infectious diseases, and health costs. Some of the recommendations to deal with this public health problem include changing antibiotic prescription patterns, changing patient attitudes about the necessity of antibiotics, increasing the worldwide surveillance of drug-resistant bacteria, improving techniques for susceptibility testing, banning the use of antibiotics as animal feed additives, and investing in research and development of new antimicrobial agents.

Gene therapy is regarded as one promising solution to antibiotic resistance. In gene therapy, the genes expressing part of the pathogen's cell are injected into a patient and stimulate a heightened immune response. Some old technologies are also being revisited. There is increasing interest in using serum treatments, in which antibodies raised against a pathogen are injected into a patient to cause an immediate immune response. Previous serum treatment techniques have yielded mixed results. However, with the advent of monoclonal antibodies and the techniques for producing them, serum treatments can now be made much more specific and the antibodies delivered in much higher concentrations.

There have been numerous reports from Russia about virus treatment for pathogenic infections. Viruses attack cells in all living organisms, including bacteria, but are extremely specific, so that they will not infect more than one type of cell. In essence, virus treatments are a form of *biocontrol*. In virus treatments, the patient is injected with viruses raised against specific pathogens. Once injected, the viruses begin specifically attacking the pathogenic bacteria. Although this technology has not been widely used, it is the subject of growing research.

Mark S. Coyne

See also: Bacteria; Biopesticides.

Sources for Further Study

Amábile-Cuevas, Carlos F., ed. *Antibiotic Resistance: From Molecular Basics to Therapeutic Options*. New York: Chapman and Hall, 1996. Explains the latest research in antibiotic resistance, drawing largely on material presented at a December, 1995, conference held in Cuernavaca, Mexico, and concentrating on the natural forces behind the rapid emergence and spread of antibiotic resistance.

"Antibacterial Overkill." *Tufts University Health and Nutrition Letter* 16, no. 8 (October, 1998): 1-3. Explains why using bacteria-fighting versions of household products is misguided. Suggests strategies to lessen bacterial ability to develop resistance.

Biddle, Wayne. *A Field Guide to Germs*. New York: Anchor Books, 1996. A guide to understanding the infectious diseases that antibiotics are supposed to stop. An entertaining, irreverent, informative book on disease.

DeKruif, Paul. *The Microbe Hunters*. San Diego: Harcourt Brace, 1996. The stories of Paul Ehrlich's discovery of Salvarsan and Alexander Fleming's discovery of penicillin are enthusiastically portrayed.

Levy, Stuart B. *The Antibiotic Paradox: How Miracle Drugs Are Destroying the Miracle*. New York: Plenum Press, 1992. A discussion of the growing problem of resistance to antibiotics. The history and explanation of the cellular mechanisms of antibiotic function

and a bacteria strain acquiring resistance, are in nontechnical language.
Witt, Steven. *Biotechnology, Microbes, and the Environment*. San Francisco: Center for Science Information, 1990. Outlines the adaptability of microbes. An easy book to read, in which the mechanisms of microbial genetic transformation are clearly illustrated.

BACTERIOPHAGES

Categories: Diseases and conditions; genetics; microorganisms

Viruses that attack bacterial cells are known as bacteriophages. Many results gained from studying bacteriophages have universal implications. For example, the physical properties of DNA and RNA are remarkably identical in all organisms, and these are perhaps easiest to study in bacteriophage systems.

Bacteriophages, or phages for short, are viruses that parasitize bacteria. Viruses are an extraordinarily diverse group of ultramicroscopic particles, distinct from all other organisms because of their noncellular organization. Composed of an inert outer protein shell, or *capsid*, and an inner core of nucleic acid—either *deoxyribonucleic acid* (DNA) or *ribonucleic acid* (RNA) but never both—viruses are obligate intracellular parasites, depending to a great extent on host cell functions for the production of new viral particles.

There is considerable variation in size and complexity among viruses. Some have fewer than ten genes and depend almost entirely on host functions. Others are known to contain from thirty to one hundred genes and rely more on proteins encoded by their own DNA. Even the largest viruses are too small to be seen under the light microscope, so studies on viral structure rely heavily on observation with the transmission electron microscope.

The Study of Bacteriophages

Because scientists know more about the molecular and cell biology of the common bacterium *Escherichia coli* than about any other cell or organism, it is perhaps not surprising that the best-known phages are those that require *E. coli* as a host (*coliphage*). It is not possible to observe phage growth directly (as bacterial growth can be detected by the appearance of colonies on an agar plate), but phage growth can be indirectly observed by the formation of plaques, small clear areas in an otherwise continous lawn of host bacteria growing on a solid growth medium in a petri dish.

Reproductive Cycles

Bacteriophages can multiply by two different mechanisms, termed the *lytic cycle* and the *lysogenic cycle*. Some phages are capable only of lytic growth, while others retain the ability to reproduce by either lytic growth or entry into the lysogenic cycle. In the lytic cycle, phages first attach themselves to specific receptor sites on the host cell wall. The phage nucleic acid (DNA or RNA) is injected inside the host, while the protein capsid of the infecting particle remains outside of the host cell at all times. Once the DNA or RNA is inside, transcription of phage genes begins, and phage-encoded proteins begin to be made. Some of these proteins serve to inactivate and destroy the host cell DNA, ensuring that the cell's energy resources will be directed exclusively toward the production of phage proteins and the replication of phage nucleic acid. Phage DNA or RNA replication ensues quickly and is followed by the packaging of this genetic material into the newly synthesized capsids of the progeny phage particles. The final step is host cell lysis—the bursting of the host cell to release the completed and infective phage progeny. The number of phages released in each burst varies with growth conditions and species, but ideal conditions often result in a burst size of one hundred to two hundred per host cell.

For temperate bacteriophages, those capable of entering the lysogenic cycle, infection of the host cell only rarely causes lysis. Injection of the phage DNA into the host is followed by a brief period of *messenger RNA* (mRNA) synthesis, necessary to direct the production of a phage repressor protein, which inhibits the production of phage proteins involved with lytic functions. A DNA-insertion enzyme is also made, allowing the phage DNA to be physically inserted into the DNA of the host. The cell then can continue to grow and multiply, and new copies of the phage genes are replicated every cell generation as part of the bacterial chromosome. The host cell is said to be lysogenic, for it retains the potential to be lysed if the prophage pops out of the host DNA and enters the lytic cycle. The integrated prophage does confer a useful property on the host cell, however, for the cell will now be immune to further infection from the same phage species.

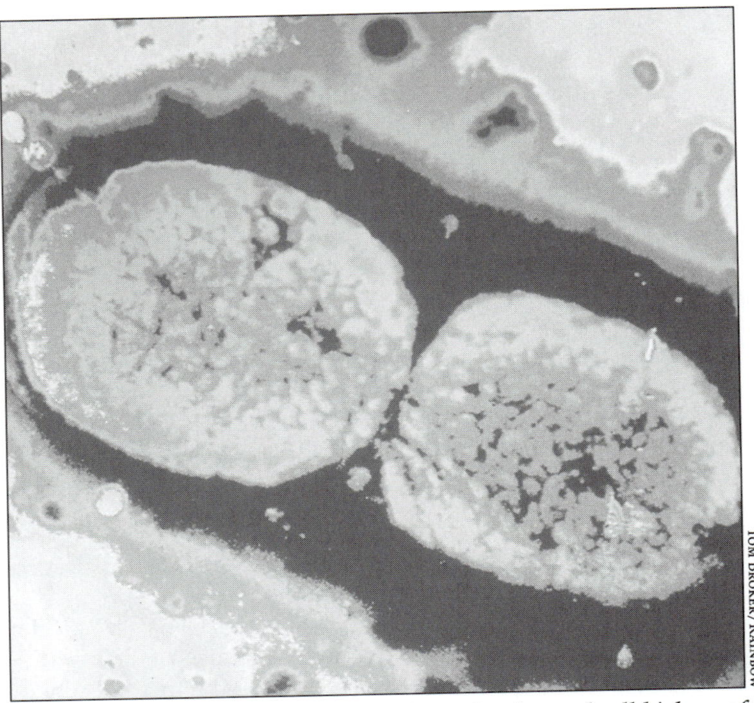

TOM BROKER/RAINBOW

Because scientists know more about the molecular and cell biology of the common bacterium Escherichia coli *than about any other cell or organism, it is perhaps not surprising that the best-known bacteriophages are those that require* E. coli *as a host.*

T4 Coliphage

One of the best-known lytic phages, which is often used in genetic studies, is the coliphage T4. Its protein capsid consists of three major sections—the head, the tail, and the tail fibers. The double-stranded circular DNA molecule of T4 is packaged into the icosahedral-shaped head, and during the infection process it is forced through the hollow core of the cylindrical tail and then directly into the host cell. Contact with the cell is established and maintained throughout the infection process by the tail fibers.

Self-assembly of progeny phages occurs in at least three distinct cellular locations, as complete heads, tails, and tail fibers are first assembled separately and then pieced together in one of the last phases of the infection cycle. Packaging of the replicated T4 DNA is an integral part of the head assembly process. Each of the three subassemblies involves a reasonably complex and highly regulated sequence of assembly steps. For example, head assembly is known to require the activity of eighteen genes, even though only eleven different proteins are found as structural components of mature heads. Identification of the number and sequence of genes involved with each subassembly process has been facilitated by the analysis of artificial lysates from t^s mutants.

For those temperate phages capable of entering the lysogenic cycle, many additional strategies for genetic control and regulation have evolved. The most thoroughly studied of the temperate coliphages is phage lambda (λ). Genes controlling phage DNA integration, excision, and recombination, and those involved with repressor functions, have been identified in phage λ as well as structural genes involved with lytic functions that are similar to those studied in T4.

Research Tool

One of the most important conclusions to be drawn from studies on bacteriophages, and viral genetics in general, is that many of the results have universal implications. For example, the physical properties of DNA and RNA are remarkably identical in all organisms, and these are perhaps easiest to study in bacteriophage systems. The experiment that provided the final proof that DNA was the genetic material was performed using a coliphage very similar to T4. Studies on the origin of spontaneous mutations, first performed in phage, have extended to higher forms of life as well. Some of the most basic questions concerning protein-DNA interactions are best addressed in viral systems, and the principles that emerge seem to hold for all other experimental systems. There is every reason to believe that many basic questions in cell and molecular biology will continue to be best studied in viruses such as bacteriophages, and that some of these investigations will spawn applications that can directly benefit humankind.

It is certain that advances in molecular biology that have revolutionized the understanding of cell biology and the molecular architecture of cells will continue to expand the frontiers of knowledge in the study of viral genetics. Applications in human medicine, veterinary medicine, and plant breeding are sure to follow, as scientists continue to unravel the complexities of these simplest of organisms.

Jeffrey A. Knight

See also: Bacterial genetics; Viruses and viroids.

Sources for Further Study

Birge, Edward A. *Bacterial and Bacteriophage Genetics: An Introduction*. 4th ed. New York: Springer-Verlag, 2000. An excellent supplementary text for a college student studying viral genetics for the first time. Some useful illustrations. References at end of each chapter are designated either "general" or "specialized."

Maloy, Stanley R., John E. Cronan, Jr., and David Freifelder. *Microbial Genetics*. 2d ed. Boston: Jones and Bartlett, 1994. An intermediate-level college textbook focusing on the genetics of bacteria and their bacteriophages. Provides an excellent summary of the properties and life cycles of phages that is accessible to readers with a limited background in biology. The text is well illustrated throughout, with many references for each chapter.

Russell, Peter J. *Genetics*. 5th ed. Menlo Park, Calif.: Benjamin Cummings, 1998. An introductory college text that is particularly well suited to the problem-solving approach to genetics. Presents a fine overview of gene regulation in phages. Well illustrated with a carefully conceived glossary and reference list.

Stahl, Franklin W. *Genetic Recombination: Thinking About It in Phage and Fungi*. San Francisco: W. H. Freeman, 1979. Probably the most technical of the references listed, this book might nevertheless be useful to interested readers. Chapter 4 provides a nice treatment of phage replication and the single burst experiment. Some illustrations, glossary, references.

Suzuki, D., A. Griffiths, J. Miller, and R. Lewontin. *An Introduction to Genetic Analysis*. 4th ed. New York: W. H. Freeman, 1989. Chapters 10-12 of this intermediate-level college text provide a historical perspective on the development of phage genetics and its applications to DNA structure and mutation. The summary of genetic fine structure and complementation is well described and illustrated. Includes glossary and references.

BASIDIOMYCETES

Categories: Fungi; taxonomic groups

The Basidiomycetes *constitute the largest of the three classes of the* Basidiomycota *(basidiosporic fungi), a very large class of about fourteen thousand species of the most diverse terrestrial fungi.*

The largest fungi belong in the *Basidiomycetes* class, as do some of the most unusual. All members of *Basidiomycetes* produce a *basidium* from hyphal cells and not from spores. (The basidium is a cell produced at the end of a dikaryotic hypha.) The basidium will produce either two or four spores as the result of meiosis. The basidium may be either a nonseptate cell, with two or four sterigmata (the *basidiospore* is produced on the end of the sterigma) at the apex, or it may be septate. The septa can be either horizontal or vertical. When observed from the apex, the vertical septa will produce a crosslike pattern. In either case, septate basidia will have one sterigma per cell. Basidiospores are thin-walled and may be released either actively or passively.

Basidiocarps

The *basidiocarp* is the fruiting body of the fungus. The fungus grows as a dikaryotic mycelium through the substrate. When the fungus has acquired sufficient energy, and environmental conditions are adequate, the fungus will produce a basidiocarp. The basidiocarp often appears overnight and may reach a meter in height. Some basidiocarps are tiny, less than a centimeter in height. The basidiocarp may look like a mushroom or may have the appearance of a golf ball or any variation in between. The basidiocarp may be edible or deadly poisonous. It often serves as food for wild animals and insects.

Classification

The *Basidiomycetes* are divided into two groups based on septa in the sterigma. Those that have a septate sterigma are classified in *Phragmobasidiomycetidae*, while those without septa in the sterigma are classified in *Holobasidiomycetidae*. *Phragmobasidiomycetidae* is a small group that includes some smaller fungi whose basidiocarps often have a gelatinous appearance. *Holobasidiomycetidae* is a large group of fungi and is easily divided into two major groups based on the release of the basidiospore. The hymenomycetes release spores actively, while the gasteromycetes release spores passively.

Hymenomycetes

The *hymenomycetes* are the most familiar fleshy fungi. These are the ones that resemble the common mushroom and can be seen when the weather is warm and damp. These fungi may have gills or pores on the underside of the *pileus* (the cap). The gills or pores are lined with a layer of basidia that produce spores. Some of these fungi are produced on the sides of trees or fallen wood and may be acentric. Many of them appear to arise from the ground. Colors of these fungi can be found anywhere in the rainbow, but most appear in earth tones.

Some of the major orders within the hymenomycetes are the *Agaricales* and the *Boletales*. The *Agaricales* contain the fungi that produce gills on the underside of the pileus. The *Boletales* contains the fungi that produce pores on the underside of the pileus. For the most part, the *Agaricales* are fleshy fungi that are supple at maturity and last for no more than a week or two in nature. The *Boletales* are also fleshy and can be confused with other fungi that have pores and are hard. Some of these hard fungi are common parasites of trees; the "shelves" that they produce can grow for years.

Gasteromycetes

The gasteromycetes are a much more diverse group of fungi. The fungi in this group are often associated with soil or decomposing organic matter, although some may be mycorrhizal (in a symbiotic association with plant roots). There is tremendous diversity in this group, and many scientists believe that this group is artificial (not based on evolutionary relationships).

Some of the more interesting fungi of this group are shaped like balls that lie on the soil. Members of *Lycoperdales* and *Sclerodermatales* produce ball-like basidiocarps that form at the soil line. The size can range from that of a small marble up to that of a soccer ball. These are called *puffballs*, as they release spores in small clouds when kicked. In nature, the upper layers of the basidiocarp crack, and spores are released as drops of water, hitting the outer layer of the basidiocarp. Many of these are edible when properly identified.

Other interesting fungi are found in the order *Nidulariales*. These are called the *bird's nest fungi*, as the basidiocarp resembles a small bowl containing two or three small "eggs," which contain the basidia. When a droplet of water lands in the "nest" the "eggs" are thrown upward and outward. As they are released, a small thread is pulled behind, and the thread sticks onto some part of a plant, such as a blade of grass. The "egg" will then degrade and release the spores to be disseminated in the wind.

Among the most bizarre fungi are the *stinkhorns*. These fungi produce basidiocarps from structures that look like chicken eggs. The elongate structure produces the basidia on the end in a mass of slimy, smelly mucus. Flies are attracted to the smell, land on the mucus, and fly away. The basidiospores adhere to their feet and drop off, thereby disseminating the fungus.

J. J. Muchovej

See also: Ascomycetes; Basidiosporic fungi; Fungi; Mushrooms; Mycorrhizae; Rusts; *Ustomycetes*.

Sources for Further Study

Alexopoulos, Constantine J., C. W. Mims, and M. Blackwell. *Introductory Mycology*. 4th ed. New York: John Wiley, 1996. The text gives detail into the biology, anatomy, and physiology of fungi. Includes illustrations, bibliographic references, and index.

Deacon, J. W. *Modern Mycology*. 3d ed. Malden, Mass.: Blackwell Science, 1997. An introduction to fungi for botanists and biologists. Emphasizes behavior, physiology, and practical significance of fungi. Contains numerous photographs, line drawings, and diagrams.

Manseth, James D. *Botany: An Introduction to Plant Biology*. 2d ed. Sudbury, Mass.: Jones and Bartlett, 1998. General botany text with a section on fungi, their importance, and their biology. Includes illustrations, bibliographic references, and index.

Raven, Peter H., Roy F. Evert, and Susan E. Eichhorn. *Biology of Plants*. 6th ed. New York: W. H. Freeman, 1999. General botany text with section on fungi, their importance in the environment, and their biology. Includes illustrations, bibliographic references, and index.

Smith, Alexander H., and Nancy Smith Weber. *The Mushroom Hunter's Field Guide*. Ann Arbor: University of Michigan Press, 1996. This illustrated field guide gives an overview of fleshy fungi found in the United States. Includes illustrations, photographs, bibliographic references, and index.

BASIDIOSPORIC FUNGI

Categories: Economic botany and plant uses; fungi; microorganisms; pests and pest control; taxonomic groups

Basidiosporic fungi (also known as the Basidiomycota *or* Basidiomycotina*) are fungi that produce sexual spores on a specialized cell called a basidium.*

The basidiosporic fungi are the most diverse phylum of the fungi world, with more than 22,300 species described. Some of the fungi in this phylum are microscopic, while the larger members of this group produce fruiting structures that are basketball-sized and weigh in excess of 10 pounds. This phylum contains fungi that fall into three classes: mushroom, rusts, and smuts—and range widely in appearance, from the common mushroom to weblike fungi with an odor that can be detected at several feet.

Taxonomy

The basidiosporic fungi are divided into three classes: *Basidiomycetes* (mushrooms); *Teliomycetes* (rusts); and *Ustomycetes* (smuts). The *Basidiomycetes* are the higher basidiosporic fungi, which are normally fleshy. They produce true basidiocarps, and the only spore formed is the basidiospore. The other two classes both have more than one spore form and do not have extensive mycelium. The *Teliomycetes* are commonly called rusts and are serious biotrophic parasites of plants. The rusts are able to complete their life cycle only in the presence of living plant host tissue. The *Ustomycetes* are commonly called smuts and are mostly minor pathogens of plants, especially monocots. Some smuts have been cultured in axenic culture, where they form a "yeastlike" phase. The yeastlike phase has no true mycelium but rather individual cells.

Basidium

The *basidium* is a single cell on which basidiospores are produced externally. The basidium forms either as the terminal cell of a dikaryotic mycelium or from a resting spore that initially is dikaryotic. The dikaryotic mycelium or spore contains two haploid nuclei, one donated by each of the parent strains. As the basidium begins to form, the two nuclei migrate into the center of the cell and fuse, forming a diploid nucleus. This nucleus then undergoes meiosis, forming four haploid nuclei. As this is occurring, the cell wall of the basidium begins to produce little extensions called *sterigmata*, upon which the basidiospores will form. The tips of the sterigmata then inflate, and one nucleus migrates into each forming basidiospore. The basidiospore is haploid and has a very thin cell wall. The spore is normally transmitted in air currents. Upon germination, the basidiospore produces a haploid mycelium which will fuse with a compatible hyphae, producing a dikaryotic mycelium.

Spore release from the basidium can be either active or passive. Passive release occurs when the junction of the sterigma and basidiospore separates, releasing the spore. Active release is more specialized. When the basidiospore is forming, a small segment of the spore wall at the junction with the sterigma loosens and fills with either gas or liquid. At the time of release, the fluid or gas escapes, propelling the basidiospore away from the basidium. The distance traveled is not great, just enough to make sure that the basidiospore is able to enter into air currents for dissemination.

Hyphal Structure

The hyphae of the *Basidiomycetes* are septate and have special modifications at the septa. When a cell divides, a crosswall forms between the two daughter cells. With the dikaryotic hyphae of the *Basidiomycetes*, as the cell divides, the nuclei migrate toward the apex of the hyphae. The nuclei then undergo mitosis, with one of the nuclei migrating into a small outgrowth of the hyphae and the other migrating backward. Septa form, creating a new dikaryotic cell near the apex and two haploid cells, one in line and the other as the outgrowth. The outgrowth then turns and fuses with the haploid cell, and the nucleus migrates back to form a dikaryotic cell. The outgrowth remains visible with a microscope and is called a clamp connection.

The reproductive structure of the *Ustomycetes* is called a sorus. The sorus is a mass of dikaryotic spores that are normally dark brown or black in color. The sorus is formed in meristematic regions of the plants. The spores are called probasidia, because they form basidia when they germinate.

With the *Teliomycetes*, there are up to five distinct spore forms. The basidiospore lands on a susceptible plant and germinates, producing a haploid mycelium that infects the plant. The infection results in the formation of a haploid spermagonium that produces both spermatia and receptive hyphae. When a compatible spermatia and receptive hypha combine, a dikaryotic hypha is produced, which initiates formation of an aecium. The aecium produces dikaryotic spores that are transmitted by air currents and infect another plant. The resultant infection produces a subcuticular or subepidermal mass of thin-walled spores. These dikaryotic spores are called urediniospores and are formed in the uredinium. The urediniospores are blown by air currents and produce reinfection of the same species of plant. At the end of the growing season, infections by urediniospores will result in the formation of a subcuticular or subepidermal mass of thick-walled spores called teliospores which are formed in the telium. These spores are initially dikaryotic but then become diploid and finally germinate by formation of the basidium.

Basidiocarps

The *basidiocarp* is the fruiting body of the higher *Basidiomycetes*. This structure is multicellular and composed of hyphae. The basidiocarp resembles the familiar image of the mushroom. The *mushroom* consists of a stalk (*stipe*) which has a cap (*pileus*) on top. The stipe can be as tall as a meter (40 inches), and the pileus as long as a meter in diameter. Alternatively, both parts could be less than a centimeter in size. The pileus has pores or gills on the underside, where the basidia are produced. The layer of basidia is called a hymenium or "fertile layer."

Other kinds of basidiocarps may be found in nature. Some are totally enclosed and remain on the ground, looking much like a golf ball. These are called *puffballs*. As the puffball matures, the other layers begin to crack at the apex. When drops of rain fall, the force of the impact causes spores to puff out of the opening. Another kind of puffball is the earthstar. In these unique fungi, the outer layers pull away from central part of the puffball and form a starlike pattern on the ground.

Ecological Importance

The basidiosporic fungi all play important roles in ecosystems. The rusts and the smuts are important plant pathogens, capable of great destruction of crops. These fungi have been known for thousands of years and are some of the most devastating fungi around.

The mushrooms are part of the natural cycle of decay. They are found on the ground or on wood and are the later stages of decay of organic matter. Some mushrooms are found on living plants, where they can be serious pathogens. Others are edible and are excellent sources of digestible protein. Still others are toxic or poisonous and can be fatal when eaten.

Stinkhorns and the *bird's nest fungi* are unique basidiosporic fungi. The stinkhorns are basidiocarps that form on the soil and produce the basidia in a mass of putrid cells. The stench from the cells draws flies, which walk over the spores and then disseminate them. These can be found in wooded areas and can be detected by smell at distances of up to several meters.

The bird's nest fungi look like small birds' nests. The outer part of the basidiocarp resembles a small nest, up to an inch in diameter. On the inside, several small puffball-like structures can be found, with basidia on the inside. These look like small eggs. When a drop of water enters the nest, the force thrusts the "egg" upward and extends a small cord from the back. The small cord catches hold of a plant and suspends the egg in the air. As the egg dries, it turns into a powdery mass, which is blown about by the wind.

J. J. Muchovej

See also: Ascomycetes; *Basidiomycetes*; Fungi; Mushrooms; Mycorrhizae; Rusts; *Ustomycetes*; Yeasts.

Sources for Further Study

Alexopoulos, Constantine J., and Charles E. Mims. *Introductory Mycology*. New York: John Wiley, 1995. The text gives detail into the biology, anatomy, and physiology of fungi. Includes illustrations, bibliographic references, and index.

Manseth, James D. *Botany: An Introduction to Plant Biology*. 2d ed. Sudbury, Mass.: Jones and Bartlett, 1998. General botany text with section on fungi, their importance and their biology. Illustrations, bibliographic references, index.

Raven, Peter H., Roy F. Evert, and Susan E. Eichhorn. *Biology of Plants*. 6th ed. New York: W. H. Freeman/Worth, 1999. General botany text with section on fungi, their importance in the environment, and their biology. Includes illustrations, bibliographic references, and index.

Smith, Alexander H., and Nancy Smith Weber. *The Mushroom Hunter's Field Guide*. Ann Arbor: University of Michigan Press, 1996. This illustrated field guide gives an overview of fleshy fungi found in the United States. Includes illustrations, photographs, bibliographic references, and index.

BIOCHEMICAL COEVOLUTION IN ANGIOSPERMS

Categories: Angiosperms; animal-plant interactions; evolution; physiology; *Plantae*; poisonous, toxic, and invasive plants

Flowering plants, or angiosperms, produce many compounds that are not directly related to growth and development. These secondary metabolites arise from primary metabolic pathways and act as antiherbivory mechanisms, allelochemicals, or attractants.

Secondary metabolites are biochemicals produced by plants in response to selection pressures. These pressures may be from herbivory, competition, or the need for pollination. As plants produce compounds to enhance their survival, predators, competitors, and pollinators react and evolve means of adjusting to the plant's efforts. Chemically simple secondary metabolites may be widespread throughout angiosperm (flowering plant) families, whereas more complex chemicals are often restricted to a single species. Secondary metabolites are often under high selection pressures, causing individual compounds to have very limited distributions and making them useful in determining the evolutionary relationships between taxonomic groups. Presence of secondary compounds influences the activities of organisms interacting with the plants and, over long periods of time, influences evolution of those species.

Antiherbivory Mechanisms

Antiherbivory chemicals may have a wide range of effects on herbivores (plant-eating animals). Many compounds merely deter grazing. Crystals produced from calcium oxalate (raphides) may be ejected from the vacuoles of cells along with proteinaceous toxins, causing tissue swelling in the mouth of an offending herbivore. Many monocotyledonous plant families, such as *Liliaceae*, *Heliconiaceae*, *Rubiaceae*, and *Arecaceae*, produce this type of antiforaging device. This type of defense is especially notable in young tissues that have not developed the toughness found in mature leaves as a herbivory deterrent. Red oaks produce tannins in response to gypsy moth attacks, reducing further herbivory. Continued feeding on plants containing tannins would lead to slow starvation of a herbivore, as its digestive system could not absorb proteins. The same tannins are not deterrents to squirrels. Squirrels harvest acorns and bury them for later consumption, providing a food source for the squirrel and a dispersal mechanism for the acorn.

Beavers provide another example of the interaction between plant and animal evolution. Some of the beaver's preferred foods include species that are unpalatable or toxic to other mammals, such as bracken fern, nettles, thistles, and skunk cabbage. This gives the beaver a largely uncontested food source that may involve a metabolic "cost" to the animal.

Other antiherbivory chemicals result in effects more severe than mere deterrence of feeding. Alkaloids such as caffeine, nicotine, and strychnine are potent antiherbivory mechanisms, causing convulsions, comas, and even death in herbivores. These effects may not occur in all herbivores. Strychnine, for example, is produced by the fruit of some plants that may be eaten by birds without ill effects, but in mammals the same fruit causes failure of the central nervous system and induces seizures. The plant reduces herbivory by mammals, and the seeds get dispersed by birds that are able to detoxify the strychnine. Grains and seed crops, such as wheat and peanuts, which are particularly attractive to animal and insect herbivores, often produce cyanogenic glycosides that release hydrogen cyanide as the tissues are digested. This compound inhibits cellular respiration, thus killing the herbivore. In each case members of a plant population are consumed by the herbivore, but future generations are spared by the loss.

Allelochemicals

Allelochemicals are compounds produced by an organism that interfere with the growth or development of another organism. Many phenolic acids act as allelochemicals, inhibiting root growth of competing species. Many grains are known to release ferulic acid and caffeic acid into the soil, thus inhibiting the germination of weed species. Phenolics may also act as antifungal compounds, increasing in concentration with fungal infection, thus protecting the plant from further attack. Phenolic compounds produced in tobacco and tomato leaves reduce the growth of these plants' natural predator, tobacco hornworm, without affecting the growth or activity of the hornworm's natural predator.

Allelochemicals produced in response to injury by herbivores may also attract predators of the herbivore. Wastes from many species of caterpillars induce the release of terpenoids from green leaves that attract parasitoid insects. Production of specific combinations of volatiles on the part of the plant signals the predator, which will then reduce further herbivory. The plants have evolved the signal in response to herbivory, and the predators have evolved the ability to detect the signal indicating the location of their host.

Lectins are widely distributed carbohydrate-binding proteins, most commonly found in the *Leguminoseae* (legume) family. When found in the seeds, these compounds act as broad-spectrum insecticides, whereas in the roots of legumes they maintain bacterial relationships in nitrogen-fixing nodules, providing the plant with a source of nitrogen unavailable to plants not producing nodules.

Attractants

Terpenoids and aliphatic compounds are often the components of essential oils of plants. The volatile nature of these compounds produces a distinctive odor that attracts pollinators. Composition of the volatile compounds often closely matches the natural pheromones produced by the pollinator, mimicking the chemical scent of a female insect in an attempt to attract male pollinators. Pheromone mimicry is found primarily in members of the *Orchidaceae*, which are often dependent on single species of wasp for pollination. Other plants may mimic the odor of food. The smell of rotting flesh, attractive to flies, is produced via ammonia and alkylamines, such as cadaverine and putrescine. Methylesters may attract moth pollinators by mimicking the sweet smell of fruit.

Flavonoids often provide color to fruits and flowers and act as visual cues for pollination. Reds, blues, or yellows in varying patterns stand out against a background of green leaves, helping pollinators locate the flower. Species may have minor chemical differences in their flavonoids that allow for the determination of identity, hybridization between species, and possible coevolution with pollinators. For example, tropical flowers tend to have a more intense red color from anthocyanins than do temperate flowers. This difference correlates with differences in pollinator preferences, indicating a role by natural selection. Birds, such as hummingbirds, prefer red to yellow, whereas bees are not able to discern reds but are attracted to yellows. Carotenoids, such as xanthophyll and beta-carotene, give fruits and flowers distinctive yellow and orange colors.

Color patterns are also important in attracting pollinators. Butterflies are attracted to red/yellow color patterns. Flavonoid compounds not only impart color but also may modify color patterns by absorbing ultraviolet (UV) light. Bees are capable of seeing in the UV range, so the presence of flavonoids may alter the bees' perception of the flower. The patterns may also create cues as to the location of nectaries within the flower, guiding the pollinator to its reward.

Cheryld L. Emmons

See also: Allelopathy; Angiosperm evolution; Animal-plant interactions; Coevolution; Flowering regulation; Hormones; Pheromones; Metabolites: primary vs. secondary; Pigments in plants; Pollination.

PhotoDisc

Color patterns are important in attracting pollinators; for example, butterflies are attracted to red and yellow colors. Flavonoids not only provide color to fruits and flowers but also may modify color patterns by absorbing ultraviolet light.

Sources for Further Study

Dodson, C. H. "Coevolution of Orchids and Bees." In *Coevolution of Animals and Plants*, edited by L. E. Gilbert and P. H. Raven. Austin: University of Texas Press, 1975. Discusses coevolution of plant and animal species.

Harborne, J. B., ed. *Phytochemical Phylogeny*. Amsterdam: Elsevier Science, 1970. Describes secondary metabolites and their evolution.

Jolivet, P. *Insects and Plants: Parallel Evolution and Adaptation*. 2d ed. Gainesville, Fla.: Sandhill Crane Press, 1992. Covers coevolution of flowering plants and insects.

Stowe, M. K. "Chemical Mimicry." In *Chemical Mediation of Coevolution*, edited by K. C. Spencer. London: Academic Press, 1988. Discusses secondary metabolites and their role in evolution.

Taiz, Lincoln, and Eduardo Zeiger. "Surface Protection and Secondary Defense Compounds." In *Plant Physiology*. New York: Benjamin Cummings, 1991. Explains the role of secondary metabolites in defense.

BIOFERTILIZERS

Categories: Agriculture; bacteria; biotechnology; economic botany and plant uses; nutrients and nutrition; soil

The use of biofertilizers, biological systems that supply plant nutrients such as nitrogen to agricultural crops, could reduce agriculture's dependency on chemical fertilizers, which are often detrimental to the environment.

Plants require an adequate supply of the thirteen mineral nutrients necessary for normal growth and reproduction. These nutrients, which must be supplied by the soil, include both *macronutrients* (nutrients required in large quantities) and *micronutrients* (nutrients required in smaller quantities). As plants grow and develop, they remove these essential mineral nutrients from the soil. Because normal crop production usually requires the removal of plants or plant parts, the nutrients are continuously removed from the soil. Therefore, the long-term agricultural utilization of any soil requires periodic fertilization to replace lost nutrients.

Nitrogen is the plant nutrient that is most often depleted in agricultural soils, and most crops respond to the addition of nitrogen fertilizer by increasing their growth and yield. Therefore, more nitrogen is applied to cropland than any other fertilizer component. In the past, nitrogen fertilizers have been limited to either *manures*, which have low levels of nitrogen, or *chemical fertilizers*, which usually have high levels of nitrogen. However, the excess nitrogen in chemical fertilizers often runs off into nearby waterways, causing a variety of environmental problems.

Less Harmful Alternatives

Biofertilizers offer a potential alternative: They supply sufficient amounts of nitrogen for maximum yields yet have a positive impact on the environment. Biofertilizers generally consist of either naturally occurring or *genetically modified* microorganisms that improve the physical condition of soil, aid plant growth, or increase crop yield. Biofertilizers provide an environmentally friendly way to increase plant health and yields with reduced input costs, new products and additional revenues for the agricultural biotechnology industry, and cheaper products for consumers.

Nitrogen Fixing

While biofertilizers could potentially be used to supply a number of different nutrients, most of the interest is focused on nitrogen. The relatively small amounts of nitrogen found in soil come from a variety of sources. Some nitrogen is present in all organic matter in soil; as this organic matter is de-

AP/Wide World Photos

These red wiggler worms are used to consume, digest, and expel hog waste into tiny pellets called worm castings which are refined, packaged, and sold to retail stores in bulk as organic fertilizer.

graded by microorganisms, it can be used by plants. A second source of nitrogen is *nitrogen fixation*, the chemical or biological process of taking nitrogen from the atmosphere and converting it to a form that can be used by plants. Bacteria such as members of *Rhizobium* can live symbiotically in the roots of certain plants, such as legumes. Rhizobia and plant root tissue form root nodules, which house the nitrogen-fixing bacteria; once inside the nodules, the bacteria use energy supplied by the plant to convert atmospheric nitrogen to ammonia, which nourishes the plant. Natural nitrogen can also be supplied by free-living microorganisms, which can fix nitrogen without forming a symbiotic relationship with plants. The primary objective of biofertilizers is to enhance any one or all of these processes.

One of the major goals for the genetic engineering of biofertilizers is to transfer the ability to form nodules and establish effective symbiosis to nonlegume plants. The formation of nodules in which the *Rhizobia* live requires plant cells to synthesize many new proteins, and many of the genes required for the expression of these proteins are not found in the root cells of plants outside the legume family (*Fabaceae*). If transfer of the appropriate genes could be accomplished, *Rhizobia* could be used as a biofertilizer for a variety of plants.

There is also much interest in using the free-living, soil-borne organisms that fix atmospheric nitrogen as biofertilizers. These organisms, including types bacteria and algae, live in the *rhizosphere* (the region of soil in immediate contact with plant roots) or thrive on the surface of the soil. Because the exudates from these microorganisms contain nitrogen that can be used by plants, increasing their abundance in the soil could reduce the dependency on chemical fertilizers. Numerous research efforts have been designed to identify and enhance the abundance of nitrogen-fixing bacteria in the rhizosphere. Soil microorganisms primarily depend on soluble root exudates and decomposed organic matter to supply the energy necessary for fixing nitrogen. Hence, there is also an interest in enhancing the *biodegradation* of organic matter in the soil. This research has primarily centered on inoculating the soil with cellulose-degrading fungi and nitrogen-fixing bacteria or applying organic matter, such as straw that has been treated with a combination of the fungi and bacteria to the soil.

D. R. Gossett

See also: Fertilizers; Nitrogen cycle; Nitrogen fixation.

Sources for Further Study

Black, C. A. *Soil-Plant Relationships*. 2d ed. Malabar, Fla.: Krieger, 1984. Provides an in-depth discussion of the relationship between soil microbes and soil fertility.

Crispeels, M. J., and D. E. Sadava. *Plants, Genes, and Agriculture*. Boston: Jones and Bartlett, 1994. Offers a discussion on the potential genetic modifications necessary to increase the use of microorganisms to supply nitrogen fertilizer.

Legocki, Andrezej, Hermann Bothe, and Alfred Pühler, eds. *Biological Fixation of Nitrogen for Ecology and Sustainable Agriculture*. New York: Springer, 1997. From the proceedings of the NATO Advanced Research Workshop held in Poznań, Poland, in 1996. Includes bibliographical references.

Lynch, J. M. *Soil Biotechnology: Microbiological Factors in Crop Productivity*. Boston: Blackwell Scientific, 1983. Contains some excellent information on the potential for genetically engineering microorganisms to improve crop production.

Salisbury, F. B., and C. W. Ross. *Plant Physiology*. 4th ed. Belmont, Calif.: Wadsworth, 1992. Contains excellent chapters on plant nutrition and nitrogen metabolism.

BIOLOGICAL INVASIONS

Categories: Ecology; ecosystems; environmental issues; poisonous, toxic, and invasive plants

Biological invasions are the entry of a type of organism into an ecosystem outside its historic range. In a biological invasion, the "invading" organism may be an infectious virus, a bacterium, a plant, an insect, or an animal.

Species introduced to an area from somewhere else are referred to as alien or exotic species or as invaders. Because an exotic species is not native to the new area, it is often unsuccessful in establishing a viable population and disappears. The fossil record, as well as historical documentation, indicates that this is the fate of many species in new environments as they move from their native habitats. Occasionally, however, an invading species finds the new environment to its liking. In this case, the invader may become so successful in exploiting its new habitat that it can completely alter the ecological balance of an ecosystem, decreasing *biodiversity* and altering the local *biological hierarchy*. Because of this ability to alter *ecosystems*, exotic invaders are considered major agents in driving native species to extinction.

Biological invasions by notorious species constitute a significant component of earth's history. In general, large-scale climatic changes and geological crises are at the origin of massive exchanges of flora and fauna. On a geologic time scale, migrations of invading species from one continent to another are true evolutionary processes, just as speciation and extinction are. On a smaller scale, physical barriers such as oceans, mountains, and deserts can be over-

AP/Wide World Photos

This Salvinia molesta *plant floats on water surfaces, multiplying rapidly. Found over an increasingly wide range in the United States, it is known to clog waterways and crowd out native plants.*

come by many organisms as their populations expand. Organisms can be carried by water in rivers or ocean currents, transported by wind, or carried by other species as they migrate seasonally or to escape environmental pressures. Humans have transplanted plants since the beginning of plant cultivation in pre-Columbian times. The geological and historical records of the earth suggest that biological invasions contribute substantially to an increase in the rate of extinction within ecosystems.

Invasive Plants

In modern times, most people are not aware of the distinction between *native plants* and exotic species growing in their region. Recent increases in intercontinental invasion rates by exotic species, brought about primarily by human activity, create important ecological problems for the recipient lands. Invasive plants in North America include eucalyptus trees, morning glory, and pampas grass.

It would seem logical to assume that invading species might add to the biodiversity of a region, but many invaders have the opposite effect. In all ecosystems the new species are often opportunistic, driving out native species by competing with them for resources. For example, *Pueraria lobata*, or kudzu, is a vine native to Japan. Introduced in the United States at the 1876 Philadelphia Exposition, kudzu was planted to control erosion on hillsides and for livestock forage. By the end of the twentieth century, it could be found from Connecticut to Missouri, extending south to Texas and Florida. Kudzu covers everything in its path and grows as much as 1 foot (0.3 meter) per day. Similarly, English ivy (*Hedera helix*), a native of Eurasia, is considered a serious problem in West Coast states. It forms "ivy deserts" in forests and crowds out native trees and shrubs that make up essential wildlife habitat.

The invasion of an ecosystem by an exotic species can effectively alter ecosystem processes. An invading species does not simply consume or compete with native species but can actually change the rules of existence within the ecosystem by altering processes such as primary productivity, decomposition, hydrology, geomorphology, nutrient cycling, and natural disturbance regimes.

Invasive Insects and Microorganisms

The invasion of native forests alone by nonnative insects and microorganisms has been devastating on many continents. The white pine blister rust and the balsam woolly adelgid have invaded both commercial and preserved forest lands in North America. Both exotics were brought to North America in the late 1800's on nursery stock from Europe. The balsam woolly adelgid attacks fir trees and causes their death within two to seven years from chemical damage and by feeding on the trees' vascular tissue. The adelgid has killed nearly every adult cone-bearing fir tree in the southern Appalachian Mountains. The white pine blister rust attacks five-needle pines; in the western United States fewer than 10 pine trees in 100,000 are resistant. Because white pine seeds are an essential food source for bears and other animals, the loss of the trees is having severe consequences across the food chain.

Since the 1800's the deciduous trees of eastern North America have been attacked numerous times by waves of invading exotic species and diseases. One of the most notable invaders is the gypsy moth, which consumes a variety of tree species. Other invaders have virtually eliminated the once-dominant American chestnut and the American elm. Tree species that continue to decline because of new invaders include the American beech, mountain ash, white birch, butternut, sugar maple, flowering dogwood, and eastern hemlock. It is widely accepted that the invasion of exotic species is the single greatest threat to the diversity of deciduous forests in North America.

Effects on Humans and Humans as Invaders

Some introduced exotic species are beneficial to humanity. It would be impossible to support the present world human population entirely on species native to their regions. Humans, the ultimate biological invaders, have been responsible for the extinction of many species and will continue to be in the future. At the beginning of the twenty-first century, the United States was spending $4 billion annually to eradicate invasive plant species, a figure that does not take into account loss of biodiversity or wildlife habitat.

Randall L. Milstein, updated by Elizabeth Slocum

See also: Food chain; Invasive plants.

Sources for Further Study

Bright, Chris. *Life out of Bounds: Bioinvasion in a Borderless World*. New York: Norton, 1998. A study of the global spread of alien, exotic organisms and how they are undermining the world's ecosystems and societies. Bright shows that this "biological pollution" is now beginning to corrode the world's economies as well and outlines scientific research on the threat, the social and economic implications if these invasions are allowed to continue unchecked, and steps that can be taken to contain the spread of exotic species.

Cox, George W. *Alien Species in North America and Hawaii: Impacts on Natural Ecosystems*. Washington, D.C.: Island Press, 1999. Provides a comprehensive overview of the invasive species phenomenon, examining the threats posed and the damage that has already been done. Offers a framework for understanding the problem and provides a detailed examination of species and regions.

Crosby, Alfred. *Ecological Imperialism: The Biological Expansion of Europe, 900-1900*. New York: Cambridge University Press, 1994. Focusing on ecological aspects, this study reveals how Europeans were able to conquer the people of temperate lands through the successful environmental adaptation of the plants, animals, and germs that they brought with them.

Drake, J. A., et al. *Biological Invasions: A Global Perspective*. New York: Wiley, 1989. Surveys the state of knowledge concerning the dramatic rearrangement of the earth's biota over the nineteenth and twentieth centuries. An international group of contributors examines animals, plants, and microorganisms that have been successful invaders in nonagricultural regions, with emphasis on those that have disrupted ecosystem function.

Hengeveld, Rob. *Dynamics of Biological Invasions*. New York: Chapman and Hall, 1989. Introduces dynamic concepts into biogeography and spatial concepts into ecology. By using mathematical models from epidemiology and human geography, generalizations can be made, and it is shown that apparently static species ranges contain dynamic internal parameters.

BIOLOGICAL WEAPONS

Categories: Economic botany and plant uses; medicine and health; microorganisms; poisonous, toxic, and invasive plants

Biological weapons are biological agents that can be used to destroy living organisms. This general definition includes the use of virtually any kind of microorganism (bacterium or fungus) or biological agent (mycoplasmalike organism, virus, viroid, or prion) to destroy any biologically important plant or animal.

There are two basic ways of using biological weapons against humanity. The first is to attack the food or water supply. This would produce hardship and the possible death from starvation of many individuals. In developed countries such as the United States, total devastation due to such an attack would likely be avoided, as there are considerable stores of food, in dispersed locations, that would mitigate against crop failure of moderate proportions. In addition, most experts agree that the amount of biological contaminant required to overcome the effects of dilution and time in most water reservoirs makes poisoning of water supplies impractical, although not impossible.

The other way of using biological weapons against humanity is to attack individuals directly, using pathogens. Numerous known pathogens could be used, but the most effective biological weapons would creep upon the population with stealth. Naturally occurring agents that have promise as biological weapons are pathogens, which can infect and colonize a host. A pathogen causes *disease*, an alteration in the metabolism of a host.

In order for disease to occur, it is necessary to have a *virulent pathogen*, that is, one that is capable of causing disease; a *susceptible host*, able to be infected; and an environment that is favorable to both the host and the pathogen. When these occur together, disease occurs. For the disease cycle to begin, it is necessary for *propagules* of the pathogen to come in contact with the susceptible host. Propagules include those that cause initial infections, called the *primary inoculum*. As the disease cycle progresses, the host may release other propagules, considered the *secondary inoculum*. When only one progression of disease occurs during a prolonged span of time, the disease is termed *monocyclic*. When more than one cycle of disease takes place in a growing season, the disease is called *polycyclic*.

Monocyclic Diseases

An example of the monocyclic disease cycle occurs with corn smut. The primary inoculum is released in the spring, as the tassel and silk of the corn appear. Infection occurs, symptoms of large black galls develop on the ears, and new inoculum is produced for the following growing season. This kind of disease produces a single cycle during a growing season. The potential for problems depends on the amount of inoculum present at the beginning of the disease cycle. In order to use corn smut as a biological weapon, the initial amount of inoculum must be sufficient to reach all of the plants that are targeted.

This is the same general case as with *anthrax*. Anthrax is a disease of animals that is caused by the bacterium *Bacillus anthracis*. The bacterium is able to survive in nature as small spores that are able to resist extremes of climate and time. These microscopic spores are produced at any time when environmental conditions are unfavorable for growth of the bacterium. The spores are less than 2 microns in diameter, and if they become suspended in the air, they will hang indefinitely. Such small particles are called PM10 (particulate matter less than 10 microns in size). These particles are of extreme importance, as they are not filtered out by common filters and must be removed by HEPA (high-efficiency particulate-arresting) filters, which can remove particles smaller than a micron. When these particles are inhaled, they cause disease of the lung tissue, which will result in destruction of the lungs and the resultant death of the host.

The reasons that *Bacillus anthracis* could be a biological weapon are that it is easily concealed and disseminates quickly as an invisible, airborne powder. The small particle size allows it to be inhaled, after which it lodges in the lungs and begins the process of infection. The infection begins like that of a common chest cold. By the time the host determines that anthrax might have been contracted, the condition is normally fatal. One of the features of anthrax as a biological weapon is that it is not contagious. Spores are not produced until the host is dead, at which time the spores are not placed back into the air currents. This permits anthrax to be used against a specific target.

Polycyclic Diseases

Some kinds of biological weapons can be termed *weapons of mass destruction*. These biological weap-

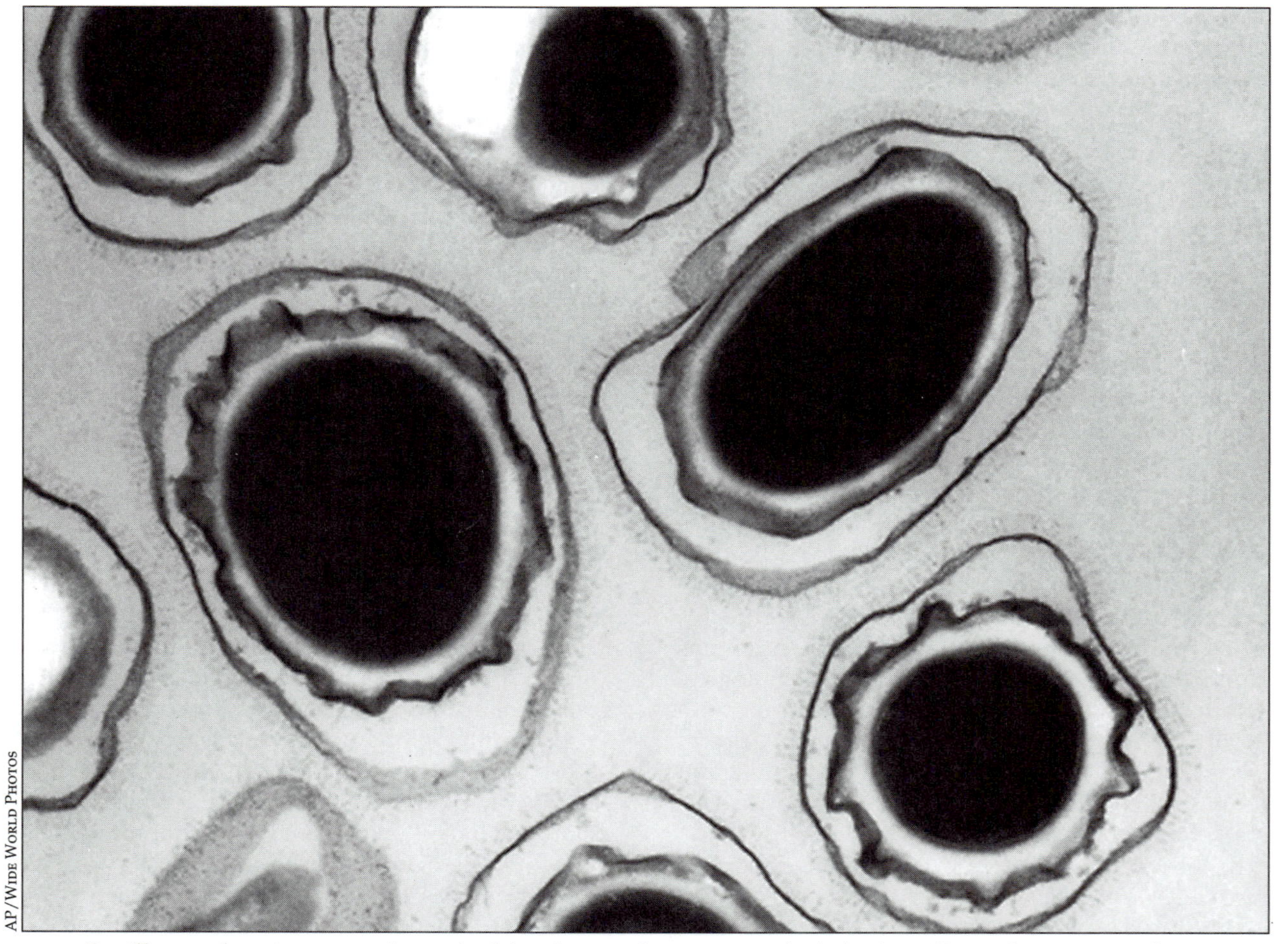

AP/Wide World Photos

Bacillus anthracis *spores, shown in this micrograph, can cause the infectious disease known as anthrax.*

ons are able to self-perpetuate and thereby create great amounts of destruction. They are polycyclic in nature.

A plant-based example of this is the epidemic of *potato late blight*, which caused the Irish Potato Famine of the mid-1840's. Most of the people living in Ireland at that time depended on their own harvests for food. During a two-year period, the climate became especially cool and damp, conditions favorable for the spread of potato late blight. Because potatoes are planted from vegetative parts, there was genetic uniformity of the crop, which was highly susceptible. The pathogen is extremely virulent, with the capacity to destroy a single plant overnight. Total crop failure resulted in the reduction of Ireland's population from about six million to about two million people. About one-third of the people emigrated to other lands, and about one-third of the population died from starvation. All of this destruction could have started from a single fungal propagule that infected a plant and then spread. This fungus is still problematic worldwide. However, regular sprays with fungicide reduce the risk of crop destruction.

Pathogens that occur at low levels in any given area are called *endemic* and may, when conditions are favorable, produce a widespread surge of disease in that area. This is called an *epidemic*. Epidemics that occur over wide geographic ranges, such as two or more continents, are called *pandemics*. The endemic potato late blight fungus caused an epidemic, and because it also spread into Europe and North America, the situation could be called a pandemic.

A similar concern is expressed with regard to the disease *smallpox*. Smallpox is caused by a virus that is easily passed from an infected individual to a healthy one. The smallpox virus no longer occurs in nature (is not endemic) anywhere in the world. The last reported case of smallpox in nature occurred in the late 1970's. As a result of a concerted vaccination effort among many nations, this normally fatal disease was eradicated. The last stores of smallpox virus are housed at the Centers for Disease Control (CDC) in Atlanta, the U.S. Army Biological Weapons Research Unit in Fort Detrick, Maryland, and the Russian Academy of Sciences in Moscow. Some people fear some of the inoculum from one of these storehouses could be transported into unconfined areas of the outside world. As such, smallpox would become a very formidable biological weapon.

Smallpox is highly contagious and spreads easily from person to person. Should an infected person walk the streets of a highly populated urban area, it would be possible to infect millions of people within the span of a week or two. Within several weeks, especially in this age of global travel, the infection could spread around the world.

Vectors

One method of spreading disease is through the use of vectors. *Vectors* are arthropods or other invertebrates that have the ability to transmit a pathogen from one host to another. The most famous vector is the mosquito, which can transmit malaria, dengue fever, West Nile virus, and many other diseases. The corresponding vector in plants is the aphid, which can transmit a host of viruses and mycoplasmas (bacteria that lack cell walls). In order for transmission to occur, it is necessary for a vector to first come in contact with an infected host, feed upon that host, and then pass the pathogen to an uninfected host. The most notable vector-borne disease is plague, which is passed from infected animals to humans by fleas. This disease, which was once responsible for hundreds of thousands of deaths, is now readily controlled by the use of insecticides to kill off the vector.

Designer Weapons

With twentieth century advances in technology, the manipulation of the genetics of a pathogen to make it a "super pathogen" is possible. The creation of a genetic code which could be pathogenic is also possible. So-called designer weapons are those that may be envisioned by someone who feels he or she has a need to make a more destructive or more "targeted" weapon. These weapons could then be used on a specific population or area. The creator of the designer weapon would have an advantage, as he or she would also be able to create a vaccine against the designer weapon, which could be administered to a specific group of individuals. Designer weapons, like all biological weapons, work best when there is an abundant supply of susceptible hosts and when the genetics of these hosts do not vary greatly.

Pathogen Dispersal

There is considerable opportunity for the use of any pathogen as a biological weapon. Most pathogens can be cultured with standard laboratory equipment; however, the difficulty is in creating a system of *dispersal* that will effectively spread the inoculum over an area of susceptible hosts. There are innumerable systems that could be used; these include spraying particles into the air, mailing them in an envelope, or placing them into a bomb. One of the less sophisticated methods of dissemination is the use of aircraft that are used for agricultural spraying. These aircraft could spray inoculum over large areas, with the potential of infecting large populations.

J. J. Muchovej

See also: Bacteria; Biological invasions; Genetically modified bacteria; Viruses and viroids.

Sources for Further Study

Agrios, George N. *Plant Pathology*. 4th ed. Orlando, Fla.: Harcourt/Academic Press, 1997. General plant pathology text with sections on important plant pathogens and their biology. Includes illustrations, bibliographic references, and index.

Forbes, Betty A., Daniel F. Sahm, and Alice S. Weissfeld. *Bailey and Scott's Diagnostic Microbiology*. 10th ed. St. Louis: Mosby, 1998. General microbiology text with sections on important pathogens and their biology. Includes illustrations, bibliographic references, and index.

Miller, Judith, Stephen Engelberg, and William J. Broad. *Germs: Biological Weapons and America's Secret War*. New York: Simon & Schuster, 2001. A look at biological weapons and how they affected society in the late twentieth century.

Murray, Patrick R., Ken S. Rosenthal, George S. Kobayashi, and M. A. Pfaller. *Medical Microbiology*. 3d ed. St. Louis: Mosby, 1998. General microbiology text with sections on important pathogens and their biology. Includes illustrations, bibliographic references, and index.

BIOLUMINESCENCE

Categories: Algae; cellular biology; fungi; physiology; water-related life

Bioluminesence is the production of light by living organisms, including algae and phytoplankton in the oceans and fungi on land.

Bioluminescence is a specific form of chemiluminescence in which the chemical energy that is produced in a chemical reaction is converted into radiant energy. In bioluminescence the reaction originates in a wide variety of living organisms, including a small number of plants. It should not be confused with fluorescence or phosphorescence, both of which do not involve a chemical reaction. In either of the former cases the energy from a source of light, not from a chemical reaction, is basically absorbed and then re-emitted in some form of another photon. The chemical reactions that lead to bioluminescence release energy in the form of light. Unlike the light bulb, in which electrical energy is converted into light, with some of this energy lost in the form of heat, a bioluminescent reaction is 100 percent efficient and converts all the emitted energy into light. Because there is no heat released, bioluminescence is also known as "cold light."

Species and Habitats

Bioluminescence is primarily marine in nature and is the only source of light in the deep ocean, which is the largest habitable biome of the earth. The phenomenon rarely occurs in any source of fresh water. Bioluminescent organisms include ctenophores, annelid worms, mollusks, insects, and fish. The most common manifestation of this phenomenon on land is seen as a glowing fungus on wood or in the few families of luminous insects. This property can be used as a means of species recognition in the darkness as well as for courtship, preying, and mating.

There are several bioluminescent fungi that are not marine in nature, occurring primarily in the tropics. These fungi appear in different colors. The most common is *Panellus stiptucus*, which is a small decay fungus that is mostly restricted to North America. The jack-o'-lantern mushroom (*Omphalotus olearius*) glows brightly, especially when fresh. A few *Armillaria* species are also reported to glow mildly. No luminous tree or plant is known, however.

Mechanisms of Bioluminescence

Bioluminescence occurs only when two different species are in contact and, almost exclusively, when oxygen is present. The two species are *luciferin*, which produces the light, and *luciferase*, a protein that triggers and catalyzes the reaction. The mechanism involves the loss of two electrons, also known as *oxidation*, by luciferin, a process achieved only through the intervention of luciferase to yield oxyluciferin. Occasionally luciferin, luciferase, and a cofactor such as oxygen are bound together in a single moiety called *photoprotein*, which leads to light formation upon contact with a positively charged species, such as the calcium cation. The mechanism appears to involve a peroxide decomposition with free radical intervention.

Dinoflagellates

Dinoflagellates known as *Pyrrhophyta*, or fire plants, are the most common sources of bioluminescence at the surface of the ocean. They are a group of marine algae that produce light upon mechanical, chemical, or temperature changes. The phenomenon was first observed in the genus *Noctiluca* in the nineteenth century and has since been observed to occur within other species.

Generally, three types of stimuli can cause bioluminescence in dinoflagellates: mechanical, chemical, and temperature stimulation.

Mechanical forms of stimulation, such as the stirring of water from a moving boat, a swimming fish, or a breaking wave, are prevalent in many *Pyrrhophyta*. The light appears to serve as a "burglar alarm" against grazing predators, which are then being seen through the flash by a larger second predator. For example, as a copepod approaches the dinoflagellate, agitation of the seawater stimulates light flashes which a small fish, the secondary predator, uses to pinpoint the position of the copepod and eventually consume it. It appears that the mechanical stimulation deforms the cell membrane to create a short flash as little as one one-hundredth of a second.

Dinoflagellate luciferin is thought to derive from the similarly structured chlorophyll, which is found in most plants. The molecule is protected from luciferase at slightly basic medium by a luciferin-binding protein. However, once the acidity increases, the free luciferin reacts, and light is emitted. The light produced by a single dinoflagellate is only six to eight photons in energy, and the flashing may last only one-tenth of a second. Larger organisms, such as jellyfish, provide flashes that may last up to tens of seconds. Temperature lowering in some dinoflagellate species also creates bioluminescence.

Purpose and Applications

The disappearance of the flash, once oxygen is consumed, has suggested that the bioluminescent reaction was originally used to remove toxic oxygen from primitive types of bacteria that developed at a time when oxygen was not available. Bioluminescence has also played a crucial role in the direct studies of several cellular and biochemical processes, such as in the formation of ultimate carcinogens from benzoapyrene. The phenomenon has served scientists in many ways. Calcium levels are monitored via the jellyfish biochemical system, adenosine triphosphate (ATP) measurements are achieved through the firefly, and the gene activity of organisms can be detected by splicing known bioluminescent proteins.

Soraya Ghayourmanesh

See also: Dinoflagellates; Phytoplankton.

Sources for Further Study

Burkenroad, M. D. "A Possible Function of Bioluminescence." *Journal of Marine Research* 5 (1943): 161-164. This article was one of the first published giving the early history and facts about bioluminescence.

Fleisher, J. Kellie, and F. James Case. "Cephalopod Predation Facilitated by Dinoflagellate Luminescence." *Biological Bulletin* 189 (1995): 263-271. Discusses the effect of mechanical

stimulation that leads to the stimulation of dinoflagellates and subsequent bioluminescence.
Ganeri, Anita. *Creatures That Glow.* New York: Harry N. Abrams, 1995. For younger readers, introduces bioluminescent creatures, including bioluminescent fungi. Illustrated; includes a pull-out, glow-in-the-dark poster of the luminous creatures of the sea.
Hamman, J. P., and H. H. Seliger. "The Mechanical Triggering of Bioluminescence in Marine Dinoflagellates." *Journal of Cellular Physiology* 80 (1972): 397-408. This article discusses the various types of mechanical stimulation that lead to bioluminescence of dinoflagellates.

BIOMASS RELATED TO ENERGY

Categories: Ecology; ecosystems; environmental issues

The relationship between the accumulation of living matter resulting from the primary production of plants or the secondary production of animals (biomass) and the energy potentially available to other organisms in an ecosystem forms the basis of the study of biomass related to energy.

Biomass is the amount of organic matter, such as animal and plant tissue, found at a particular time and place. The rate of accumulation of biomass is termed *productivity*. *Primary production* is the rate at which plants produce new organic matter through photosynthesis. *Secondary production* is the rate at which animals produce their organic matter by feeding on other organisms. Biomass is an instantaneous measure of the amount of organic matter, while primary and secondary production give measures of the rates at which biomass increases. Plant and animal biomass consists mostly of carbon-rich molecules, such as sugars, starches, proteins, and lipids, and other substances, such as minerals, bone, and shell. The carbon-rich organic molecules are not only the building blocks of life but also the energy-rich molecules used by organisms to fuel their activities.

Solar Energy and Photosynthesis

Ultimately, all energy used by organisms to produce the building blocks of life and to drive life processes originated as solar energy captured by plants. Only a small fraction, less than 2 percent, of the total solar light energy received by a plant is absorbed and transformed by photosynthesis into energy-containing organic molecules. The rest of the sun's energy passes out of the plant as heat. The rate at which plants capture light energy and transform it into chemical energy is called primary production. Because plants do not rely on other organisms to provide their energy needs, they are referred to as primary producers, or *autotrophs* (meaning "self-feeding"). In addition to light energy, plants must absorb water, carbon dioxide gas, and simple nutrients, such as nitrate and phosphate, to produce various organic molecules during photosynthesis. Oxygen gas is also produced.

Sugars are the first energy-containing organic molecules produced in photosynthesis, and they can be changed to other, more complex, molecules, such as starches, proteins, and fats. The energy in the sugar molecules can be used immediately by the plants to maintain their own respiration needs, stored as starches and fats, or can be converted to new plant tissue. It is the stored organic matter plus new tissue that contributes to the growth of plants and to biomass.

Because the energy-containing products of photosynthesis can be used either immediately in respiration or in the formation of new plant biomass, two types of primary production can be distinguished. *Gross production* refers to the total amount of energy produced by photosynthesis. It includes both the energy used by the plant for respiration and the energy that goes into new biomass. *Net production* refers only to the amount of energy that accumulates as new biomass. It is only the energy in net production that is potentially available to animal consumers as food.

The rate of primary production varies directly with the rate of photosynthesis; therefore, factors in the environment that affect the rate of photosynthesis affect the rate of primary production. These factors most often include light intensity, temperature, nutrient concentrations, and moisture conditions. Each species of plant has a specific combination of these factors that promotes maximum rates of primary production. If one or more of these factors is in excess or is in short supply, then the rate of primary production is slowed.

Primary Production

On land, the rate of primary production by plants is determined largely by light, temperature, and rainfall. The favorable combination of intense sunlight for twelve hours per day, warm temperatures throughout the year, and considerable rainfall make the tropical rain forests the most productive ecosystems on land. In contrast, Arctic tundra vegetation is exposed to reduced light intensity, very cold winters, and cool summers. Primary production there is very low. In deserts, the lack of water severely limits primary production even though light and temperature are otherwise favorable.

In aquatic habitats, rates of primary production by algae, such as phytoplankton, are determined by nutrient concentration and light intensity. As sunlight penetrates water, it is quickly absorbed by the water molecules and by small suspended particles. Thus, all primary production occurs near the surface, as long as nutrients are available. Although the waters of the open ocean are very clear, and sunlight can penetrate to great depths, the scarcity of nutrients reduces the rate of primary production to less than one-tenth that of coastal bays.

Secondary Production

The energy and material needs of some organisms are met by consuming the organic materials produced by others. These consumer organisms are called *heterotrophs*; there are two types. Those that obtain their food from other living organisms are called *consumers* and include all animals. Those that obtain their energy from dead organisms are called *decomposers* and include mostly the fungi and bacteria.

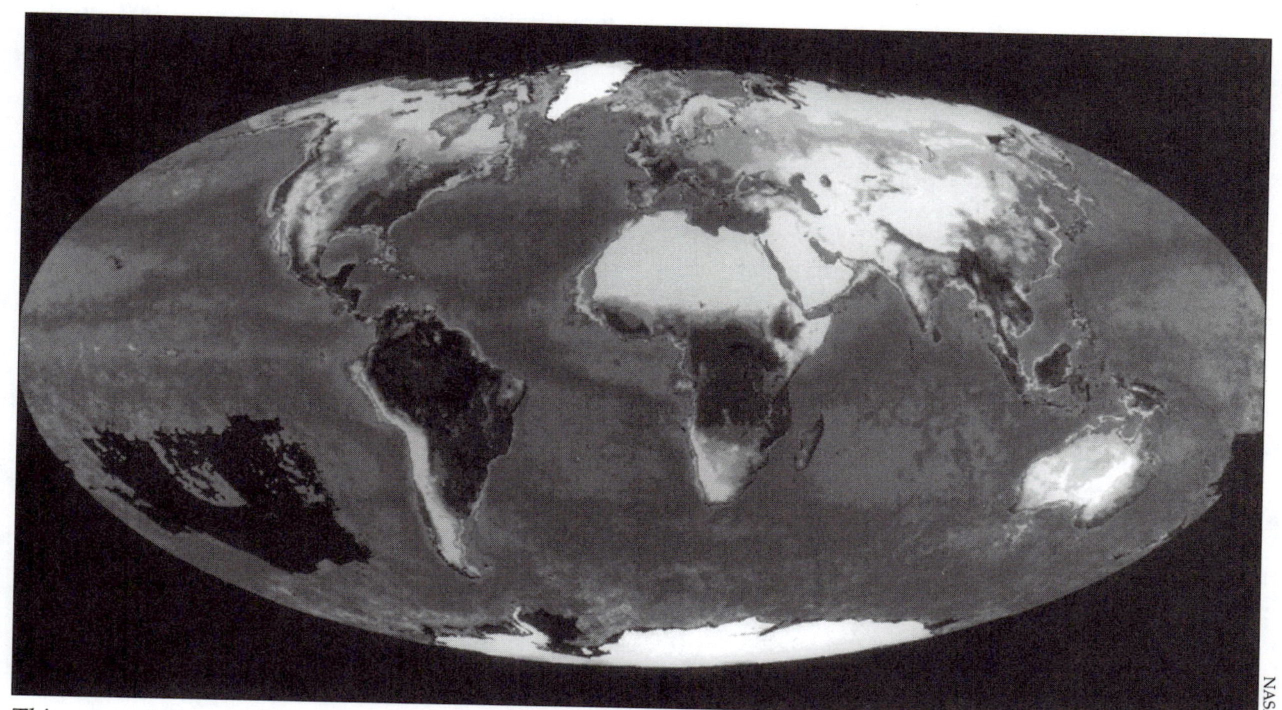

NASA

This composite image of the earth's biosphere shows the planet's heaviest vegetative biomass in the dark sections, known to be rain forests. The combination of intense sunlight for twelve hours per day, warm temperatures throughout the year, and considerable rainfall makes tropical rain forests the most productive ecosystems on land.

The energy available to each type of consumer becomes progressively less at each level of the *food chain*. Each consumer level uses most of its food energy, about 90 percent, to fuel its respiratory activities. In this energy-releasing process, most of the food energy is actually converted to heat and is lost to the environment. Only 10 percent or less of the original food energy is used to form new biomass. It is only this small amount of energy that is available for the next consumer level. The result is that food chains are limited in their number of links or levels by the reduced amount of energy available at each higher level.

Generally, the greater the amount of primary production, the larger the number of consumer organisms and the longer the food chain. Most food chains consist of three levels; rarely are there examples of up to five levels. It should be noted that the food chain concept is a simplified view of a more complex network of energy pathways, known as *food webs*, that occur in nature. Another outcome of the reduction in energy flow up the food chain is a progressive decrease in production and biomass. The most productive level, and the one with the greatest biomass, is therefore the primary producers, or plants.

Human Threats to Primary Production

The total natural primary production of the earth is limited, and human efforts to increase total world primary production much beyond its present levels may be futile. One reason for this is that much of the earth's surface lacks optimal conditions for plant growth. The open ocean, which covers about 71 percent of the earth's surface, has very little plant growth. On land, the Arctic, subarctic, and Antarctic regions are very unproductive most of the year. Human attempts to increase primary production in the form of food or fuel crops usually involve changing the characteristics of the land, converting forests into croplands, for example, and adding large quantities of nutrients and water. It has been estimated that humans are currently utilizing most of the easily workable croplands and that the development of additional lands for agriculture would require major changes to currently unworkable habitats, changes that would be expensive and demand much fuel energy.

The study of production processes is vitally important in understanding the ecology of natural ecosystems. Such information is necessary to manage and conserve *habitats* and their organisms in the face of human pressures. These processes provide insight into the general health of ecosystems. Pollutants, such as acid rain or industrial toxic wastes, are known to reduce the primary and secondary productivity of forests and lakes.

Throughout the world, humans are reducing the biomass of the world's primary producers through *deforestation*. This is particularly true in the tropics, where high population pressures have necessitated that land be cleared for agriculture and development. There is a worldwide demand for lumber. One obvious consequence is the dramatic reduction in the primary and secondary production of these areas. The *clear-cutting* (removal of all the trees) of tropical forests allows unprotected soils to wash away quickly during the heavy tropical rains. It will take hundreds of years for new soils to develop and for the forest to return—if it can return at all.

Deforestation is also harmful in that tropical forests form a major part of the world's life-support system. For millions of years these forests have buffered the earth's atmosphere by producing the oxygen gas needed by animals and by removing carbon dioxide and other toxic gases. The low level of carbon dioxide in the atmosphere is believed to have moderated the earth's temperature, counteracting the so-called *greenhouse effect*. It is therefore of great importance to understand and preserve these forests and other primary producers of the world.

Ray P. Gerber

See also: Deforestation; Food chain; Trophic levels and ecological niches.

Sources for Further Study

Brower, James E., and Jerrold H. Zar. *Field and Laboratory Methods for General Ecology*. 4th ed. Boston, Mass.: WCB/McGraw-Hill, 1998. In this manual, no previous ecological knowledge is assumed. Terms are defined and concepts developed as they appear. Principles and procedures explain how to make ecological studies. Examples of data and its analysis are given. Discusses analysis of production, including biomass measurements and aquatic productivity. Index, several appendices, and selected references are included.

Nybakken, James W. *Marine Biology: An Ecological Approach*. 5th ed. Menlo Park, Calif.: Benjamin Cummings, 2001. Providing an ecological approach to the entire marine environment, this undergraduate-level text emphasizes processes and interactions. The discussions of primary and secondary productivity are presented as each habitat is introduced and therefore appear throughout the text. Each chapter has fine illustrations and a useful bibliography. Includes glossary and index.

Odum, Howard T. *Ecological and General Systems: An Introduction to Systems Ecology*. Rev. ed. Niwot: University Press of Colorado, 1994. A thorough introduction to the movement of energy within ecosystems. Chapters cover energy systems, ecosystems and energy hierarchy, storage and flow, food webs, producers, and consumers. Includes references and author and subject indexes.

Pasztor, Janos, and Lars A. Kristoferson, eds. *Bioenergy and the Environment*. Boulder, Colo.: Westview Press, 1990. The results of a study by the Stockholm Environment Institute addressing the increasing need for biomass-based energy systems, with recommendations for energy planners.

Ricklefs, Robert E. *Ecology*. 4th ed. New York: W. H. Freeman, 1999. Stimulating text is written in an easy-to-follow format that combines theory with experimental work and field observations. Section 3 deals with energy and materials in the ecosystem. Chapter summaries, detailed glossary, bibliography, and index are included.

Smith, Robert L. *Ecology and Field Biology*. 5th ed. Menlo Park, Calif.: Addison Wesley Longman, 1996. This college-level text is easily readable by senior-level high school students. It has many illustrations with high information content, and the extensive appendices make the book a useful source. Chapters discuss basic concepts of energy flow and production in terms of specific ecosystems. Excellent glossary and index.

BIOMES: DEFINITIONS AND DETERMINANTS

Categories: Biomes; ecology; ecosystems

The concept of biomes is similar to plant ecologists' classification of plant formations and classification of life zones. However, the biomes usually refer to ecological communities of both plants and animals, whereas plant formations concern plant communities only. Worldwide, there are six major types of biomes on land: forest, grassland, woodland, shrubland, semidesert scrub, and desert.

One who travels latitudinally from the equator to the Arctic will cross tropical forests, deserts, grasslands, temperate forests, coniferous forest, tundra, and ice fields. Those major types of natural vegetation at regional scales are called biomes. A biome occurs wherever a particular set of climatic and edaphic (soil-related) conditions prevail with similar physiognomy. For example, prairies and other grasslands in the North American Middle West and West form a biome of temperate grasslands, where moderately dry climate prevails. Tropical rain forests in the humid tropical areas of South and Central America, Africa, and Southeast Asia create a biome where rainfall is abundant and well-distributed through the year.

In general, biomes are delineated by both physiognomy and environment. There are six major physiognomic types on land: *forest*, *grassland*, *woodland*, *shrubland*, *semidesert scrub*, and *desert*. Each of the six types occurs in a wide range of environments. Therefore, more than one biome may be defined within each physiognomic type according to major differences in climate. Tropical forests, temperate deciduous forests, and coniferous forests are, for example, separate biomes, although forests dominate all of them. On the other hand, some

Biomes and Their Features

Biome	*Annual Mean Rainfall*[1]	*Climate and Temperature*[2]
Desert	250 mm or less	Arid, with extremes of heat and cold
Grasslands	250-750 mm	Cold winters, warm summers; dry periods
Mediterranean scrub	Low to moderate	Cool winters, hot summers; latitudes 30° to 45°; includes chaparral, maquis
Rain forest (tropical)	2,500-4,500 mm	20-30°
Savanna, deciduous tropics	1,500-2,500 mm	Hot summers; 3-6 months dry; seasonal fires
Taiga (boreal forest)	1,000 mm	Cold, long winters; mild, short summers; seasonal fires
Tundra	Very low year-round	Very cold (3° or less); soil characterized by permafrost; Arctic tundra occurs in Arctic Circle; alpine tundra in other high elevations

1. In millimeters
2. Degrees Celsius

biome types, such as the tundra, are dominated by a range of physiognomic types and are in one prevailing environmental region.

Classification of Biomes

There are many ways to classify biomes. One system, which designates a small number of broadly defined biomes, divides global vegetation into nine major terrestrial biomes: tundra, taiga, temperate forest, temperate rain forest, tropical rain forest, savanna, temperate grasslands, chaparral, and desert. Other systems more narrowly define biomes, designating a larger total number. In those cases, some of the broadly defined biomes are divided into two or more biomes. For example, the biome called temperate forest in a broad classification may be separated into temperature deciduous forest and temperate evergreen forest in a fine classification. The biome of desert in the broad classification may be broken into warm semidesert, cool semidesert, Arctic-alpine semidesert, Arctic-pine desert, and true desert in the fine classification.

Description of Biome Distributions

Naturalists, geographers, and ecologists have tried to correlate world major types of biomes to climatic patterns in both descriptive and quantitative approaches. For example, in northern North America, the tundra and boreal forests are two broad belts of vegetation that stretch from east to west. The distribution of the two biomes is primarily influenced by temperature. South of those two belts are biome types that are mostly controlled by precipitation and evaporation. From east to west in North America, available moisture decreases, influencing biome distribution. Humid regions along the East Coast support forest biomes, including temperate coniferous forests and temperate deciduous forest. West of the eastern forests is a biome type of grasslands, including tall-grass prairie and short-grass steppe. In this zone, there is less precipitation than evaporation. The ratio of precipitation to evaporation is about 0.6 to 0.8 in the land that supports a tall-grass prairie and 0.2 to 0.4 farther west, where a short-grass steppe is supported. Beyond the short-grass steppe are shrubland and the deserts of the West. Western North America is a mountainous country in which vegetation zones reflect climatic changes on an altitudinal gradient. The vegetation in the lowlands is characteristic of the regions (short-grass steppe on the east side of Rocky Mountains, sagebrush cold semideserts in the Great Basin between the Rocky Mountains and the Sierra Nevada, and grasslands in California's Central Valley west of the Sierra Nevada). Above the base regions, the vegetation changes from shrub, woodland, or deciduous forest to montane coniferous forest or alpine tundra. In Central America, from Mexico to Panama where precipitation becomes ample and temperatures are high, tropical rain forests and tropical seasonal forests occur.

Similar distributions of biomes along latitude

and altitude can be found in South America, Africa, and Eurasia. In general, the climate-induced patterns of vegetation are influenced by latitude; the location of regions within a continent, which affects the amount of moisture they receive; and altitude, in which mountains modify the climate patterns. In addition, other factors, such as fire and human disturbance, may influence distributions of biomes. For example, most grasslands require periodic fires for maintenance, renewal, and elimination of incoming woody growth. Grasslands at one time covered about 42 percent of the land surface of the world. Humans have converted much of that area into croplands.

Quantitative Relationships

Descriptive relationships can provide pictures of world vegetation distributions along latitudinal and altitudinal gradients of temperature and moisture. Ecologists in the past several decades have also sought quantitative relationships between distributions of biomes and environmental factors. For example, when R. H. Whittaker plotted various types of biomes on gradients of mean annual temperature and mean annual precipitation in 1975, a global pattern emerged relating biomes to climatic variables. It was shown that tropical rain-forest biomes are distributed in regions with annual mean precipitation of 2,500 to 4,500 millimeters and annual mean temperatures of 20 to 30 degrees Celsius. Tropical seasonal forest and savannas also occur in warm regions with precipitation of 1,500-2,500 millimeters and 500-1,500 millimeters per year, respectively. Temperate forests occupy regions with annual temperature of 5 to 20 degrees Celsius and precipitation exceeding 1,000 millimeters per year.

PhotoDisc

Humid regions along the East Coast of North America support forest biomes, including temperate coniferous forests and temperate deciduous forests.

This thermal zone can support temperate rain forest when annual precipitation is more than 2,500 millimeters and temperate grassland when annual precipitation is below 750 millimeters. Temperate woodland occurs between temperate forests and grasslands. Tundra and taiga are distributed in regions with an annual mean temperature below 3 degrees Celsius, whereas deserts occupy areas with annual precipitation below 250 millimeters.

These relationships between climatic variables and biomes provide a reasonable approximation of global vegetation patterns. Many types of biomes intergrade with one another. Soil, exposure to fire, and regional climate can influence distributions of biomes in a given area.

Yiqi Luo

See also: Biomes: types; Climate and resources; Ecology: concept; Ecology: history; Ecosystems: overview; Ecosystems: studies.

Sources for Further Study

Archibold, O. W. *Ecology of World Vegetation*. London: Chapman & Hall, 1995. An advanced book of world major biomes.

Smith, R. L., and T. M. Smith. *Ecology and Field Biology*. 6th ed. San Francisco: Benjamin Cummings, 2001. An introductory ecology book with general descriptions and great illustrations of biomes.

Whittaker, R. H. *Communities and Ecosystems*. 2d ed. New York: Macmillan, 1975. A comprehensive book on community and ecosystem ecology with excellent discussions on biomes.

BIOMES: TYPES

Categories: Biomes; ecology; ecosystems

The major recognizable life zones of the continents are divided into biomes, characterized by their plant communities.

Temperature, precipitation, soil, and length of day affect the survival and distribution of biome species. Species diversity within a biome may increase its stability and capability to deliver natural services, including enhancing the quality of the atmosphere, forming and protecting the soil, controlling pests, and providing clean water, fuel, food, and drugs. Major biomes include the *temperate*, *tropical*, and *boreal forests*; *tundra*; *deserts*; *grasslands*; *chaparral*; and *oceans*.

Temperate Forests

The temperate forest biome occupies the so-called temperate zones in the midlatitudes (from about 30 to 60 degrees north and south of the equator). Temperate forests are found mainly in Europe, eastern North America, and eastern China, and in narrow zones on the coasts of Australia, New Zealand, Tasmania, and the Pacific coasts of North and South America. Their climates are characterized by high rainfall and temperatures that vary from cold to mild.

Temperate forests contain primarily *deciduous* trees—including maple, oak, hickory, and beechwood—and, secondarily, *evergreen* trees—including pine, spruce, fir, and hemlock. Evergreen forests in some parts of the Southern Hemisphere contain eucalyptus trees. The root systems of forest trees help keep the soil rich. The soil quality and color are due to the action of earthworms. Where these forests are frequently logged, soil runoff pollutes streams, which reduces spawning habitat for fish. Raccoons, opossums, bats, and squirrels are found in the trees. Deer and black bears roam forest floors. During winter, small animals such as marmots and squirrels burrow in the ground.

Tropical Forests

Tropical forests exist in frost-free areas between the Tropic of Cancer and the Tropic of Capricorn.

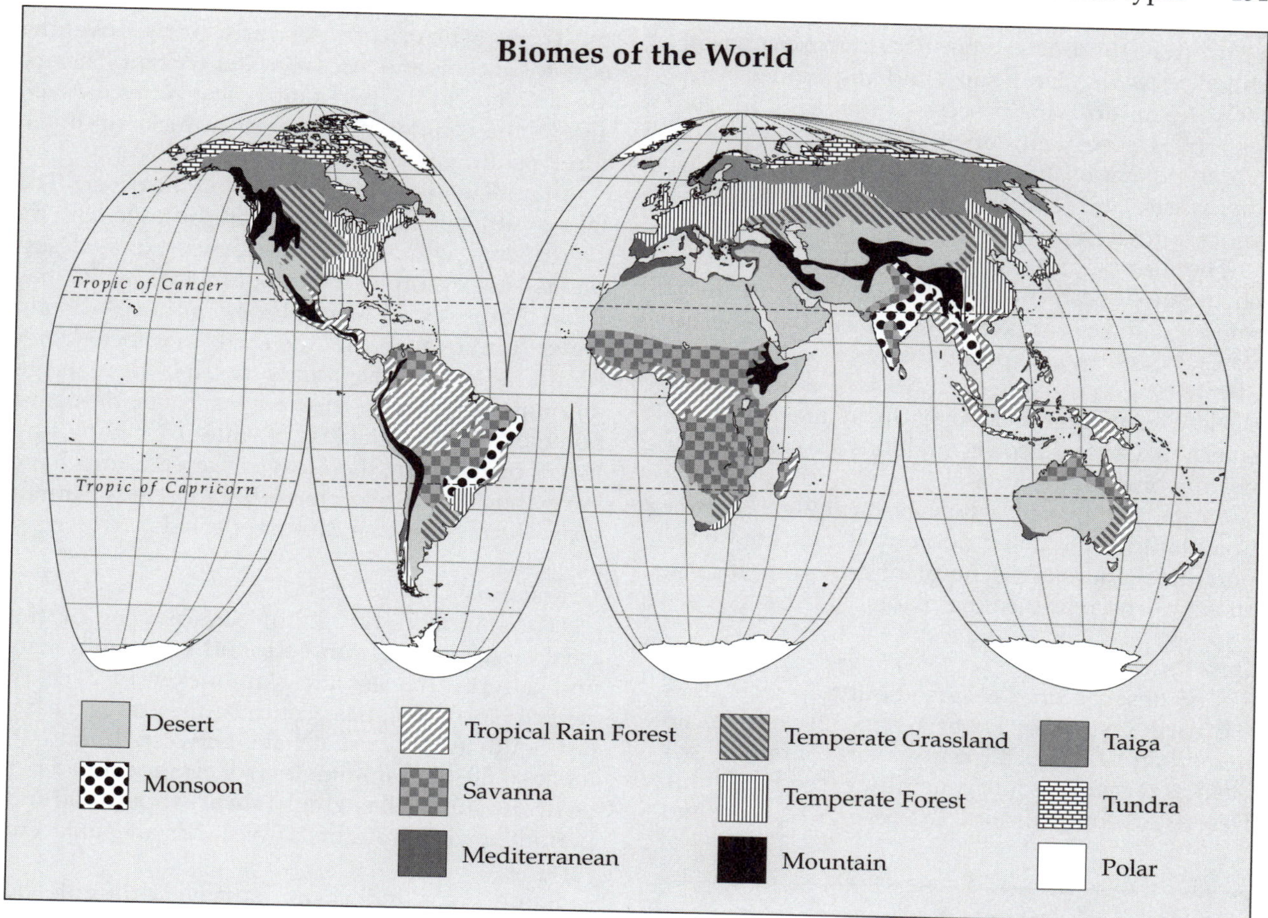

Temperatures range from warm to hot year-round. These forests are found in northern Australia, the East Indies, Southeast Asia, equatorial Africa, and parts of Central America and northern South America.

Tropical forests have high biological diversity and contain about 15 percent of the world's plant species. Animal life thrives at each layer of tropical forests. Nuts and fruits on the trees provide food for birds, monkeys, squirrels, and bats. Monkeys and sloths feed on tree leaves. Roots, seeds, leaves, and fruit on the forest floor feed larger animals. Tropical forest trees produce rubber and hardwood, such as mahogany and rosewood. Deforestation for agriculture and pastures has caused reduction in plant and animal diversity in these forests.

Boreal Forests

The boreal forest is a circumpolar Northern Hemisphere biome spread across Russia, Scandinavia, Canada, and Alaska. The region is very cold. Evergreen trees such as white spruce and black spruce dominate this zone, which also contains larch, balsam, pine, fir, and some deciduous hardwoods such as birch and aspen. The acidic needles from the evergreens make the leaf litter that is changed into soil humus. The acidic soil limits the plants that develop.

Animals in boreal forests include deer, bears, and wolves. Birds in this zone include red-tailed hawks, sapsuckers, grouse, and nuthatches. Relatively few animals emigrate from this habitat during winter. Conifer seeds are the basic winter food.

Tundra

About 5 percent of the earth's surface is covered with *Arctic tundra* and 3 percent with *alpine tundra*. The Arctic tundra is the area of Europe, Asia, and North America north of the boreal coniferous forest zone, where the soils remain frozen most of the year. Arctic tundra has a permanent frozen subsoil, called *permafrost*. Deep snow and low temperatures slow the soil-forming process. The area is bounded by a 50 degrees Fahrenheit (122 degrees Celsius) circumpolar *isotherm*, known as the summer isotherm. The cold temperature north of this line prevents normal tree growth.

The tundra landscape is covered by mosses, lichens, and low shrubs, which are eaten by caribou, reindeer, and musk oxen. Wolves eat these herbivores. Bears, foxes, and lemmings also live there. The most common Arctic bird is the old squaw duck. Ptarmigans and eider ducks are also very common. Geese, falcons, and loons are some of the nesting birds of the area.

The alpine tundra, which exists at high altitude in all latitudes, is acted upon by winds, cold temperatures, and snow. The plant growth is mostly cushion- and mat-forming plants.

Deserts

The desert biome covers about one-seventh of the earth's surface. Deserts typically receive no more than 10 inches (25 centimeters) of rainfall per year, and evaporation generally exceeds rainfall. Deserts are found around the Tropic of Cancer and the Tropic of Capricorn. As warm air rises over the equator, it cools and loses its water content. The dry air descends in the two subtropical zones on each side of the equator; as it warms, it picks up moisture, resulting in drying the land.

Rainfall is a key agent in shaping the desert. The lack of sufficient plant cover contributes to soil *erosion* during wind- and rainstorms. Some desert plants—for example, the mesquite tree, which has roots that grow 40 feet (13 meters) deep—obtain water from deep below the earth's surface. Other plants, such as the barrel cactus, store large amounts of water in their leaves, roots, or stems. Some plants slow the loss of water by having tiny leaves or shedding their leaves. Desert plants have very short growth periods, because they cannot grow during the long drought periods.

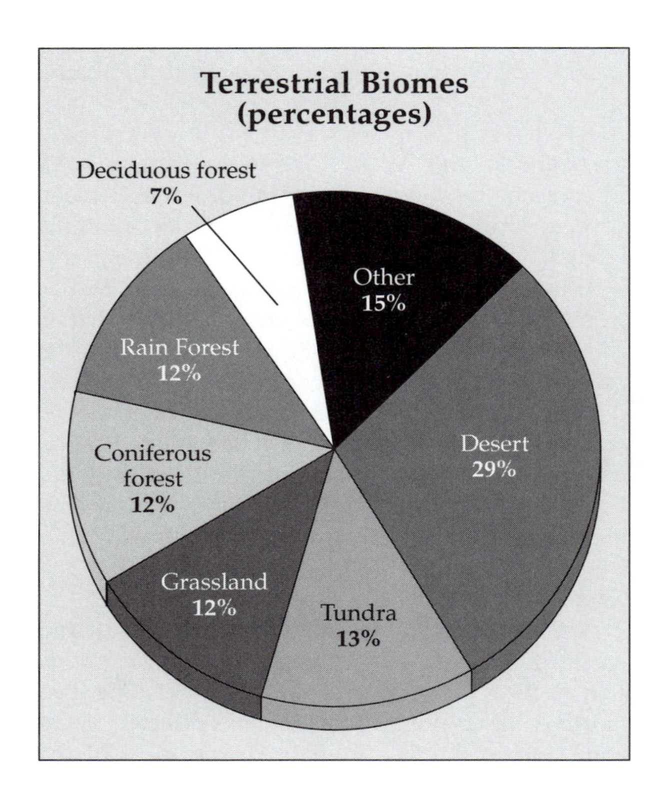

Grasslands

Grasslands cover about one-quarter of the earth's surface and can be found between forests and deserts. Treeless grasslands exist in parts of central North America, Central America, and eastern South America that have between 10 and 40 inches (250-1,000 millimeters) of erratic rainfall per year. The climate has a high rate of evaporation and periodic major droughts. Grasslands are subject to fire.

Some grassland plants survive droughts by growing deep roots, while others survive by being dormant. Grass seeds feed the lizards and rodents that become the food for hawks and eagles. Large animals in this biome include bison, coyotes, mule deer, and wolves. The grasslands produce more food than any other biome. Overgrazing, inefficient agricultural practices, and mining destroy the natural stability and fertility of these lands, resulting in reduced carrying capacity, water pollution, and soil erosion. Diverse natural grasslands appear to be more capable of surviving drought than are simplified manipulated grass systems. This may be due to slower soil mineralization and nitrogen turnover of plant residues in the simplified system.

Savannas are open grasslands containing deciduous trees and shrubs. They are near the equator and are associated with deserts. Grasses there grow in clumps and do not form a continuous layer.

Chaparral

The chaparral, or *mediterranean*, biome is found in the Mediterranean Basin, California, parts of

Australia, middle Chile, and the Cape Province of South America. This region has a climate of wet winters and summer drought. The plants have tough, leathery leaves and may have thorns. Regional fires clear the area of dense and dead vegetation. The seeds from some plants, such as the California manzanita and South African fire lily, are protected by the soil during a fire and later germinate and rapidly grow to form new plants. Vegetation *dwarfing* occurs as a result of the severe summer drought and extreme climate changes.

Oceans

The ocean biome covers more than 70 percent of the earth's surface and includes 90 percent of its volume. Oceans have four zones. The *intertidal zone* is shallow and lies at the land's edge. The *continental shelf*, which begins where the intertidal zone ends, is a plain that slopes gently seaward. The *neritic zone* (*continental slope*) begins at a depth of about 600 feet (180 meters), where the gradual slant of the continental shelf becomes a sharp tilt toward the ocean floor, plunging about 12,000 feet (3,660 meters) to the ocean bottom, which is known as the abyss. The *abyssal zone* is so deep that it does not have light.

Plankton are animals that float in the ocean. They include algae and copepods, which are microscopic crustaceans. Jellyfish and animal larva are also considered plankton. The nekton are animals that move freely through the water by means of their muscles. These include fish, whales, and squid. The benthos are animals that are attached to or crawl along the ocean's floor. Clams are examples of benthos. Bacteria decompose the dead organic materials on the ocean floor.

The circulation of materials from the ocean's floor to the surface is caused by winds and water temperature. Runoff from the land contains pollutants such as pesticides, nitrogen fertilizers, and animal wastes. Rivers carry loose soil to the ocean, where it builds up the bottom areas. Overfishing has caused fisheries to collapse in every world sector.

Human Impact on Biomes

Human interaction with biomes has increased *biological invasions*, reduced species *biodiversity*, changed the quality of land and water resources, and caused the proliferation of toxic compounds. Managed care of biomes may not be capable of undoing these problems.

Ronald J. Raven

See also: Arctic tundra; Biological invasions; Biomes: definitions and determinants; Deserts; Ecosystems: overview; Forests; Grasslands; Marine plants; Mediterranean scrub; Rain-forest biomes; Taiga; Tundra and high-altitude biomes; Wetlands.

Sources for Further Study

Food and Agriculture Organization of the United Nations. *State of the World's Forests, 2001*. Rome: Author, 2001. Discussion of forest policy and management. Includes a map titled "The World's Forests in 2000."

Gawthorp, Daniel, and David Suzuki. *Vanishing Halo: Saving the Boreal Forest*. Seattle: Mountaineers, 1999. Describes why the largest forest ecosystem in the world is so important and why so many people know so little about it.

Linsenmair, K. E., ed. *Tropical Forest Canopies: Ecology and Management*. London: Kluwer Academic, 2001. Summary of the understanding of canopy ecology; maps a path to a greater understanding of tropical forest ecology and management of this "last biological frontier."

Prager, Ellen J., with Cynthia A. Earle. *The Oceans*. New York: McGraw-Hill, 2000. Overview of oceanic knowledge, written by scientists and divers. Covers the same topics found in an introductory oceanography text but is directed at the general reader.

Solbrig, Otto Thomas, E. Medina, and J. F. Silva, eds. *Biodiversity and Savanna Ecosystem Processes: A Global Perspective*. New York: Springer, 1996. Addresses the role of species in the function of the most widespread ecosystem in the tropics. The intense human pressure on savannas may result in massive soil degradation.

BIOPESTICIDES

Categories: Agriculture; bacteria; biotechnology; economic botany and plant uses; environmental issues; microorganisms; pests and pest control

Biopesticides are biological agents, such as viruses, bacteria, fungi, mites, and other organisms used to control insect and weed pests in an environmentally and ecologically friendly manner.

Biopesticides allow biologically based, rather than chemically based, control of pests. A pest is any unwanted animal, plant, or microorganism. When the environment provides no natural resistance to a pest and when no natural antagonists are present, pests can run rampant. For example, spread of the fungus *Endothia parasitica*, which entered New York in 1904, caused the nearly complete destruction of the American chestnut tree because no natural control was present. Viruses, bacteria, fungi, protozoa, mites, insects, and flowers have all been used as biopesticides.

AP/Wide World Photos

This leaf-eating beetle, one of a species imported from Europe, is making headway in controlling invasive plant populations in New Hampshire and Vermont. Here, it prepares to dine on a leaf of the invasive plant purple loosestrife.

Advantages of Biopesticides

Many plants and animals are protected from pests by passive means. For example, plant rotation is a traditional method of insect and disease protection that is achieved by removing the host plant long enough to reduce a region's pathogen and pest populations. Biopesticides have several significant advantages over commercial pesticides. They appear to be ecologically safer than commercial pesticides because they do not accumulate in the food chain. Some biopesticides provide persistent control, as more than a single mutation is required to adapt to them and because they can become an integral part of a pest's life cycle. In addition, biopesticides have slight effects on ecological balances because they do not affect nontarget species. Finally, biopesticides are compatible with other control agents. The major drawbacks to using biopesticides are the time required for them to kill their targets and the inefficiency with which they work; also, if the organism being used as a biopesticide is a nonnative species, it may cause unforeseen damage to the local ecosystem.

Viruses and Bacteria

Viruses have been developed against insect pests such as *Lepidoptera* (butterflies and moths), *Hymenoptera* (bees, wasps, and ants), and *Dipterans* (flies). Gypsy moths and tent caterpillars, for example, periodically suffer from epidemic virus infestations, which could be exploited and encouraged.

Many *commensal* microorganisms (microorganisms that live on or in other organisms causing no direct benefit or harm) that occur on plant roots and leaves can passively protect plants against microbial pests by *competitive exclusion* (that is, simply crowding them out). *Bacillus cereus* has been used as an inoculum on soybean seeds to prevent infection by fungal pathogens in the genus *Cercospora*. Some microorganisms used as biopesticides produce antibiotics, but the major mechanism in most cases seems to be competitive exclusion. For example, *Agrobacterium radiobacter* antagonizes *Agrobacterium tumefaciens*, which causes the disease crown gall. Species of two bacterial genera—*Bacillus* and *Streptomyces*—when added as biopesticides to soil help control the damping-off disease of cucumbers, peas, and lettuce caused by *Rhizoctonia solani*. *Bacillus subtilis* added to plant tissue also controls stem rot and wilt rot caused by species of the fungus *Fusarium*. *Mycobacteria* species produce cellulose-degrading enzymes, and their addition to young seedlings helps control fungal infection by species of *Pythium*, *Rhizoctonia*, and *Fusarium*. Species of *Bacillus* and *Pseudomonas* produce enzymes that dissolve fungal cell walls.

***Bacillus thuringiensis* and *Bacillus popilliae* as Microbial Biocontrol Agents**

	Bacillus thuringiensis	*Bacillus popilliae*
Pest controlled	*Lepidoptera* (many)	*Coleoptera* (few)
Pathogenicity	low	high
Response time	immediate	slow
Formulation	spores and toxin crystals	spores
Production	in vitro	in vivo
Persistence	low	high
Resistance in pests	developing	reported

Source: Data adapted from J. W. Deacon, *Microbial Control of Plant Pests and Diseases* (1983).

***Bacillus thuringiensis* Toxins**

The best examples of microbial insecticides are *Bacillus thuringiensis* (*B.t.*) toxins, which were first used in 1901. They have had widespread commercial production and use since the 1960's and have been successfully tested on 140 insects, including mosquitoes. Insecticidal endotoxins are produced by *B.t.* during sporulation, and exotoxins are contained in crystalline parasporal protein bodies. These protein crystals are insoluble in water but readily dissolve in an insect's gut. Once dissolved, the proteolytic enzymes paralyze the gut. Spores that have been consumed germinate and kill the insect. *Bacillus popilliae* is a related bacterium that produces an insecticidal spore that has been used to control Japanese beetles, a corn pest.

The gene for the *B.t.* toxin has also been inserted into the genomes of cotton and corn, producing genetically modified, or GM, plants that produce their own *B.t.* toxin. GM cotton and *B.t.* corn both express the gene in their roots, which provides them with protection from root worms. Ecologists and environmentalists have expressed concern that constantly exposing pests to *B.t.* will cause insects to develop resistance to the toxin. In such a scenario, the effectiveness of traditionally applied *B.t.* would decrease.

Fungi and Protozoa

Saprophytic fungi can compete with pathogenic fungi. There are several examples of fungi used as biopesticides, such as *Gliocladium virens*, *Trichoderma hamatum*, *Trichoderma harzianum*, *Trichoderma viride*, and *Talaromyces flavus*. For example, *Trichoderma* species compete with pathogenic species of *Verticillium* and *Fusarium*. *Peniophora gigantea* antagonizes the pine pathogen *Heterobasidion annosum* by three mechanisms: It prevents the pathogen from colonizing stumps and traveling down into the root zone, it prevents the pathogen from traveling between infected and uninfected trees along interconnected roots, and it prevents the pathogen from growing up to stump surfaces and sporulating.

Nematodes are pests that interfere with commercial button mushroom (*Agaricus bisporus*) production. Several types of nematode-trapping fungi can be used as biopesticides to trap, kill, and digest the nematode pests. The fungi produce constricting and nonconstricting rings, sticky appendages, and spores, which attach to the nematodes. The most common nematode-trapping fungi are *Arthrobotrys oligospora*, *Arthrobotrys conoides*, *Dactylaria candida*, and *Meria coniospora*.

Protozoa have occasionally been used as biopesticide agents, but their use has suffered because of slow growth and the complex culture conditions associated with their commercial production.

Mites, Insects, and Flowers

Well-known "terminator" bugs include praying mantis and ladybugs as well as decollate snails, which eat the common brown garden snail. Fleas, grubs, beetles, and grasshoppers often have natural nematode species that prey on them, which can be used as biocontrol agents. Predaceous mites are used as a biopesticide to protect cotton from other insect pests such as the boll weevil. Parasitic wasps of the genus *Encarsia*, especially *E. formosa*, prey on whiteflies, as does *Delphastus pusillus*, a small, black ladybird beetle.

Dalmatian and Persian insect powders contain pyrethrins, which are a toxic insecticidal compounds produced in *Chrysanthemum* flowers. Synthetic versions of these naturally occurring compounds are found in products used to control head lice.

Mark S. Coyne, updated by Elizabeth Slocum

See also: Bacterial resistance and super bacteria; Herbicides; Pesticides.

Sources for Further Study

Carozzi, Nadine, and Michael Koziel, eds. *Advances in Insect Control: The Role of Transgenic Plants*. Bristol, Pa.: Taylor & Francis, 1997. Provides a rich source of information about technologies which have been proven successful for engineering of insect-tolerant crops as well as an overview of the new technologies for future genetic engineering.

Deacon, J. W. *Microbial Control of Plant Pests and Diseases*. Washington, D.C.: American Society for Microbiology, 1983. Good but brief monograph that gives a broad perspective of biocontrol agents.

Hall, Franklin R., and Julius J. Menn, eds. *Biopesticides: Use and Delivery*. Totowa, N.J.: Humana Press, 1999. A guide to development, application, and use of biopesticides as a complementary or alternative treatment to chemical pesticides. Reviews their development, mode of action, production, delivery systems, and future market prospects and discusses current registration requirements for biopesticides as compared with conventional pesticides.

Hokkanen, Heikki M. T., and James M. Lynch, eds. *Biological Control: Benefits and Risks*. New York: Cambridge University Press, 1995. A discussion and debate of the benefits and risks associated with biological control.

BIOSPHERE CONCEPT

Categories: Ecology; ecosystems; environmental issues

The term "biosphere" was coined in the nineteenth century by Austrian geologist Eduard Suess in reference to the 20-kilometer-thick zone extending from the floor of the oceans to the top of mountains, within which all life on earth exists. Thought to be more than 3.5 billion years old, the biosphere supports nearly one dozen biomes, regions of climatic conditions within which distinct biotic communities reside.

Compounds of hydrogen, oxygen, carbon, nitrogen, potassium, and sulfur are cycled among the four major spheres, one of which is the biosphere, to make the materials that are essential to the existence of life. The other spheres are the *lithosphere*, the outer part of the earth; the *atmosphere*, the whole mass of air surrounding the earth; and the *hydrosphere*, the aqueous vapor of the atmosphere, sometimes defined as including the earth's bodies of water.

The Water Cycle

The most critical of these compounds is water, and its movement among the spheres is called the *hydrologic cycle*. Dissolved water in the atmosphere condenses to form clouds, rain, and snow. The annual precipitation for any region is one of the major factors in determining the terrestrial biome that can exist. The precipitation takes various paths leading to the formation of lakes and rivers. These flowing waters interact with the lithosphere (the outer part of the earth's crust) to dissolve chemicals as they flow to the oceans. Evaporation of water from the oceans then supplies most of the moisture in the atmosphere. This cycle continually moves water among the various terrestrial and oceanic biomes.

Solar Energy

The biosphere is also dependent upon the energy that is transferred from the various spheres. Solar energy is the basis for almost all life. Light enters the biosphere as the essential energy source for photosynthesis. Plants take in carbon dioxide, water, and light energy, which is converted via *photosynthesis* into chemical energy in the form of sugars and other organic molecules. Oxygen is generated as a by-product. Most animal life reverses this process during *respiration*, as chemical energy is released to do work by the oxidation of organic molecules to produce carbon dioxide and water.

Incoming solar energy also interacts dramatically with the water cycle and the worldwide distribution of biomes. Because of the earth's curvature, the equatorial regions receive a greater amount of solar heat than the polar regions. Convective movements in the atmosphere—such as winds, high- and low-pressure systems, and weather fronts—and the hydrosphere—such as water currents—are generated during the redistribution of this heat. The weather patterns and climates of earth are a response to these energy shifts. Earth's various climates are defined by the mean annual temperature and the mean annual precipitation.

Toby R. Stewart and Dion Stewart

See also: Biomes: types; Carbon cycle.

Sources for Further Study

McNeely, Jeffrey A. *Conserving the World's Biological Diversity*. Washington, D.C.: International Union for Conservation of Nature and Natural Resources, 1990. Covers strategies being used to conserve the biosphere around the world.

Smith, Vaclav. *Cycles of Life: Civilization and the Biosphere*. New York: W. H. Freeman, 2000. Introduction to biogeochemical cycles. Explains the interrelationship of carbon, nitrogen, sulfur, and living organisms as agents of change in the environment.

Vernadskii, V. I. *The Biosphere*. Translated by Mark A. S. McMenamin. New York: Copernicus, 1998. Reviewed as required reading for all students in earth and planetary sciences. Describes life as a cosmological phenomenon and a means by which energy is stored and transformed on a planetary scale.

Weiner, Jonathon. *The Next One Hundred Years: Shaping the Fate of Our Living Earth*. New York: Bantam Books, 1991. Discusses threats to the earth's biosphere.

Wilson, E. O., ed. *Biodiversity*. Washington, D.C.: National Academy Press, 1988. Contains articles from noted biologists on topics about biodiversity and problems facing biodiversity in biomes.

BIOTECHNOLOGY

Categories: Agriculture; bacteria; biotechnology; disciplines; economic botany and plant uses; environmental issues; genetics; history of plant science

Biotechnology is the use of living organisms, or substances obtained from those organisms, to produce processes or products of value to humanity, such as foods, high-yield crops, and medicines.

Modern biotechnological advances have provided the ability to tap into a natural resource, the world gene pool, with such great potential that its full magnitude is only beginning to be appreciated. Theoretically, it should be possible to transfer one or more genes from any organism in the world into any other organism. Because genes ultimately control how any organism functions, gene transfer can have a dramatic impact on agricultural resources and human health in the future.

History of Biotechnology

Although the term "biotechnology" is relatively new, the practice of biotechnology is at least as old as civilization. Civilization did not evolve until humankind learned to produce food crops and domestic livestock through the controlled breeding of selected plants and animals. Eventually humans began to utilize microorganisms in the production of foods such as cheese and alcoholic beverages. During the twentieth century, the pace of modification of various organisms accelerated. Through carefully controlled breeding programs, plant architecture and fruit characteristics of crops have been modified to facilitate mechanical harvesting. Plants have been developed to produce specific drugs or spices, and microorganisms have been selected to produce antibiotics and other medicinal or food products.

Developments in Biotechnology

Since the mid-twentieth century, the ability to utilize artificial media to propagate plants has led to the development of a technology called tissue culture. The earliest form of tissue culture involved using the culture of meristem tissue to produce numerous tiny shoots that can be grown into full-size plants, referred to as clones because each plant is genetically identical. More than one thousand plant species have been propagated by tissue culture techniques. Plants have been propagated via the culture of other tissues, including the stems and roots. In some of these techniques, the plant tissue is treated with hormones to produce callus tissue, masses of undifferentiated cells. The callus tissue can be separated into single cells to establish a cell suspension culture. Callus tissue and cell suspensions can be used to produce specific drugs and other chemicals. Entire plants can also be generated from the callus tissue or from single cells by addition of specific combinations of hormones.

A far more complex method of cloning of plants and animals from the deoxyribonucleic acid (DNA) of a single cell is a more recent development. Proponents of this method of producing copies of organisms have suggested that cloning technology might be used to improve agricultural stock and to regenerate endangered species. These ideas have had their detractors, however, as critics have noted the potential dangers of narrowing a species' gene pool. The July, 1996, birth in Scotland of Dolly, a sheep cloned and raised to adulthood, demonstrated that the cloning of animals had left the realm of science fiction and become a matter of scientific fact.

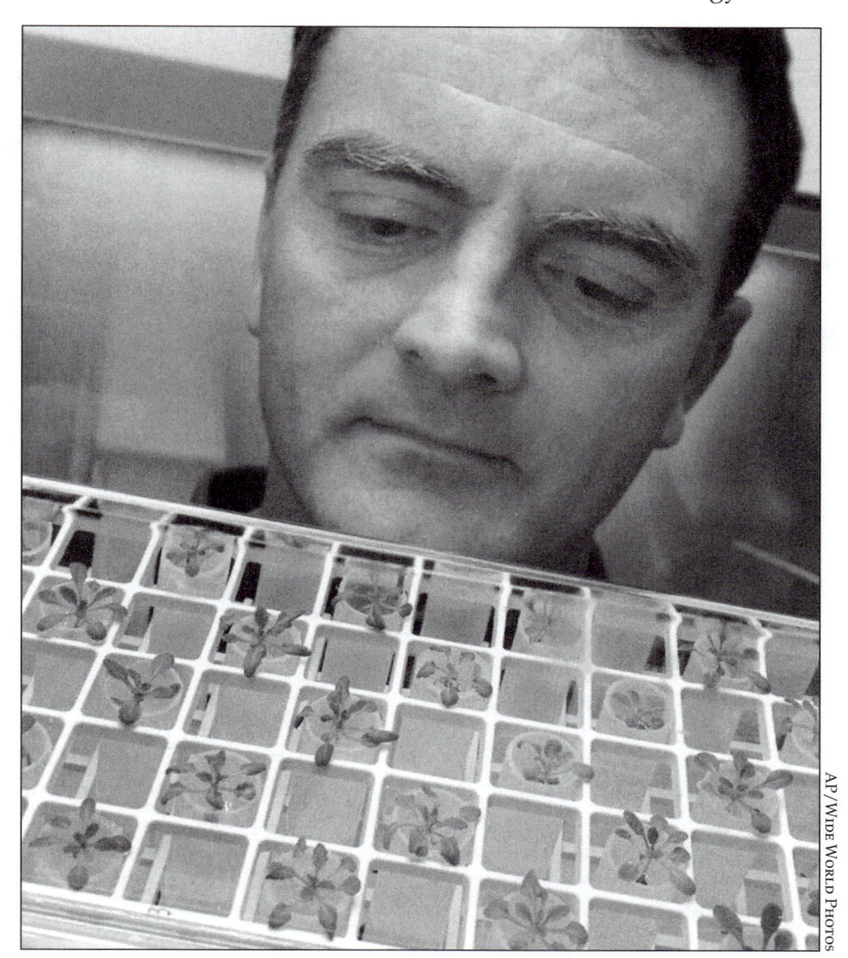

AP/Wide World Photos

These genetically modified mustard plants could be sent on a mission to Mars to provide NASA scientists with information about soil conditions on the red planet. Plants in distress have been bred to emit a fluorescent glow.

Recombinant DNA Technology

In practice, recombinant DNA methodology is complex, but in concept, it is fairly easy to comprehend. The genes in all living cells are composed of the same chemical, DNA. The DNA of all cells, whether from bacteria, plants, or animals including humans, is very similar. When DNA from a foreign species is transferred into a different cell, it functions exactly as the native DNA functions; that is, it "codes" for protein.

The easiest way to manipulate genes is using bacterial cells (most often *Escherichia coli*) and a *vector*, an agent that can be used to pass the gene from one cell to another. *Plasmids*, small extra circular DNA molecules found in many bacterial cells, are commonly used for this purpose. Plasmids are replicated along with the bacterial cell's own DNA every time the cell reproduces. Plasmids can be easily isolated from bacterial cells. When a specific gene has been isolated, it can be fused, using *restriction endonucleases*, with a plasmid to produce a recombinant plasmid. These *recombinant plasmids* can then be put into bacterial cells by a process called *transformation*. Special plasmids called *expression vectors* allow expression of inserted genes once they are inside a bacterial cell.

Although expressing foreign genes in bacterial cells is relatively simple, inserting them into plants and getting them expressed is more complicated. The *Ti plasmid* is a widely used vector that works well in dicots but has never worked for monocots. Consequently, the first successful transgenic plants were dicots, while success with the most important food crops, monocots such as rice and corn, took more time and effort.

Many alternative methods for inserting genes into plant cells have been developed that work on both monocots and dicots. *Microinjection* can be used to insert a gene into individual cells. A less laborious method is called *biolistic*, for biological ballistic, where millions of copies of the gene are attached to tiny projectiles that are then fired into groups of plant cells. Using these and other methods, genetic modification of plants is becoming more routine.

Future of Biotechnology in Agriculture

This new technology could have a tremendous impact on agriculture. As the human population grows, biotechnology will most likely play an important role in producing an increase in food production. Such an increase will require developments such as crop plants that will produce higher yields under normal conditions and crops that will produce higher yields when grown in marginal environments. Biotechnology provides a means of developing higher-yielding crops in much less time than it takes to develop them though traditional plant-breeding programs. Genes for the desired characteristics can be inserted directly into the plant without having to go through repeated controlled selection and breeding cycles to establish the trait.

There are also economic advantages in diversifying agriculture production in a given area. A producer might wish to grow a particular high-value cash crop in an area where soil or climate conditions would prevent such a crop from thriving. Biotechnology can help solve these types of problems. For example, high value crops can be developed to grow in areas that heretofore would not have supported such crops. Plants also can be developed to produce new products such as antibiotics, drugs, hormones, and other pharmaceuticals. Crop plants bioengineered to produce novel products mean that pharmaceuticals and other valuable products could be grown in farm environments rather than in laboratories.

While there will be a growing pressure for agriculture to produce more food in the future, there will also be pressure for crop production to be more friendly to the environment. Biotechnology has the potential to play a major role in the development of a long-term, sustainable, environmentally friendly agricultural system. For example, the development of crop varieties with improved resistance to pests will reduce the reliance on pesticides. Methods of crop production and harvest with less environmental impact will also have to be developed. Because agriculture will continue to have an impact on the environment, the need to remediate polluting agents will continue to exist. Hence biotechnology will play an important role in the development of bioremediation systems for agriculture as well as other industrial pollutants.

Ownership Issues

There will be many difficult ethical and economic issues surrounding the use of this new biotechnology. One of the major questions concerns ownership. Patent laws in the United States read that ownership over an organism can be granted if the organism has been intentionally genetically modified through the use of recombinant DNA techniques. In addition, processes that utilize genetically modified organisms can be patented. Therefore one biotechnology firm may own the patent to an engineered organism, but another firm may own the rights to the process used to produce it.

D. R. Gossett, updated by Bryan Ness

See also: Microbial nutrition and metabolism.

Sources for Further Study

Bruno, Kenny. "Say It Ain't Soy, Monsanto." *Multinational Monitor* 18, no. 1-2 (January-February, 1997): 27-30. Discusses labeling and other biotechnology industry issues in the context of Roundup Ready Soybeans.

Callow, J. A., B. V. Ford-Lloyd, and H. J. Newbury, eds. *Biotechnology and Plant Genetic Resources: Conservation and Use*. New York: CAB International, 1997. Illustrates how new techniques in biotechnology are being applied to plant genetic resources, especially genetic diversity and conservation.

Rissler, Jane. *The Ecological Risks of Engineered Crops*. Cambridge, Mass.: MIT Press, 1996. Identifies and categorizes the environmental risks presented by commercial uses of transgenic plants. Presents a practical, feasible method of precommercialization evaluation to balance the needs of ecological safety with those of agriculture and business.

Santaniello, V., R. E. Evenson, and D. Zilberman, eds. *Agriculture and Intellectual Property Rights: Economic, Institutional, and Implementation Issues in Biotechnology*. New York: CAB International, 2000. Covers issues in plant breeding patents, the ownership of biological innovation, and associated intellectual property rights. Includes perspectives of policymakers and economists.

Shargool, Peter, ed. *Biotechnological Applications of Plant Cultures*. Boca Raton, Fla.: CRC Press, 1994. Plant culture techniques play a pivotal role in supplying uniform and reproducible materials for biotechnological experimentation. This book presents reviews of techniques in plant culture work.

Tant, Carl. *Awesome Green*. Angleton, Tex.: Biotech, 1994. Explores myths, fears, and facts about plant biotechnology. Down-to-earth descriptions include flow charts and diagrams. Offers a glimpse of what is coming soon, such as tomatoes that taste like tomatoes; eat a banana to get your next vaccination; new drugs for cancer; and biodegradable plastics from plants. Questions are raised about the social responsibility of high-tech industries.

BOTANY

Categories: Disciplines; history of plant science

Botany is the branch of science that studies plant life.

Botany is a very old branch of science that began with early people's interest in the plants around them. Plant science now extends from that interest to cutting-edge biotechnology. Any topic dealing with plants, from the level of their *cellular biology* to the level of their economic production, is considered part of the field of botany.

History and Subdisciplines

The origins of this branch of biology are rooted in human beings' attempts to improve their lot by raising better food crops around 5000 B.C.E. This practical effort developed into intellectual curiosity about plants in general, and the science of botany was born. Some of the earliest botanical records are included with the writings of Greek philosophers, who were often physicians and who used plant materials as curative agents. In the second century B.C.E. Aristotle had a botanical garden and an associated library.

As more details became known about plants and their functions, particularly after the discovery of the microscope, a number of subdisciplines arose. Plant *anatomy* is concerned chiefly with the internal structure of plants. Plant *physiology* delves into the living functions of plants. Plant *taxonomy* has as its interest the discovery and systematic classification of plants. Plant geography, also known as geobotany or *phytogeography*, deals with the global distribution of plants. Plant *ecology* studies the interactions between plants and their surroundings. Plant *morphology* studies the form and structure of plants. Plant *genetics* attempts to understand and work with the way that plant traits are inherited. Plant *cytology*, often called cell biology, is the science of cell structure and function. *Economic botany*, which traces its interest back to the origins of botany, studies those plants that play important economic roles. These include major crops such as wheat, rice, corn, and cotton.

Ethnobotany is a rapidly developing subarea in which scientists communicate with indigenous peoples to explore the knowledge that exists as a part of their folk medicine. Several new drugs and the promise of others have developed from this search.

At the forefront of botany today is the field of *genetic engineering*, including the *cloning* of organisms. New or better crops have long been developed by the technique of *crossbreeding*, but genetic engineering offers a much more direct course. Using its techniques, scientists can introduce a gene carrying a desirable trait directly from one organism to another. In this way scientists hope to protect crops from frost damage, to inhibit the growth of weeds, to provide insect repulsion as a part of the plant's own system, and to increase the yield of food and fiber crops.

The role that plants play in the energy system of the earth (and may someday play in space stations or other closed systems) is also a major area of study. Plants, through photosynthesis, convert sunlight into other useful forms of energy upon which humans have become dependent. During the same process carbon dioxide is removed from the air, and oxygen is delivered. Optimization of this process and discovering new applications for it are goals for botanists.

Kenneth H. Brown

See also: History of plant science; Plant science.

Sources for Further Study

Raven, Peter H., Ray F. Evert, and Susan E. Eichhorn. *Biology of Plants*. 6th ed. New York: W. H. Freeman/Worth, 1999. The authors devote a significant portion of the introductory material identifying and defining the field of botany, including why in modern contexts it often includes the study of life-forms other than plants and how it overlaps with disciplines such as cellular biology, genetics, and ecology.

Stannard, Jerry, Katherine E. Stannard, and Richard Kay, eds. *Pristina Medicamenta: Ancient and Medieval Medical Botany*. Brookfield, Vt.: Ashgate, 1999. Articles on premodern texts on plants.

BROMELIACEAE

Categories: Angiosperms; economic botany and plant uses; *Plantae*; taxonomic groups

The family Bromeliaceae *comprises a group of perennial, monocotyledon herbs or trees that often age slowly.*

Important ornamentals (called bromeliads) as well as sources of food and medicines, *Bromeliaceae* have substantial economic value and are widely cultivated. The colors of the leaves offer decorative foliage, and the flowers are of astonishing hues due to the rich content of pigment-forming substances known as anthocyanins. Based on ovary position, habit, and floral and pollen morphology, the family *Bromeliaceae* has been split into three subfamilies: subfamily *Pitcairnioideae*, subfamily *Tillandsioideae*, and subfamily *Bromelioideae*. There are fifty-six genera and approximately twenty-six hundred species, growing mostly in the neotropical regions of the world, from Virginia to southern Argentina. One species, *Pitcairnia feliciana*, originated in Africa. This interesting family can nevertheless occupy a variety of ecologically diverse environments, ranging from the dry deserts in Peru to the highest montane forest in the Andes Mountains.

Appearance and Structure

The *Bromeliaceae* family shares a basic ground plan of construction that consists of branches (ramets) and an inflorescence that follows a repetitive pattern when growing. However, modifications, in the form of reductions, of this basic plan have evolved in different subfamilies. The basic pattern consists of sympodial branching, a rhythmic type of growth in which the axis is built up by a linear series of shoot units, each distal unit developing from an axillary bud located on the previous shoot unit. This pattern of development leads to a series of condensed ramets with terminal flowers. Roots, when present, usually emerge from the lower half of each ramet.

Growing Habit

Bromeliaceae range from small plants, such as some miniature *Tillandsia*, to very tall individuals, such as *Puya raimomndii*, reaching up to 32 feet (10 meters) in height. They can be epiphytes, that is, plants that use other species as support without harming them, or terrestrial. Some grow on top of rocks, and some are carnivorous.

Those species whose leaves are born from a common place in the stem (in a rosulate shape) can develop the tank form, also known as *phytotelma*, that is common in genera such as *Aechmea* and *Brocchinia*. These phytotelma harbor a variety of insects and small vertebrates that grow in small pools

PhotoDisc

The pineapple, Ananas comosus, *is the species of bromeliad most widely used as food.*

of water and old leaves that collect at the bottom of the "tank." The tanks accumulate water and partially dissolved organic matter, creating a nutrient-rich substrate as a continuous supply of moisture. Other *Bromeliaceae* do not form tanks; instead, they have fully functional roots and specialized hairs for water absorption.

Scales

Physiological adaptations to different environments among some species correlate with the presence of a highly evolved type of foliar hair (or trichome) known as a *scale*. The scales may cover the entire surface of the leaf, sometimes appearing in different locations and patterns; they absorb atmospheric water through capillary action, like blotting paper, and the water is later transported to the leaf tissue, where it is stored in the parenchyma. Division of the scale in two parts—known as the shield, or *trichome covering*, and the water absorption cells—is what makes *Bromeliaceae* unique. When water is scarce the scale shrinks, and when water is present the shield cells expand. Scales protect the leaves against transpiration and reduce water evaporation during the dry periods.

Flowers and Pollination

The flowers of *Bromeliaceae* are generally hermaphroditic (functionally unisexual). Their shape can be radial or slightly asymmetric, and the number of floral parts known as sepals and petals is always three. The stamen arrangement is in two whorls, with three stamens in each one. The ovary can be superior or inferior, and the placentation (position of the ovules) is mostly axial. Septal nectaries are always present at the base of the flower. The sepals are distinguished from the petals by their color and size. The petals show bright colors, while the sepals may remain mostly in green hues. Fruits are usually a capsule or a berry, and the seeds are winged.

The bloom of *Bromeliaceae* flowers is usually odorless, although some species may have scented flowers, indicating pollination by nocturnal moths or butterflies. However, their abundant secretion of nectar indicates that the plants are pollinated primarily by birds.

Uses

The main uses of *Bromeliaceae* are as textile fiber, food, medicine, and ornamental plantings. In the food category the pineapple, *Ananas comosus*, is the most widely used species. The medicinal properties of pineapple are based on the presence of bromelain, a proteolitic (protein-breaking) enzyme that is widely used to treat inflammation and pain. Serotonin, a neurotransmitter, is also present, and steroids from the leaves possess estrogenic activity. Thirteen species of *Bromeliaceae* are used as a source of textile fibers; for example, hammocks are made from the fibers of *Aechmea bracteata* and of pineapple.

Miriam Colella

See also: South American flora.

Sources for Further Study

Benzing, David H. *Bromeliaceae: Profile of an Adaptive Radiation*. New York: Cambridge University Press, 2000. Synthesis of available information on the biology of *Bromeliaceae*, including reproductive and vegetative structure, ecology, and evolution, rather than floristics and taxonomy.

Parkhurst, Ronald W. *The Book of Bromeliads and Hawaiian Tropical Flowers*. Honolulu: Pacific Isle, 2001. Covers commercial, collector, and hybrid bromeliads as well as growing and care, diseases and pests, landscaping, and more. Includes glossary and index.

Paull, R. E., ed. *The Pineapple: Botany, Production, and Uses*. New York: Oxford University Press, 2001. Aimed at researchers and horticulturalists, the book covers topics of botany, taxonomy, genetics, breeding, production, disease, and postharvest techniques.

BROWN ALGAE

Categories: Algae; *Protista;* taxonomic groups; water-related life

Seaweeds that are brown to olive-green in color belong to the phylum Phaeophyta, *or brown algae, which includes between fifteen hundred and two thousand species.*

Brown algae (phylum *Phaeophyta*) are familiar to most people as brown or dark green seaweeds. Some brown algae are microscopic in size, but many are relatively large: One giant kelp measured 710 feet in length. All brown algae are multicellular.

Appearance and Distribution

Brown algae have a body, called a *thallus*, which is a fairly simple, undifferentiated structure. Some thalli consist of simple branched filaments. Some brown algae have more complex structures called *pseudoparenchyma* because they superficially resemble the more complex tissues of higher plants.

Giant kelp have a thallus that is differentiated into a *holdfast*, a *stipe*, and one or more flattened, leaflike *blades*. The holdfast functions as the name implies, and holds the rest of the organism to the substrate. It is a tough, sinewy structure resembling a mass of intertwined roots. The stalk that constitutes the stipe is often hollow, with a meristem (a zone of growing tissue) either at its base or at the blade junctions. Because the meristem produces new tissue at the base, the oldest parts of the blades are at the tips.

The blades, which, like most of the rest of the giant kelp body, are photosynthetic, may have gas-filled floats called bladders toward their bases, which may contain carbon monoxide gas. The function of this particular gas has not yet been determined.

The vast majority of species are marine, living in cold, shallow ocean waters, and may be the dominant plant life on rocky coastlines. The giant kelp can be found in waters around 100 feet deep. Only 4 of the 260 identified genera occur in fresh water. Brown algae of the order *Fucales* are commonly

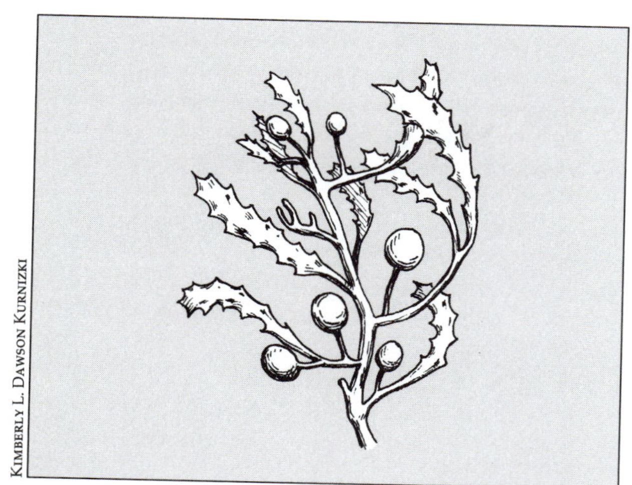

Brown algae, which range in color from olive green to golden, are among the largest algae, including the giant kelps, and often have a thallus (body) that is differentiated into a holdfast, a stipe (stalk), and one or more flattened, leaflike blades.

called rockweeds; kelp belong to the order *Laminariales.*

Brown algae are less common in tropical and subtropical areas. However, in the Caribbean region, *sargassum* (large masses of brown algae having a branching thallus with lateral outgrowths differentiated into leafy segments, air bladders, or spore-bearing structures) make up large floating mats; they gave their name to the Sargasso Sea.

Pigments and Food Reserves

The color of the brown algae can vary from light yellow-brown to almost black. Its color reflects the presence of varying amounts of the brown xanthophyll pigment fucoxanthin, a carotenoid pigment, in addition to chlorophylls *a* and *c*. The main food reserve is a carbohydrate called laminarin, although giant kelp can also translocate mannitol. Algin (alginic acid) can be found in or on the cell walls and may comprise as much as 40 percent of the dry weight of some kelps.

Reproduction

Reproductive cells of brown algae are unusual in that their two flagella are located laterally, instead of at the ends. The only motile cells in the brown algae are the gametes or reproductive cells. In the common genus *Fucus*, separate male and female thalli are produced. Fertile areas called *receptacles* develop at the tips of the lobes of the thallus. Each receptacle has pores on the surface. These pores open into special spherical, hollow chambers called *conceptacles*, in which the gametes are formed. Eight eggs are produced in the female structure, while sixty-four sperm cells are produced in the male structure. Eventually, both eggs and sperm are released into the water, where fertilization takes place and the resulting zygotes develop into mature thalli.

Economic Uses

Brown algae have several uses and applications for humans. Giant kelp is eaten, and one species found in the Pacific Ocean has been used, in chopped-up form, as a poultice applied to cuts. Algin, a colloidal substance produced by brown algae, is used as a thickener or stabilizer in commercially produced ice cream, salad dressing, beer, jelly beans, latex paint, penicillin suspensions, paper, textiles, toothpastes, and floor polish. Brown algae, with its high concentration of the element iodine, has been used to treat goiter, an iodine-deficiency disease. Kelp, also high in nitrogen and potassium, has been used as fertilizer and as livestock feed.

Some types of brown algae, such as *Fucus*, contain either phenols or terpenes. Botanists believe these substances may discourage herbivory. These substances also have been shown to possess microbe- and cancer-fighting properties. Brown algae is the subject of continuing research in these areas of medicine.

Carol S. Radford

See also: Agriculture: marine; Algae; Marine plants; Medicinal plants; Red algae.

Sources for Further Study

Graham, Linda E., and Lee W. Wilcox. *Algae*. Upper Saddle River, N.J.: Prentice Hall, 2000. Comprehensive textbook focuses on diversity and relationships among the major algal types, algal roles in food webs, global biogeochemical cycling, the formation of harmful algal blooms, and ways people use algae. Also provides broad coverage of freshwater, marine, and terrestrial algae.

Meinesz, Alexandre. *Killer Algae*. Chicago: University of Chicago Press, 1999. French scientist recounts the epidemic spread of a tropical green algae, *Caulerpa taxifolia*, along the Mediterranean coastline. Recommended for environmentalists and students of biology or ecology.

Sze, Philip. *A Biology of the Algae*. Boston: WCB/McGraw-Hill, 1998. Basic textbook presents an overview of different algal groups.

BRYOPHYTES

Categories: Nonvascular plants; paleobotany; *Plantae*; taxonomic groups

Bryophytes comprise three phyla of nonvascular plants, which generally lack the specialized conductive tissues (xylem and phloem) that are found in the vascular plants, are small in size, and are distributed worldwide in moist, shady habitats.

Bryophytes (from the Greek word *bryon*, meaning "moss") were once grouped together into one large phylum. Many botanists today recognize that these organisms belong to at least three distinct phyla: phylum *Hepatophyta* (the liverworts), phylum *Anthocerophyta* (the hornworts), and phylum *Bryophyta* (the mosses).

Origin and Relationships

Bryophytes are thought to have originated more than 430 million years ago, during the Silurian period. Many botanists speculate that bryophytes arose from an ancestor in the green algal order *Charales* or *Coleochaetales* based on biochemical, morphological, and life history comparisons. For example, *Chara* has a flavonoid biosynthesis pathway that is similar to that of higher plants, while *Coleochaete* retains its zygote inside parental tissue, similar to higher plants. These characteristics, along with similarities in cell division patterns, photosynthetic pigment contents, and the use of starch as a storage material, all suggest ancestry in the *Charales* or *Coleochaetales* orders.

Historically, bryophytes were thought to represent a group that formed a separate lineage from that of vascular plants. By the late 1990's a growing body of evidence suggested that bryophytes and vascular plants were derived from a common green algal ancestor. Some botanists suggest that the earliest land plants may have been members of the phylum *Anthocerophyta*. One of the key arguments in this theory is that the structure of some hornwort chloroplasts is virtually identical to the chloroplast structure of the presumed algal ancestors.

Studies conducted in the 1990's involving the presence or absence of certain portions of noncoding deoxyribonucleic acid (DNA) called *introns* in the genetic information of several groups of algae, bryophytes, and vascular plants revealed that members of the *Hepatophyta*, the liverworts, were likely among the first land plants. Like the algae, they lack the introns that are found in groups that are presumed to be more derived. Thus, based on the assumption that introns are derived characters, ancestors of modern liverworts may have given rise to vascular plants.

Anatomy

The dominant phase of the bryophyte life cycle is the haploid gametophyte phase. The gametophyte is photosynthetic and is usually small because of the lack of efficient vascular tissues.

Bryophytes possess rootlike *rhizoids* that anchor the plant to the soil and aid in nutrient uptake. A waxy *cuticle*, which helps prevent water loss, covers the body. Liverworts have pores for gas exchange, while hornworts and mosses have stomata to regulate gas movement. Some liverworts and hornworts have a thalloid body type, which is not differentiated into leaf and stem. The *thallus* may be simple, composed of a ribbonlike, flattened body of relatively undifferentiated tissues, or complex, in which there is a distinct differentiation of tissues. The flat body may aid in the uptake of water and minerals and in gas exchange. The bodies of some liverworts and the mosses are divided into leaf and stem. These terms are used for convenience even though xylem and phloem are not present.

Some mosses possess tissues that have functions similar to xylem and phloem. *Hydroids* are water-conducting cells that make up a tissue called hadrom. *Leptoids* are food-conducting cells that make up a tissue called leptom. These tissues appear similar to the conducting tissues in a group of fossil plants called *protracheophytes*, which are thought to be an intermediate group between the bryophytes and the vascular plants.

The diploid sporophyte of liverworts and mosses consists of a *foot*, which is attached to a stalklike *seta*. The seta connects the foot to the spore-producing organ called the *sporangium*, or *capsule*. The hornwort sporophyte, however, lacks a seta and possesses a long, cylindrical sporangium. The foot of the bryophyte sporophyte contains specialized transfer cells, which bring materials from the maternal gametophyte to the sporophyte. The sporophyte is totally dependent on the maternal gametophyte for its survival. A layer of sterile tissue called the *calyptra* covers the capsules of liverworts and mosses. When the spores are mature, the sporophyte may die, allowing the release of spores as the capsule decays (as in some thalloid liverworts). Alternatively, the capsule may rupture, allowing spores to be released through pores (as in mosses and leafy liverworts), or the capsule may split along the side to release the spores (as in the hornworts). Liverworts and hornworts often have specialized structures in the capsules called *elaters* that aid in dispersing spores from the capsules.

Reproduction and Life Cycle

Bryophytes may reproduce either asexually or sexually. Asexual reproduction primarily occurs by *fragmentation*. Some of the liverworts also repro-

Rob and Ann Simpson/Photo Agora

Common liverworts, male and female plants. Liverworts make up one of three phyla of the bryophytes.

duce asexually by the production of small masses of vegetative tissue called *gemmae* in special structures called *gemma cups*. Water drops disperse the gemmae.

Bryophytes exhibit a typical plant life-cycle pattern called *alternation of generations*. There are distinct male and female gametophytes in some species, while other species produce both male and female organs in one plant. The reproductive organs are all multicellular. Male organs are called *antheridia*. Special cells within an antheridium undergo mitotic cell division to produce flagellated haploid sperm cells. The sperm cells are the only flagellated cells produced by bryophytes. As with many other plant groups, the presence of flagella on the sperm indicates that these cells require liquid water to swim to the egg.

The female organs are called *archegonia*. The archegonium is composed of a slender neck, within which is a canal. The base of the archegonium has a swollen region called the venter, which contains the egg. Special cells within an archegonium undergo mitotic cell division to produce a haploid egg.

If one gametophyte produces both antheridia and archegonia, the organs usually develop at different times, to reduce to likelihood of self-fertilization. When the sperm and eggs are mature, sperm are released from the antheridia in the presence of liquid water. Water drops transfer sperm from an antheridium to an archegonium. Sperm cells swim through the neck canal of the archegonium where fertilization occurs in the venter. The resulting zygote develops into an embryo, which then grows into the diploid sporophyte.

Sporogenous tissues in the sporangium undergo meiosis to produce haploid *spores*. The spore walls contain a substance called *sporopollenin*, which is resistant to chemicals and decay. After release, spores germinate and grow into new haploid gametophytes. The early threadlike stage of mosses and some liverworts is called the *protonema*. Protonemata are very similar to the body form of some algae.

Phylum *Hepatophyta*

There are between six thousand and eight thousand species of *hepatophytes* (from the Greek word *hepar*, meaning "liver"), which are commonly called liverworts. Hepatophytes are divided into three general groups: the simple thalloid liverworts, the complex thalloid liverworts, and the leafy liverworts. More than 85 percent of all hepatophyte species are leafy. Liverworts are usually terrestrial, although some species may be semiaquatic. Thalloid types are found worldwide. Leafy liverworts, which are often similar in appearance to mosses, are abundant in tropical jungles and fog belts. However, they are typically found in habitats that are more moist than those preferred by mosses.

Phylum *Anthocerophyta*

This phylum, the hornworts, consists of some one hundred species and represents the smallest group of bryophytes. The best-known genus, *Anthoceros* (from the Greek words *anthos*, meaning "flower" and *keras*, meaning "horn"), is found in temperate regions. The gametophyte is similar to thalloid liverworts. The cavities of the gametophyte body are filled with *mucilage*, a slimy secretion, in which grow nitrogen-fixing cyanobacteria, such as the genus *Nostoc*.

Phylum *Bryophyta*

Phylum *Bryophyta*, the mosses, consists of more than ninety-five hundred species. There are three important classes: class *Sphagnidae*, which includes the globally distributed, and economically as well as ecologically important genus *Sphagnum*; class *Andreaeidae*, which consists of a small group of blackish green to reddish brown tufted rock mosses growing on granitic or calcareous rocks in northern latitudes; and the class *Bryidae*, which consists of true mosses.

Economic Uses

Bryophytes are ecologically important members of terrestrial ecosystems. They are primary producers, providing food and habitat for animals. Humans have used bryophytes for many purposes. For example, *Sphagnum* deposits in peat bogs have been used for centuries as fuel for heating and cooking. Dried *Sphagnum* also has the ability to absorb large amounts of liquid, which makes it ideal to act as a soil conditioner for planting. American Indians used mosses as compresses to dress wounds. The antiseptic quality of *Sphagnum*, along with its absorptive properties, made its use attractive as bandage material for the British when cotton supplies were low during World War I.

Darrell L. Ray

See also: Hornworts; Liverworts; Mosses.

Sources for Further Study

Bell, Peter R., and Alan R. Hemsley. *Green Plants: Their Origin and Diversity*. 2d ed. New York: Cambridge University Press, 2000. Contains an introduction to major groups of bryophytes. Includes photos, diagrams.

Raven, Peter H., Ray F. Evert, and Susan E. Eichhorn. *Biology of Plants*. 6th ed. New York: W. H. Freeman/Worth, 1999. A standard text for introductory botany courses at the university level. Discusses structure and significance of major groups of bryophytes. Includes photos and diagrams.

Shaw, A. Jonathan, and Bernard Goffinet, eds. *Bryophyte Biology*. New York: Cambridge University Press, 2000. A scholarly study of bryophyte origins, systematics, and biology. Includes tables, photos, and diagrams.

BULBS AND RHIZOMES

Categories: Anatomy; physiology; reproduction and life cycles

Bulbs and rhizomes are modified stems, stem bases, or other underground organs used by plants for food (or energy) storage and in asexual reproduction.

Plants reproduce both sexually and asexually. Although sexual reproduction is part of the typical life cycle of plants, for a variety of reasons a plant may reproduce asexually. Exact duplicates of a plant, called *clones*, are formed by asexual reproduction.

Asexual Reproduction

Asexual reproduction involves the production of offspring through the formation of *propagules* by mitosis (the process of nuclear cell division). Because genetic recombination does not occur in mitosis, the offspring are genetically identical to the parent plant. Asexual reproduction does not occur in all plants; some reproduce asexually only when humans intervene. Asexual reproduction occurs when a single plant produces a vegetative propagule that develops into a separate free-living plant. Many of the propagules that support asexual reproduction are actually highly modified branches. Others are modified roots. In rare instances, the tissues of leaves may be modified by nature to support asexual reproduction. The propagules of asexual reproduction vary enormously. They are often found in catalogs describing "bulbs," but technically they include *true bulbs*, *corms*, *stolons*, *tubers*, *rhizomes*, *turions*, *pseudobulbs*, and *fleshy roots*.

True Bulbs

Bulbs, corms, stolons, tubers, rhizomes, and turions are all modified stems. Bulbs are modified stem bases that develop underground. The stem is shortened and thickened to produce a mass of tissue shaped like a coin or like a child's toy top. Scalelike leaves with thickened bases are attached to the base of the bulb. Starch is stored in the thickened bases, a food supply that allows the bulb to survive through a dormant season and to produce adventitious roots. Roots are often absent when the bulb is dormant. The starch can also support a period of rapid stem and leaf growth in the growing season and may support the flowering and fruiting of the plant.

In *tunicate* bulbs, a cloak of dried leaves surrounds the outside of the bulb. These dried leaves provide a barrier to desiccation and allow the tunicate bulbs to be stored aboveground for weeks or even months. Onions (*Allium cepa*) and daffodils (*Narcissus*) are examples of tunicate bulbs. Other bulbs have no cloak and usually have shorter, less cylindrical leaves. These scaly bulbs dry quickly when kept aboveground and usually develop flowers only after a more normal, aerial branch system forms. Lilies (*Lilium*) have scaly bulbs.

Stems of both tunicate and scaly bulbs can branch. Belowground, branches appear at first as

AP/Wide World Photos

*The mountain ramp plant (*Allium tricoccum*) is a relative of the onion. Its bulb can be seen at the base of the plant. In late winter or early spring, the ramp bulb sends up several broad, ovate leaves.*

miniature bulbs (*bulbils* or *bulblets*). Bulblets take their energy from the parent bulb but eventually produce aerial stems or leaves and can be separated from the parent. Profuse branching can be stimulated by wounding the stem of the parent bulb. All the bulbs produced by this technique are clones, identical to the parent in their genetic makeup and physical characteristics.

Corms and Tubers

Corms are similar to bulbs in many ways. They have a disk-shaped or top-shaped stem mass that is shorter and wider than most typical stems. They are often cloaked in a tunic of dried leaves that are thinner and smaller than those on bulbs. Corms do not store significant amounts of starch; it is instead stored primarily in the basal plate of the stem. Branches of the corm stem produce new, miniature corms (*cormels*). Wounding the parent stem stimulates greater branching. Gladiolus and crocus are two common garden plants that produce corms.

Tubers are thick, starchy stems that form usually at the tip of a stolon, runner, or tiller. Tubers may form on the soil surface or belowground. A familiar example is the white or Irish potato (*Solanum tuberosum*). The leaves on most tubers are much smaller than the leaves on other parts of the stem, but above each leaf on the tuber is a well-developed axillary bud, commonly called an *eye*. The axillary bud has the potential to elongate, forming a complete and fully developed branch. If the stolon connecting the tuber to the parent plant dies, the branch from the eye of the tuber becomes an independent clone of the parent plant.

Most tubers contain many eyes. If the tuber is cut into smaller pieces, each containing an eye, each piece develops a rhizome, which in turn develops a new tuber. By this technique, a significant increase in the number of plants can be obtained. The cut pieces of tuber are initially prone to decay, but after they have dried for a few days, they heal over with a layer of callus which protects them like a skin. Cutting tubers into small "seed" pieces is a common method for propagating tuber-forming species.

Rhizomes, Stolons, and Fleshy Roots

Rhizomes are specialized, underground stems. Unlike most areal stems, the rhizomes are normally oriented horizontally. Just as pieces of tuber can provide the tissue and energy source for the formation of a new plant, so too can pieces of a rhizome. Many ferns and fern allies propagate naturally by rhizomes. Large stands of these plants can form from a single individual as the rhizomes grow and branch. Eventually, older pieces of the rhizome die, leaving a population of individuals that all have identical genetic characteristics.

Stolons are long, thin, horizontal stems, also called *runners* or *tillers*, which grow along the surface of the ground. When the stolon has grown far enough from the parent plant, the growth pattern changes, and a *crown*, or tuber, forms. A crown is a compressed stem mass with leaves arranged close to one another, also called a *rosette*. Within the crown, roots form at the points of attachment of the leaves to the stems. If the stolon is broken or dies, the crown becomes an independent clone of the parent plant. In this way, a large number of offspring can be produced from a single plant. This is a mechanism of reproduction of the strawberry (*Fragaria*) and is also a common reproductive mechanism for grasses, including crab grass (*Digitaria sanguinalis*) and quack grass (*Agropyron repens*).

Fleshy roots store energy that can be useful for asexual propagation. Most require at least a small amount of stem tissue to support cell growth and differentiation. The true yam (*Dioscorea*) is a tuber, but the sweet potato (*Ipomoea batatas*) is a fleshy root which can easily be propagated asexually. Many buttercups (*Ranunculus*) are also propagated by breaking up the clumps of their fleshy roots.

Craig R. Landgren, updated by Bryan Ness

See also: Roots; Stems.

Sources for Further Study

Hartmann, Hudson T., and Dale E. Kester. *Plant Propagation*. 6th ed. Englewood Cliffs, N.J.: Prentice Hall, 1997. A classic text covering plant propagation, asexual, and sexual reproduction. Illustrated.

Hartmann, Hudson T., A. M. Kofranek, V. E. Rubatzky, and William J. Flocker. *Plant Science: Growth, Development, and Utilization of Cultivated Plants*. 2d ed. Englewood Cliffs, N.J.: Prentice Hall, 1988. A readable introductory botany text that places plant growth in context with the biology of plants. The chapter on propagation of plants describes asexual reproduction. Heavily illustrated with black-and-white drawings and photographs.

Kaufman, Peter B., et al. *Plants: Their Biology and Importance*. San Francisco: Harper & Row, 1989. Illustrated with black-and-white and color drawings and photographs, this botany text is easily understood. The chapter on modifications of plant organs provides a good review of bulbs but is not as well illustrated as it could be.

Mogie, Michael. *The Evolution of Asexual Reproduction in Plants*. New York: Chapman and Hall, 1992. A study of the "cost of sex" in plant evolution, concluding that plants that reproduce asexually never developed sexual reproduction, rather than having developed sexual reproduction and then selecting against it. Includes bibliography and index.

Raven, Peter H., and George B. Johnson. *Biology*. 5th ed. Boston: WCB/McGraw-Hill, 1999. Beautifully illustrated with many color drawings and photographs, this college-level biology text places plant reproduction in context of the biology of plants and that of the plants in context of the living and nonliving world. Chapter 40, "Sex and Reproduction," provides a good explanation of the consequences of asexual reproduction.

Richards, A. J. *Plant Breeding Systems*. 2d ed. New York: Chapman and Hall, 1997. A unified and comprehensive treatment of plant reproduction. Mostly focuses on sexual reproduction but considers asexual modes as well. Includes glossary, references, and index.

C_4 AND CAM PHOTOSYNTHESIS

Categories: Photosynthesis and respiration; physiology

Alternative forms of photosynthesis are used by specific types of plants, called C_4 and CAM plants, to alleviate problems of photorespiration and excess water loss.

Photosynthesis is the physiological process whereby plants use the sun's radiant energy to produce organic molecules. The backbone of all such organic compounds is a skeleton composed of carbon atoms. Plants use carbon dioxide from the atmosphere as their carbon source.

The overwhelming majority of plants use a single chemical reaction to attach carbon dioxide from the atmosphere onto an organic compound, a process referred to as *carbon fixation*. This process takes place inside specialized structures within the cells of green plants known as *chloroplasts*. The enzyme that catalyzes this fixation is ribulose bisphosphate carboxylase (Rubisco), and the first stable organic product is a three-carbon molecule. This three-carbon compound is involved in the biochemical pathway known as the *Calvin cycle*. Plants using carbon fixation are referred to as C_3 plants because the first product made with carbon dioxide is a three-carbon molecule.

C_4 Photosynthesis

For many years scientists thought that the only way photosynthesis occurred was through C_3 photosynthesis. In the early 1960's, however, researchers studying the sugarcane plant discovered a biochemical pathway that involved incorporation of carbon dioxide into organic products at two different stages. First, carbon dioxide from the atmosphere enters the sugarcane leaf, and fixation is accomplished by the enzyme phosphoenolpyruvate carboxylase (PEP carboxylase). This step takes place within the cytoplasm, not inside the chloroplasts. The first stable product is a four-carbon organic compound that is an acid, usually malate. Sugarcane and other plants with this photosynthetic pathway are known as *C_4* plants.

In C_4 plants, this photosynthetic pathway is tied to a unique leaf anatomy known as *Kranz anatomy*. This term refers to the fact that in C_4 plants the cells that surround the water- and carbohydrate-conducting system (known as the *vascular system*) are packed very tightly together and are called *bundle sheath cells*. Surrounding the bundle sheath is a densely packed layer of *mesophyll* cells. The densely packed mesophyll cells are in contact with air spaces in the leaf, and because of their dense packing they keep the bundle sheath cells from contact with air. This Kranz anatomy plays a major role in C_4 photosynthesis.

In C_4 plants the initial fixation of carbon dioxide from the atmosphere takes place in the densely packed mesophyll cells. After the carbon dioxide is fixed into a four-carbon organic acid, the malate is transferred through tiny tubes from these cells to the specialized bundle sheath cells. Inside the bundle sheath cells, the malate is chemically broken down into a smaller organic molecule, and carbon dioxide is released. This carbon dioxide then enters the chloroplast of the bundle sheath cell and is fixed a second time with the enzyme Rubisco and continues through the C_3 pathway.

Advantages of Double-Carbon Fixation

The double-carbon fixation pathway confers a greater photosynthetic efficiency on C_4 plants over C_3 plants, because the C_3 enzyme Rubisco is highly inefficient in the presence of elevated levels of oxygen. In order for the enzyme to operate, carbon dioxide must first attach to the enzyme at a particular location known as the *active site*. However, oxygen is also able to attach to this active site and prevent carbon dioxide from attaching, a process known as *photorespiration*. As a consequence, there is an ongoing competition between these two gases for attachment at the active site of the Rubisco enzyme. Not only does the oxygen outcompete carbon dioxide; when oxygen binds to Rubisco, it also destroys some of the molecules in the Calvin cycle.

At any given time, the winner of this competition is largely dictated by the relative concentrations of these two gases. When a plant opens its *stomata* (the pores in its leaves), the air that diffuses in will be at equilibrium with the atmosphere, which is 21 percent oxygen and 0.04 percent carbon dioxide. During hot, dry weather, excess water vapor diffuses out, and under these conditions plants face certain desiccation if the stomata are left open continuously. When these pores are closed, the concentration of gases will change. As photosynthesis proceeds, carbon dioxide will be consumed and oxygen generated.

When the concentration of carbon dioxide drops below 0.01 percent, oxygen will outcompete carbon dioxide at the active site, and no net photosynthesis occurs. C_4 plants, however, are able to prevent photorespiration, because the PEP carboxylase enzyme is not inhibited by oxygen. Thus, when the stomata are closed, this enzyme continues to fix carbon inside the leaf until it is consumed. Because the bundle sheath is isolated from the leaf's air spaces, it is not affected by the rising oxygen levels, and the C_3 cycle functions without interference. C_4 photosynthesis is found in at least nineteen families of flowering plants. No family is exclusively composed of C_4 plants. Because C_4 photosynthesis is an adaptation to hot, dry environments, especially climates found in tropical regions, C_4 plants are often able to outcompete C_3 plants in those areas. In more temperate regions, they have less of an advantage and are therefore less common.

CAM Photosynthesis

A second alternative photosynthetic pathway, known as *crassulacean acid metabolism* (CAM), exists in succulents such as cacti and other desert plants. These plants have the same two carbon-fixing steps as are present in C_4 plants, but rather than being spatially separated between the mesophyll and bundle sheath cells, CAM plants have both carbon dioxide-fixing enzymes within the same cell. These enzymes are active at different times, PEP carboxylase during the day and Rubisco at night. Just as Kranz anatomy is unique to C_4 plants, CAM plants are unique in that the stomata are open at night and largely closed during the day.

The biochemical pathway of photosynthesis in CAM plants begins at night. With the stomata open, carbon dioxide diffuses into the leaf and into mesophyll cells, where it is fixed by the C_4 enzyme PEP carboxylase. The product is malate, as in C_4 photosynthesis, but it is transformed into malic acid (a nonionic form of malate) and is stored in the cell's *vacuoles* (cavities within the cytoplasm) until the next day.

AP/Wide World Photos

The photosynthetic pathway known as crassulacean acid metabolism (CAM) is used by this saguaro cactus and by many other desert plants.

Although the malic acid will be used as a carbon dioxide source for the C_3 cycle, just as in C_4 photosynthesis, it is stored until daylight because the C_3 cycle requires light as an energy source. The vacuoles will accumulate malic acid through most of the night. A few hours before daylight, the vacuole will fill up, and malic acid will begin to accumulate in the cytoplasm outside the vacuole. As it does, the pH of the cytoplasm will become acidic, causing the enzyme to stop functioning for the rest of the night.

When the sun rises the stomata will close, and photosynthesis by the C_3 cycle will quickly deplete the atmosphere within the leaf of all carbon dioxide. At this time, the malic acid will be transported out of the vacuole to the cytoplasm of the cell. There it will be broken down, and the carbon dioxide will enter the chloroplast and be used by the C_3 cycle; thus, photosynthesis is able to continue with closed stomata.

Crassulacean acid metabolism derives its name from the fact that it involves a daily fluctuation in the level of acid within the plant and that it was first discovered to be common in species within the stonecrop family, *Crassulaceae*. The discovery of this photosynthetic pathway dates back to the 1960's. The observation that succulent plants become very acidic at night, however, dates back to at least the seventeenth century, when it was noted that cactus tastes sour in the morning and bitter in the afternoon.

CAM Plant Ecosystems

There are two distinctly different ecological environments where CAM plants may be found. Most are terrestrial plants typical of deserts or other harsh, dry sites. In these environments, the pattern of stomatal opening and closing provides an important advantage for surviving arid conditions: When the stomata are open, water is lost; however, the rate of loss decreases as the air temperature decreases. By restricting the time period of stomatal opening to the nighttime, CAM plants are extremely good at conserving water.

The other ecological setting where CAM plants are found is in certain aquatic habitats. When this environment was first discovered, it seemed quite odd, because in these environments conserving water would be of little value to a plant. It was found, however, that there are aspects of the aquatic environment which make CAM photosynthesis advantageous. In shallow bodies of water, the photosynthetic consumption of carbon dioxide may proceed at a rate in excess of the rate of diffusion of carbon dioxide from the atmosphere into the water, largely because gases diffuse several times more slowly in water than in air. Consequently, pools of water may be completely without carbon dioxide for large parts of the day. Overnight, carbon dioxide is replenished, and aquatic CAM plants take advantage of this condition to fix the plentiful supply of carbon dioxide available at night and store it as malic acid. Hence, during the day, when the ambient carbon dioxide concentration is zero, these plants have their own internal supply of carbon dioxide for photosynthesis. Thus, two very different ecological conditions have selected for the identical biochemical pathway.

These two modified photosynthetic pathways adequately describe what happens in most terrestrial plants, although there is much variation. For example, there are species that appear in many respects to have photosynthetic characteristics intermediate to C_3 and C_4 plants. Other plants are capable of switching from exclusively C_3 photosynthesis to CAM photosynthesis at different times of the year. Photosynthesis by aquatic plants appears to present even more variation. C_3-C_4 intermediate plants seem to be relatively common compared to the terrestrial flora, and several species have C_4 photosynthesis but lack Kranz anatomy.

Jon E. Keeley, updated by Bryan Ness

See also: Cacti and succulents; Carbon 13/carbon 12 ratios; Vacuoles.

Sources for Further Study

Hall, D. O., and K. K. Rao. *Photosynthesis*. 6th ed. New York: Cambridge University Press, 1999. A concise, easy-to-understand introduction to all aspects of photosynthesis. Written for undergraduate-level students.

Lawlor, D. W. *Photosynthesis: Molecular, Physiological, and Environmental Processes*. 3d ed. New York: Springer-Verlag, 2001. A completely revised and updated textbook for upper-level undergraduate and graduate students. Provides an introduction to the subject and its literature, incorporating new research on the molecular basis of photosynthesis and the effects of environmental change.

Pessarakli, Mohammad, ed. *Handbook of Photosynthesis*. New York: Marcel Dekker, 1997. A comprehensive reference book, containing sixty-three essays on all aspects of photosynthesis.

Winter, Klaus, and J. A. C. Smith, eds. *Crassulacean Acid Metabolism: Biochemistry, Ecophysiology, and Evolution*. New York: Springer, 1996. Advanced and up-to-date coverage of CAM photosynthesis research, including regulation of gene expression and the molecular basis of CAM, the ecophysiology of CAM plants from tropical environments, the productivity of agronomically important cacti and agaves, the ecophysiology of CAM in submerged aquatic plants, and the taxonomic diversity and evolutionary origins of CAM.

Yunus, Mohammed, Uday Pathre, and Prasanna Mohanty, eds. *Probing Photosynthesis: Mechanism, Regulation, and Adaptation*. London: Taylor & Francis, 2000. A comprehensive treatise featuring the state of the art in photosynthesis research. Divided into four main sections—evolution, structure, and function; biodiversity metabolism and regulation; stress and adaptations; and techniques. A useful reference for students and professionals in botany, plant biochemistry, and related subjects.

CACTI AND SUCCULENTS

Categories: Angiosperms; *Plantae*

Succulents are fleshy plants that store water in natural reservoirs such as stems or leaves. Cacti are a group of flowering plants; all cacti are succulents.

The *Cactaceae* family includes about 1,650 to 3,500 species of *cacti* and *succulents* classified in 130 genera. Because they live in harsh, arid environments, these fleshy, spiny perennial plants have developed a variety of unique characteristics for protection and to retain water, reduce evaporation, and resist heat.

Cacti

The word "succulent" is derived from the Latin term *sucus*, meaning sap. All cacti are succulents. The word "cactus" is derived from the Greek term *kaktos*, describing thistles. Botanists estimate cacti first existed during the Mesozoic era, about 130 million years ago. Limited cacti fossil evidence exists (the earliest known specimen is about forty thousand years old).

Cacti vary in size. The *Copiapoa laui* is a spherical plant several millimeters in diameter, while the *Pachycereus weberi* is cylindrical, stands more than 20 meters tall, and can weigh more than 25 tons. Cacti often develop bizarre shapes to cope with arid conditions. Some stems are flat, and others are puffy. Many consist of jointed segments, while others have one round stem. Cacti stems swell when storing water. Surface ridges and grooves gather water. Roots extend in a wide area near the soil surface, to capture any moisture. The pincushion, barrel, saguaro, prickly pear, night-blooming cereus, and Christmas cactus are some of the most familiar cacti.

Unlike other plants, cacti have *areolas* on stems where branches, spines, *glochids* (bristles), leaves, and flowers grow. Spines protect and shade the plant and its seedlings from predators and ultraviolet radiation and serve as condensation sites. Known as *crassulacean acid metabolism* (CAM), photosynthesis in cacti is reversed from the process in other plants. Stems have chlorophyll because leaves are either absent or tiny. At night, instead of day, cacti open the stomata on their stems to collect carbon dioxide and expel oxygen. The carbon diox-

DIGITAL STOCK

Cacti often develop bizarre shapes to cope with arid conditions. Stems of these plants swell when storing water, surface ridges and grooves gather water, and roots extend in a wide area near the soil surface to capture any available moisture.

ide is stored as organic acids for conversion to sugar during the day. Because the temperature is cooler when the stomata are opened, less water is lost. During the day, the closed stomata prevent evaporation from occurring.

Life Cycles of Cacti

Cacti grow slowly and can live more than a century. Flowers usually bloom in late spring and vary in color, size, and shape. Seeds are inside the fruits that blossoms produce. Some cacti grow from seeds if they are shaded and not consumed by predators. Other cacti emerge from stems and take root where they fall. Artificially, cacti can also take root from cuttings. Diverse insects and animals are attracted to the flowers and assist in their pollination. Some birds nest in holes in cacti stems.

Distribution of Cacti

Cacti grow in deserts, prairies, mountains, and tropical climates and have developed a tolerance for extreme conditions. Cacti are indigenous to North, Central, and South America. The *Epiphyllum* species live in tropical trees. Other cacti grow in rocky places. Some Chilean cacti in the Atacama Desert secure water from sea fog. The largest and most diverse population of cacti is in Mexico. The prickly pear is the most widely distributed cactus, ranging from near the Arctic circle to southern South America.

Uses of Cacti

Cactus fruits are edible by humans and animals and used as livestock forage, as a water source, for fuel, and to erect organic barriers. Spines are used as needles and fishhooks, and fibers are twisted into rope. Historically, peyote is a ceremonial hallucinogenic, and other cacti have medicinal purposes. While no cacti are poisonous, some species have unpleasant chemicals that discourage predators.

Some hybrids have naturally occurred, and cactus segments can be detached at joints to graft to artificially unique plants. Because of poaching, cacti are considered endangered plants, with some species threatened by extinction, and are federally protected at Saguaro National Park (in Arizona's Tucson Basin) and Organ Pipe Cactus National Monument (in Arizona's Sonoran Desert).

Succulents

Although they share many traits with their close relatives the cacti, other succulents do not have areolas. Succulents vary in shape and size. Some are as tiny as peas, while others are large as livestock. Succulents take many forms, including that of the string of beads (*Senecio rowleyanus*). Yucca and jade plants are two of the most familiar succulents.

Because of evolutionary adaptation to endure climatic extremes, succulents have small leaves and spongy tissues that keep water for prolonged durations. Succulents retain water to withstand such environmental stresses as drought, scorching wind, shallow or salty topsoil, steep locations, and overcrowding by other plants. Succulents keep flower stalks and fruit until all the water is depleted from them. They have a thick skin, which is waxy, and sometimes alter their shape while adjusting to differences of light and moisture. Most succulents are gray, although a few are colored lilac, pink, light green, beige, or ivory, often in patterns that may serve as camouflage.

The greatest quantity and most diverse succulents can be found in Mexico and South Africa, which have thousands of species. New species are still being discovered because of variations arising from environmentally triggered adaptations. Some succulents, such as the *Argyroderma*, are abundant, growing in thick clumps. The rarer succulents include *Conophytum burgeri*, which lives on only one South African hill. Succulents are threatened by overgrazing and industrial and agricultural development of habitats. Some succulents, particularly aloe, have healing juices to soothe burns.

Elizabeth D. Schafer

See also: Angiosperm evolution; C_4 and CAM photosynthesis; Culturally significant plants; Deserts; Drought; Endangered species; Evolution of plants; Growth habits; Hybridization; Leaf anatomy; Liquid transport systems; Photosynthesis; Plant fibers; Water and solute movement in plants.

Sources for Further Study

Anderson, Edward F., with Roger Brown. *The Cactus Family*. Foreword by Wilhelm Barthlott. Portland, Oreg.: Timber Press, 2001. Comprehensive discussion of the global distribution of cacti naturally and in botanical gardens. Includes illustrations, maps, bibliographical references, appendices, and indexes.

Grantham, Keith, and Paul Klaassen. *The Plantfinder's Guide to Cacti and Other Succulents*. Portland, Oreg.: Timber Press, 1999. Useful overview of species and how to tend and conserve them. Includes illustrations, maps, bibliographical references, appendices, and index.

Hewitt, Terry. *The Complete Book of Cacti and Succulents*. New York: D. K. Publishing, 1997. Describes how to identify, collect, propagate, graft, and display desert plants. Includes illustrations, charts, maps, lists, glossary, addresses, and index.

Nobel, Park S. *Remarkable Agaves and Cacti*. New York: Oxford University Press, 1994. Discusses how cacti can be used by humans. Includes illustrations, bibliographical references, and index.

Pizzetti, Mariella, and Stanley Schuler, eds. *Simon and Schuster's Guide to Cacti and Succulents*. New York: Simon and Schuster, 1985. Descriptions of international specimens. Includes illustrations, bibliography, and indexes.

CALVIN CYCLE

Categories: Biogeochemical cycles; photosynthesis and respiration; physiology

The Calvin cycle is the principal mechanism that leads to the conversion of carbon dioxide into sugars by plants, algae, photosynthetic bacteria, and certain other bacteria that use chemicals as an energy source instead of light.

The Calvin cycle, also known as the Calvin-Benson cycle, is an integral part of the process of *photosynthesis* in plants, algae, and photosynthetic bacteria. Named after its discoverer, Melvin Calvin of the University of California at Berkeley, its principal product is a three-carbon compound called *glyceraldehyde 3-phosphate*, or PGAL. Sugars are synthesized using PGAL as a starting material. Light, absorbed by *chlorophyll*, is used to synthesize the high-energy compounds *adenosine triphosphate* (ATP) and reduced *nicotinamide adenine dinucleotide phosphate* (NADPH). Chlorophyll and the enzymes that are used for synthesis of ATP and NADPH are associated with internal membranes in all photosynthetic cells. The ATP and NADPH, once formed, are released from the membrane-bound enzymes and diffuse into the surrounding solution inside the cell. The Calvin cycle takes place in this solution, using the ATP and NADPH molecules as a source of energy to drive the conversion of carbon dioxide into PGAL.

Enzymes

All the steps in the Calvin cycle and sugar biosynthesis are catalyzed by specific enzyme molecules. The carbon dioxide molecules react with a five-carbon sugar-phosphate molecule called *ribulose bisphosphate* (RuBP) to form a six-carbon intermediate. The reaction is catalyzed by the enzyme *ribulose bisphosphate carboxylase/oxygenase* (Rubisco). The six-carbon intermediate reacts with water and decomposes into two identical three-carbon molecules called phosphoglycerate. These, in turn, react with ATP and NADPH to produce PGAL molecules. Some of these leave the Calvin cycle and are used for the formation of sugars. The ADP and NADP molecules, produced when PGAL is formed, diffuse back to the chlorophyll-containing membranes, where they can be used to regenerate a supply of ATP and NADPH for the next round of the Calvin cycle. The remaining PGAL molecules are used for the regeneration of sufficient amounts of RuBP to permit the reactions of the Calvin cycle to be repeated.

The regeneration of RuBP from PGAL involves the rearrangement of the carbon atoms in three-carbon containing molecules to form five-carbon molecules. For example, if there are ten three-carbon molecules remaining after two three-carbon molecules have been removed from the cycle, then six five-carbon molecules are synthesized by the Calvin cycle. This is accomplished by no less than nine separate enzyme-catalyzed steps, involving intermediate compounds containing two, three, four, five, or seven carbon atoms derived from the PGAL molecules. At the end of this complex process, ATP is used to add another phosphate group to each five-carbon molecule, thus regenerating the required amount of RuBP, the original organic starting material for the cycle. A continuous supply of phosphate must be made available in order to continue running the Calvin cycle. Ultimately, all this phosphate must be supplied to the organism from the environment; in the short term, however, it is gleaned from other biochemical reactions, such as sugar biosynthesis.

PGAL

The principal product of the Calvin cycle is not sugar but PGAL. In higher plants, the Calvin cycle takes place inside *chloroplasts*, and the PGAL molecules are transported across the membranes of the chloroplasts and released into the solution between the chloroplast membranes and the cell's outer membrane. In this solution, called the *cytosol*, the PGAL molecules react to form six-carbon sugar phosphates. These six-carbon sugar phosphates then react to form *sucrose*, a twelve-carbon molecule (ordinary table sugar).

A sucrose molecule consists of one molecule each of the six-carbon sugars *fructose* and *glucose*.

Phosphate is released from the sugar phosphates during the formation of sucrose. The phosphate can then be returned to the chloroplast, where it is needed for the formation of ATP. Most of the sucrose is transported out of the cell and flows to various parts of the plant, such as the fruits or the roots. The transport of sucrose out of the cell requires energy derived from ATP. The accumulation of sucrose in the water outside the cell causes the hydrostatic pressure to rise. This pressure drives the flow of the water and sucrose (sap) through the phloem away from the leaves and toward the fruit or roots. The accumulation of sucrose in the fruit accounts for a large part of the nutritional value of plants.

When conditions do not favor the formation of sucrose, the triose phosphates may remain inside the chloroplast. These can react to form six-carbon sugar bisphosphates that, in turn, can react in several steps to form an insoluble carbohydrate storage compound called *starch*. The conversion of six-carbon sugar bisphosphate into starch releases phosphate. The phosphate released can then participate in the synthesis of more ATP, permitting continued operation of the Calvin cycle even when sucrose is not being formed. The accumulation of starch is another major source of nutritional value in plants.

In the morning, when plants begin receiving light, the amounts of phosphoglycerate and six-carbon sugar phosphates increase dramatically. The amounts of other intermediates in the cycle do not change as much. This suggests that some (but not all) steps in the Calvin cycle shut down in the dark and are activated in the light. The Calvin cycle also operates in nonphotosynthetic bacteria that use environmental chemicals as an energy source for the synthesis of ATP and other high-energy molecules. Although these organisms are responsible only for a minor proportion of the total carbon dioxide converted to organic form every year, their existence is interesting because it suggests that the Calvin cycle evolved before the origin of photosynthesis. The starch that builds up in the chloroplasts during the day is converted to sucrose at night and is then exported from the leaf.

Photorespiration

A significant complication must be taken into account when discussing the Calvin cycle: Oxygen can also react with RuBP, because the active site of Rubisco has affinity for both oxygen and carbon dioxide. Under normal conditions in many higher plants, three out of ten RuBP molecules react with oxygen instead of reacting with carbon dioxide. Under conditions where carbon dioxide levels are lower than normal and oxygen levels are higher than normal, oxygen may even react more frequently than carbon dioxide. This has deleterious consequences, because each RuBP molecule that reacts with oxygen is cleaved into two parts. One part is PGA, identical to that produced by the reaction of RuBP with carbon dioxide. The other part, however, is a two-carbon compound called phosphoglycolate. The latter molecule subsequently is cleaved to produce carbon dioxide; only one of the two carbons in phosphoglycolate is salvaged by the cell in a complex series of reactions called *photorespiration*. Because the photorespiratory reactions use energy, the chlorophyll-containing membranes must produce more ATP and NADPH than would otherwise be needed for the Calvin cycle.

Evolution of the Cycle

The Calvin cycle is believed to have originated more than 3.5 billion years ago in marine bacteria that were using very simple carbon compounds as an energy source. Some of the descendants of these bacteria later acquired the ability to synthesize ATP and NADPH (or their equivalents), using light as an energy source. As long as one billion years ago, some of these photosynthetic bacteria are believed to have established mutually beneficial, or *symbiotic*, relationships with other cells. These symbiotic relationships became stabilized and led to the evolution of algae and higher plants. The photosynthetic organelles of plants and algae, the chloroplasts, are thought to be the direct descendants of the symbiotic photosynthetic bacteria.

The early atmosphere of the earth probably had significantly less oxygen than it does now, so the existence of the oxygenase activity of Rubisco would not have been a problem. When the oxygen concentration in the atmosphere rose to its present level about 1.7 billion years ago, however, the losses of energy caused by the oxygenase reaction became significant. Some organisms evolved mechanisms to prevent these losses: In algae, for example, there are molecular pumps which, in effect, concentrate carbon dioxide in the cell so that the oxygenase reaction is inhibited. In sugarcane, corn, and certain other plants that are specialized to live in hot, dry climates, a similar effect is achieved by C_4 photosynthesis.

Not content with the results of evolution, biotechnologists are interested in altering Rubisco. They reason that if this enzyme can be genetically engineered to lower its oxygenase activity, the net photosynthetic rates of some plants could be improved. Possibly a 30 percent increase in plant productivity could be expected if such strategies prove successful, provided other materials (nitrogen, phosphorus, and other nutrients) are present in sufficient supply to permit the extra growth. Thus, although the Calvin cycle is the major route of entry of inorganic carbon into the biosphere, it is also something of a bottleneck. It remains to be seen whether the Calvin cycle can be made to function more efficiently.

Harry Roy, updated by Bryan Ness

See also: ATP and other energetic molecules; C_4 and CAM photosynthesis; Photosynthesis; Photosynthetic light absorption; Photosynthetic light reactions.

Sources for Further Study

Hall, D. O., and K. K. Rao. *Photosynthesis*. 6th ed. New York: Cambridge University Press, 1999. A concise, easy-to-understand introduction to all aspects of photosynthesis. Written for undergraduate-level students.

Hoober, J. Kenneth. *Chloroplasts*. New York: Plenum Press, 1984. A senior-level college text which covers photosynthesis, photorespiration, C_4 photosynthesis, and other aspects of chloroplast biology. Well illustrated with diagrams and data from the original research literature. Very readable.

Lawlor, D. W. *Photosynthesis: Molecular, Physiological, and Environmental Processes*. 3d ed. New York: Springer Verlag, 2001. Textbook for upper-level undergraduates and graduate students. Provides an introduction to the subject and its literature, incorporating new research on the molecular basis of photosynthesis and the effects of environmental change.

Pessarakli, Mohammad, ed. *Handbook of Photosynthesis*. New York: Marcel Dekker, 1997. A comprehensive reference book, containing sixty-three essays on aspects of photosynthesis.

Solomon, Eldra Pearl, Claude A. Villee, and P. William Davis. *Biology*. 5th ed. Fort Worth, Tex.: Saunders, 1999. An introductory college-level text with a good chapter on photosynthesis that presents the Calvin cycle in outline and discusses light-dependent reactions, the C_4 pathway, and photorespiration. Good illustrations.

Stryer, Lubert. *Biochemistry*. 4th ed. New York: W. H. Freeman, 2000. A textbook that is popular with college students who have a year in organic chemistry in their backgrounds. The Calvin cycle, photorespiration, and C_4 photosynthesis are discussed in a chapter on photosynthesis. Illustrated with diagrams, molecular structures of some key intermediates, and a variety of data from the literature.

CARBOHYDRATES

Categories: Cellular biology; nutrients and nutrition

Common organic chemicals found in all living organisms, important in energy metabolism and structural polymers, carbohydrate molecules are made up of carbon, hydrogen, and oxygen.

Carbohydrates are made of carbon, hydrogen, and oxygen molecules in a 1:2:1 ratio, respectively. This is often simplified using the formula nCH_2O, where n represents the number of CH_2O subunits in a carbohydrate. This formula should make it clear how the name carbohydrate was derived, as nCH_2O is essentially carbon and water. The simplest carbohydrates are the *monosaccharides*, or simple sugars. Individual monosaccharides can be joined together to make *disaccharides* (composed of two monosaccharides), *oligosaccharides* (short polymers composed of two to several monosaccharides), and *polysaccharides* (longer polymers composed of numerous monosaccharides).

Monosaccharides

The common monosaccharides found in plants have from three to six carbon atoms in a straight chain with one oxygen atom. Most of the oxygen atoms also have a hydrogen atom attached, making them hydroxyl groups (^{-}OH). One of the oxygen atoms is connected to a carbon by a double covalent bond, while the hydroxyl groups are attached to carbon atoms by single covalent bonds. If the double-bonded oxygen is on a terminal carbon (as an aldehyde group), the monosaccharide is called an *aldose*. If the double-bonded oxygen is on an internal carbon, the monosaccharide is called a *ketose*.

The simplest monosaccharides are the three-carbon sugars, or *trioses*. *Pentoses*, with five carbons, are also important in plants. *Ribose* and *deoxyribose* are found in RNA (ribonucleic acid) and DNA (deoxyribonucleic acid), respectively. *Ribulose bisphosphate* is an important intermediate in the incorporation of carbon dioxide into carbohydrates during photosynthesis. *Xylose* and *arabinose* are found as components of some plant polysaccharides.

Hexoses, six-carbon monosaccharides such as *glucose*, *fructose*, and *galactose*, are the most common monosaccharides in plants. These sugars all have the same formula, $C_6H_{12}O_6$ (note the 1:2:1 ratio of C:H:O), but their atoms are arranged differently. Glucose is the primary carbon-containing product of photosynthesis and reverse glycolysis and later can be metabolized through glycolysis and the Krebs cycle to release energy or can be converted to other carbohydrates needed by the plant.

Oligosaccharides

Oligosaccharides are made by joining two or more monosaccharides. The smallest are the disaccharides, formed from two monosaccharides that are joined together by a condensation reaction. Condensation reactions get their name from the fact that when the two monosaccharides are joined together, a molecule of water is released. *Sucrose* (glucose-fructose) is the most common plant disaccharide and is the principal molecule of short-term energy storage and of translocation (transport) in the phloem. Many plants, including sugarcane (*Saccharum officinarum*) and sugar beets (*Beta saccharifera*), have high concentrations of sucrose, which can be extracted and refined for use as table sugar.

Other disaccharides found in plants are *maltose*, which is a glucose disaccharide formed from the hydrolysis (the reverse of a condensation reaction, wherein water is used to "split" the bond between the monosaccharides) of starch, and *trehalose*, also a glucose disaccharide, which is the primary molecule of translocation in species of *Selaginella* and is seen in cyanobacteria (blue-green algae or blue-green bacteria), red algae, and fungi. *Cellobiose*, another glucose disaccharide, is formed by the hydrolysis of cellulose.

The trisaccharide *raffinose* (galactose-glucose-fructose) is a storage molecule in sugar beets and in cotton and legume seeds. *Stachyose* (galactose-galactose-glucose-fructose) and *verbascose* (galactose-galactose-galactose-glucose-fructose) are also storage oligosaccharides, seen mainly in *Fabaceae* (the legume or pea family).

Polysaccharides

The two main functions of polysaccharides in plants are long-term energy storage and structure. Glucose is the most common subunit in plant polysaccharides. The glucose molecules in these polymers are joined together in different ways. The carbon atoms in glucose molecules are numbered from one to six. In some, the 1-carbon of a glucose is attached to the 4-carbon of the next, and this linkage is repeated throughout the molecule. At other times, an additional bond is formed between the 1-carbon and the 6-carbon of adjacent glucoses, which results in a branched polysaccharide.

Starch is the most common storage polysaccharide of plants. Two forms of this glucose polymer exist. *Amylose* is a linear polymer made up of between one hundred and several thousand glucose units. *Amylopectin* is very similar, but it is a branched polymer. In most plants, starch is 15-25 percent amylose and 75-85 percent amylopectin. However, starch in some waxy varieties of corn is nearly 100 percent amylopectin and in some wrinkled varieties of peas is as high as 80 percent amylose. Phytoglycogen found on corn (*Zea mays*) is an even more branched glucose polymer.

Fructosans are another type of storage polysaccharide in plants. They are branched or unbranched fructose polymers with a terminal glucose subunit. *Inulin* is found in the tubers or rhizomes of plants in *Campanulaceae* (the bellflower family) and *Asteraceae* (the sunflower or aster family) and usually has thirty to fifty fructose subunits. *Levans,* used for temporary storage by several monocots, especially in *Poaceae* (the grass family), range from seven to eight fructose subunits in the unbranched levans to seventy-two fructose subunits in some highly branched ones.

Structural Polysaccharides

Structural polysaccharides form the fibrous material in plant cell walls. *Cellulose,* an unbranched glucose polymer that averages about eight thousand glucose subunits per molecule, is the main cell wall component of plants, a few fungi, and some algae. Cellulose molecules form microfibrils, many individual cellulose molecules held together by hydrogen bonds. Other microfibrillar cell wall polysaccharides are sometimes called the hemicelluloses. Examples are *mannans* and *glucomannans,* found in the primary cell walls of several green algae and simple vascular plants and in the secondary cell walls of some conifers; *xylans* are found in other algae and in the secondary cell walls of many hardwoods. *Chitin,* a polymer of N-acetylglucosamine, is the main substance forming the cell walls of fungi. (Chitin is the same substance that forms the exoskeletons of most insects.)

Pectins are matrix polysaccharides found in plant cell walls. The most common pectin in higher plants is unbranched polygalacturonic acid (galacturan). Branched and unbranched rhamnogalacturans and arabinans are also present in smaller quantities. Pectin is commercially important as a gelling agent in the production of jams and jellies. A similar pectinlike polysaccharide found in brown algae is alginic acid, a mixture of mannuronic and guluronic acids. It is used as a thickener and a stabilizer in many prepared foods.

PhotoDisc

Glycosides are formed when carbohydrates are attached to various plant chemicals and include the cardiac glycosides of the foxglove plant (Digitalis purpurea)*, which has strong physiological effects on heart muscle.*

Other Plant Carbohydrates

Carbohydrates are often found attached to other cell components. In both cell membranes and cell walls, there are many glycoproteins, proteins with short oligosaccharides attached. *Glycosides* are interesting carbohydrate-containing secondary metabolites found in many plants. Glycosides are formed when carbohydrates are attached to various plant chemicals. Anthocyanins, which give red to blue color to flowers, fruits, and autumn leaves, are glycosides. Other glycosides include the cardiac glycosides of the foxglove (*Digitalis purpurea*) and milkweed (*Asclepias*) species, which have strong physiological effects on heart muscle, and the cyanogenic glycosides of the almond (*Prunus amygdalus*), which liberate cyanide.

Richard W. Cheney, Jr.

See also: Algae; Calvin cycle; Cell wall; Fungi; Glycolysis and fermentation; Krebs cycle; Metabolites: primary vs. secondary; Photosynthesis; Sugars.

Sources for Further Study

Goodwin, T. W., and E. I. Mercer. *Introduction to Plant Biochemistry*. 2d ed. Woburn, Mass.: Butterworth-Heinemann, 1990. This text contains a thorough look at plant carbohydrates.

Raven, Peter H., Ray F. Evert, and Susan E. Eichhorn. *Biology of Plants*. 6th ed. New York: W. H. Freeman/Worth, 1999. This text has a basic description of carbohydrates, especially related to plants.

Robyt, John F. *Essentials of Carbohydrate Chemistry*. New York: Springer, 1998. Broad treatment of carbohydrates, presenting their structures, reactions, modifications, and properties. Written for students as well as scientists.

CARBON CYCLE

Categories: Biogeochemical cycles; ecology; environmental issues; photosynthesis and respiration

The carbon cycle is the movement of the element carbon through the earth's rock and sediment, the aquatic environment, land environments, and the atmosphere. Large amounts of organic carbon can be found in both living organisms and dead organic material.

An enormous reservoir of carbon, on the order of 20×10^{15} tons, may be found on the surface of the earth. Most of this reservoir is found in rock and sediment. The carbon cycle therefore represents the movement of this element through the biosphere in a process mediated by photosynthetic plants on land and in the sea. The process involves the fixation of carbon dioxide (CO_2) into organic molecules, a process called *photosynthesis*. Energy used in the process is stored in chemical form, such as that in carbohydrates (sugars such as glucose). The organic material is eventually oxidized, as occurs when a photosynthetic organism dies. Through the process of respiration, the carbon is returned to the atmosphere in the form of carbon dioxide. Because the "turnover" time of such forms of carbon is so slow (on the order of thousands of years), the entrance of this material into the carbon cycle is insignificant on the human scale.

Photosynthesis

Organisms that use carbon dioxide as their source of carbon are known as *autotrophs*. Many of these organisms also use sunlight as the source of energy for reduction of carbon dioxide; hence, they are frequently referred to as *photoautotrophs*. This process of carbon dioxide fixation is carried out by phytoplankton in the seas, by land plants (particularly trees), and by many microorganisms. Most of the process is carried out by the land plants.

The process of photosynthesis can be summarized by the following equation:

$$CO_2 + \text{water} + \text{energy} \rightarrow \text{carbohydrates} + \text{oxygen}$$

The process requires energy from sunlight, which is stored in the form of the chemical energy in carbohydrates. While most plants produce oxygen in the process—the source of the oxygen in the earth's atmosphere—some bacteria may produce products other than oxygen. Organisms that carry out carbon dioxide fixation, using photosynthesis to synthesize carbohydrates, are often referred to as *producers*. Approximately 20 billion to 30 billion tons of carbon are fixed each year by the process—clearly a large amount but only a small proportion of the total carbon found on the earth. Approximately 450 billion tons of carbon are contained within the earth's forests; some 700 billion tons exist in the form of atmospheric carbon dioxide.

Much of the organic carbon on the earth is found in the form of land plants, including forests and grasslands. When these plants or plant materials die, as when leaves fall to the earth in autumn, the dead organic material becomes *humus*. Much of the carbon initially bound during photosynthesis is in the form of humus. Degradation of humus is a slow process, on the order of decades. However, it is the decomposition of humus, particularly through the process called respiration, that returns much of the carbon dioxide to the atmosphere. Thus, the carbon cycle represents a dynamic equilibrium between the carbon in the atmosphere and carbon fixed in the form of organic material.

Respiration

Respiration represents the reverse of photosynthesis. All organisms that use oxygen, including humans, carry out the process. However, it is primarily humic decomposition by microorganisms that returns most of the carbon to the atmosphere. Depending on the particular microorganism, the carbon is in the form of either carbon dioxide or methane (CH_4). Respiration is generally represented by the equation:

Carbohydrate + oxygen →
carbon dioxide + water + energy

Energy released by the reaction is used by the organism (that is, the consumer) to carry out its own metabolic processes.

Carbon Sediment

Despite the enormous levels of carbon cycled between the atmosphere and living organisms, most carbon is found within carbonate deposits on land and in ocean sediments. Some of this originates in marine ecosystems, where organisms use dissolved carbon dioxide to produce carbonate shells (calcium carbonate). As these organisms die, the shells sink and become part of the ocean sediment. Other organic deposits, such as oil and coal, originate from

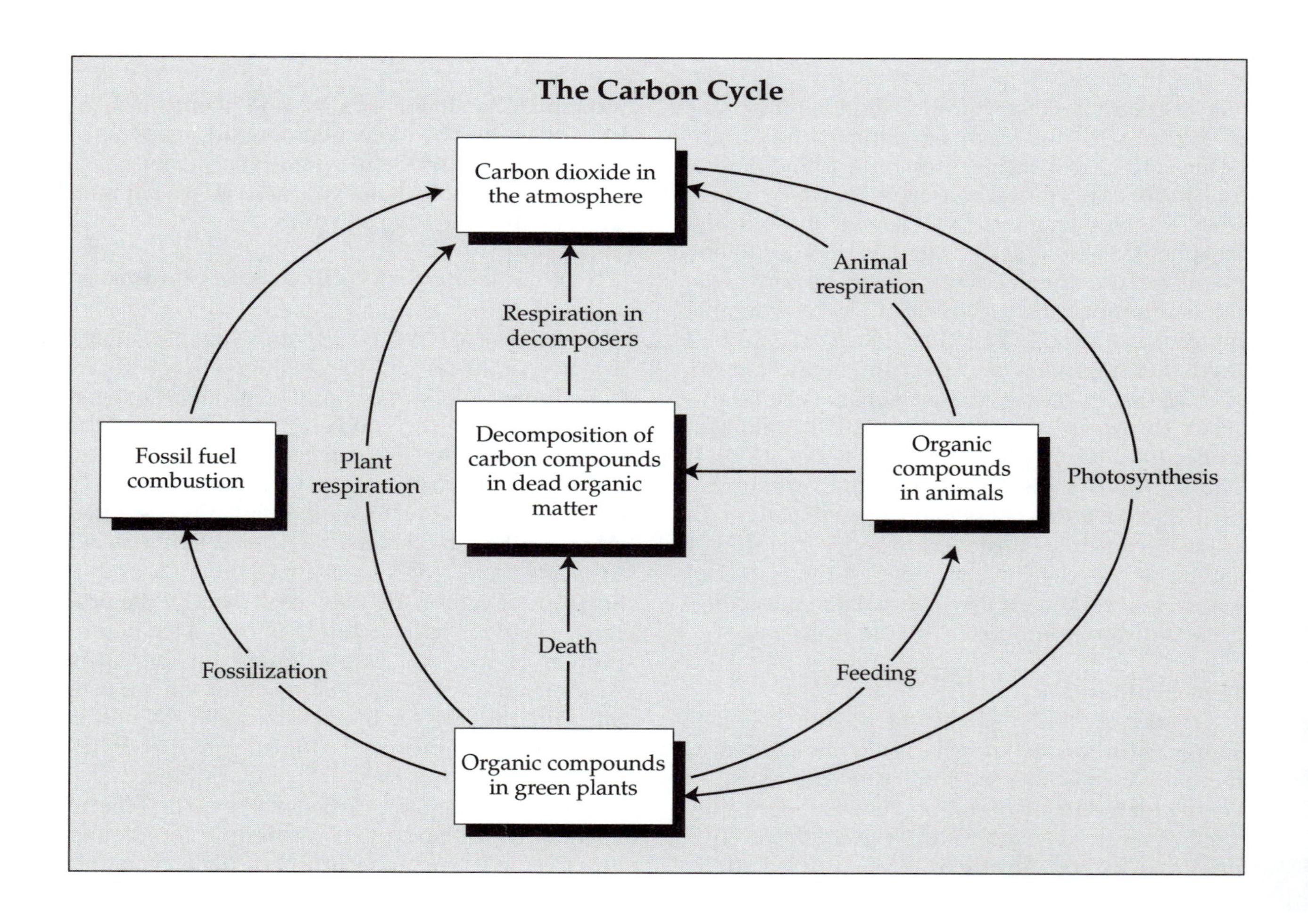

fossil deposits of dead organic material. The recycling time for such sediments and deposits is generally on the order of thousands of years; hence their contribution to the carbon cycle is negligible on a human time scale. Some of the sediment is recycled naturally, as when sediment dissolves or when acid rain falls on carbonate rock (limestone), releasing carbon dioxide. However, when such deposits are burned as fossil fuels, the levels of carbon dioxide in the atmosphere may increase at a rapid rate.

PhotoDisc

The biogeochemical carbon cycle is the movement of carbon through the earth's ecosystems. An enormous reservoir of carbon may be found on the surface of the earth, mostly within rock and sediment.

Environmental Impact of Human Activities

Carbon dioxide gas is only a small proportion (0.036 percent) of the volume of the atmosphere. However, because of its ability to trap heat from the earth, carbon dioxide acts much like a thermostat, and even small changes in levels of this gas can significantly alter environmental temperatures. Around 1850, humans began burning large quantities of fossil fuels; the use of such fuels accelerated significantly after the invention of the automobile. By the end of the twentieth century, between 5 billion and 6 billion tons of carbon were being released into the atmosphere every year from the burning of fossil carbon. Some of the released carbon probably returns to the earth through biological carbon fixation, with a possible increase in the land biomass of trees or other plants. (Whether this is so remains a matter of dispute.) Indeed, large-scale deforestation could potentially remove this means by which levels of atmospheric carbon dioxide could be naturally controlled.

Richard Adler

See also: Greenhouse effect; Photosynthesis; Respiration; Trophic levels and ecological niches.

Sources for Further Study

Bolin, Bert. "The Carbon Cycle." *Scientific American* 223 (September, 1970). This special issue is devoted to the carbon cycle.

Breymeyer, A. I., ed. *Global Change: Effects on Coniferous Forests and Grasslands*. New York: Wiley, 1996. Reviews extant forest and grasslands models and develops foundations for the design of diagnostic and predictive models as well as identifying plans for future research on ecosystem response to global change. Carbon flow and storage are emphasized.

Kasischke, Eric S., and Brian J. Stocks, eds. *Fire, Climate Change, and Carbon Cycling in the Boreal Forest*. New York: Springer, 2000. Discusses the direct and indirect mechanisms by which fire and climate interact to influence carbon cycling in North American boreal forests.

Madigan, Michael, John Martinko, and Jack Parker, eds. *Brock Biology of Microorganisms*. 9th ed. Upper Saddle River, N.J.: Prentice Hall, 2000. A textbook for microbiology students. Includes an overview of the carbon cycle in diagrammatic form.

Max, A., ed. *Carbon Sequestration in the Biosphere: Processes and Prospects*. New York: Springer, 1995. Discusses the possibilities for limiting the release of carbon dioxide into the atmosphere by the long-term storage of carbon in soils, vegetation, wetlands, and oceans. Each storage medium is analyzed in detail to elucidate those processes responsible for the uptake and release of carbon.

Wallace, Robert, Gerald Sanders, and Robert Ferl. *Biology: The Science of Life*. 3d ed. New York: HarperCollins, 1991. Presents the important concepts in cellular biology and heredity, evolution, diversity, plant and animal functions, and animal behavior in relation to ecology. Includes a discussion of the greenhouse effect and the carbon cycle.

CARBON 13/CARBON 12 RATIOS

Categories: Cellular biology; photosynthesis and respiration; physiology

The carbon 13 to carbon 12 ratio is the ratio of two stable carbon isotopes in a given organic sample, often used to determine the photosynthetic pathway being used by a plant.

Although the atom is the smallest unit having the properties of its element, atoms are composed of subatomic particles, of which protons, neutrons, and electrons are the most important. All atoms of a given element (in their non-ionic form) have the same number of protons and electrons, but some atoms may have more neutrons than other atoms of the same element. These different atomic forms are called *isotopes* of the element, and in nature there is a mixture of isotopes for most elements.

Carbon Isotopes

Carbon has an atomic number of 6, meaning that it has six protons. Most carbon atoms also have six neutrons. Because the atomic weight is largely determined by the mass of the protons plus neutrons, this isotope is called carbon 12 (^{12}C); it is the most common form of carbon, accounting for about 99 percent of the carbon in nature. Most of the remaining 1 percent of carbon consists of atoms of the isotope carbon 13 (^{13}C), with seven rather than six neutrons; thus, this isotope is heavier than ^{12}C. A third isotope, carbon 14 (^{14}C), is present in the environment in minute quantities but is not very stable. Consequently, ^{14}C decays spontaneously, giving off radiation, and thus it is a *radioactive isotope*. Both ^{12}C and ^{13}C are considered stable isotopes.

Stable isotopes are measured using a mass spectrometer, an instrument that separates atoms on the basis of their mass differences. Initially when plant material is combusted, the carbon dioxide (CO_2) given off is analyzed by the mass spectrometer for the ratio of carbon 13 to carbon 12 isotopes. This ratio is compared to the ratio of carbon 13 to carbon 12 in an internationally accepted standard of known ^{13}C to ^{12}C ratio and is expressed as the difference between the sample and the standard, minus one. This number is multiplied by one thousand and expressed as a "per mil" (per million parts). In plant matter, this number is always negative. The more negative the ratio of carbon 13 to carbon 12, the less carbon 13 there is present.

C_3, C_4, and CAM Plants

There are three biochemical pathways of carbon acquisition used by different plant species, and the ratio of carbon 13 to carbon 12 in plant tissues is often a useful means of distinguishing the *photosynthetic pathway* being used. Most plants photosynthesize by attaching CO_2 obtained from the atmosphere onto an organic compound in a single carbon fixation step that is a part of the Calvin cycle. This reaction is catalyzed by an enzyme known as ribulose bisphosphate carboxylase (Rubisco), and the first stable organic product is a three-carbon molecule. Some of this three-carbon compound enters a biochemical pathway leading to sugar formation, and the remainder is used to maintain the Calvin cycle. Such plants are referred to as *C_3 plants*.

Some plants, such as corn and sugarcane, have two carbon fixation steps. Atmospheric CO_2 is fixed initially by the enzyme phosphoenolpyruvate carboxylase (PEP carboxylase), and the first product is a four-carbon organic acid; these plants are known as *C_4 plants*. This product is moved to the interior of the leaf and broken down, releasing CO_2 into specialized cells known as Kranz-type bundle sheath cells. Within these cells the CO_2 is fixed a second time by the same process used in C_3 plants.

A third group of plants, known as *CAM plants* (for crassulacean acid metabolism), open their stomatal pores (located on the leaf surfaces) primarily at night. CO_2 enters and is fixed with the enzyme PEP carboxylase. The organic acid produced is stored in the cell overnight. During the day, the stomata are closed, and the acid is broken down. The CO_2 released is then fixed by the same method as in C_3 plants.

Discrimination Against Carbon 13

In terrestrial plants, carbon isotope ratios of photosynthetic tissues vary from $^{-}8$ per mil to $^{-}15$ per mil in C_4 plants and from $^{-}20$ per mil to $^{-}35$ per mil in C_3 plants. CAM plants may range from C_4-like to C_3-like in their carbon isotope ratios. For atmospheric carbon dioxide, the ratio of carbon 13 to carbon 12 is about $^{-}8$ per mil, and thus C_4 plant tissues have slightly less carbon 13 than the air. C_3 plants have much less carbon 13 than the air. In other words, during photosynthesis plants tend to discriminate against $^{13}CO_2$ molecules and more readily fix $^{12}CO_2$. This discrimination against carbon 13 is even more pronounced in C_3 plants.

Discrimination against carbon 13 is attributable to the greater mass of this isotope. One consequence is that $^{13}CO_2$ does not diffuse as readily to the site of photosynthesis as does $^{12}CO_2$, which accounts for a small component of discrimination against carbon 13 in all plants. The major difference in carbon isotope ratio between C_3 and C_4 plants, however, results from a difference in discrimination among the initial carbon-fixing enzymes. In C_3 plants, the carbon-fixing enzyme Rubisco results in a $^{-}27$ per mil discrimination against carbon 13. In C_3 plants, this enzyme is present in the cells adjacent to the stomatal pores and thus obtains CO_2 more or less directly from the atmosphere. Because carbon 13 is discriminated against, $^{13}CO_2$ will tend to accumulate, but it readily diffuses out of the leaf when stomatal pores are open.

In C_4 plants, on the other hand, the initial carbon-fixing enzyme, PEP carboxylase, discriminates very little against carbon 13. In C_4 photosynthesis, the secondary carbon-fixing enzyme, Rubisco, is sequestered in the interior of the leaf in the bundle sheath cells, and the CO_2 it fixes is derived from the breakdown of the C_4 fixation product. As Rubisco discriminates against $^{13}CO_2$, this heavier CO_2 accumulates within the bundle sheath cells and diffuses out very slowly. As $^{13}CO_2$ accumulates in the bundle sheath cells, the higher concentration of $^{13}CO_2$ will overcome the discrimination by the enzyme; in effect, the enzyme will be forced to fix $^{13}CO_2$, and thus discrimination is minimal. C_4 plant tissues consequently have less negative $^{13}CO_2/^{12}CO_2$ ratios.

In typical CAM photosynthesis, the atmospheric CO_2 is fixed at night by the enzyme PEP carboxylase, and, as in C_4 plants, this enzyme discriminates very little against $^{13}CO_2$. During the day, the C_4 fixation product is broken down, and the CO_2 that is released is fixed by Rubisco. This enzyme will discriminate against carbon 13, but because the stomata are closed during the day, $^{13}CO_2$ will accumulate within the leaf and eventually be fixed. Consequently, little discrimination occurs. Such CAM plants have ratios similar to those of C_4 plants. Some CAM plants, however, will open their stomatal pores for varying lengths of time during the day or switch to strictly C_3 photosynthesis during certain times of the year. In these plants, the $^{13}CO_2/^{12}CO_2$ ratio will be more similar to that observed for C_3 plants.

In aquatic plants, the $^{13}CO_2/^{12}CO_2$ ratio does not indicate the photosynthetic pathway used. C_3 aquatic plants frequently will have carbon isotope ratios very similar to that of the source carbon from the water: Because the enzyme Rubisco discriminates against carbon 13, $^{13}CO_2$ tends to accumulate in the layer of water around the leaf. Because the diffusion of gases in water is very slow, the plant will eventually be forced to fix the $^{13}CO_2$. Other aspects of the aquatic environment also influence the carbon isotope ratio of aquatic plant tissues.

Jon E. Keeley, updated by Bryan Ness

See also: C_4 and CAM photosynthesis; Calvin cycle; Photosynthesis.

Sources for Further Study

Beerling, D. J. "Interpreting Environmental and Biological Signals from the Stable Carbon Isotope Composition of Fossilized Organic and Inorganic Carbon." *Journal of the Geological Society* 154, no. 2 (March, 1997): 303-309. Highlights recent developments in the use of stable carbon isotopes of organic and inorganic carbon sources, particularly with respect to interpreting environmental and biological signals.

Coleman, David C., and Brian Fry, eds. *Carbon Isotope Techniques*. San Diego: Academic Press, 1991. A reference book with protocols for using carbon isotope tracers in experimental biology and ecology.

Ehleringer, James R., Anthony E. Hall, and Graham D. Farquhar, eds. *Stable Isotopes and Plant Carbon/Water Relations*. San Diego: Academic Press, 1993. A wide-ranging collection of essays on the theory and uses of stable isoptopes, written from both agricultural and ecological perspectives.

Erez, Jonathan, Anne Bouevitch, and Aaron Kaplan. "Carbon Isotope Fractionation by Photosynthetic Aquatic Microorganisms." *Canadian Journal of Botany* 76, no. 6 (June, 1998): 1109-1118. Concludes that the natural fractionation of carbon isotopes is mainly due to the discrimination of ribulose-1,5-bisphosphate carboxylase-oxygenase against carbon 13 during photosynthesis.

Rundel, Phillip W., James R. Ehleringer, and Kenneth A. Nagy, eds. *Stable Isotopes in Ecological Research*. New York: Springer-Verlag, 1988. Includes twenty-eight chapters describing all aspects of how stable isotopes can be used in ecological studies. The use of stable isotopes other than carbon is discussed. Each chapter includes an extensive bibliography.

CARIBBEAN AGRICULTURE

Categories: Agriculture; economic botany and plant uses; food; world regions

Agriculture in the Caribbean islands, from the Bahamas to Trinidad, is concentrated in sugarcane, bananas, coffee, tobacco, and some citrus and cacao.

The Caribbean Sea is an extension of the western Atlantic Ocean that is bounded by Central and South America to the west and south and the islands of the Antilles chain on the north and east. At the end of the twentieth century, agriculture was basic to the economies of nearly every island. Two fundamentally different types of agriculture dominate: large-scale commercial, or plantation, agriculture and small-scale semisubsistence, or peasant, farming. Plantation farming provides the most exports, by value, whereas peasant farming involves far more human labor.

Caribbean agriculture operates under various natural and cultural restraints. Most of the islands have rugged terrain, restricting productive agriculture to river valleys and coastal plains. Typically, less than one-third of an island's land area is suitable for crops. The windward portions of islands are commonly very wet, whereas their leeward areas suffer drought, necessitating irrigation. Various hazards also impact agriculture, including the damaging winds of hurricanes, flooding, accelerated erosion, and landslides. In addition, some crops (notably bananas) have suffered from diseases. On the human side, most peasant farms are restricted to steep, unproductive slopes, while plantations control most of the productive lowland soils. Population pressures have led to the loss of some of the best lands and have caused fragmentation of farmland. Farm labor shortages, climbing wages, and foreign competition have added to the burden.

PhotoDisc

Among the Caribbean islands, coffee is raised for export mainly in Haiti, Jamaica, and the Dominican Republic. Its production is largely for European, Japanese, and U.S. markets.

Commercial Agriculture

Modern plantations own large tracts of land and specialize in one crop, commonly sugarcane, bananas, coconuts, coffee, rice, or tobacco. They are more mechanized and better managed than colonial plantations, although they are still largely British-, French-, or American-owned. The largest plantations are found on the largest islands, especially Hispaniola, Jamaica, and Puerto Rico. Cuba also has large-scale farming, but the operations are government-owned. Plantations always have been smaller in the Lesser Antilles, where relatively little land is available.

Sugar dominates the export economies of Cuba, the Dominican Republic, Guadeloupe, and Saint Kitts. Among traditional sugar producers in the Caribbean, notably Jamaica, Puerto Rico, Trinidad, and Barbados, sugar exports are exceeded by those of other commodities. Haiti, a leading sugar producer as a French colony, now produces little. Overall, sugar production in the Caribbean has been on the decline since the 1960's as a result of the variety of problems noted above.

Other commercial export crops grown in the Caribbean region include bananas, coffee, tobacco, and ganja. Bananas, introduced in the sixteenth century by Spanish missionaries, became an important export in the late nineteenth century as markets developed in Europe and the United States. Sweet bananas are significant exports of Guadeloupe, Martinique, the Dominican Republic, Jamaica, Granada, St. Lucia, and St. Vincent. Overall production is not significant on the world scale. Coffee is raised for export mainly in Haiti, Jamaica, and the Dominican Republic. Jamaica's famous Blue Mountain coffee, grown in the Blue Mountains northeast of Kingston, is among the most prized and expensive coffees of the world. Its production and export is largely for European, Japanese, and U.S. markets.

Tobacco was important before the sugar era and has seen a recent resurgence in the Greater Antilles, especially in Cuba, Puerto Rico, Jamaica, and the Dominican Republic, mostly for cigar production. Ganja, marijuana prepared especially for smoking, is illegal throughout the Caribbean region. The product is nevertheless of considerable commercial importance. Its chief producer is Jamaica, and its main destination is the United States. Other significant export crops include cacao (for chocolate) and citrus.

Peasant Farming

Peasant farming in the Caribbean began after emancipation in the nineteenth century, when freed slaves sought out the only land available, in the hills and mountains. Unfortunately, this land is unsuitable for crop agriculture, having thin and erodible soils. Individual peasant farms average less than 5 acres (2 hectares) in area, often in disconnected plots. A variety of crops are raised, including fruits such as mangoes, plantains, akee, and breadfruit; vegetables such as yams, potatoes, and okra; sugarcane; and coffee.

P. Gary White

See also: African agriculture; Asian agriculture; Australian agriculture; North American agriculture; South American agriculture.

Sources for Further Study

Baud, Michiel. *Peasants and Tobacco in the Dominican Republic, 1870-1930*. Knoxville: University of Tennessee Press, 1995. As well as being a history of tobacco agriculture in the northern valley of the Dominican Republic, known as the Cibao, this book analyzes the place of the tobacco sector in the national and international economies.

Dunn, Richard S. *Sugar and Slaves: The Rise of the Planter Class in the English West Indies, 1624-1713*. Chapel Hill: University of North Carolina Press, 2001. A vivid portrait of English life in the Caribbean more than three centuries ago and what made those colonies the richest, but in human terms the least successful, in English America.

Harrison, Michele. *King Sugar: Jamaica, the Caribbean, and the World Sugar Industry*. New York: New York University Press, 2001. Describes life on a sugar plantation at the end of the twentieth century. Examines the world sugar business, how the industry works, and how ordinary people fit into this global industry.

McIntyre, Arnold M. *Trade and Economic Development in Small Open Economies*. Westport, Conn.: Praeger, 1995. Analyzes trade and economic data relating to Trinidad and Tobago, Barbados, Jamaica, and Guyana from 1968 to 1990 to explain the poor performance of exports during that time.

Rosset, Peter. *The Greening of the Revolution: Cuba's Experiment with Organic Farming*. Melbourne, Victoria, Australia: Ocean Press, 1994. Detailed account of Cuba's turn to a system of organic agriculture prepared on an international scientific delegation on low-input sustainable agriculture in 1992.

CARIBBEAN FLORA

Category: World regions

The Caribbean region is noted for its diverse and varied vegetation. Flowers thrive in the moist, tropical environment found on many islands. Hibiscus, bougainvillea, and orchids are just a few of the varieties found there. The heavily touristed region faces challenges in balancing development with preserving its flora.

The Caribbean region comprises the islands from the Bahamas to Trinidad and is noted for its tropical vegetation and flowers. This region has been significantly affected by human activities. Deforestation began with the development of sugarcane culture in the seventeenth century. When forests are cut for farmland, soil erosion and depletion often occur. Jamaica, Haiti, and many of the smaller islands have suffered acute ecological degradation. Thorn scrub and grasses have replaced native forests that were cleared for farming. This new vegetation does not protect the ground from the sun and provides little protection against moisture loss during drought. Livestock overgrazing also has contributed to ecological degradation.

However, these complex and fragile island ecosystems are finally being appreciated and protected. Most islands' governments recognize that they must balance development with protection of the natural environment. Many have established active conservation societies and national wildlife trusts for this purpose.

Rain Forest

The only rain forest left in the Caribbean Islands is a small area in Guadeloupe. However, many of the islands still have large stands of good secondary forest that are being harvested selectively by commercial lumber companies. Gommier, balata, and blue mahoe are some of the valuable species of trees cut commercially. Such tropical woods as cedar, mahogany, and palms grow on many islands. Martinique has some of the largest tracts of forest (rain forest, cloud forest, and dry woodland) left in the Caribbean.

While the Caribbean rain forest is not as diverse as those of Central and South America, it supports numerous plant species. Several of these plants are endemic to the region. For example, Jamaica has more than three thousand species with eight hundred endemics. Orchids and bromeliads are stunning examples of climbing and hanging vegetation in the rain forests of the Caribbean. Huge tree ferns, giant elephant ear plants, figs, and balsam trees are also found in these tropical island rain forests.

Plantations of commercial timber (blue mahoe, Caribbean pine, teak, and mahogany) have been established in many locations. These plantations reduce pressure on natural forest, help protect watersheds and soil, and provide valuable wildlife

AP/Wide World Photos

Tree ferns in the Caribbean National Forest in Puerto Rico, a tropical rain forest.

habitat. Much of the original forests of the Caribbean have been cut down to make room for sugar plantations and for use as fuel. Haiti, once covered with forests, is now on the verge of total deforestation. Soil erosion and desertification have severely impacted the country.

Saint Kitts, on the other hand, is one of the few places in the world where the forest is actually expanding. It provides abundant habitat for exotic vines, wild orchids, and candlewoods.

Closer to the coasts, dry scrub woodland predominates. Some trees lose their leaves during the dry season. The turpentine tree is common to the dry scrub forests. It is sometimes called the tourist tree because of its red, peeling bark. The trees' bark and leaves are often sold in marketplaces and used for herbal remedies and bush medicine.

Along the marshy coastal waters of many of the islands are dense mangrove swamps. Mangroves grow thick roots that stand above the waterline. These wetlands provide habitat for the endangered manatee, the American crocodile, and for huge numbers of migratory birds and resident birds, such as the egret and heron.

Dominica, one of the windward islands, is known as the "nature island" of the Caribbean. Most of the southern part of the island (17,000 acres, or approximately 7,000 hectares) has been designated the Morne Trois Pitons National Park, which became a World Heritage Site in 1998. Elfin forests of dense vegetation and low-growing plants cover the highest volcanic peaks and receive an abundance of rain. The trees, stunted by the wind, have leaves adapted with drip tips to cope with the excess moisture. Lower down, the slopes are covered with rain forest. Measures are being taken to protect this forest from cutting for farmland or economic gain because it is a valuable water source and a unique laboratory for scientific research.

Carol Ann Gillespie

See also: African flora; Antarctic flora; Asian flora; Australian flora; North American flora; Rain-forest biomes; South American flora; Wetlands.

Sources for Further Study

Knight, Franklin W., and Colin A. Palmer, eds. *The Modern Caribbean*. Chapel Hill: University of North Carolina Press, 1989. Collection of thirteen articles on the diverse elements, such as politics and environment, that have shaped the Caribbean.

Littler, Diane Scullion, et al. *Marine Plants of the Caribbean: A Field Guide from Florida to Brazil*. Washington, D.C.: Smithsonian Institution Press, 1992. The text covers common species as well as locally abundant or unusual flora. Includes color illustrations.

Nellis, David W. *Seashore Plants of South Florida and the Caribbean: A Guide to Identification and Propagation of Xeriscape Plants*. Sarasota, Fla.: Pineapple Press, 1994. Handbook includes many illustrations as well as bibliographical references and an index.

CARNIVOROUS PLANTS

Categories: Angiosperms; *Plantae*; poisonous, toxic, and invasive plants

Carnivorous plants have the ability to capture and digest insects and other small animals.

Carnivorous plants are a diverse group. All live in nutrient-poor, often boggy or aquatic environments where minerals from digested prey are essential for survival, or at least for good health. None use captured prey as an energy source, as animals do; rather, carnivorous plants are photosynthetic, like other plants. The mechanisms for capturing prey are varied, and carnivory is found in at least nine distinct plant families, having evolved a number of times independently.

Pitchers

One widespread carnivorous mechanism is the formation of *pitchers*, specialized leaves or portions of leaves that have developed into hollow tubular or pitcherlike structures that contain fluid. The fluid is mostly water but may contain digestive enzymes and mild narcotics. The mouth, or entrance, to the pitcher is generally marked with bright colors and may produce droplets of nectar to attract insects. The throat of the pitcher is usually lined with wax or downward-pointing hairs so that insects will slowly slide downward toward the pool of water but will have great difficulty climbing back out.

Pitchers are found in a number of genera, in several unrelated families. *Sarracenia* is the genus of North American pitcher plants found in bogs, seeps, and marshy areas, primarily in the southeastern United States. The pitchers arise directly from an underground rhizome. A related species, *Darlingtonia californica*, is found in cold-water bogs in northern California and southern Oregon. The genus *Heliamphora* includes many species in South America. The Asian pitcher plants are in the unrelated genus *Nepenthes*. These pitchers hang from the ends of leaves on plants that are generally *epiphytic* (growing on tree branches) or grow on rocky surfaces. The genus *Cephalotus* is not related to any of the foregoing but produces pitchers similar to those of *Nepenthes*. The single species of *Cephalotus* is found in seeps and bogs in southwestern Australia. Finally, in South America there are species in two genera of *bromeliads* (pineapple family), *Catopsis* and *Brocchinia*, that form crude pitchers that have been found to be carnivorous.

Digital Stock

Specialized leaves or portions of leaves of the pitcher plant form hollow, tubular structures which contain fluid, mostly water, that traps insects.

Leaf Traps

The most famous of all carnivorous plants is Venus's flytrap, *Dionaea muscipula*, which is native to bogs in the coastal plain of North and South Carolina. In this species, the two sides of the leaf snap together quickly when trigger hairs on its surface are brushed twice within a short span of time. The closing mechanism involves a wave of rapid cell elongation on the outer surface of the leaf trap that pushes the two sides of the leaf together. The edges of the trap leaves are lined with long spikes which serve as bars to prevent quick-responding insects from escaping during closing. A reddish coloration and drops of nectar lure insects to the trap. Digestive enzymes are secreted from the inner surface of the trap. After the meal is digested, slower growth of cells on the inner surface reopens the trap.

Related to the fly traps are the *sundews*, in the genus *Drosera*, and similar plants in the genera *Drosophyllum*, *Byblis*, and *Tripohylophylum*. Leaves of these plants are covered with tiny glandular hairs, each producing a drop of sticky, nectarlike fluid that attracts the insects and then traps them. Once an animal is tangled in the gluey exudate, digestive enzymes are secreted, and in some, the leaf slowly folds around the prey.

There are many species of sundew, nearly two hundred in Australia alone, with many species in South America, southern Africa, North America, and Europe. The species vary in the size and shape of their leaves and also in their flowers, which come in a variety of colors, sizes, and shapes. They in-

Carnivorous Plants

Common Name	*Genus*	*Species*	*Location*
Australian pitcher plant	*Cephalotus*	1	Southwest Australia
Bladderworts	*Utricularia*	220	Worldwide
Butterworts	*Pinguicula*	80	Asia, Europa, North America
Cobra lily	*Darlingtonia*	1	California, Pacific Northwest
Corkscrew or forked trap	*Genlisea*	20	South America
North American pitchers	*Sarracenia*	10	United States, Southeast
Portuguese sundew	*Drosophyllum*	1	Portugal, western Spain
Rainbow plants	*Byblis*	5	Northwest Australia
Sun pitchers	*Heliamphora*	6	South America
Sundews	*Drosera*	150	Worldwide
Tropical pitcher vines	*Nepenthes*	90	Australia, Indonesia, Madagascar
Venus's flytrap	*Dionaea*	1	North Carolina, South Carolina
Waterwheel plant	*Aldrovanda*	1	Eurasia

habit seeps and bogs or are found in sandy soil that is damp for only a short time. Most of them are tiny rosettes of paddle-shaped to linear leaves, but some are erect, herbaceous tufts 1-2 feet high, or even vines.

One genus of the unrelated family *Lentibulariaceae* also employs leaf traps. This is *Pinguicula*, found in North America in moist, nutrient-poor habitats. Its leaves are flat, somewhat curled at the edges, and moderately sticky, without conspicuous glandular hairs. They do not close around prey. The many species of *Pinguicula* have variously colored, snapdragon-like flowers on long stalks.

Underwater and Underground Traps

The final category of carnivorous plants produces traps under water, or sometimes in damp soil, where they capture tiny crustaceans and other invertebrate animals. *Aldrovanda* is a submerged plant related to Venus's flytrap but is native to Australia. Its miniature underwater traps snap shut when animals swim or bump into them. Another genus with tiny underwater traps is *Utricularia*, related to *Pinguicula*. These traps operate by suddenly expanding to suck water and small animals inside when triggered. The traps are borne on specialized leaves that grow downward into the water or damp soil. Often one only sees the brightly colored, snapdragon-like flowers above the surface. There are hundreds of species of *Utricularia*, found all over the world, particularly in Australia and North America. A third carnivorous genus in the *Lentibulariaceae* is *Genlisea*, which grows in Africa and South America, and which has traps similar to those of *Utricularia*.

Frederick B. Essig

See also: Animal-plant interactions; Coevolution; Leaf anatomy; Metabolites: primary vs. secondary; Nastic movements; Nutrient cycling.

Sources for Further Study

Cheers, Gordon. *A Guide to Carnivorous Plants of the World*. New York: Angus and Robertson, 1992. A well-illustrated guide to the different groups of carnivorous plants.

Lowrie, Allen. *Carnivorous Plants of Australia*. 3 vols. Nedlands, Western Australia: University of Western Australia Press, 1987-1998. A truly spectacular, comprehensive, and fully illustrated guide to the hundreds of species of carnivorous plants in Australia.

Schnell, Donald. *Carnivorous Plants of the United States and Canada*. Winston-Salem, N.C.: John F. Blair, 1976. Illustrated reference on the carnivorous plants of North America.

Slack, Adrian. *Carnivorous Plants*. Cambridge, Mass.: MIT Press, 1980. An extensive, well-illustrated guide to carnivorous plants worldwide, considered to be one of the best.

CELL CYCLE

Categories: Cellular biology; physiology

In plants, as in all eukaryotic life-forms, the cell cycle comprises the processes of cell division, constituted by three preparatory phases (G_1, G_2, and S), followed by mitosis (nuclear division) and cytokinesis (cytoplasm division).

One of the fundamental characteristics of living organisms is their ability to grow and reproduce. At the cellular level, growth is accomplished by a gain of mass, followed by division into two daughter cells. In unicellular species, such as bacteria and green algae, this division results in the production of new organisms. In multicellular organisms, such as plants, cells must divide many times to produce new individuals, and additional processes of differentiation into mature cell types must occur.

Eukaryotic Cell Division

Plant cells are eukaryotic, meaning that they contain a *nucleus* and membrane-bound cellular compartments called *organelles*, such as chloroplasts, the sites of photosynthesis, and the central *vacuole* used for storage. In order for eukaryotic cell division to occur, cells must accomplish several steps. First, a cell must reach a sufficient size to ensure that the daughter cells will be large enough to survive. The cell achieves this status by making needed molecules, using synthetic biochemical reactions, and monitoring the outside environment to ensure continued favorable conditions for reproduction. Next, a cell must replicate all required macromolecules, including the deoxyribonucleic acid (DNA), as well as the organelles. Finally, the cell must be able to distribute appropriately the cellular contents, including the genetic material, to the daughter cells. The cyclical repetition of these steps comprises the eukaryotic cell division cycle or, more simply, the cell cycle.

Cell Cycle Phases

The general arrangement of the cell cycle was first determined through microscopic observations of living cells undergoing cell division. Originally, four stages were identified: G_1 phase, S phase, G_2 phase, and M phase. G_1 and G_2 phases were named as gap phases, because it was presumed that the cell did not appear to accomplish any specific events during these phases. G_1 phase occurs after the previous cell division and is the phase where the cell grows, metabolizes, and prepares to undergo another round of cell division. During G_1 phase, the cell also transcribes and translates the genes required for DNA synthesis. In S phase, the DNA synthetic phase, the cell concentrates its efforts on accurately replicating its genetic material, the DNA genome. In G_2 phase, the cell coordinates both transcription and translation to synthesize proteins that are required specifically for asexual cell division, or mitosis. During M phase, the DNA chromosomes condense and are divided into the two daughter cells. G_1, S, and G_2 phases together are referred to as *interphase*.

Mitosis itself is composed of four phases: (1) *prophase*, in which chromosomes are condensed, homologous chromosomes are paired together, and the spindle apparatus made of microtubules forms; (2) *metaphase*, in which the paired chromosomes are lined up across the center of the cell on the metaphase plate; (3) *anaphase*, in which the homologous chromosomes are pulled to separate poles in the dividing cell by the attached spindle apparatus; and (4) *telophase*, in which the daughter cell chromosomes are collected together at the poles. Mitosis is followed by cytokinesis, or the process by which the two daughter cells are physically separated. Because plant cells have cell walls, the division of one cell into two daughter cells requires the formation of a cell plate to complete cytokinesis. This cell plate grows outward between the two new nuclei. Once the cell plate reaches the walls of the dividing cell, it forms the cell wall that separates the two new cells.

Checkpoints

The cell will die if it goes through any cell cycle phase out of order. Therefore, the cell has evolved sophisticated *checkpoints* to ensure that critical events, such as DNA replication and cell division, occur in the correct order and that each required step is completed prior to movement to the next phase. Additionally, external conditions, such as a change in nutrient availability, can cause a cellular checkpoint response. There are two main points in the cell cycle which are regulated by checkpoints, the G_1-to-S transition and the G_2-to-M transition. The G_1-to-S transition is important because once the cell commits to divide and starts to copy its DNA, it must complete cell division or die. After DNA replication is complete, the G_2-to-M transition is important for the cell to make sure that all of the DNA has been copied correctly, or the daughter cells will not have the full complement of genetic material.

CDKs and Cyclins

Enzymes are proteins that catalyze chemical reactions in the cell. Without enzymes, chemical reactions would occur too slowly to sustain life. Kinases are enzymes that place a chemical group, called a phosphate group, onto other proteins in the cell. Because of the nature of the phosphate chemical group, the addition of these groups causes conformational changes in the three-dimensional structure of the proteins that accept them. The cell cycle is controlled by the actions of kinases called cyclin-dependent kinases, or CDKs. Each CDK has a partner called a cyclin, which is required for the kinase activity to place phosphate groups on target proteins. The CDK/cyclin pairs perform kinase reactions, which activate proteins leading to completion of cell cycle phases. Opposing the action of the CDK and cyclin pairs are cyclin-dependent kinase inhibitors, or CKIs, which act as brakes to prevent CDK/cyclin activity until the cell is ready to go on to the next step in the cell cycle.

Many of the cell cycle control mechanisms have been conserved throughout all eukaryotic cells, including the plant cell. *Arabidopsis* is a genus that has been widely characterized and used as an experimental model for plant growth and differentiation in scientific laboratories worldwide. In plants such as *Arabidopsis*, two major groups of CDKs have been studied, the A-type and B-type CDKs. The A-type CDKs show kinase activity during the S, G_2, and M phases of the cell cycle and regulate the transitions from G_1-to-S and G_2-to-M phases. The B-type CDKs are active during mitosis and regulate only the G_2-to-M transition in plants.

Jennifer Leigh Myka

See also: Angiosperm cells and tissues; Cell theory; Chromosomes; DNA replication; Eukaryotic cells; Mitosis and meiosis; Nucleic acids; Reproduction in plants.

Sources for Further Study

Alberts, Bruce, et al. *Molecular Biology of the Cell*. New York: Garland, 1994. Survey of cell biology for an introductory university course. Beautifully illustrated in color throughout the book.

Darnell, James, Harvey Lodish, and David Baltimore. *Molecular Cell Biology*. 4th ed. New York: W. H. Freeman, 2000. Detailed undergraduate textbook combines molecular biology with cell biology and genetics. Coverage is applied to problems such as cell development.

Lewin, Benjamin. *Genes VII*. New York: Oxford University Press, 2000. Explains the structure and function of the gene and offers an integrated approach to prokaryotes and eukaryotes. The gene is covered from all perspectives.

Raven, Peter H., Ray F. Evert, and Susan E. Eichhorn. *Biology of Plants*. 6th ed. New York: W. H. Freeman/Worth, 1999. Introductory college-level textbook. Chapter 8, "The Reproduction of Cells," covers the cell cycle and includes photographs, drawings, and diagrams.

CELL THEORY

Categories: Cellular biology; history of plant science; physiology

The notion that the cell is the smallest division of life and its attendant principles have been developed over the past three centuries and are collectively known as the cell theory.

Before the invention of the microscope, people studying living organisms saw whole and complete organisms and did not imagine that life was subdivided into smaller compartments. It is now known that the cell is the fundamental unit of life and that all living organisms are composed of cells. Because cells are microscopic, their existence was not discovered until the seventeenth century.

Discovery of the Cell

The discovery of the cell did not come about until the last half of the seventeenth century, after the Dutch inventor Antoni van Leeuwenhoek built the first light microscope. When looking at pond water using his light microscope in 1674, Leeuwenhoek saw many tiny creatures which were invisible to the naked eye. Leeuwenhoek assumed that these tiny "animalcules" were alive because he could see them moving.

However, the first description of the cell is attributed to the English scientist Robert Hooke. In 1665 Hooke first published *Micrographia*, a work devoted to observations made with his compound light microscope. Hooke examined the structure of cork, a dead plant tissue, by cutting cork into very thin slices and observing the slices under his light microscope. Hooke saw the dead cells of the cork outlined by the thickened cell walls of that tissue and determined that cork was composed of a pattern of spaces which resembled small rooms, or "cells." Interestingly, Hooke did not recognize the significance of the cells that he described and thought that the cork cells were merely channels for fluid conduction in the plant.

Cells as Globules

Partly because of problems with chromatic aberrations in early microscopes, scientists thought that living organisms were made of "globules." As early as 1682, plant tissues were described as bladders clustered together. In 1771 William Hewson performed one of the first cell biology experiments by confirming Leeuwenhoek's earlier observation of blood and brain globules and by showing that the blood globules swelled and shriveled in different solutions. By 1812 Johann Jacob Paul Moldenhawer had shown that plant tissue was composed of independent cells, and in 1826 Henri Milne-Edwards determined that all animal tissues were formed from globules. Finally, in 1824 Henri Dutrochet proposed that animals and plants had a similar cellular structure. Ironically, Thomas Hodgkin and Joseph Jackson Lister showed in 1827 that many of the previously observed globules were likely to have been optical artifacts that disappeared when the new achromatic microscope was used. However, the idea that life was made of tiny cells remained.

Cells Compose All Organisms

Matthias Schleiden, a German botanist, suggested in 1838 that all of the structural elements found in plant tissues were composed of cells or cellular products. In 1839 Theodor Schwann, a zoologist, reported that animal tissues were also composed of cells and suggested that all living organisms were actually composed of cells. Based on their contributions, Schleiden and Schwann are considered to have established the official "cell theory." Schwann also contributed to the description of the cell by defining a cell as having three essential components: a cell wall, a nucleus, and a fluid content. The Scottish botanist Robert Brown first described the nucleus of the cell as an essential component of living cells in 1831.

The work of both Schleiden and Schwann contested the notion of *vitalism*, the belief that no single part of an organism was alive but that somehow the substances composing the whole organism together shared the characteristics of life. The idea of vitalism made sense prior to the invention of the microscope (because cells cannot be seen with the naked eye) but did not hold up when cells were observed in all plant and animal tissues. Therefore, the idea that living organisms were made of cells was becoming widespread at this time. An 1830 plant anatomy textbook by Franz Julius Ferdinand Meyen contained a chapter on cell structure and discussed the idea that cells unite to form cellular tissues.

Cell Origins in Organisms

Interestingly, while Schleiden and Schwann realized that all living organisms were made of cells, they did not realize that all cells came from preexisting cells. Indeed, Schwann mistakenly thought that cells had an extracellular origin and could arise de novo. Schwann originally observed plant cells in the endosperm of seeds, where the nuclei multiply before the cell walls form, and he generalized from this unusual system. This idea persisted for a time, even though Hugo von Mohl first described cell division in green algae in 1837. Essentially, von Mohl investigated his hypothesis that all cells must start out very small and gradually grow to full size by observing a filamentous green alga. By studying the alga, Mohl discovered cell division as the algal cells divided and formed partitions between the newly formed daughter cells.

By 1858 the pathologist Rudolf Virchow had expressed the idea that all cells arise from preexisting cells. This idea was important because until that time, the idea of *spontaneous generation* was also popular. Spontaneous generation was the mistaken idea that living organisms could arise spontaneously from nonliving material and was thought to explain such phenomena as the fact that frogs are found in ponds and maggots in rotten meat. In contrast, Virchow wrote: "Where a cell exists, there must have been a preexisting cell, just as the animal arises only from an animal and the plant only from a plant."

Modern Cell Theory

The modern cell theory consists of several principles. As stated by Schleiden and Schwann, all living organisms are composed of one or more cells. Because cells are the smallest units of life, cells must be the site of the chemical reactions that sustain living organisms. Finally, it is clear that all cells arise from preexisting cells, as first stated by Virchow. In order for cells to perform this replication, it is now apparent that all cells contain hereditary information and pass this information from parent cell to daughter cell. Additionally, this hereditary information is contained in the deoxyribonucleic acid (DNA) found in the nuclei of all cells, with this DNA being copied for each daughter cell to use.

Jennifer Leigh Myka

See also: Angiosperm cells and tissues; Cell cycle; Chromosomes; DNA replication; Eukaryotic cells; History of plant science; Mitosis and meiosis; Nucleic acids; Reproduction in plants.

Sources for Further Study

Darnell, James, Harvey Lodish, and David Baltimore. *Molecular Cell Biology*. 4th ed. New York: W. H. Freeman, 2000. Detailed undergraduate textbook combines molecular biology with cell biology and genetics. Coverage includes cell development.

Raven, Peter H., Ray F. Event, and Susan E. Eichhorn. *Biology of Plants*. 6th ed. New York: W. H. Freeman/Worth, 1999. Introductory college-level textbook. Cell theory is discussed in chapter 3, "Introduction to the Plant Cell."

CELL-TO-CELL COMMUNICATION

Category: Cellular biology

Cell-to-cell communication involves the various stimuli to which plants respond, whether biotic, such as hormones and disease, or abiotic, such as water status, heat, cold, and light.

Throughout their lives, plants and plant cells continually respond to both external and internal signals, which they use to alter their physiology, morphology, and development. The manner in which plants respond to a stimulus is determined by developmental age, previous environmental experience, and internal biological clocks that specify the time of year and time of day.

Chemical Messengers

In complex multicellular eukaryotes, the coordination of responses to environmental and developmental stimuli requires an array of signaling mechanisms. Animals have evolved two systems, the nervous system and the endocrine (hormone) system, for responding to stimuli. While plants lack a nervous system, they did evolve *hormones* and other chemicals, such as phytochrome, as chemical messengers.

The major groups of plant hormones include the auxins, gibberellins, cytokinins, ethylene, and abscisic acid. These hormones serve as signals for a wide range of physiological, biochemical, and developmental responses. These signals may impact different cells at different times, or they may impact the same cells at different sites. A variety of pathways are available for information flow from any one signal. Information from the same signal may travel to different areas of the cell under different environmental conditions. Alternatively, information might travel to the same site after traveling by different pathways. Most signals appear to induce changes in gene expression, and this altered gene expression is responsible for the observed response.

There are two classes of chemical messengers, based on their ability to cross cellular membranes. *Lipophilic messengers* readily diffuse across membranes and combine with intracellular receptor proteins. When activated by the messenger, these proteins function as transcription factors, thereby inducing the transcription of new proteins. In other words, the messenger-receptor complex signals the activation of genes which encode the proteins that produce the response to the stimulus. While animals have numerous lipophilic messengers, such as the steroid hormones in animals, only one, brassosteroid, has been demonstrated in plants.

Most plant messengers are *hydrophilic*, or water-soluble (rather than lipophilic) and are unable to enter the target cell because they cannot diffuse across the hydrophobic ("water-hating") interior of the membrane. These messengers must first bind with a membrane receptor molecule. This messenger-receptor complex then communicates with other molecules inside the cell to initiate a cascade of events referred to as a *signal transduction pathway*. Most signal transduction pathways cause the activation of other chemicals, referred to as second messengers.

Second Messengers

Signal transduction pathways using a variety of second messengers have been well documented in animal systems. Some of the most common second messengers are 3′,5′-cyclic AMP (cAMP), G-proteins, 1,2 diacylglycerol (DAG), inositol 1,4,5-triphosphate (IP_3), and Ca^{2+}, and many of these have been shown to be active in plants. When a chemical messenger such as a hormone binds to a membrane receptor, one or more of these second messengers are elevated. The elevated level of the second messengers results in the activation of regulatory proteins such as protein kinases or phosphatases. Activated protein kinases will *phosphorylate* transcription factors (that is, add a phosphate group), and activated phosphatases will dephosphorylate (remove a phosphate group from) transcription factors.

A typical signal transduction cascade is presented in the following scenario. An environmental stress causes an elevation in the level of the hormone abscisic acid (ABA), which is responsible for leaf fall. ABA binds to receptors in the membranes of the target cells. The ABA-receptor complex activates a G-protein, which then activates the enzyme phospholipase C. This enzyme catalyzes the conversion of a substrate to DAG and IP_3. The IP_3 stimulates the opening of Ca^{2+} channels in the endoplasmic reticulum or tonoplasts (the membranes surrounding vacuoles). The release of Ca^{2+} from these organelles activates protein kinases, which then activate transcription factors by phosphorylation. The activated transcription factors induce transcription of genes, which encode the proteins necessary for the plant to respond to the environmental stress.

Transport of Messengers

As discussed above, cells communicate primarily via chemical messengers. In most instances, particularly in the case of plant hormones, the messengers are produced in one cell and transported to other cells. Plant cells are usually in contact with others around them, and cell communication (transport of messengers) can occur by transport through either the *apoplast* or the *symplast*. The apoplast refers to the free space between cells and cell wall materials. Water moves freely through the apoplast, and certain chemicals can be found moving with the water. Some important developmental molecules, such as the auxins, have been shown to move through the apoplast. The symplast is composed of the living cytoplasm of the cells, and many substances are transported symplastically.

Rapid transport through the symplast is possible because most living plant cells are connected to neighboring cells by *plasmodesmata* that pass through the adjoining cell walls and provide some degree of cytosolic continuity between them. Plasmodesmata are tubelike cytoplasmic extensions that are divided into eight to ten micro-channels. Although the exact pathway of communication has not been determined, some molecules can pass from cell to cell through plasmodesmata, probably by flowing through the micro-channels. Plasmodesmata appear to be gated, which means that they allow the passage of some molecules and restrict the passage of others.

D. R. Gossett

See also: Active transport; Circadian rhythms; Gene regulation; Growth and growth control; Hormones; Liquid transport systems; Osmosis, simple diffusion, and facilitated diffusion; Oxidative phosphorylation; Photoperiodism; Tropisms; Vesicle-mediated transport; Water and solute movement in plants.

Sources for Further Study

Dey, P. M., and J. B. Harbone, eds. *Plant Biochemistry*. San Diego: Academic Press, 1997. An excellent source on the biochemistry of signaling molecules, for the more advanced student. Includes bibliographical references.

Hopkins, William G. *Introduction to Plant Physiology*. 2d ed. New York: John Wiley and Sons, 1999. This general text for the beginning plant physiology student contains a very clear discussion of chemical signaling; diagrams, illustrations, index.

Raven, Peter H., Ray F. Evert, and Susan E. Eichhorn. *Biology of Plants*. 6th ed. New York: W. H. Freeman/Worth, 1999. A very good general botany text with a simplified discussion of cellular communication. Includes diagrams, illustrations, and index.

Salisbury, Frank B., and Cleon W. Ross. *Plant Physiology*. 4th ed. Belmont, Calif.: Wadsworth, 1992. One of the standard texts for the more advanced plant physiology student, with an excellent discussion on chemical messengers. Includes biographical references, diagrams, illustrations, and index.

Taiz, Lincoln, and Eduardo Zeiger. *Plant Physiology*. 2d ed. Sunderland, Mass.: Sinauer Associates, 1998. A text written for the advanced plant physiology student with an outstanding discussion of signal transduction. Includes biographical references, diagrams, illustrations, and index.

CELL WALL

Categories: Anatomy; cellular biology; physiology

The cell wall is the outer, rigid wall of a cell, dividing the protoplast (the interior, including the cytoplasm and nucleus) from the cell's external environment. The plant cell wall is both unique to and a major feature of plants, perhaps second only to the plant's photosynthetic ability.

The primary functions of the cell wall in plant cells include are to provide protection for the enclosed *cytoplasm* and give mechanical support to the entire plant structure. Plant cell walls are part of the *extracellular matrix*, a complex mixture of extracellular materials found between cells. These materials are synthesized by the intracellular contents and transported through the plasma membrane. All plant cell types consist of at least a *primary cell wall*, and many also produce a *secondary cell wall*. In addition, certain cells also secrete specialized substances into the extracellular matrix.

Primary Cell Wall

The primary cell wall is first synthesized by young, actively growing cells. It is thin and is composed of *cellulose* embedded within a *noncellulose matrix*. Cellulose is a polysaccharide polymer composed entirely of glucose molecules joined together end on end to form long, unbranched chains. These chains may consist of up to several thousand glucose molecules. The chemical bond joining the glucose molecules in cellulose is slightly different from that found in starch, another polysaccharide composed entirely of glucose. It is this difference in bonds that makes cellulose a structural polysaccharide.

Microfibrils are groupings of about 50 to 60 cellulose chains that are parallel to one another and held together by hydrogen bonds. The tensile strength of microfibrils is comparable to steel wire of the same thickness. The cellulose microfibrils make up approximately 25 percent of the primary cell wall and are arranged somewhat randomly (more parallel in fast growing cells) within the noncellulose matrix.

Noncellulose Matrix

The noncellulose matrix of the primary cell well accounts for about 60 percent to 80 percent of the primary cell wall and is composed of *hemicelluloses*, *pectins*, and *extensins*. Hemicelluloses are highly branched polysaccharide structures composed of heterogeneous mixtures of sugars, some of which are chemically modified. Pectins are also heterogeneous mixtures of sugars but are particularly rich in galacturonic acid. Pectic substances are also a major component of the outermost *middle lamella*, which can be thought of as the cement that holds adjacent cells together. Pectins, extracted from unripe fruits, have been used commercially as thickening agents for jellies and jams. The softness of ripe fruits is due to the enzymatic breakdown of pectins within the middle lamella. The hemicelluloses and various pectins are thought to coat and reinforce the cellulose microfibrils. Extensins are protein components of the matrix. They are glycoproteins that contain a high amount of a modified amino acid called hydroxyproline. Extensins make up about 10 percent of the matrix material, add strength to the cell wall, and are involved in cell growth.

Secondary Cell Wall

Cell types that are directly involved in support of the plant body, such as *sclerechyma* and particularly the *tracheids* and *vessels* of the xylem, undergo secondary wall formation. This wall is typically much thicker and is synthesized by the intracellular contents to the inside of the primary cell wall. Secondary cell wall formation occurs as the cell reaches its mature size. In some cell types there may even be two or more distinct layers of secondary cell wall deposition. Its composition is somewhat similar to the primary cell wall but also differs significantly.

Cellulose microfibrils can make up to 45 percent of this wall, while hemicelluloses and pectins can make up to about 20 percent.

However, what most distinguishes the secondary cell wall is the deposition of *lignin*. Lignin is an extremely tough and durable complex compound characterized by interlocking phenolic groups. Lignin is stronger than cellulose microfibrils, and together they are responsible for the superior strength of many types of wood. In cells that undergo secondary cell wall formation, lignin may also be deposited in the preexisting middle lamella and primary cell wall areas.

Specialized Cell Wall Substances

Certain cell types, especially the epidermis, which is exposed directly to the outer environment, secrete a variety of highly waterproof and protective substances, such as *cutin*, *suberin*, and a variety of *waxes*. These substances are deposited on the outside of the primary cell wall. Cutin and suberin are polymers composed of long-chain fatty acids that are linked or esterified at the acid ends. Suberin differs in that it contains dicarboxylic fatty acids and various phenolic compounds. For the most part, cutin is associated with and is the main constituent of the *cuticle* of the aboveground epidermis, while suberin is mostly associated with the belowground epidermis.

Suberin is also the major component of the *Casparian strip* found in the *endodermis* of the root. This important root tissue forces water and dissolved minerals to move intracellularly into the vascular tissue. Additionally, suberin constitutes scar tissue that is formed when cells are injured or otherwise wounded. Waxes are a family of extremely waterproof substances characterized by long-chain alcohols linked with long-chain fatty acids or hydrocarbons. They are typically secreted as droplets on the exterior of the cuticle and crystallize in a variety of geometric patterns.

Plasmodesmata

Plant cells contain cytoplasmic channels called *plasmodesmata* that connect adjacent cells. As cells divide and the *cell plate* is formed, there are areas where new cell wall material is not deposited because of the extension of the endoplasmic reticulum (ER) from the mother cell to the daughter cell. This ER channel is referred to as the *desmotubule*. Thus, each plasmodesma contains an inner desmotubule. Some cells contain areas called *primary pit fields*, in which numerous plasmodesmata are found.

The purpose of plasmodesmata involves cell-to-cell communication via transport of small molecules. Movement is not thought to occur inside the desmotubule, as it is too narrow, but rather through the cytoplasmic channel between the desmotubule and the plasmodesma itself. The presence of plasmodesmata allows for a continuous cytoplasmic connection within plant tissues called the *symplast*.

In cells which form thick secondary cell walls, particularly the xylem vessels and tracheids, which contain no living protoplast at maturity, numerous pit pairs among adjacent cells are found that allow multiple pathways for the flow of water.

Thomas J. Montagno

See also: Angiosperm cells and tissues; Carbohydrates; Cell-to-cell communication; Cytoplasm; Endomembrane system and Golgi complex; Endoplasmic reticulum; Plant fibers; Plant tissues; Plasma membranes; Vesicle-mediated transport; Water and solute movement in plants; Wood.

Sources for Further Study

Hopkins, William G. *Introduction to Plant Physiology*. 2d ed. New York: John Wiley & Sons, 1999. Basic textbook with general information on plant cell walls. Includes illustrations, bibliographical references.

Taiz, Lincoln, and Eduardo Zeiger. *Plant Physiology*. 2d ed. Sunderland: Sinauer Associates, 1998. Advanced textbook with a chapter devoted to plant cell walls. Includes illustrations, photographs, and extensive bibliographic references.

CELLS AND DIFFUSION

Categories: Cellular biology; transport mechanisms

Plant cells, like all other living cells, are surrounded by a semipermeable membrane, and any particle moving into or out of the cell must cross this membrane. There are three basic processes by which particles move across plant cell membranes: diffusion, facilitated diffusion, and active transport.

The process of *active transport* requires the direct input of energy to move particles across the cell membrane. Diffusion and *facilitated diffusion* can occur without the direct expenditure of cellular energy.

Diffusion

If one were to drop a sugar cube into glass of water and immediately use a straw to sip a little water from the top of the glass, the water would not have a sweet taste. However, after a few hours, a sip of water from the top would taste sweet. The reason for the change in the taste of the water is diffusion, the net movement of particles down a *concentration gradient* (that is, from an area of higher concentration to an area of lower concentration). Concentration is the number of particles or amount of substance per unit volume, and a gradient occurs when some factor such as concentration changes from one volume of space to another. Hence, the sugar molecules move more frequently from around the cube where they were highly concentrated to other parts of the glass where they were less concentrated. There is always some movement in both directions, but the net movement is down the concentration gradient.

Diffusion is possible because molecules in a liquid or gaseous phase are not static; they are in constant motion as a result of kinetic energy, which exists at temperatures above absolute zero (⁻273.16 degrees Celsius, or ⁻459.69 degrees Fahrenheit). As the concentration of a substance increases, its free energy also increases. When molecules move, they collide with one another and exchange kinetic energy, and there is a random but progressive movement from regions of high free energy (high concentration) to regions of low free energy (low concentration). Diffusion can occur quite rapidly over short distances but can be extremely slow over long distances. For example, a molecule of glucose can diffuse across a typical 50-micrometer diameter cell in 2.5 seconds, but it takes thirty-two years for it to diffuse a distance of 1 meter.

Role in Plants

Diffusion is an important process in the lives of plants. Water is an important component of all cells, and water moves into plant cells by the process of *osmosis*. Osmosis is the diffusion of water across a semipermeable membrane. Many plant nutrients reach the root surface via diffusion through the soil solution. Some nutrient molecules diffuse across root cell membranes into the *cytosol* (cell sap or cytoplasm) or from the cytosol of the endodermal cells into the xylem tissue. Carbon dioxide diffuses from the atmosphere through the stomata and into the air spaces of leaves. Water vapor evaporates from the surface of a leaf by diffusion through the open stomata.

Diffusion also plays a role in the movement of photosynthetic products such as sugars into the phloem for transport throughout the plant. Because cellular membranes are composed of a lipid bilayer, lipid-soluble materials use simple diffusion to cross the membrane surface. Substances with low lipid solubility can move across membranes via facilitated diffusion. In this process, the substance binds to a transporter molecule, generally called an ionophore, which transports the substance across the membrane and down its concentration gradient.

Bulk Flow

As previously mentioned, diffusion occurs rapidly over short distances. In order to move substances such as water over long distances—for example from the roots to the leaves, plants use a process referred to as *bulk flow*. Bulk flow is the concerted *en masse* movement of groups of molecules, usually in response to a pressure gradient. A moving stream, water flowing through a garden hose, and wind currents are examples of bulk flow. Pressure-driven bulk flow is the major mechanism behind the movement of water over long distances through the xylem. This process is different from diffusion because it is independent of solute concentration.

Active Transport

Whereas diffusion and facilitated diffusion do not require the direct input of cellular energy because they involve transport down a concentration gradient, moving substances against their concentration gradient requires the expenditure of energy. This "uphill" movement across membranes is called active transport.

The most common source of energy for active transport comes from adenosine triphosphate (ATP). When this high-energy phosphate is hydrolyzed, the stored energy is released to drive cellular reactions such as active transport. More specifically, the substance is moved across the membrane by a carrier protein embedded in the membrane. The carrier protein uses energy from the hydrolysis of ATP (that is, the removal of one phosphate group). Although active transport is primarily for movement against a concentration gradient, it can also be used to move substance down their concentration gradient.

There are two important modifications of the active transport process: *cotransport* and *countertransport*, both of which involve the movement of one substance down its concentration gradient while simultaneously transporting another substance against its specialized membrane proteins. Although these proteins do not require an energy source to operate, ATP is still indirectly consumed. The substance being moved down its concentration gradient would eventually be at equal concentrations on each side of the membrane. To counteract this, active transport, with hydrolysis of ATP as the energy source, is used to pump the substance across the membrane to maintain the gradient.

D. R. Gossett

See also: Active transport; Gas exchange in plants; Liquid transport systems; Osmosis, simple diffusion, and facilitated diffusion; Plasma membranes; Root uptake systems; Vesicle-mediated transport; Water and solute movement in plants.

Sources for Further Study

Hopkins, William G. *Introduction to Plant Physiology*. 2d ed. New York: John Wiley and Sons, 1999. This general text for the beginning plant physiology student contains a very clear discussion of diffusion and water relations. Includes diagrams, illustrations, and index.

Nobel, P. S. *Physicochemical and Environmental Plant Physiology*. San Diego: Academic Press, 1991. An excellent source on the water relations and water movement in plants, for the more advanced student. Includes bibliographical references.

Raven, Peter H., Ray F. Evert, and Susan E. Eichhorn. *Biology of Plants*. 6th ed. New York. W. H. Freeman/Worth, 1999. A very good general botany text with a simplified discussion of diffusion in plants. Includes diagrams, illustrations, and index.

Salisbury, Frank B., and Cleon W. Ross. *Plant Physiology*. 4th ed. Belmont, Calif.: Wadsworth, 1992. One of the standard texts for the more advanced plant physiology student, with an excellent discussion on diffusion and water relations. Includes biographical references, diagrams, illustrations, and index.

Taiz, Lincoln, and Eduardo Zeiger. *Plant Physiology*. 2d ed. Sunderland, Mass.: Sinauer Associates, 1998. A text written for advanced plant physiology students, with an outstanding discussion of diffusion and bulk flow. Includes biographical references, diagrams, illustrations, and index.

CELLULAR SLIME MOLDS

Categories: Microorganisms; molds; *Protista;* taxonomic groups; water-related life

Cellular slime molds, or dictyostelids, were originally considered to be fungi. These microscopic, multicellular organisms are easily mistaken for some of the microfungi that commonly occur as contaminants in laboratory cultures. However, cellular slime molds are more closely related to the protozoans than to fungi.

Although once thought to be fungi, the protists of the phylum *Dictyosteliomycota* actually have more in common with the *paramecium* or *amoeba* that can be observed in a drop of pond water when viewed under the microscope than they do with mushrooms and toadstools. Cellular slime molds are essentially microscopic throughout their entire life cycle, and only rarely can they be observed directly in nature, as is the case for the plasmodial slime molds. Cellular slime molds must therefore be grown under controlled laboratory conditions in order to be studied.

Life Cycle

Since their discovery in the late nineteenth century, cellular slime molds have intrigued biologists. Their life cycle exhibits a curious alternative to the way in which most other creatures on earth grow, develop, and become multicellular, with different specialized tissues produced as a result of the process. Most plants and animals begin life as a single cell (called a zygote) that is the product of the fusion of an egg cell and sperm cell. Shortly after the two cells fuse (through a process termed fertilization), the zygote divides into two cells that stick together. These cells soon divide again to produce a cluster of four cells that in turn divide, and so on. Within hours or days (depending upon the particular plant or animal), clusters of dozens to thousands of cells form an embryo. Specialized cells begin to take form, and the basic shape of the body of the organism begins to become apparent.

Cellular slime molds approach reproduction differently. Like fungi and plasmodial slime molds, they produce *spores* as reproductive structures. When a spore germinates (no fusion of cells is required), it releases a single *amoeboid cell* that begins to engulf and digest bacteria in soil and decaying plant debris, the usual habitats for cellular slime molds. When the amoeboid cell divides, the two cells produced separate and become completely independent of each other, with each continuing to feed and undergo additional divisions for a number of hours or days. Only after the growing population of amoeboid cells depletes the local supply of bacteria is there any indication that a multicellular structure will be produced.

In response to the production of chemical attractants, thousands of amoeboid cells that have been operating as individual single-celled organisms begin to move, either singly or in streaming masses, to form multicellular clumps, or *aggregations*. Shortly thereafter, one or more cigar-shaped structures called *pseudoplasmodia* emerge from each aggregation. A pseudoplasmodium is a unified collection of thousands of what had once been separate, independent amoeboid cells. The cells remain distinct in the pseudoplasmodium but no longer act independently. Instead, they cooperate as parts of a multicellular entity. Remarkably, when amoeboid cells of two or more different species of cellular slime molds are grown together, the amoeboid cells of the different species can recognize each other, so that the cells that form any one aggregation are all of a single species rather than a mixture.

Immediately, or perhaps after the entire structure has migrated a short distance toward a light source, cells of the pseudoplasmodium begin to display different patterns of specialization. Cells that happen to have been positioned near the anterior end of the moving "cigar" begin to secrete a wall consisting of *cellulose*. These cells bind together to form a slender stalk that grows upward from the surface of the substrate upon which the pseudoplasmodium occurs. Other cells, those that happened to have been nearer the posterior end of the pseudoplasmodium, are lifted off the surface on the end of the extending stalk. These cells begin to become encapsulated and specialized as spores. Only the latter live on and produce another generation of amoeboid cells to feed on soil bacteria. The cells that produced the stalk in order to elevate the spore cluster above the substrate eventually die, dry up, and decay.

Reproduction

It appears that cellular slime molds reproduce asexually most of the time, at least under laboratory conditions. All of the cells that originate from the same spore are basically genetically identical to one another and collectively represent a genetic clone. As is the case for asexual reproduction in other lifeforms, finding a "mate" is not necessary to perpetuate the species. If amoeboid cells are equipped with the genetic characteristics necessary to survive long enough to produce spores, the same gene combinations will be passed faithfully to all offspring, thus providing the same qualities for survival.

However, a method of sexual reproduction, with its potential of introducing genetic variability, also seems to exist in cellular slime molds. Occasionally in laboratory cultures, a number of large, thick-walled cells are found that are quite different from spores or encysted amoeboid cells. These giant cells are called *macrocysts*. Macrocysts appear to form when several amoeboid cells (sometimes described as being of compatible "mating types") fuse together and rearrange their genetic libraries and those of other amoeboid cells that may be engulfed. When macrocysts germinate, the amoeboid cells that emerge seem to have different combinations of genetic information than the cells that initially formed the macrocysts. This mixing up of genetic information, along with the genetic changes resulting from mutations, provides cellular slime molds with an ability to cope with changing environments.

Distribution and Ecology

Most of what is known about cellular slime molds has been acquired from studying these organisms in laboratory culture. What about the biology of "wild" slime molds in nature? In natural ecosystems, it is quite likely that cellular slime molds play a significant role in controlling the size of bacterial populations in soil and decaying litter. Nutrients that are taken up from decaying plants and animals by bacteria are transferred to cellular slime cells when the latter feed upon these bacteria. The cellular slime molds, in turn, become food for soil protozoans, nematode worms, microscopic arthropods such as mites, and other small invertebrate animals. Because of this, cellular slime molds play an essential role in patterns of energy flow and nutrient cycles within terrestrial ecosystems.

There are about seventy-five described species of cellular slime molds. These have been assigned to one of three genera: *Dictyostelium*, *Polysphondylium*, and *Acytostelium*. Some species of cellular slime molds have been found in almost all parts of the world. Two good examples are *Dictyostelium mucoroides* and *Polysphondylium pallidum*. Numbers of species of cellular slime molds appear to be highest in the American tropics, which suggests that this region represents a center of evolutionary diversification of the group. More than thirty-five different species have been found in the small area around the Mayan ruins at Tikal in Guatemala. In general, numbers of species of cellular slime molds decrease with increasing elevation and with increasing latitude.

Some species have restricted habitat associations. One species (*Dictyostelium caveatum*) has been found only in a single cave system in Arkansas. Another species (*Dictyostelium rosarium*), known from a number of localities worldwide but rarely aboveground, also seems to have an affinity for the type of conditions found in caves. Of the thirty-five species that occur at Tikal, many appear to be restricted to tropical or subtropical locations. *Dictyostelium discoideum* is the most intensively studied cellular slime mold and the one most widely used in research on developmental biology and genetics. Any search for information about cellular slime molds would probably turn up numerous references to this particular form.

Dispersal of Spores

Unlike most spore-producing organisms (including plasmodial slime molds), cellular slime molds produce spores that do not seem to be carried appreciable distances by wind. Instead, dispersal of cellular slime mold spores seems to depend more upon their accidental transport on the body surface or within the digestive tract of some animal. Viable spores of cellular slime molds have been recovered from the droppings of a number of animals, including rodents, amphibians, bats, and even migratory birds that travel great distances between winter and summer homes. In tropical forests, many living plants and considerable amounts of organic material are found high above the ground in the forest canopy. Cellular slime molds have been isolated from the mass of organic material (literally a "canopy soil") found at the bases of epiphytic plants growing on the trunks and branches of trees in these forests. It seems likely that they are introduced to such habitats by being carried up from the ground by birds, insects, or other animals that move between the forest floor and the canopy above it.

John C. Landolt and Steven L. Stephenson

See also: Bacteria; Plasmodial slime molds; *Protista*.

Sources for Further Study

Bonner, John T. *The Cellular Slime Molds*. Princeton, N.J.: Princeton University Press, 1967. A bit dated but a good general reference to the biology of cellular slime molds.

Hagiwara, Hiromitsu. *The Taxonomic Study of Japanese Cellular Slime Molds*. Tokyo: National Science Museum, 1989. This monograph is available in English. The majority of the species described are found in many parts of the world other than Japan.

Raper, Kenneth B. *The Dictyostelids*. Princeton, N.J.: Princeton University Press, 1984. A comprehensive monograph on all aspects of the biology, ecology, and taxonomy of cellular slime molds.

CENTRAL AMERICAN AGRICULTURE

Categories: Agriculture; economic botany and plant uses; food; world regions

Agriculture is generally understood to be concerned with the production of food; however, in Central America, ornamental plants and flowers, forest products, and fibers are also important agricultural commodities.

The nations of Central America are generally considered to be Belize, Costa Rica, El Salvador, Guatemala, Honduras, Nicaragua, and Panama. At the end of the twentieth century, the agricultural sector employed about 46 percent of the available labor force in Central America, most of which was engaged in subsistence agriculture. This percentage is higher than that of the neighboring developing countries of Mexico (28 percent) and Colombia (30 percent). The Central American percentage is higher than those in more developed countries, such as the United States and Canada, each of which is below 4 percent. The percentage of suitable land in Central America is about equal to that in Mexico (12 percent) but significantly more than in Colombia (4 percent). Arable land in the United States is about 19 percent.

Early Agriculture

Considerable archaeological evidence supports the existence of sedentary agriculture in the region for more than two thousand years. The early Maya farmed raised fields in lowland swamp areas and constructed irrigation systems in areas with a dry season. In highland areas, steep slopes were terraced. The most prominent terrace agriculture in the Americas was in the Andean cultures, but Central Americans also used this practice. Agriculture was based mainly on corn, but other crops were widely grown, including squash, beans, and chile peppers. Nonfood crops such as cotton and tobacco were grown for both domestic use and trade. These two crops continue to be important.

Exactly what group of Central Americans established the various agricultural practices, or when, is debatable. However, it is known that agriculture supported large communities of people early in the first millennium. The cities of Tikal, Copán, Caracol, and others had populations of thirty-five thousand or more.

Raised field agriculture had several benefits. Sediment dredged from channel bottoms was added to the fields, raising the surface above the surrounding swamp, creating dry land. This mate-

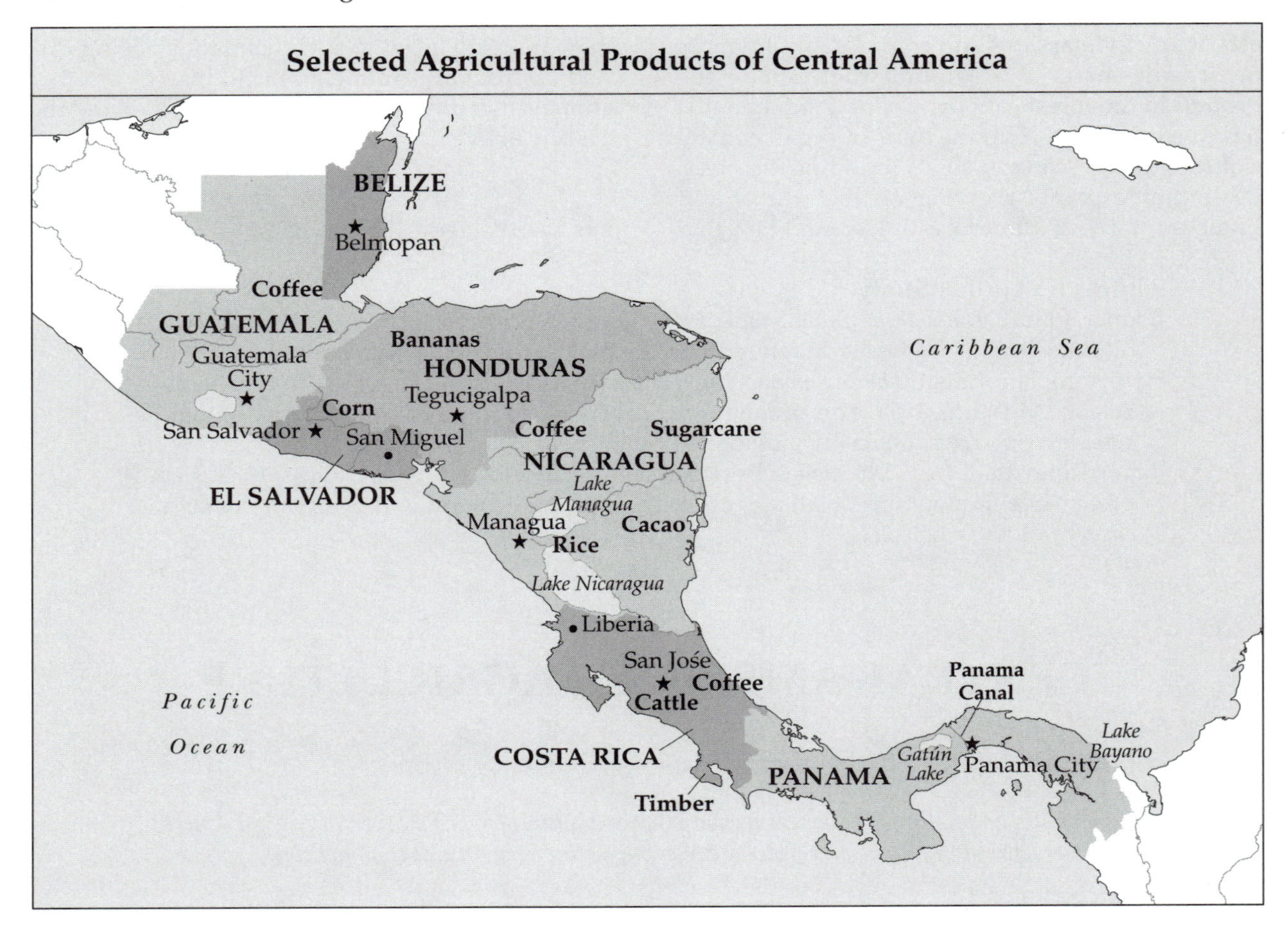

rial was rich in nutrients from decaying plant matter and wastes from aquatic creatures. Channels of water dividing the dry land provided habitat for fish and turtles, which were a protein-rich food source for people.

Slash-and-burn agriculture was practiced. The process involved stripping forests and burning the debris in place. Trees too large to be cut with primitive stone implements were *girded*; that is, a circle of bark was removed from around the tree, and the tree died afterward. Burned debris added nutrients to the topsoil. Because the soil was generally poor, the fields, usually known as *milpas*, or cornfields, but sometimes referred to as *swidden*, were abandoned after two or three years of production and left *fallow* for up to twenty years. This process is still practiced.

Intercropping, or *polyculture*, was a practice that helped ensure a harvest. The planting of several crops and different varieties provided a harvest even if one crop failed. This practice is also in use today.

Traditional Crops

Since the nineteenth century, certain crops have been raised in Central America as export crops and others principally for domestic consumption. Many of the traditional crops grown are not native to the region. Many of the most widely grown crops are termed *exotics*, that is, plants not native to the region that were introduced by European settlers. Bananas, coffee, and sugarcane are three principal exotic crops, with corn being a fourth. Most of the production of introduced plants is grown for export, although native corn is for local use.

Bananas are grown extensively in the Caribbean and Pacific lowlands but most prominently in the Sula Valley of Honduras, a leading world exporter of this crop. The banana industry flourished under the control of North American growers, especially the United Fruit Company. In the later part of the nineteenth century, the banana export business grew and enjoyed large markets in the United States and Europe. The United Fruit Company also exerted strong influence over governmental policies in the region.

Coffee is grown extensively in the highland areas of all seven Central American countries. A slow-ripening crop, coffee requires as much as two months to harvest. Small-scale growers who sell their product through cooperatives produce much of the area's coffee. The best-quality coffee is shade-grown, and so banana trees often are planted throughout the small fields.

Sugarcane, first introduced by Christopher Columbus to the island of Cuba, is another plant grown throughout a wide area. Sugarcane is labor-intensive during harvest but requires little attention at other times. The harvest of sugarcane begins with the burning of the fields. This practice reduces the volume of foliage and leaves only the stalks, or canes, which are the source of sugar. After the burning—which has the side benefit of chasing out the snakes that inhabit the cane fields—teams of workers with machetes march through the fields cutting the cane.

Corn (maize) is not grown for export. Along with regionally grown rice, it is for domestic consumption. Corn meal is used in the preparation of tortillas, which are eaten at nearly every meal. Rice is commonly served with red or black beans.

AP/Wide World Photos

Sugarcane field near Santa Lucia Cotzulmalguapa, Guatemala.

Leading Agricultural Crops of Central American Countries

Country	*Products*
Belize	Bananas, cacao, citrus, sugarcane, lumber
Costa Rica	Coffee, bananas, sugar, corn, rice, beans, potatoes, timber
El Salvador	Coffee, sugarcane, corn, rice, beans, oilseed, cotton, sorghum
Guatemala	Sugarcane, corn, bananas, coffee, beans, cardamom
Honduras	Bananas, coffee, citrus, timber
Nicaragua	Coffee, bananas, sugarcane, cotton, rice, corn, cassava, citrus, beans
Panama	Bananas, corn, sugarcane, rice, coffee, vegetables

Source: Data are from *The Time Almanac 2000*. Boston: Infoplease, 1999.

Export crops have varied in their economic value to the region. A banana disease nearly ruined the industry in the 1930's. The Great Depression in those same years sharply reduced exports to North America. Import quotas imposed by the United States on sugar and the U.S. trade embargo on all Cuban products imposed in the early 1960's provided both a low and a high for Central American sugar producers. Overproduction of coffee by South American producers has led to depressed prices several times. During the late 1990's, the European Union's agricultural import practice of favoring former colonies reduced the value of bananas to growers. In Central America, only Belize (formerly British Honduras) benefits from European tariff regulations.

Nontraditional Crops

Vegetables, high-value crops, and ornamental plants and flowers are being grown at an increased rate. The leading crops are broccoli, cauliflower, snow peas, melons, strawberries, and pineapples.

Palm oil from a nonnative tree is another high-value farm product. Nontraditional crops are labor-intensive and affect the environment because of the heavy requirements for chemical pesticides. Workers face health risks due to these chemical applications, but employment is high. In Costa Rica, the government encourages investment in reforestation using teak from Southeast Asia.

Donald Andrew Wiley

See also: Agriculture: traditional; Central American flora.

Sources for Further Study

Grigg, David. *An Introduction to Agricultural Geography*. 2d ed. New York: Routledge, 1995. Introduces environmental, human, and economic issues in agriculture in developed and developing nations. This edition includes two new chapters on modernization and the environment.

Janzen, Daniel H., ed. *Costa Rican Natural History*. Chicago: University of Chicago Press, 1983. Interesting essays are divided into six sections: agricultural organisms, plants, herpetofauna, mammals, birds, and insects.

Kricher, John, William E. Davis, and Mark Plotkin. *A Neotropical Companion: An Introduction to the Animals, Plants, and Ecosystems of the New World Tropics*. 2d ed. Princeton, N.J.: Princeton University Press, 1999. Covers a wide variety of topics including rain forests, regeneration, ecological succession, evolutionary theory, and tropical medicinal plants.

Schwartz, Norman B. *Forest Society: A Social History of Peten, Guatemala*. Philadelphia: University of Pennsylvania Press, 1990. Examines the social history of a region in the lowlands of Northern Guatemala, including relationships between ecology and society.

CENTRAL AMERICAN FLORA

Category: World regions

Central America—comprising the nations Belize, Costa Rica, El Salvador, Guatemala, Honduras, Nicaragua, and Panama—is a land bridge that connects North America and South America, and many of its plants are similar to plants found on both those continents.

Lowlands

Tropical *rain forests* lie on the eastern half of Central America and typically have many tall, broad-leaved evergreen trees 130 feet (40 meters) or more in height, and 4-5 feet (1.2-1.5 meters) in diameter that form a dense canopy. Shade-seeking plants, such as palms, figs, ferns, vines, philodendrons, and orchids, form the forest undergrowth beneath the trees.

Epiphytes, such as orchids, ferns, bromeliads, and mosses, cling to the branches of the trees in a dense mat of vegetation—these plants have no roots but grow by clinging to the trunks of trees and drawing moisture and nourishment from the air. Rain-forest trees that are harvested for their commercial value include mahogany, kapok, cedarwood, tagua, ebony, and rosewood for making furniture; breadfruit, palm, and cashew; sapodilla, used to make latex; and the rubber tree. Many brilliantly colored flowers also grow in Central America. The most common of these are orchids (with close to a thousand species), heliconias, hibiscus, and bromeliads.

In the Caribbean lowlands, where the soil is porous and dry, extensive savanna *grasslands* with sparse forests of pines, palmettos, guanacastes, cedars, and oaks are found. Along the Caribbean coast (called the Mosquito Coast), mangroves and coconut palms flourish in swamps and lagoons.

Highlands

The central mountains and highlands of Central America are cooler than the coastal lowlands, and

Cacao: The Chocolate Bean

Cacao (cocoa) beans (*Theobroma cacao*), from which chocolate is made, have been cultivated in Central America for centuries. Although now a major African crop as well, cacao originated in the Americas. Once considered the drink of the gods, chocolate was reserved for royalty. Today, millions around the world enjoy chocolate, especially mixed with sugar and milk.

The beans are actually the berries of the small cacao tree, which grows in shade and is rarely more than 20 feet (6 meters) tall. The tree's football-shaped pods are 6-8 inches (15-20 centimeters) in length. When ripe, the pods can be red, yellow, or orange, depending on the variety. The tree's flowers are tiny, inconspicuous white blossoms that emerge singularly from the lower branches or trunk, not from stem ends. The cacao seeds, or beans, are surrounded by a whitish, gelatinous mass. Cacao beans must be fermented, dried, and cleaned before the chocolate aroma and taste develop.

the vegetation there is mainly deciduous hardwood trees such as walnut, pine, and oak. The eastern slopes of the mountains have abundant rainfall. "Cloud forests" that are 5,000 feet (1,525 meters) above sea level are thick with evergreen oak, sweet gum, pine, and laurel, which grow to a height of about 65 feet (20 meters) and are festooned with ferns, bromeliads, mosses, and orchids.

On the western side of the mountains, facing away from the moist Caribbean winds and receiving rain only seasonally, vegetation is sparse and semiarid, and soils are poor and unproductive. Deciduous tropical forests dominate there, and vegetation is characterized by evergreen herbs and shrubs, plumeria (frangipani), eupatorium pines, myrtles, and sphagnum mosses.

Helen Salmon

See also: African flora; Antarctic flora; Asian flora; Australian flora; Caribbean flora; North American flora; South American flora.

Sources for Further Study

Beletsky, Les. *Belize and Northern Guatemala: The Eco-Traveller's Wildlife Guide*. San Diego: Academic Press, 1999. An encyclopedic introduction to Central America's flora and fauna and a primer on the principles of ecotourism.

Kricher, John, William E. Davis, and Mark J. Plotkin. *A Neotropical Companion: An Introduction to the Animals, Plants, and Ecosystems of the New World Tropics*. 2d ed. Princeton, N.J.: Princeton University Press, 1999. Covers a wide variety of topics, including rain forests, regeneration, ecological succession, evolutionary theory, and tropical medicinal plants.

Parker, Edward. *Central America*. Austin, Tex.: Raintree Steck-Vaughn, 1999. A survey, with clear color photographs on every page. Includes historical time line.

CHAROPHYCEAE

Categories: Algae; microorganisms; *Protista*; taxonomic groups; water-related life

It is almost impossible not to see Spirogyra *floating on the surface of a pond on a hot summer day, but most people dismiss it as pond scum. Few realize that what they are looking at is a member of the* Charophyceae, *a class in the phylum Chlorophyta, or green algae, and a cousin of the ancestor of the* Embryophyta, *or bryophytes and vascular plants.*

Most *Charophyceae*, like *Spirogyra*, live in freshwater habitats, but some also occur in moist soil in terrestrial habitats. *Charophyceae* can live as single cells, colonies, or branched and unbranched filaments and come in a variety of shapes. The characteristics that unite members of the class—and which link them with the embryophytes—include flagellated cells (similar to sperm cells in vascular plants), a nuclear envelope that breaks down during mitosis, mitotic spindles that persist as phragmoplasts (a type of cytoskeletal scaffolding) through cell division either by furrowing or by forming a cell plate, the presence of chlorophylls *a* and *b* and phytochrome, and the storage of starch inside plastids.

Charophytes possess decay-resistant cell walls made of phenolic compounds as well as lignins or ligninlike compounds. Cell walls made of similar compounds are found in bryophytes and vascular plants as well. Likewise, all three groups of plants also contain sporopollenin, the substance in the walls of spores and pollen grains that makes them virtually indestructible. Communication channels between cells are similar, too. Plasmodesmata similar to that seen in embryophyte cells allow between-cell communications in charophytes and embryophytes.

Life Cycle

Charophytes have a two-stage life cycle involving a dominant haploid stage, upon which develops the sex organs; antheridia, which produce sperm cells; and oogonia, which produce egg cells. Typically, individual charophytes produce both antheridia and oogonia, but in some species an individual will produce only one or the other. Fertilization—which in one group, the *Zygnematales*, takes place via conjugation—produces a diploid zygote, which quickly undergoes meiosis. If the environment is unfavorable, the zygote will go dormant and remain so for a long period of time. Dormancy ends when the environment improves.

Classification

Genetic analysis supports the recognition of six orders within the *Charophyceae*: the *Mesostigmatales*, *Chlorokybales*, *Klebsormidiales*, *Zygnematales*, *Coleochaetales*, and *Charales*.

Of these, the most abundant group is the *Zygnematales*, a large order which consists of more than three thousand species, including *Spirogyra* and the desmids, a group of mostly single-celled organisms with a constriction across the middle which nearly divides the cells in two. *Zygnematales* live primarily in freshwater habitats as phytoplankton, as benthic dwellers, or attached to other aquatic plants. Some species live on snow and ice.

The *Charales*, commonly called stoneworts or brittleworts, are a large group of filamentous charophytes that feature complex branching patterns. Some can reach lengths of more than a meter. The branching pattern—branches reach out from nodes along the filament—is similar to that of higher plants. *Charales* reside primarily in freshwater habitats, but some can be found in brackish water as well as on land. The stems of stoneworts and brittleworts can be encrusted with calcium and magnesium carbonates. As a result, their hard bodies are well known from the fossil record. The lineage extends back to more than 400 million years ago. Two current genera, *Chara* and *Nitella*, date back about 200 million years.

The *Coleochaetales* are a small group of complex, microscopic filamentous algae that can be found only in freshwater habitats. *Klebsormidiales* are a small group of unbranched, filamentous charophytes that occur in both freshwater and terrestrial environments. *Mesostigmatales* and *Chlorokybales* are two groups of rare algae. The *Coleochaetales* and *Charales* are more closely related to bryophytes and vascular plants than the other groups.

Evolutionary Significance

For decades, structural similarities led plant biologists to suspect that the *Embryophyta* evolved from charophytes. Recently, cladistic analyses of chemical, structural, and genetic characteristics have opened up research on the topic. In cladistics, characteristics among a number of organisms are analyzed statistically in the hopes of developing a classification system for the group which will reveal evolutionary relationships. Cladistic analyses support the notion that embryophytes are monophyletic; in other words, bryophytes and vascular plants descend from a common ancestor.

Furthermore, several recent analyses support the notion that a member of the *Charophyceae* gave rise to embryophytes. Cladistic analyses were somewhat unclear, however, about the relationships of the charophyte orders with one another and with other green algae (*Chlorophyta*), bryophytes, and land plants.

One of the latest analyses of mitochondrial, chloroplast, and nuclear genes helps resolve some of the confusion. The research indicates that the *Mesostigmatales* were probably the most ancient group of charophytes, followed by the *Chlorokybales*, *Klebsormidiales*, *Zygnematales*, *Coleochaetales*, and *Charales*.

The work also supports earlier suggestions that the *Charales* are the closest living relatives to extant embryophytes and that the charophytes descended from other green algae.

David M. Lawrence

See also: Algae; Aquatic plants; Bryophytes; *Chlorophyceae*; Cladistics; Eukaryotic cells; Evolution of plants; Green algae; Phytoplankton; *Ulvophyceae*.

Sources for Further Study

Graham, Linda E., and Lee W. Wilcox. *Algae*. New York: Prentice-Hall, 2000. An overview of the biology of algae, with emphases on systematics, biogeochemistry, and environmental and economic effects.

Raven, Peter H., Ray F. Evert, and Susan E. Eichhorn. *Biology of Plants*. 6th ed. New York: W. H. Freeman/Worth, 1999. Includes chapter on protists, such as green algae, with plantlike characteristics.

CHEMOTAXIS

Categories: Cellular biology; microorganisms; movement; physiology

Chemotaxis is the ability of a cell to detect certain chemicals and to respond by movement, such as microbial movement toward nutrients in the environment.

Many microorganisms possess the ability to move toward a chemical environment favorable for growth. They will move toward a region that is rich in nutrients and other growth factors and away from chemical irritants that might damage them. Among the organisms that display this *chemotactic behavior,* none is simpler than bacteria. Bacteria are single-celled *prokaryotic* microorganisms, which means that their deoxyribonucleic acid (DNA) is not contained within a well-defined nucleus surrounded by a nuclear membrane, as in *eukaryotic* (plant and animal) cells. *Prokaryotes* lack many of the cellular structures associated with more complex eukaryotic cells; nevertheless, many species of bacteria are capable of sensing chemicals in their environment and responding by movement.

Bacterial Flagella

Bacteria capable of movement are called *motile bacteria*. Not all bacteria are motile, but most species possess some form of motility. Although there are three different ways in which bacteria can move, the most common means is by long, whiplike structures called *flagella*.

Bacterial flagella are attached to cell surfaces and rotate like propellers to push the cells forward. A bacterial cell must overcome much resistance from the water through which it swims. In spite of this, some bacteria can move at a velocity of almost 90 micrometers per second, equivalent to more than one hundred bacterial cell lengths per second.

A flagellum is composed of three major structural components: the *filament*, the *hook*, and the *basal body*. The filament is a hollow cylinder composed of a protein called *flagellin*. A single filament contains several thousand spherically shaped flagellin molecules bound in a spiral pattern, forming a long, thin cylinder. A typical filament is between 15 and 20 micrometers long but only 0.02 micrometer thick. The filament is attached to the cell by means of the hook and basal body. The hook is an L-shaped structure composed of protein and slightly wider than the filament. One end of the hook is connected to the filament, and the other end is attached to the basal body. The basal body, also known as the *rotor,* consists of a set of protein rings embedded in the cell wall and plasma membrane. Inside these rings is a central rod attached to the hook. The central rod of the basal body rotates inside the rings, much like the shaft of a motor. As it rotates, it causes the hook and the filament to turn.

Bacteria in Motion

While they are moving, bacteria change direction by reversing the rotation of their flagella. As a bacterium swims forward in a straight line, its flagella spin in a counterclockwise direction. Because of their structure, the flagella twist together when they rotate counterclockwise and act cooperatively to push the cell forward. The forward movement is referred to as a *run*. Every few seconds, a chemical change in the basal body of each flagellum causes it to reverse its spin from counterclockwise to clockwise. When the flagella spin clockwise, they fly apart and can no longer work together to move the cell forward. The cell stops and tumbles randomly until the flagella reverse again, returning to counterclockwise spin and a forward run. This type of movement, in which the cell swims forward for a short distance and then randomly changes its direction, is called *run and tumble* movement.

Certain eukaryotic microorganisms, such as *Euglena* and some other protozoa, are also motile by means of flagella. The structure and activity of eukaryotic flagella are, however, completely different from those of bacteria. Eukaryotic flagella are composed of protein fibers called *microtubules*, which move back and forth in a wavelike fashion to achieve movement. The rotation of bacterial flagella and the run and tumble movement they produce are unique to bacteria.

Attractants and Repellants

Bacteria respond by chemotaxis to two broad classes of substances, *attractants* and *repellants*. They move toward high concentrations of attractants (*positive chemotaxis*) and away from high concentrations of repellants (*negative chemotaxis*). Attractants are most often nutrients and growth factors, such as monosaccharides (simple sugars), amino acids (the building blocks of protein), and certain vitamins required for bacterial metabolism. Repellants include waste products given off by the bacteria as well as other toxic substances found in the environment.

Bacteria respond to attractants and repellants by altering the time between tumbles in their run and tumble movement. When a bacterial cell detects an attractant, the time between tumbles and the time of the runs increase. As long as the cell is moving toward a higher concentration of attractant, its runs will be longer. The opposite effect occurs when a cell encounters a repellant. A repellant causes the time between tumbles to decrease, resulting in shorter runs as the cell changes direction more frequently while trying to avoid the repellant. The net result is that the cell tends to move toward a lower concentration of the repellant.

Chemotactic Receptors

Bacteria recognize attractants and repellants through specialized proteins called *chemotactic receptors*, also called *methyl-accepting chemotactic proteins* (MCPs), which are embedded in their plasma membranes just inside the cell wall. Biologists have identified roughly twenty different receptors for attractants and some ten for repellants. Each receptor protein is believed to respond to only a single type of attractant or repellant.

When an attractant molecule binds to its chemotactic receptor, two separate events occur. First, there is a rapid activation of the receptor. The attractant molecule binds to a special site on the receptor protein to form an activated receptor. This binding is not permanent, however, so a cell must remain in an area with attractant molecules for its receptors to remain activated. The activated receptor sends a chemical signal to the basal bodies of flagella, which causes them to spin in a counterclockwise direction, producing continuous swimming in one direction.

At the same time, there is adaptation of the activated receptors to the attractant. Adaptation is important because it keeps the cell from swimming too long in one direction. It is accomplished by *methylation* of the receptors, a process in which methyl groups are attached to the protein by an enzyme in the cell. (A methyl group consists of an atom of carbon attached to three atoms of hydrogen.) Methylated receptors do not stimulate the basal bodies for counterclockwise rotation as effectively as nonmethylated receptors. After a cell has been in the presence of an attractant for a short while, its receptors adapt to the attractant, and it returns to the original pattern of run and tumble movement. Adaptation is reversed by *demethylation*, the removal of methyl groups from the receptor by a separate enzyme. Together, the balance between methylation and demethylation makes the receptors very sensitive to small changes in attractant concentration, so that cells remain in the region with the greatest concentration of attractant.

The action of repellants appears to be similar to that of attractants. Repellant molecules bind to sites on their chemotactic receptors, activating the receptors. The activated receptors signal the flagella to spin clockwise instead of counterclockwise, causing the cell to tumble and to change direction. Repellant receptors also adapt through methylation and demethylation, much like attractant receptors.

It is not entirely understood how an activated chemotactic receptor can signal flagella to rotate. Four different proteins inside the bacterial cell have been identified as a possible link between the chemotactic receptors and the basal bodies of flagella. These proteins are believed to regulate flagellar rotation using a process called *phosphorylation*. Phosphorylation, the attachment of phosphate molecules to a protein, is used in all types of cells as a kind of "on and off" switch to regulate protein activity.

Jerald D. Hendrix

See also: Bacteria; Flagella and cilia; Prokaryotes.

Sources for Further Study

Alberts, Bruce, et al. *Molecular Biology of the Cell*. 4th ed. New York: Garland, 2002. An intermediate-level college text with many informative diagrams and photographs. The chapter on cell signaling includes a section on bacterial motility and chemotaxis which explains the movement of flagella, the adaptation of receptor proteins, the way that receptor activation is coupled to flagellar movement, and includes a list of references to the research literature.

Diagram Group, The. *Genetics and Cell Biology on File*. New York: Facts On File, 1997. A useful reference book summarizing cell biology and genetics. Discusses topics including parts of cells and how they work, how cell division occurs and what happens when cancer develops, units of measurement, how a centrifuge works, types of viruses, and concepts of evolution to aspects of genetic counseling.

Gerhardt, Philipp, et al., eds. *Methods for General Bacteriology*. Washington, D.C.: American Society for Microbiology, 1994. This manual presents a wealth of information on bacteriological techniques, including methods for the growth, manipulation, and analysis of bacterial cells. The chapter on solid culture contains a simple procedure for detecting chemotaxis using a capillary tube technique.

Kurjan, Janet, and Barry L. Taylor, eds. *Signal Transduction: Prokaryotic and Simple Eukaryotic Systems*. San Diego: Academic Press, 1993. Sixteen essays reviewing current knowledge of the molecular mechanisms responsible for signal transduction in a variety of eukaryotic and prokaryotic microorganisms. Intended for advanced students and researchers in the field.

Prescott, Lansing, John P. Harley, and Donald A. Klein. *Microbiology*. 5th ed. Boston: McGraw-Hill, 2002. An intermediate-level college textbook that covers the structure, activities, and environmental significance of bacteria and other microorganisms. The chapter on prokaryotic cell structure and function has an extensive section on the structure and movement of bacterial flagella and an informative discussion on chemotaxis. The text is supplemented with useful diagrams and photographs; in particular, there is an excellent electron micrograph showing the filament, hook, and basal body assembly of the flagellum. There is also an interesting discussion of magnetotactic bacteria, which respond to the earth's magnetic field.

Raven, Peter H., and George B. Johnson. *Understanding Biology*. 3d ed. Dubuque, Iowa: Wm. C. Brown, 1995. Introductory-level college text presents basic concepts of cell biology in a fashion easily comprehended by nonscientists. The chapter on cells contains a thorough comparison of prokaryotic and eukaryotic cell types and describes the differences between prokaryotic and eukaryotic flagella.

CHLOROPHYCEAE

Categories: Algae; microorganisms; *Protista;* taxonomic groups; water-related life

Chlorophyceae *(from the Greek word* chloros, *meaning "green") make up an extremely large and important class of green algae. Members may be unicellular, colonial, or filamentous. Cells of unicellular and colonial chlorophyceans may have two or more flagella.*

There are about 2,650 living species of chlorophyceans. The main features of the class (and most plants) are the use of starch as the principal food reserve and the green chloroplasts with chlorophylls *a* and *b*. In spite of plant characteristics, this algal group is not directly related to early land plants. Chlorophyceans are almost entirely restricted to freshwater and terrestrial habitats. Some members of this class have adapted to life on snow as *snow algae*. Snow algae cause snow to appear red-burgundy or orange in color because of high levels of unusual carotenoid pigments within the algal cells.

There are a variety of asexual and sexual reproduction modes among members of this class. Sexual reproduction is characterized by the formation of a zygote produced by gametic fusion. Chlorophyceans show differences during cell division compared to other green algal groups. For example, they produce a set of microtubules, the *phycoplast*, that is parallel to the plane of cell division.

Diversity

The *Chlorophyceae* include some familiar green algae. Perhaps the most famous chlorophyceans are *Chlamydomonas* (from the Greek word *chlamys*, meaning "cloth") and *Volvox* (from the Latin *volvo*, meaning "to roll"). Both are important research models in laboratories. Chlorophyceans fall into several orders, including *Volvocales*, *Chlorococcales*, *Chaetophorales*, and *Oedogoniales*.

Volvocales

Members of the order *Volvocales* include both unicellular organisms, such as those in the genus *Chlamydomonas* with their two equal flagella, and colonial forms. The *Chlamydomonas* are a large genus of chlorophyceans. More than six hundred species have been described worldwide. The *Chlamydomonas* probably represent the most primitive structure among chlorophyceans. Nevertheless, their basic cell features may be found among other representatives of this order.

A cell wall made of glycoproteins, rather than cellulose, surrounds each *Chlamydomonas* cell. Inside the cell, there is a single large chloroplast and a pyrenoid, which forms starch.

Other cytoplasmic structures include the contractile vacuole rather than a central vacuole. The contractile vacuole is responsible for the removal of water from the cell. Cells of *Chlamydomonas* are capable of phototaxis: They swim toward moderate light but away from high-intensity light. Rhodopsin-like pigment is their primary light-sensing photoreceptor. Under dry conditions, *Chlamydomonas* form a palmelloid stage, in which nonflagellate cells are held together by common mucilage.

Chlamydomonas reproduce asexually via cell division. Also, cells of this alga can become gametes. In most species of *Chlamydomonas*, the male and female gametes appear the same; they are designated (+) and (–). Colonial flagellates of the order *Volvocales* range from simple colonies of *Gonium* to visible-without-magnification spheres of *Volvox* with up to several thousands of cells and some sort of cellular specialization.

Volvox are one of the most structurally advanced colonial forms of green algae. Only specialized cells participate in reproduction. During asexual reproduction, some cells of *Volvox* divide and bulge inward, forming new daughter colonies, which are held for some time within the parent colony. *Volvox* are also capable of sexual reproduction.

They produce gametes that differentiate into sperm and eggs.

Chlorococcales

Members of the order *Chlorococcales* include nonmotile unicellular and colonial algae. Typical representatives of the unicellular nonmotile form are found in *Chlorococcum*. They occur as spherical single cells or cell aggregates and produce flagellated zoospores. Examples of colonial representatives of *Chlorococcales* are *Hydrodictyon*, commonly known as the "water net"; *Pediastrum*, famous for their distinctive, starlike shape; and *Scenedesmus*, widespread inhabitants of the freshwater phytoplankton. The order *Chlorococcales* has now been divided on the basis of small subunit ribosomal ribonucleic acid (RNA) sequence data into several groups, including the *Sphaeropleales*, *Tetracystis* clade, and *Dunaliella* clade.

Chaetophorales and *Oedogoniales*

The most complex of the class *Chlorophyceae* are the filamentous members in orders *Chaetophorales* and *Oedogoniales*, some of which exhibit features that are observed primarily in plants. The chaetophoralean green algae have plantlike bodies with a system of primary and secondary branches. The *Draparnaldia* (named for Jacques Phillipe Raymond Draparnaud, a French naturalist) from order *Chaetophorales* have a main filamentous axis with relatively large cells, primary branches with smaller cells, and secondary branches with even smaller cells. One representative of *Oedogoniales*, the green alga *Oedogonium* (from the Greek *oidos*, meaning "swelling"), has been a subject of intense study for its unusual cell division technique. The entire contents of an *Oedogonium* cell may be used in the formation of one large zoospore with multiple flagella. Members of *Bulbochaete* (from the Greek *bolbos*, meaning "bulb") resemble *Oedogonium* in cell division but differ in being branched and having a distinctive hair cell at the end of each branch.

Technological Uses

A few chlorophycean green algae have commercial value. These algae are good candidates for the industrial production of hydrogen gas because they are able to release the gas from water using solar energy. Hydrogen gas is an environmentally desirable fuel because the burning of hydrogen produces water, and it can be converted effectively to electricity.

Another "commercial" organism is *Dunaliella salina*, a saltwater alga that accumulates massive amounts of beta-carotene, a vital antioxidant also used in food coloring and in pharmaceuticals. *Selenastrum capricornutum* are the most widely used algal biomonitors in the detection of water pollution. Chlorophyceans are used in freshwater aquaculture systems as food for fish. One alga with possible potential for salmon feeds is *Haematococcus*. Algae contain large amounts of the pigment astaxanthin, which is responsible for the red coloration typical of salmon flesh. *Chlorella* (formerly classified in the order *Chlorococcales*) are famous both as the experimental systems in the discovery of the photosynthetic Calvin cycle and as health food in Asia.

Sergei A. Markov

See also: Algae; Brown algae; Calvin cycle; *Charophyceae*; Chrysophytes; Cryptomonads; Diatoms; Dinoflagellates; Green algae; Haptophytes; Red algae; *Ulvophyceae*.

Sources for Further Study

Graham, Linda E., and Lee W. Wilcox. *Algae*. Upper Saddle River, N.J.: Prentice Hall, 2000. A comprehensive textbook on major algal groups, including *Chlorophyceae*.

Sze, Philip. *A Biology of the Algae*. Boston: WCB/McGraw-Hill, 1998. A basic textbook presenting an overview of different algal groups.

CHLOROPLAST DNA

Categories: Cellular biology; genetics; photosynthesis and respiration; reproduction and life cycles

Plants are unique among higher organisms in that they meet their energy needs through photosynthesis. The specific location for photosynthesis in plant cells is the chloroplast, which also contains a single, circular chromosome composed of DNA. Chloroplast DNA contains many of the genes necessary for proper chloroplast functioning.

A better understanding of the genes in chloroplast deoxyribonucleic acid (cpDNA) has improved the understanding of photosynthesis, and analysis of the deoxyribonucleic acid (DNA) sequence of these genes has been useful in studying the evolutionary history of plants.

Discovery of Chloroplast Genes

The work of nineteenth century Austrian botanist Gregor Mendel showed that the inheritance of genetic traits follows a predictable pattern and that the traits of offspring are determined by the traits of the parents. For example, if the pollen from a tall pea plant is used to pollinate the flowers of a short pea plant, all the offspring are tall. If one of these tall offspring is allowed to self-pollinate, it produces a mixture of tall and short offspring, three-quarters of them tall and one-quarter of them short. Similar patterns are observed for large numbers of traits from pea plants to oak trees. Because of the widespread application of Mendel's work, the study of genetic traits by controlled mating is often referred to as Mendelian genetics.

In 1909 German botanist Karl Erich Correns discovered a trait in the four-o'clock plants (*Mirabilis jalapa*) that appeared to be inconsistent with Mendelian inheritance patterns. He discovered that four-o'clock plants had a mixture of leaf colors on the same plant: Some were all green, many were partly green and partly white (variegated), and some were all white. If he took pollen from a flower on a branch with all-green leaves and used it to pollinate a flower on a branch with all-white leaves, all the resulting seeds developed into plants with white leaves. Likewise, if he took pollen from a flower on a branch with all-white leaves and used it to pollinate a flower on a branch with all-green leaves, all the resulting seeds developed into plants with green leaves. Repeated pollen transfers in any combination always resulted in offspring whose leaves resembled those on the branch containing the flower that received the pollen, that is, the maternal parent. These results could not be explained by Mendelian genetics.

Since Correns's discovery, many other such traits have been discovered. It is now known that the reason these traits do not follow Mendelian inheritance patterns is that their genes are not on the chromosomes in the nucleus of the cell where most genes are located. Instead, the gene for the four-o'clock leaf color trait is located on the single, circular chromosome found in chloroplasts. Because chloroplasts are specialized for photosynthesis, many of the genes on the single chromosome produce proteins or ribonucleic acid (RNA) that either directly or indirectly affect synthesis of chlorophyll, the pigment primarily responsible for trapping energy from light. Because chlorophyll is green and because mutations in many chloroplast genes cause chloroplasts to be unable to make chlorophyll, most mutations result in partially or completely white or yellow leaves.

Identity of Chloroplast Genes

Advances in molecular genetics have allowed scientists to take a much closer look at the chloroplast genome. The size of the genome has been determined for a number of plants and algae and ranges from 85 to 292 kilobase pairs (one kb equals one thousand base pairs), with most being between 120 kb and 160 kb. The complete DNA sequences for several different chloroplast genomes of plants and algae have been determined. Although a simple sequence does not necessarily identify the role of each gene, it has allowed the identity of a number of genes to be determined, and it has allowed scientists to estimate the total number of genes. In terms of genome size, chloroplast genomes are relatively small and contain slightly more than one hundred genes.

Roughly half of the chloroplast genes produce either RNA molecules or polypeptides that are important for protein synthesis. Some of the RNA genes occur twice in the chloroplast genomes of almost all land plants and some groups of algae. The products of these genes represent all the ingredients needed for chloroplasts to carry out transcription and translation of their own genes. Half of the remaining genes produce polypeptides directly required for the biochemical reactions of photosynthesis. What is unusual about these genes is that their products represent only a portion of the polypeptides required for photosynthesis. For example, the very important enzyme ATPase, the enzyme that uses proton gradient energy to produce the important energy molecule adenosine triphosphate (ATP), comprises nine different polypeptides. Six of these polypeptides are products of chloroplast genes, but the other three are products of nuclear genes that must be transported into the chloroplast to join with the other six polypeptides to make active ATPase. Another notable example is the enzyme ribulose biphosphate carboxylase (RuBP carboxylase), which is composed of two polypeptides. The larger polypeptide, called rbcL, is a product of a chloroplast gene, whereas the smaller polypeptide is the product of a nuclear gene.

HULTON ARCHIVE

The work of nineteenth century Austrian botanist Gregor Mendel showed that inheritance of genetic traits follows a predictable pattern and that the traits of offspring are determined by the traits of the parents, leading to the discovery of chloroplast genes.

The last thirty or so genes remain unidentified. Their presence is inferred because they have DNA sequences that contain all the components found in active genes. These kinds of genes are often called "open reading frames" (ORFs) until the functions of their polypeptide products are identified.

Impact and Applications

The discovery that chloroplasts have their own DNA and the further elucidation of their genes have had some impact on horticulture and agriculture. Several unusual, variegated leaf patterns and certain mysterious genetic diseases of plants are now better understood. The discovery of some of the genes that code for polypeptides required for photosynthesis has helped increase understanding of the biochemistry of photosynthesis. The discovery that certain key chloroplast proteins, such as ATPase and RuBP carboxylase, are composed of a combination of polypeptides coded by chloroplast and nuclear genes also raises some as yet unanswered questions. For example, why would an important plant structure like the chloroplast have only part of the genes it needs to function? Moreover, if chloroplasts, as evolutionary theory suggests, were once free-living bacteria-like cells, which must have had all the genes needed for photosynthesis, why and how did they transfer some of their genes into the nuclei of the cells in which they are now found?

Of greater importance has been the discovery that the DNA sequences of many chloroplast genes are highly conserved; that is, they have changed very little during their evolutionary history. This fact has led to the use of chloroplast gene DNA sequences for reconstructing the evolutionary history of various groups of plants. Traditionally, plant systematists (scientists who study the classification and evolutionary history of plants) have used structural traits of plants, such as leaf shape and flower anatomy, to try to trace the evolutionary history of plants. Unfortunately, there are a limited number of structural traits, and many of them are uninformative or even misleading when used in evolutionary studies. These limitations are overcome when gene DNA sequences are used.

A DNA sequence of a few hundred base pairs in length provides the equivalent of several hundred traits, many more than the limited number of structural traits available (typically much fewer than one hundred). One of the most widely used sequences is the rbcL gene. It is one of the most conserved genes in the chloroplast genome, which in evolutionary terms means that even distantly related plants will have a similar base sequence. Therefore, rbcL can be used to retrace the evolutionary history of groups of plants that are very divergent from one another. The rbcL gene, along with a few other very conservative chloroplast genes, has already been used in attempts to answer some basic plant evolution questions about the origins of some of the major flowering plant groups. Less conservative genes and ORFs show too much evolutionary change to be used at higher classification levels but are extremely useful in answering questions about the origins of closely related species, genera, or even families. As analytical techniques are improved, chloroplast genes show promise of providing even better insights into plant evolution.

Bryan Ness

See also: Chloroplasts and other plastids; DNA in plants; DNA replication; Genetics: Mendelian; Genetics: post-Mendelian; Mitochondrial DNA; Pollination; RNA.

Sources for Further Study

Doyle, Jeff J. "DNA, Phylogeny, and the Flowering of Plant Systematics." *Bioscience* 43 (June, 1993). Introduces the reader to the basics of using DNA to construct plant phylogenies and discusses the future of using DNA in evolutionary studies on plants.

Palmer, Jeffry D. "Comparative Organization of Chloroplast Genomes." *Annual Review of Genetics* 19 (1985). One of the best overviews of chloroplast genome structure, from algae to flowering plants.

Svetlik, John. "The Power of Green." *Arizona State University Research Magazine* (Winter, 1997). Provides a review of research at the Arizona State University Photosynthesis Center, with good background for understanding the genetics of chloroplasts.

CHLOROPLASTS AND OTHER PLASTIDS

Categories: Cellular biology; photosynthesis and respiration; physiology

Plastids are highly specialized, double membrane-bound organelles found within the cells of all plants and algae. A type of plastid called the chloroplast is the cellular location of the process of photosynthesis.

Plastids exhibit remarkable diversity with respect to their development, morphology, function, and physiological and genetic regulation. Chloroplasts, a type of plastid, are arguably largely responsible for the maintenance and perpetuation of most of the major life-forms on earth through *photosynthesis*. The process of photosynthesis uses visible light as an energy source to power the conversion of atmospheric carbon dioxide into organic molecules that can be used by living organisms. As a by-product of photosynthesis, oxygen is released into the atmosphere and is used by living organisms in the energy-obtaining process of cellular respiration. Other plastid types are specialized for synthesis and storage of pigments, starch, and other secondary metabolites.

Plastid Structure

The typical plastid from the cell of a flowering plant is surrounded by a double membrane system consisting of an inner and outer membrane, with an intermembrane space between the two. In chloroplasts, the photosynthetic pigments that are responsible for absorbing sunlight are located in the *thylakoid* membrane system. This continuous internal membrane system is found throughout the chloroplast *stroma*, an internal fluid matrix analogous to the cellular cytosol. *Granal thylakoids* are organized into stacks, and the *stromal thylakoids* are unstacked and exposed to the stromal matrix. The internal space within the thylakoid membrane system is called the *lumen*. The pigments and proteins involved in the light reactions of photosynthesis, the processes whereby light energy is converted into chemical energy, are embedded in the thylakoid membrane system. The *dark reactions*, or *Calvin cycle*, which is the carbon fixation pathway that leads to the formation of simple carbohydrates, occurs in the stroma. Small starch granules and oil bodies, termed *plastoglobuli*, are often found in chloroplasts. These serve as energy storage reserves for the plant cell. Plastids other than chloroplasts typically lack thykaloids.

Proplastids

The developmental precursor to all plastid types is the *proplastid*. Proplastids are relatively undifferentiated plastids typically found in young, undifferentiated meristematic cells and tissues. Under the appropriate cellular and environmental conditions, proplastids can undergo development and differentiation to any of three main plastid types: chloroplasts, *chromoplasts*, or *leucoplasts*.

Chloroplasts

Chloroplasts typically contain one or more of the three types of plastid chlorophylls (chlorophyll *a*, *b*, or *c*) and, often, members of the two classes of photosynthetic accessory pigments: *carotenoids* and *phycobilins*. The most obvious and essential physiological process unique to chloroplasts is photosynthesis. In the energy transduction reactions (the light reactions), radiant energy in the form of visible light (mostly of the violet, blue, and red wavelengths) is harnessed primarily by the green pigment chlorophyll. The harnessed energy is then used to phosphorylate adenosine diphosphate (ADP) to produce adenosine triphosphate (ATP) in a process termed *noncyclic photophosphorylation* and reduce the electron carrier nicotinamide adenine dinucleotide phosphate (NADP) to NADPH. Oxygen is liberated through the light-dependent oxidative splitting of water.

In the carbon-fixation reactions (often called the dark reactions, although they can occur in the presence of light) the ATP is used as an energy source for the attachment of atmospheric carbon dioxide to the simple sugar ribulose 1,5-bisphosphate (RuBP). The NADPH is used to facilitate the reduction of RuBP through a series of simple sugars in a biochemical set of reactions known as the Calvin cycle. One of the products of this cycle, glyceraldehydes-3-phosphate (G3P), is used by the chloroplast to make glucose and other carbohydrates. G3P is also needed to perpetuate the Calvin cycle, so only one of every three produced is used for carbohydrate synthesis.

Chloroplasts are also the site of synthesis for the three aromatic amino acids: phenylalanine, tyrosine, and tryptophan. The precursor compound aspartate is imported into chloroplasts from the cell cytosol and is used for the synthesis of the amino acids lysine, threonine, and isoleucine. An intermediate in the synthesis of threonine, called homoserine 4-phosphate, is exported from the chloroplast into the cytosol as a precursor for methionine. Thus, there is a strong integration of function among the chloroplast, cytosol, and nucleus, in that the enzymes involved in these amino acid biosynthetic pathways are nuclear-encoded, their mRNAs are translated using cytosolic ribosomes, and most of the biosynthetic enzymes are imported into the chloroplast.

Fatty acid biosynthesis is another biochemical function that occurs in chloroplasts. Fatty acids, as lipid precursors, might be either incorporated directly into chloroplast lipids via a plastid-localized biochemical pathway or exported into the cytoplasm for conversion into endoplasmic reticulum lipids. Lipids found in the inner plastid membrane are plastid-synthesized, whereas those of the outer plastid membrane are synthesized in the endoplasmic reticulum.

Other Plastids

Other plastid types include *chromoplasts*, which typically contain carotene or xanthophyll pigments and are responsible for the colors of many fruits, flowers, and roots. Under some conditions chromoplasts can differentiate into chloroplasts. Leucoplasts are colorless and lack complex inner membranes. One type of leucoplast, the *amyloplast*, synthesizes and stores starch. Other leucoplasts synthesize a wide range of products, including oils and proteins. Proplastids that are arrested during their normal development into chloroplasts are termed *etioplasts*. These typically are formed when developing plant tissues are deprived of light.

Evolutionary History

Plastids possess a number of features that provide insights into their remarkable evolutionary history. The chloroplasts of eukaryotic cells photosynthesize in a manner similar to the more ancient prokaryotic cyanobacteria by using membrane-bound chlorophyll to capture radiant energy. Some plastids even bear a strong morphological similarity to cyanobacteria, being similar in size and having similar internal structures. Plastids divide by binary fission in a manner similar to bacterial reproduction.

Plastids also have a certain degree of autonomy in terms of their genetic system. Typically, the majority of flowering plant plastids contain multiple copies (50-100) of a circular chromosome, ranging in size from 130 to 180 kilobase pairs (kb) in higher plants. Chromosome size in algae is much more variable, ranging all the way from 57 kb to 1,500 kb. The plastid chromosome contains genes for RNAs, such as rRNA (ribosomal RNA) and tRNA (transfer RNA), and structural genes that code for polypeptides involved in photosynthesis, transcription, protein synthesis, energy transduction, and several other functions. Many of the genes on the chloroplast chromosome are organized into clusters termed *operons* in a manner similar to that found in eubacteria.

The nucleotide sequences of many plastid genes, especially the ribosomal RNA genes, are highly similar to those in eubacteria, and the ribosomes found within plastids have a similar composition and size to eubacterial ribosomes. The plastid-encoded genes are transcribed either by a nuclear-encoded or plastid-encoded RNA polymerase, and the resultant mRNAs are translated by plastid ribosomes found within the stroma. The majority of plastid biochemical processes rely on both nuclear- and plastid-encoded genes. Some proteins, such as RuBP carboxylase/oxygenase (Rubisco), are composed of both nuclear- and plastid-encoded protein subunits, again demonstrating the remarkable coordination of biogenesis and development between organelle and cytosol.

This evidence lends strong support to the *endosymbiotic theory* of the origin of plastids. This theory, in essence, states that plastids were once free-living, *autotrophic* (having the ability to obtain carbon from carbon dioxide), prokaryotic cells that were engulfed through *phagocytosis* by an ancestral *heterotrophic* nucleated cell (a cell having a metabolism where carbon must be obtained from organic molecules) termed a *protoeukaryote*. Typically this engulfment would result in the ingestion and subsequent destruction of the engulfed prokaryotic cell. However, in one—or perhaps several—independent incidents, a symbiotic relationship was gradually established between the engulfed photosynthetic bacterium and the protoeukaryote. The captured bacterium provided an internal source of food production for the heterotrophic eukaryotic cell through photosynthesis. The bacterium, in turn, was provided with protection and a stable external environment in the cytosol of the eukaryotic cell.

To coordinate further the physiological and genetic interactions between the two, massive transfer of genes took place over time from the genome of the photosynthetic bacterium to the nuclear genome of the protoeukaryote, leading to the genetic capture and control of the photosynthetic endosymbiont. Recent investigations have shown that this gene transfer event is an ongoing process, with examples of transfer documented in recent evolutionary time for both plastids and mitochondria in several different evolutionary lineages of flowering plants.

Pat Calie

See also: Algae; Anaerobic photosynthesis; Bacteria; Brown algae; C_4 and CAM photosynthesis; Calvin cycle; Cell theory; Cytoplasm; Diatoms; Eukaryotic cells; Evolution of plants; Extranuclear inheritance; Gas exchange in plants; Green algae; Mitochondria; Photorespiration; Photosynthesis; Photosynthetic light absorption; Photosynthetic light reactions; Plant cells: molecular level; Plant tissues; Prokaryotes.

Sources for Further Study

Alberts, Bruce, et al. *Molecular Biology of the Cell*. 4th ed. New York: Garland, 2002. An advanced undergraduate or graduate text, with comprehensive treatments of all major aspects of plastid biology. Includes tables, diagrams, photographs, bibliographies, and index.

Buchanan, Bob B., Wilhelm Gruissem, and Russell L. Jones. *Biochemistry and Molecular Biology of Plants*. Rockville, Md.: American Society of Plant Physiologists, 2000. A comprehensive treatise of plant biochemistry and molecular biology, with discussion of plastid morphology, function, biochemistry, development, and evolutionary origins. Includes tables, diagrams, photographs, bibliographies, and index.

Gillham, Nicholas W. *Organelle Genes and Genomes*. New York: Oxford University Press, 1994. An extensive review of the biology of plastids and mitochondria, with a focus on organelle genome evolution and organization, organelle genetics, regulation of gene expression, and organelle biogenesis. Includes tables, diagrams, bibliographies, and index.

Raven, Peter H., Ray F. Evert, and Susan E. Eichhorn. *Biology of Plants*. 6th ed. New York: W. H. Freeman/Worth, 1999. A standard college textbook with chapters devoted to plastids, chloroplasts, and photosynthesis. Includes tables, diagrams, photographs, and index.

Smith, John Maynard, and Eors Szathmary. *The Origins of Life*. New York: Oxford University Press, 1999. A review of the leading theories of the origins of life, with a chapter dedicated to the origin of the eukaryotic cell and its organelles. Includes tables, diagrams, and index.

CHROMATIN

Categories: Cellular biology; genetics; reproduction and life cycles

Chromatin is an inclusive term referring to DNA and the proteins that bind to it, located in the nuclei of eukaryotic cells. The huge quantity of DNA present in each cell must be organized and highly condensed in order to fit into the discrete units of genetic material known as chromosomes. Gene expression can be regulated by the nature and extent of this DNA packaging in the chromosome, and errors in the packaging process can lead to genetic disease.

Scientists have known for many years that the hereditary information within plants and other organisms is encrypted in molecules of *deoxyribonucleic acid* (DNA) that are themselves organized into discrete hereditary units called *genes* and that these genes are organized into larger subcellular structures called *chromosomes*. James Watson and Francis Crick elucidated the basic chemical structure of the DNA molecule in 1952, and much has been learned since that time concerning its replication and expression. At the molecular level, DNA is composed of two parallel chains of building blocks called *nucleotides*, and these chains are coiled around a central axis to form the well-known *double helix*.

Each nucleotide on each chain attracts and pairs with a complementary nucleotide on the opposite chain, so a DNA molecule can be described as consisting of a certain number of these nucleotide base pairs.

The entire human genome consists of more than six billion base pairs of DNA, which, if completely unraveled, would extend for more than 2 meters. It is a remarkable feat of engineering that in each human cell this much DNA is condensed, compacted, and tightly packaged into chromosomes within a nucleus that is less than 10^{-5} meters in diameter. Plants typically have larger genomes than humans; for example, wheat has fifteen billion base pairs of DNA. By contrast, the most widely studied plant among current scientists is *Arabidopsis*. The species *Arabidopsis thaliana* was selected as a model organism in plant research because of its comparatively simple structure: Its 26,000 genes make up "only" 125 million base pairs.

What is even more astounding is the frequency and fidelity with which this DNA must be condensed and relaxed, packaged and unpackaged, for replication and expression in each individual cell at the appropriate time and place during both development and adult life. The essential processes of DNA replication or gene expression (*transcription*) cannot occur unless the DNA is in an open or relaxed configuration.

Chemical analysis of mammalian chromosomes reveals that they consist of DNA and two distinct classes of proteins, known as *histone* and *nonhistone* proteins. This nucleoprotein complex is called chromatin, and each chromosome consists of one linear, unbroken, double-stranded DNA molecule that is surrounded in predictable ways by these histone and nonhistone proteins. The histones are relatively small, basic proteins (having a net positive charge), and their function is to bind directly to the negatively charged DNA molecule in the chromosome. Five major varieties of histone proteins are found in chromosomes, and these are known as H1, H2A, H2B, H3, and H4. Chromatin contains about equal amounts of histones and DNA, and the amount and proportion of histone proteins are constant from cell to cell in all higher organisms, including the higher plants. In fact, the histones as a class are among the most highly conserved of all known proteins. For example, for histone H3, which is a protein consisting of 135 amino acid "building blocks," there is only a single amino acid difference in the protein found in sea urchins as compared with the one found in cattle. This is compelling evidence that histones play the same essential role in chromatin packaging in all higher organisms and that evolution has been quite intolerant of even minor sequence variations between vastly different species.

Nonhistones as a class of proteins are much more heterogeneous than the histones. They are usually acidic (carrying a net negative charge), so they will most readily attract and bind with the positively charged histones rather than the negatively charged DNA. Each cell has many different kinds of nonhistone proteins, some of which play a structural role in chromosome organization and some of which are more directly involved with the regulation of gene expression. Weight for weight, there is often as much nonhistone protein present in chromatin as histone protein and DNA combined.

Nucleosomes and Solenoids

The fundamental structural subunit of chromatin is an association of DNA and histone proteins called a *nucleosome*. First discovered in the 1970's, each nucleosome consists of a core of eight histone proteins: two each of the histones H2A, H2B, H3, and H4. Around this histone octamer is wound 146 base pairs of DNA in one and three-quarters turns (approximately eighty base pairs per turn). The overall shape of each nucleosome is similar to that of a lemon or a football. Each nucleosome is separated from its adjacent neighbor by about fifty-five base pairs of *linker DNA*, so that in its most unraveled state they appear under the electron microscope to look like tiny beads on a string. Portions of each core histone protein protrude outside the wound DNA and interact with the DNA that links adjacent nucleosomes.

The next level of chromatin packaging involves a coiling and stacking of nucleosomes into a ribbonlike arrangement, which is twisted to form a chromatin fiber about 30 nanometers in diameter, commonly called a *solenoid*. Formation of solenoid fibers requires the interaction of histone H1, which binds to the linker DNA between nucleosomes. Each turn of the chromatin fiber contains about twelve hundred base pairs (six nucleosomes), and the DNA has now been compacted by about a factor of fifty. The coiled solenoid fiber is organized into large domains of 40,000 to 100,000 base pairs, and these domains are separated by attached nonhistone proteins that serve both to organize and to control their packaging and unpackaging.

Loops and Scaffolding

Physical studies using the techniques of X-ray crystallography and neutron diffraction have suggested that solenoid fibers may be further organized into giant supercoiled *loops*. The extent of this additional looping, coiling, and stacking of solenoid fibers varies, depending on the cell cycle. The most relaxed and extended chromosomes are found at *interphase*, the period of time between cell divisions. Interphase chromosomes typically have a diameter of about 300 nanometers. Chromosomes that are getting ready to divide (*metaphase* chromosomes) have the most highly condensed chromatin, and these structures may have a diameter of up to 700 nanometers. One major study on the structure of metaphase chromosomes has shown that a skeleton of nonhistone proteins in the shape of the metaphase chromosome remains even after all of the histone proteins and the DNA have been removed by enzymatic digestion. If the DNA is not digested, it remains in long loops (10 to 90 kilobase pairs) anchored to this nonhistone protein scaffolding.

Impact and Applications

Studies on chromatin packaging continue to reveal the details of the precise chromosomal architecture that results from the progressive coiling of the single DNA molecule into increasingly compact structures. Evidence suggests that the regulation of this coiling and packaging within the chromosome has a significant effect on the properties of the genes themselves. In fact, errors in DNA packaging can lead to inappropriate gene expression and developmental abnormalities. In humans, the blood disease thalassemia, several neuromuscular diseases, and even male sex determination can all be explained by the altered assembly of chromosomal structures.

The unifying lesson to be learned from these examples of DNA packaging and disease is that DNA sequencing studies and the construction of genetic maps will not by themselves provide all the answers to questions concerning genetic variation and genetic disease. An understanding of genetics at the molecular level depends not only on the primary DNA sequence but also on the three-dimensional organization of that DNA within the chromosome. Compelling genetic and biochemical evidence has left no doubt that the packaging process is an essential component of regulated gene expression.

Jeffrey A. Knight

See also: Chromosomes; DNA: historical overview; DNA in plants; DNA replication; Genetic code; Mitosis and meiosis; Model organisms; Nucleic acids; RNA.

Sources for Further Study

Darnell, James, Harvey Lodish, and David Baltimore. *Molecular Cell Biology*. 4th ed. New York: W. H. Freeman, 2000. Covers chromatin structure and function from a cellular and biochemical perspective.

Kornberg, Roger, and Anthony Klug. "The Nucleosome." *Scientific American* 244 (1981). Provides a somewhat dated but highly readable summary of the primary association of DNA with histone proteins.

Russell, Peter. *Genetics*. 5th ed. Menlo Park, Calif.: Benjamin Cummings, 1998. College-level textbook with an excellent discussion of chromatin structure and organization.

Wolffe, Alan. *Chromatin: Structure and Function*. 3d ed. San Diego: Academic Press, 1998. Comprehensive review of chromatin for advanced students and professionals.

CHROMATOGRAPHY

Category: Methods and techniques

Chromatography is a method of separating the components of a mixture over time. Chromatography has allowed for the discovery of many specialized pigments, including at least five forms of chlorophyll.

Chromatography was first described in 1850 by a German chemist, Friedlieb Ferdinand Runge. It was not until the early twentieth century, however, that Mikhail Semenovich Tsvet became the first to explain the phenomenon and methods of this analytical tool.

Chromatography and Photosynthesis

Tsvet's chromatography of plant leaf pigments prompted scientific investigations of photosynthesis—the all-important biochemical reaction that transforms inorganic to organic energy and therefore is at the base of most life. Chromatography has revealed that many different *pigments*, not only green ones, are simultaneously present in leaves. Each pigment absorbs only certain colors of light from sunlight, rather than absorbing all the incident light energy that falls upon it. Each pigment behaves as though it has a tiny "window" that allows the energy of certain wavelengths of light to be harvested. These little bundles of energy are *quantized*, or set, amounts of energy, and they are unique for each different type of pigment. (White sunlight is actually composed of a broad range of wavelengths, with the visible wavelengths appearing as a rainbow of colors when passed through a prism.)

Paper chromatography has allowed for the discovery of many specialized pigments, including at least five forms of *chlorophyll*. Chlorophyll pigments are now known to include chlorophylls *a* through *e*. Also, many different forms of *carotenes* and *xanthophylls* exist. Paper chromatography reveals that red and yellow pigments are always present in the leaves of deciduous trees and shrubs and not just during the fall color change. Because of the high abundance of the green chlorophyll pigments, as compared with the bright reds of carotenes or yellows of xanthophyll, only the dominant green hues are generally seen. In the fall, deciduous trees show a loss of chlorophyll pigments, thereby revealing the brilliant foliage associated with an autumn forest.

Once pigments are separated from one another, they can be chemically characterized and further studied. Carotenes and xanthophylls have been discovered to be of similar chemical composition, with each being made of forty carbon atoms covalently bonded to one another. Different arrangements of these covalent bonds produce the different colors of red and orange.

Chromatography has allowed scientists the opportunity to trace the path that carbon atoms follow through every tiny increment of the photosynthetic process. Paper chromatography, coupled with radioisotopic studies of carbon-labeled (with radioactive carbon 14) compounds, eventually led to the ability to describe the carbon-containing products of each step in the series of reactions of photosynthesis. Today this pathway is called the *Calvin cycle*.

Methodology

A classical demonstration of chromatographic principles utilizes techniques that allow plant pigments to be isolated. Spinach leaves are an excellent tool for the identification of four pigments: chlorophyll *a*, chlorophyll *b*, carotene, and xanthophyll. The stationary phase is a piece of chromatography paper with a dried spot of the plant extract near one end. The mobile phase is an acetone-ligroin mixture, a nonpolar (hydrophobic) solvent mixture. The paper is placed with a small portion of the end with the pigment spot in the solvent, the mobile phase. As the acetone-ligroin mobile phase comes into contact with the paper, *capillary action* allows the liquid to travel upward, against gravity.

The mobile phase has a migrating moisture line, or leading line of wetness, which is called the solvent front. As the solvent travels over the spot, each of the pigments will travel with the mobile phase at different rates from the original spot. Some pigments will adhere to the paper more strongly than others, and thus travel shorter distances along the paper. Yellow-green chlorophyll *b* travels the least distance with the mobile phase. Chlorophyll *b* is a more polar (water-loving) pigment than the other pigments found in spinach extracts and is therefore more strongly attracted to the polar surface of the paper than to the nonpolar solvent.

The remaining pigments travel increasing distances with respect to chlorophyll *b*, beginning with blue-green chlorophyll *a*, followed by yellow-orange xanthophyll and, finally, the orange pigment of carotene. Carotene moves the farthest because it is the most nonpolar of the pigments and it is attracted more strongly to the acetone-ligroin mixture (mobile phase) than to the paper. This stronger, nonbonded interaction with the mobile phase indicates that carotene is the most nonpolar pigment found in spinach chloroplasts.

Once the solvent front is about half an inch from the top of the paper strip, the strip is removed from the chamber. A pencil line must be drawn immediately across the top of the strip to indicate how far up the paper the mobile phase traveled. The paper strip is then referred to as a *chromatogram*.

The *Rf value* is a numerical constant that is unique for each of the four pigments identified in spinach. The ratio of the distance each pigment travels, as compared with the distance traveled by the mobile phase (from the start to finish lines), will be unique to that pigment alone. Thus, chlorophyll *b* will not switch places with carotene on the chromatogram because of the unique interactions it has with the stationary and mobile phases. For this reason, the Rf values determined by the method described above can be generated repeatedly by anyone using this method.

Types of Chromatography

As performed by Runge and Tsvet, chromatography has evolved from the days of paper, chalk, and dyes into a computerized and versatile instrumentation requiring expert training and a significantly larger budget. *Thin-layer chromatography* (TLC) is useful in protein chemistry. The stationary phase of this method consists of thin gel applied to a plastic or glass plate (strip). Various gels can be used to coat the plate. Some coatings may be polar, while others may be nonpolar.

Column chromatography can look for the amounts and types of vitamins in food or diet supplement tablets. Pigments, steroids, alkaloids, and carbohydrates all can be identified and measured using an appropriate column-chromatographic system. Many recent advances in column chromatography now allow for isolation and purification of proteins, DNA, RNA, and many other biological molecules.

High-pressure liquid chromatography (HPLC) can purify biologically important enzymes from living systems without destroying the biological activity of the enzyme. In *gas chromatography* (GC), an inert gas such as helium or nitrogen flows through several feet of a packed and coiled column. The gas acts as the mobile phase by sweeping the sample through the column. The packing is often a solid material, but liquid-coated solid particles are also used. Paper chromatography continues be a popular method for analysis of plant pigments, dyes, inks, and food colorings. It is largely used, however, in academic settings to demonstrate the principles of chromatography.

Mary C. Fields

See also: Calvin cycle; Pigments in plants.

Sources for Further Study

Bassham, J. A. "The Path of Carbon in Photosynthesis." *Scientific American* 206 (June, 1962): 40, 88-100. Describes the paper chromatography method as coupled with radioisotope tracers. It is an excellent article of historical significance, as it gives an account of the surprise intermediate products, such as amino acids, fats, carbohydrates, and other compounds, made during photosynthesis. The photographs of the chromatograms are worth scrutiny.

Braithwaite, A., and F. J. Smith. *Chromatographic Methods*. 5th ed. London: Blackie Academic & Professional, 1996. An excellent book that describes the many types of chromatography available today. Specific chapters deal with column chromatography, TLC, GC, HPLC, and combined techniques, such as mass spectroscopy and gas chromatography. The final chapter of the book provides model experiments in a cookbook style. This book is not overly technical, but it is more sophisticated than some references geared specifically to the beginner.

Clevenger, Sarah. "Flower Pigments." *Scientific American* 210 (June, 1964): 84-88. This article is a great resource to teachers because it offers options for pigments from nature that can be investigated. Technical in sections, but the introduction is quite accessible. The photographs show both one-dimensional and two-dimensional paper chromatograms, using methods that can be applied to the analysis of either plant leaf or petal extracts.

Hamilton, R. J., Sheila Hamilton, and David Kealey. *Thin Layer Chromatography: Analytical Chemistry by Open Learning*. New York: John Wiley & Sons, 1987. An excellent self-teaching approach to TLC theory and methods. This book is not as difficult to read as the title may imply; much of the material is very clearly written.

Kost, Hans-Peter, ed. *Plant Pigments*. Boca Raton, Fla.: CRC Press, 1988. A handbook that gathers together data tables on plant pigments chromatography, giving not only the results of modern separation methods but also older sources that are often difficult to access. A workbench reference book for researchers.

Morholt, E., P. Brandwein, and A. Joseph. *A Sourcebook for the Biological Sciences*. 3d ed. San Diego: Harcourt Brace Jovanovich, 1986. This reference was designed as a resource for high school science teachers and is still very useful. The diagrams of paper chromatography (which in this case is applied to amino acid analysis) and column chromatograpy are accurate and clear. Contacts and sources of audiovisual support materials can be found toward the end of the text. The best features of the book, however, are its well-written presentations and step-by-step approach.

Tocci, Salvatore. *Biology for Young Scientists*. Rev. ed. New York: Franklin Watts, 2000. This book contains excellent, brief descriptions of paper chromatography and photosynthesis. The chapter on photosynthesis and respiration makes particularly good reading for high school students.

CHROMOSOMES

Categories: Cellular biology; genetics; reproduction and life cycles

Chromosomes contain the genetic information of cells. Replication of chromosomes assures that genetic information is correctly maintained as cells divide.

The *genome* of an organism is the sum total of all the genetic information of that organism. In eukaryotic cells, this information is contained in the cell's nucleus and organelles, such as mitochondria and plastids. In prokaryotic organisms (bacteria and archaea), which have no nucleus, the genomic information resides in a region of the cell called the *nucleoid*. A chromosome is a discrete unit of the genome that carries many *genes*, or sets of instructions for inherited traits. Genes, the blueprints of cells, are specific sequences of deoxyribonucleic acid (DNA) that code for messenger ribonucleic acids (monas), which in turn direct the synthesis of proteins.

Each eukaryotic chromosome contains a single long DNA molecule that is coiled, folded, and compacted by its interaction with chromosomal proteins called *histone*. This complex of DNA with chromosomal proteins and chromosomal RNAs is *chromatin*. DNA of higher eukaryotes is organized into loops of chromatin by attachment to a *nuclear scaffold*. The loops function in the structural organization of DNA and may increase transcription of certain genes by making the chromatin more accessible.

To maintain the genetic information of a cell, it is essential that chromosomes correctly replicate and divide as a cell divides. After DNA replication, chromosomes separate in a process called mitosis. During this process, the nuclear envelope breaks down and chromosomes condense into compact structures. A cellular structure known as the mitotic spindle forms, pulling pairs of replicated chromosomes apart so that the two cells receive identical sets of chromosomes.

Chromosomes are readily visualized when they condense during cell division. All the chromosomes of a cell visualized during mitosis constitute that cell's *karyotype*. Each chromosome has a *centromere*—a constricted area of the condensed chromosome where the mitotic or meiotic spindle attaches to assure correct distribution of chromosomes during cell division—and a *telomere*, the end or tip of a chromosome, which contains tandem repeats of a short DNA sequence.

The number of chromosomes in a gamete (either egg or sperm) is the *haploid* number, n. The haploid number of chromosomes in humans is 23; in corn, 10; in peas, 7; in *Arabidopsis* (the model organism used in much botanical research), 4. Some carp and some ferns have more than 50 chromosomes in the haploid genome. Pollen grains of some plants, such as pear, contain three haploid cells: One directs the growth of the pollen tube down the style to the ovary; the other two are sperm. In flowering plants (angiosperms), there is a unique double fertilization whereby one sperm nucleus fuses with the egg nucleus to form the diploid ($2n$) zygote, and the other sperm nucleus fuses with two polar nuclei to form the triploid nutritive tissue, or *endosperm*, which will nourish the embryo in the seed. The zygote then increases in cell number by *mitosis,* a type of cell division during which chromosomes in a nucleus are replicated and then separated to form two genetically identical daughter nuclei. This is followed by *cytokinesis,* the process of cytoplasmic division, which results in two daughter cells, each having the same number of chromosomes and genetic composition as the parent cell. The mature $2n$ plant forms the haploid (n) gametes by *meiosis,* a type of cell division that reduces the number of chromosomes to the haploid number.

A distinctive feature of plant cell division is the plant cell has three genomes (the nuclear, mitochondrial, and plastid genome) to replicate and divide. The chromosomes of eukaryotes consist of unique genes among a complex pattern of repeated DNA sequences. *Arabidopsis* has only 4 chromosomes containing about 120 million base pairs. There are typically between twenty and one hundred copies of the mitochondrial genome per mitochondrion, ranging in size from two hundred to twenty-four hundred kilobase pairs (or kb; one kilobase pair equals one thousand base pairs). Plant mitochondrial genomes are much larger than the mitochondrial genomes of yeast or animals. Chloroplast genomes range in size from 130 to 150 kb, with 50 to 150 copies of that genome per plastid.

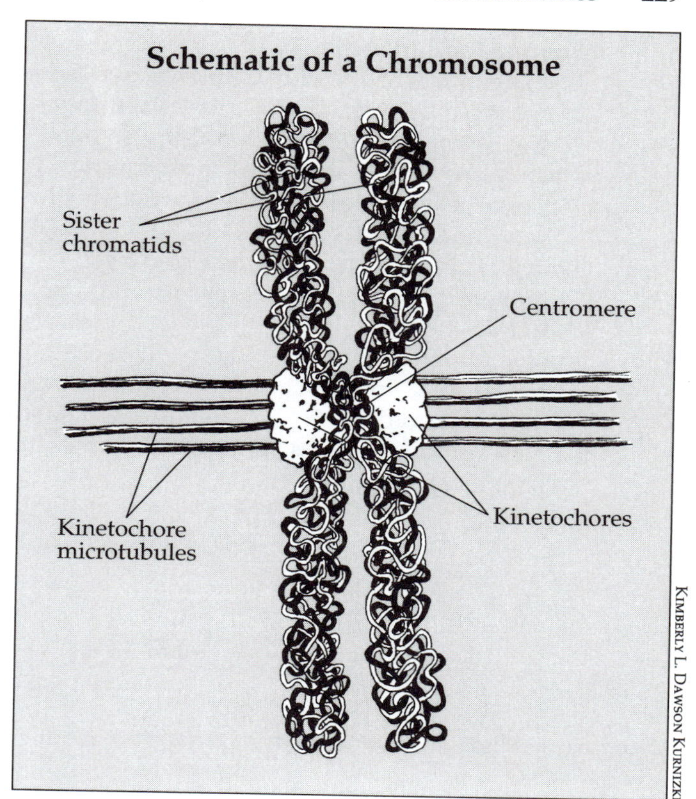

In cell division in plant cells, the two daughter nuclei are partitioned to form two separate cells by a cell plate that grows at the equator of the mother cell. In animal cells, this separation involves the constriction of the cell at a central contractile ring. DNA replication is strictly controlled during the cell cycle. DNA synthesis occurs in the synthesis (S) phase, beginning at origins of replication distributed around the genome, occurring on average every 66 kb in dicotyledonous plants and on average every 47 kb in monocotyledonous plants. *Heterochromatin* is the term for regions of chromosomes that are permanently in a highly condensed state, are not transcribed, and are late-replicating. Heterochromatin contains highly repeated DNA sequences. *Euchromatin* is the rest of the chromosomes that is extended, accessible to RNA polymerase, and at least partially transcribed.

Some plants and animals have extra chromosomes that do not seem to be essential. These are called accessory or *supernumerary chromosomes.* They have been most studied in corn where these extra chromosomes are called *B-chromosomes.* B-chromosomes are usually highly condensed heterochromatin that may or may not be present in an individual of that species.

An increase in the copy number of the genome is common in plants and animals, occurring during the development of individuals. *Polyploids* have three or more complete sets of chromosomes in their nuclei instead of the two sets found in *diploids.* For example, in *Arabidopsis,* tissues of increasing age have an increase in polyploidy, reaching up to sixteen duplications.

Susan J. Karcher

See also: Cell cycle; Chloroplasts and other plastids; Chromatin; Cytoplasm; DNA: historical overview; DNA in plants; DNA replication; Eukaryotic cells; Genetic equilibrium: linkage; Genetics: mutations; Genetics: post-Mendelian; Mitochondria; Mitosis and meiosis; Nucleic acids; Nucleus; Reproduction in plants.

Sources for Further Study

Buchanan, Bob B., Wilhalm Gruissem, and Russell L. Jones. *Biochemistry and Molecular Biology of Plants*. Rockville, Md.: American Society of Plant Physiologists, 2000. A comprehensive textbook. Illustrations, photos, sources, index.

Griffiths, Anthony J. F., et al. *An Introduction to Genetic Analysis*. 7th ed. New York: W. H. Freeman, 2000. Upper-level college textbook. Illustrations, problem sets with solutions, glossary, index. Comes with CD-ROM.

Lodish, Harvey, et al. *Molecular Cell Biology*. 4th ed. New York: W. H. Freeman, 2000. Comprehensive text with an experimental emphasis. Figures and index. Comes with CD-ROM.

Reece, Jane B., and Neil A. Campbell. *Biology*. 6th ed. Menlo Park, Calif.: Benjamin Cummings, 2002. Introductory textbook. Illustrations, problems sets, glossary, index.

Wagner, Robert P., Marjorie P. Maguire, and Raymond L. Stallings. *Chromosomes: A Synthesis*. New York: Wiley-Liss, 1993. Detailed presentation of chromosomes. Figures, extensive references, index.

CHRYSOPHYTES

Categories: Algae; microorganisms; *Protista*; taxonomic groups; water-related life

The Chrysophyceae, *classified within the kingdom* Chromista, *are mostly unicellular or colonial organisms found in fresh and salt water throughout the world.*

The *Chrysophyceae* (in some systems corresponding to the phylum *Chrysophyta*) are related to heterokont algae and include more than eight hundred described species that are classified in approximately one hundred genera. They are most closely related to the *Synurophyceae* and other pigmented heterokont algae, including the *Bacillariophyceae* (diatoms), *Eustigmatophyceae*, *Phaeophyceae* (brown algae), and *Xanthophyceae* (yellow-green algae), among others. The classification of chrysophycean species remains in a state of flux. In one system of classification primary importance is placed upon the number of flagella (zero, one, or two) that are present in the motile cell stage. A second classification organizes species based upon the predominant vegetative state of the organism. For example, in this classification amoeboid, coccoid, palmelloid, and flagellate species are assigned to separate orders.

Ecology and Diversity

Chrysophytes are predominantly found in freshwater environments, although some are marine, and a few are reported from soil or snow. Members of the group are widely distributed but are most common in cold-temperate lakes, ponds, bogs, and ditches. Some species are common members of the phytoplankton, whereas others are epibionts or are neustonic (attached to the surface film of quiet water). Other species are only rarely observed.

Most chrysophytes are free-swimming unicellular or colonial flagellates. Others are coccoid (that is, immobile, walled unicells), amoeboid, or palmelloid (with cells enveloped in a gelatinous matrix). A few species are parenchymatous.

Cell Walls

Most chrysophytes lack a cell wall, but others produce species-specific outer coverings of scales or loricae. For example, complex siliceous scales or spines that are produced in silica deposition vesicles cover the cells of *Paraphysomonas*. The scales of *Chrysolepidomonas* are organic and of two types: those that are dendritic (tree-shaped) and those that are canistrate (cylindrical). The cells of other species may be enclosed within an organic vaselike or flasklike lorica composed of cellulose and proteins or chitin (for example, *Dinobryon, Pseudokephyrion, Poteriochromonas, Lagynion*, and *Stenocalyx*). In such species the lorica is typically composed of fine, interwoven fibrils. In *Dinobryon* these fibrils are helically arranged and secreted as the cell rotates about its longitudinal axis. In contrast, the loricae of *Epipyxis* species are composed of imbricate, overlapping scales. The posterior pole of the cell is typically positioned at the base of the lorica and may be attached by a fine cytoplasmic extension; the flagella protrude externally through the lorica opening.

Flagella

Chrysophytes are heterokont, biflagellate organisms that swim with at least one flagellum forwardly directed. The two flagella of motile cells are anteriorly inserted in an apical or subapical position and are unequal in length. The flagella differ morphologically and are heterodynamic. In most species, the basal bodies from which the flagella arise are either oriented at an acute angle to one another or are perpendicular to one another.

In *Hydrurus, Chromphyton*, and *Lagynion*, the basal bodies form an obtuse (oblique) angle with respect to one another. The long (immature) flagellum is anteriorly directed and is ornamented with two rows of mastigonemes and finer lateral filaments. Each mastigoneme is composed of a base, a tubular shaft, and one to three terminal filaments; these are known as tripartite tubular hairs. Mastigonemes are produced in the perinuclear space between the two outer membranes of the chloroplast and the two surrounding membranes of the chloroplast endoplasmic reticulum (see below). The long flagellum beats in an undulatory, sine-wave-like motions that are initiated at the base of the flagellum. The relatively stiff short (mature) flagellum is directed laterally or posteriorly, lacks mastigonemes, and rotates helically. A distinct swelling associated with the eyespot is typically present at the proximal base of the smooth flagellum.

In some taxa (such as *Chromulina, Chrysococcus*, and *Sphaleromantis*), the short flagellum is highly reduced and may be nonemergent; it is therefore undetectable by light microscopy. In a handful of species the short flagellum is entirely absent, although the mature basal body may persist within the cell. Naked motile cells bearing two visible flagella are often referred to as *Ochromonas*-like (or ochromonadalean), whereas those with one visible flagellum are typically assigned to the genus *Chromulina*.

The transitional region between the basal body and flagellum contains an electron-dense transitional plate, above which lies a coiled, apparently springlike transitional helix. The functions of the transitional plate and helix, which are also found in other flagellates, are uncertain.

In heterotrophic and mixotrophic species, the flagella play a role in prey capture. Particles actively captured by the flagella that are recognized as food are pushed into a feeding basket; those not recognized as food are released. The feeding basket is formed and closed by movements of underlying microtubules (see below). Water currents produced by the undulation of the long flagellum may passively bring food particles in contact with the cells that, in some species, are collected by pseudopodia.

Cell Organization

Cells possess a single pear-shaped nucleus that is positioned at the anterior end of the cell. The narrow end of the nucleus typically lies close to the basal bodies. A prominent Golgi apparatus with distended cisternae lies against the nucleus. Contractile vacuoles (absent in some marine forms) are also found at the anterior end of the cell.

One or more mitochondria with tubular cristae are present in the cell. Because the mitochondria are usually long and coiled, the actual number of mitochondria present is difficult to discern. Fibrous bands, sometimes referred to as connecting fibers, connect the basal bodies to one another. A cross-striated band of fibers known as the rhizoplast extends from the basal apparatus and forms a connection to the nucleus. Typically four microtubular roots (R1, R2, R3, and R4) originate near the basal bodies, take characteristic paths through the cell, and proliferate beneath the plasmalemma. For example, in most species roots R3 and R4 often form a loop beneath the short flagellum. Other microtubules are nucleated from the four major roots that provide the cytoskeletal elements needed to maintain cell shape.

Muciferous bodies or discobolocysts are present in some species. Muciferous bodies are capable of extruding long threads, whereas discobolocysts forcefully eject discoid projectiles. These functions of these organelles have been little studied but may be involved in prey capture or predator avoidance.

Nutrition

The *Chrysophyceae* employ a variety of means to obtain energy. Most chrysophytes are photosynthetic but require an exogenous source of vitamins (such as vitamin B_{12}, biotin, and thiamin) for growth. It is probable that all chrysophytes are opportunistically or facultatively osmotrophic; that is, they are capable of directly absorbing small inorganic or organic molecules (such as sugars and amino acids) from the surrounding medium. Several species, particularly those with leucoplasts, are obligate heterotrophs that are bactivorous or consume small organic particles. Mixotrophic species are also well represented among the chrysophytes. This category includes photosynthetic species that, routinely or under unfavorable conditions, supplement their nutrition via phagotrophy.

Chloroplasts, Photosynthetic Pigments, and Storage Products

The chloroplasts of chrysophytes are typically golden-brown or yellow-green in color, and there are usually one to two chloroplasts per cell. Chloroplasts are peripherally located, and pyrenoids may be present or absent. Four unit membranes surround each chloroplast; the outer two are derived from the endoplasmic reticulum and are typically continuous with the nuclear envelope. Chloroplast lamellae are typically composed of three adpressed thylakoid membranes, and a girdle lamella, which completely encircles the chloroplast, is usually present. The chloroplast deoxyribonucleic acid (DNA) is ring-shaped and lies just beneath the girdle lamella.

The light-harvesting complex of chrysophytes contains chlorophylls *a* and *c*, beta-carotene, and the xanthophylls fucoxanthin, neoxanthin, violaxanthin, and zeaxanthin. Among these, fucoxanthin is dominant and is therefore responsible for the golden-brown color observed in most chrysophytes.

The major product of photosynthesis is a water-soluble o-1,3-linked glucan (known as chrysolaminarin or leucosin) that is stored in cytoplasmic vacuoles in the posterior region of the cell. Lipids may also be produced and are also stored in the cytoplasm.

Eyespots (or stigmata) are present in many, but not all, species. The eyespot takes the form of a single layer of orange or reddish colored, lipidlike droplets that are located just beneath the chloroplast membrane. These droplets lie near a swelling located at the base of the smooth (short) flagellum; together the eyespot and flagellar swelling form a photoreceptor apparatus.

Several chrysophyte genera are known that contain a vestigial chloroplast (leucoplast) that lacks pigments (including *Anthophysa*, *Monas*, *Oikomonas*, *Paraphysomonas*, and *Spumella*).

Reproduction

Asexual reproduction in amoeboid and flagellate species occurs by longitudinal division of the cell; fragmentation is common among colonial, palmelloid, and parenchymatous species. In coccoid species reproduction may proceed via cell division or the formation of autospores that rupture and exit the parent cell wall. Some taxa, such as the parenchymatous genera *Phaeodermatium* and *Hydrurus* or members of the palmelloid family *Chrysocapsaceae*, reproduce by means of flagellated swarmers (zoospores).

Under certain environmental conditions, silicified resting cysts, or statospores, are produced by many species. Statospores are formed endogenously, are roughly spherical or ellipsoidal, and have walls that may be smooth or ornamented. The stomatocyst opening (porus) may be simple, possess a thickened collar, or take the form of a narrow neck. The cyst wall is formed by the deposition of silicate on an internal membrane, and the porus is preformed or produced by resorption of a portion of the cyst wall. Depending on the species, cytoplasm located outside the cyst wall may or may not be absorbed through the porus, which at maturity is occluded by a pectic plug. During excystment the plug is lost, and one or more amoeboid or free-swimming flagellate cells emerge.

Sexual reproduction is known only in a handful of species. In those cases observed, vegetative cells behave as gametes and fuse apically. The resulting quadriflagellate cell (planozygote) will encyst forming sexually derived binucleate hypnozygotes or stomatocysts. It is presumed that karyogamy (nuclear fusion) and meiosis occur within the cyst, but these processes have yet to be studied. Depending upon the species, sexual stomatocysts may give rise to one, two, or four vegetative cells.

J. Craig Bailey

See also: Algae; Brown algae; Cryptomonads; Diatoms; Dinoflagellates; Flagella and cilia; Haptophytes; Heterokonts; Phytoplankton; Photosynthesis; *Protista*.

Sources for Further Study

Graham, Linda E., and Lee W. Wilcox. "Chrysophyceans." In *Algae*. Prentice Hall, Upper Saddle River, N.J.: Prentice Hall, 2000. A general introduction to the chrysophytes within a modern context, with illustrations.

Lee, Robert E. "Chrysophyta." In *Phycology*. 2d ed. New York: Cambridge University Press, 1989. Key characters of *Chrysophyceae* and related taxa.

Preisig, Hans R., and Robert A. Andersen. "Chrysomonada: Class Chrysophyceae Pascher, 1914." In *An Illustrated Guide to the Protozoa*, edited by John J. Lee, Gordon F. Leedale, and Phyllis Bradbury. 2d ed. Lawrence, Kans.: Society of Protozoologists, 2001. Presents historical aspects of chrysophyte classification, key taxonomic features of the class, and illustrated descriptions of many common genera. Includes important references.

Sandgren, Craig D., John P. Smol, and Jørgen Kristiansen, eds. *Chrysophyte Algae: Ecology, Phylogeny, and Development*. New York: Cambridge University Press, 1995. A collection of articles devoted to various aspects of chrysophyte biology.

CHYTRIDS

Categories: Fungi; taxonomic groups; water-related life

Chytrids are fungi in the phylum Chytridiomycota. *They have motile spores and are primarily aquatic organisms.*

Like all fungi, chytrids live in their food and have an absorptive mode of nutrition in which they secrete digestive enzymes and absorb the breakdown products. Chytrids also have cell walls made of chitin, make the amino acid lysine via the amino adipic acid (AAA) pathway, and possess a ribosomal DNA (deoxyribonucleic acid) sequence that places them more closely with other fungi than with any other group of organisms. The feature that sets them apart from other fungi is the possession of a motile zoospore. All other fungi produce spores without flagella.

Characteristics

A posteriorly oriented, whiplash-type flagellum is the feature that unites all the organisms in the division *Chytridiomycota* within the kingdom *Fungi*. As absorptive heterotrophs, they live either as *saprophytes*, growing on dead organic matter, or as parasites in living plants, other fungi, insects, or algae. The vegetative organism may take the form of a spherical structure, with or without branching rhizoids, on the surface of substrate or host or may send mycelial threads through the material in which it is living. Asexual reproduction occurs by a variety of means described below. Sexual reproduction is known to exist in several types of chytrids and in some species involves the alternation between a gamete-producing phase and a spore-producing phase.

The sporangia that produce the motile zoospores develop in a variety of ways. Two features are used to characterize development: the fate of the nucleus upon encystment of the zoospore and the number of zoosporangia produced from a single zoospore. The three most common types of thallus development are endogenous-monocentric, exogenous-monocentric, and exogenous-polycentric. Endogenous-monocentric development occurs when the zoospore nucleus stays within the encysted zoospore wall, undergoes mitosis, and produces a single zoosporangium. Exogenous-monocentric development occurs when the zoospore nucleus migrates into the germ tube, undergoes mitosis, and produces a single zoosporangium. Exogenous-polycentric development occurs when the zoospore nucleus migrates into the germ tube, undergoes mitosis, and spreads to many locations for zoosporangium production.

The phylum-defining zoospore may be one of

four basic morphological types. Though the types are determined by electron microscope, the morphological type can be recognized using light microscopy with experience. The four morphological types are the basis of classification at the ordinal level as described below.

Ecology and Habitats

Because chytrids are absorptive heterotrophs, they grow in their food, digesting complex food molecules and absorbing the simpler breakdown products. When growing in dead material, these fungi are saprophytes and are decomposing organisms in ecosystems. Because the zoospore requires water for dispersal, these fungi are found in aquatic environments. However, they also can be found in soils that are wet with soil water. Chytrids also can live within living organisms as parasites, causing major declines in populations. The gut chytrids, *Neocallimastigales*, live in the rumina (stomachal cavities) of herbivorous mammals.

Taxonomy

There are approximately eight hundred species of chytrids, arranged in five orders. Taxonomy of the different orders is based on the ultrastructure of the zoospore. Ultrastructure features used in taxonomy include the presence or absence of a connection between the nucleus and the kinetosome by microtubules; whether ribosomes are dispersed or collected into a mass surrounded by membranes; the degree of organization of the microbody-lipid complex (MLC); the location and number of lipid globules; and presence or absence of a rumposome—a honeycomblike organelle of unknown function. The main characteristics of the five orders are described below.

Chytridiales. During examination of the main features of the zoospore—lipid globule, microbody, mitochondria, and nucleus—the nucleus seems to occupy whatever space is left over within the zoospore. Rootlet microtubules are located within the plasma membrane connecting the kinetosome to the rumposome. Ribosomes are gathered in the center of the cell, enclosed within membranes. In the MLC, the posteriorly located lipid globules are in close association with the microbody, mitochondrian, and rumposome.

Spizellomycetales. The nucleus of the zoospore is close to the kinetosome or, if separated, is connected to it via microtubules or a rhizoplast. Rootlet morphology is variable, and ribosomes are scattered throughout the cytoplasm. The MLC has a loose association of the microbody and lipid at the anterior end of the zoospore with the mitochondria located toward the rear. There is no rumposome. Ribosomes are dispersed throughout the zoospore.

Neocallimastigales. *Neocallimastix* and other genera of the order are uniflagellate or multiflagellate and live in the rumen of herbivorous mammals. Because they live in this unique environment, rumen chytrids are obligate anaerobes. The zoospores lack any of the MLC organelles and the rumposome. All these anaerobic fungi are cellulolytic and digest plant cell walls of the food upon which sheep and cattle feed.

Monoblepharidales. The zoospores have a centrally located nucleus that is not connected to the kinetosome. Microtubules extend randomly into the cytoplasm from the kinetosome. The MLC has a rumposome in close association with a microbody and anteriorly located lipid globules. The ribosomes are centrally located, surrounding the nucleus. These fungi have a mycelial growth form and reproduce sexually by producing a motile male cell and a nonmotile egg cell.

Blastocladiales. A nuclear cap consisting of ribosomes encased within a membrane located anteriorly to a cone-shaped nucleus and a single large mitochondrian with a side body complex are the two most distinctive features of these fungi. Some of these fungi produce mycelial growth forms, whereas others produce the saclike zoosporangium with rhizoids.

Evolutionary History

Evolutionary history of the chytrids can be traced back to the Pennsylvanian period through fossil evidence. Sequential analysis of the small subunit ribosomal DNA gene from fifty-four chytrids indicates that the *Chytridiomycota* are related to other fungi and that there are natural groups within the division: *Blastocladiales*, *Monoblepharidales*, and *Neocallimastigales*. Despite the diversity of the data, the monophyletic nature of the *Chytridiales* and *Spizellomycetales* is not rejected. The DNA groupings closely resemble groupings based on zoospore ultrastructure.

Representative Organisms

Allomyces is a mycelial member of the *Blastocladiales*, which is interesting because it has an alternation of generations between a gamete-producing thallus (gametothallus) and a spore-producing thallus (sporothallus). In all organisms with alternation of generations, the gametothallus produces gametes by mitosis in gametangia. The gametes are distinguished by size, the male being smaller than the female. The motile male gamete is chemotactically attracted to the hormone sirenin, which is produced by the female gametes and enables fertilization. Upon fertilization, the zygote nucleus undergoes mitosis as the germ tube develops into mycelia without crosswalls. The dichotomously branched mycelia of the sporothallus produce two types of sporangia. The thin-walled sporangia produce diploid spores by mitosis. These diploid zoospores are responsible for increasing numbers of *Allomyces* in its habitat.

The sporothallus also can produce a thick-walled sporangium capable of withstanding harsh environmental conditions. Zoospores in this sporangium are produced by meiosis. When these haploid zoospores geminate, the nucleus divides by mitosis and spreads throughout the dichotomously branched mycelia. The life cycle of the fungus now is completed.

Batrachochytrium is interesting because it parasitizes frogs. Within the last decade, declines in populations of frogs around the world have been described. *Batrachochytrium dendrobatidis* is responsible for this chytridiomycosis in amphibians, including salamanders.

Blastocladiella is a developmental biology tool. The thallus has the exogenous, monocentric developmental pathway resulting in a rhizoidal system with a single thin-walled, colorless sporangium or a single thick-walled, resistant sporangium. The chemical environment of the developing thallus determines which sporangium is produced. High carbon dioxide levels favor the development of the thick-walled sporangium. This shift from a thin-walled sporangium pathway to a thick-walled sporangium pathway has been traced to a disruption of the Krebs cycle. This organism is one of a few nongreen organisms in which light promotes the growth of the organism.

Coelomomyces is a mycelial member of the *Blastocladiales* that parasitizes invertebrate animals. Coelomomyces alternates between a haploid gametothallus and a diploid sporothallus. The unique feature of *Coelomomyces* is that each phase is specific for a different host. The diploid sporothallus parasitizes mosquitoes and grows as wall-less mycelia within the hemocoel of the mosquito larvae. *Coelomomyces* has been studied as a possible mycoinsecticide against mosquitoes. Difficulty in using *Coelomomyces* as a mycoinsecticide occurred until the discovery of the fact that an alternate host was required to achieve completion of the life cycle. The zoospores produced by the thick-walled sporangium within the mosquito are produced by the process of meiosis and are haploid. The haploid zoopore must infect a microcrustacean copopod or ostracod in order for the gametes to be produced. The haploid zoospore develops into the gametothallus, which produces the motile gametes. The resulting zygote will infect mosquito larvae, completing the life cycle.

John C. Clausz

See also: Flagella and cilia; Fungi; Krebs cycle.

Sources for Further Study

Alexopoulos, C. J., C. W. Mims, and M. Blackwell. *Introductory Mycology*. 4th ed. New York: John Wiley and Sons, 1996. The primary thrust of the book is morphology and taxonomy, but it also includes topics about activities of fungi.

Hudler, George W. *Magical Mushrooms, Mischievous Molds*. Princeton, N.J.: Princeton University Press, 1998. A very readable book about fungi in human affairs. The book is intended for a nontechnical audience.

Kendrick, Bryce. *The Fifth Kingdom*. 3d ed. Newburyport, Mass.: Focus Information Group, 2001. A book dealing with classification, ecology, and genetics of fungi, fungi as plant pathogens, predatory fungi, biological control, mutualistic relationships with animals and plants, fungi as food, food spoilage and mycotoxins, poisonous and hallucinogenic fungi, medical mycology, antibiotics, and organ transplants.

CIRCADIAN RHYTHMS

Categories: Movement; physiology

Circadian rhythms in plants are phases of growth and activity that appear in regular, approximately twenty-four-hour, cycles.

Biological activities that cycle in approximately twenty-four-hour intervals are called circadian rhythms (from the Latin *circa*, meaning "about" and *dies*, meaning "a day"). Circadian rhythms allow plants to anticipate environmental cycles and to coordinate their activities with them. Circadian rhythms are not simply responses to changing external conditions, as they continue even when a plant is placed under constant conditions. This continuation indicates that circadian rhythms are controlled by endogenous (internal) timing mechanisms, collectively referred to as the *biological clock*.

Digital Stock

Many plants, such as these California poppies, open their flowers in the morning and close them at night. These cycles are important for timing pollen availability with the activity of insect, bird, and mammal pollinators. It is essential that flowers of the same species be open at the same time of day or night to promote outcrossing that results in increased genetic variation.

Plant circadian rhythms include cycles in gene regulation, enzyme activity, leaf movements, flower opening, and stomatal opening. Circadian rhythms also interact with photoperiodism in the control of major developmental processes, such as dormancy and the induction of flowering.

History

In 1729 the French astronomer Jean-Jacques Dortous de Mairan discovered the endogenous nature of circadian rhythms when he looked at the sleep movements of leaves of the sensitive plant, *Mimosa*, known as *nyctinastic leaf movements*. Mimosa leaves fold closed at night and open during the day. It had been thought that these leaf movements occurred in response to external cycles of light and darkness. De Mairan examined the plants under constant environmental conditions and discovered that the nyctinastic movements of the leaves continued. This was the first description of a biological activity with an endogenous circadian rhythm. Current models for how plants accomplish circadian rhythms are divided into three parts: entrainment, biological clock, and output pathways.

Entrainment

The synchronization of circadian rhythms to the cycles of the outside world is accomplished via input pathways and is referred to as *entrainment*. In nature, circadian rhythms are entrained primarily by light or temperature cycles to have periods of twenty-four hours. It is essential that circadian rhythms be entrained, because without synchronization of the biological clock with environmental cycles, the advantages of circadian rhythms would be lost.

Biological Clock

The biological clock is also referred to as the central oscillator and the pacemaker. It is endogenous and self-sustaining. Although circadian rhythms are entrained by external stimuli, they continue in the absence of external cycles. Under artificially constant conditions, circadian rhythms do not maintain twenty-four-hour periods but revert to free-running periods that are usually between twenty-one and twenty-seven hours. The molecular mechanisms of the biological clock remain unknown, but they are thought to include autoregulatory feedback mechanisms.

An interesting feature of the biological clock is that the free-running period is generally insensitive to changes in temperature. Most chemical and biological processes are affected by temperature changes; higher temperatures make them go faster, and lower temperatures make them go slower. That the biological clock is able to compensate for temperature changes and maintain timekeeping functions is important to plants experiencing extreme changes in temperature.

Output Pathways

The *output pathways*, or "hands" of the biological clock, are the measurable rhythms exhibited by the plant. Known circadian rhythms range from the subcellular level to the cell and tissue level to the developmental level.

Subcellular Level

Subcellular circadian rhythms include cycles in gene regulation (at the levels of transcription, transcript abundance, translation, and post-translational modification), calcium signaling, and enzyme activity. One well-characterized rhythm is the rate of carbon dioxide assimilation in plants with CAM (crassulacean acid metabolism) photosynthesis. Such plants open their stomata at night to allow for gas exchange, fixing carbon dioxide into an organic acid that is stored in the vacuole. During the day, the plants close their stomata (presumably to conserve water) and continue photosynthesis using carbon dioxide released from the organic acids. The circadian rhythm of carbon dioxide assimilation in CAM plants is controlled by rhythmic changes in the activity of the enzyme (PEP carboxylase) that fixes carbon dioxide into the organic acid.

Cell and Tissue Levels

Cell- and tissue-level circadian rhythms include those controlled by cycles in cell expansion and contraction, such as the obvious rhythms of leaf and petal movements and the opening and closing of stomata. Leaf movements are brought about by a cycling in the expansion and contraction of specialized cells in a region at the base of the leaf that is called the pulvinus. Nyctinastic leaf movements presumably allow a plant to maximize light interception for photosynthesis.

Many plants open their flowers in the morning and close them at night. Other plants open their flowers in the afternoon (such as the four o'clocks,

Mirabilis jalapa), in the evening (evening primrose, *Oenothera biennis*), or even at night (the bat-pollinated cactus *Cereus*). These cycles are important for timing pollen availability with the activity of insect, bird, and mammal pollinators. It is essential that flowers of the same species be open at the same time of day or night to promote outcrossing that results in increased genetic variation.

Circadian cycles of stomatal opening and closing allow a plant to balance carbon dioxide uptake with water loss. Plants with CAM photosynthesis open their stomata at night, in contrast to plants that carry out C_3 and C_4 photosynthesis, which open their stomata during the day. Other known tissue-level circadian rhythms include hypocotyl elongation, nectar secretion, and hormone synthesis.

Developmental Level

Developmental processes that depend on interactions with circadian rhythms and the biological clock include the photoperiodic control of flowering and dormancy. These photoperiodic responses rely on the ability of a plant to measure relative amounts of light and darkness within each twenty-four-hour period. It remains unknown whether one biological clock controls both photoperiodism and circadian rhythms.

Margaret A. Olney

See also: C_4 and CAM photosynthesis; Dormancy; Flowering regulation; Gas exchange in plants; Heliotropism; Nastic movements; Photoperiodism; Photosynthesis; Tropisms.

Sources for Further Study

Hopkins, William G. *Introduction to Plant Physiology*. 2d ed. New York: John Wiley & Sons, 1999. Plant physiology textbook with descriptions of rhythmic phenomena and photoperiodism. Includes charts and diagrams.

Moore, Randy, W. Dennis Clark, and Darrell S. Vodopich. *Botany*. 2d ed. Boston: WCB/ McGraw-Hill. 1998. Introductory textbook describing circadian rhythms and how plants respond to environmental stimuli. Includes illustrations, diagrams, and photographs.

Raven, Peter H., Ray F. Evert, and Susan E. Eichhorn. *Biology of Plants*. 6th ed. New York: W. H. Freeman/Worth, 1999. Introductory textbook describing circadian rhythms, biological clocks, photoperiodism, and dormancy. Includes illustrations, charts, diagrams, and photographs.

Wilkins, Malcolm. *Plantwatching: How Plants Remember, Tell Time, Form Relationships, and More*. New York: Facts on File, 1988. Scientifically accurate and accessible descriptions of basic plant physiology, covering plant clocks and flowering control. Includes illustrations, charts, diagrams, and photographs.

CLADISTICS

Categories: Classification and systematics; disciplines; methods and techniques

Cladistics is a quantitative method of classification of plants that attempts to recover evolutionary relationships, based on observable characters.

Since the dawn of history, humans have classified plants. In primitive cultures classifications were by economic use, such as food, clothing, medicine, and shelter. Later the form (morphology) of a plant became important, for example, trees, shrubs, or herbs. Carolus Linneaus considered the similarity of floral parts to be critical, and this formed the basis of his classification system. Each of these systems is said to be "artificial." That is, the classification was solely for a human purpose and did not attempt to indicate genetic relationships between plants. Since Charles Darwin, the goal of plant systematics has been to develop a "natural," phylogenetic classification, one that represents the natural relationships of each species to all others. Cladistics was developed as a method to construct phylogenetic classifications.

A Brief History

Three systems have evolved to aid systematists (scientists who study the phylogenetic relationships of organisms) in their work. Traditional *phylogenetics* was based on intuition and involved the "art and science" of character weighting. The scientist studied a group of plants and decided which characters he or she thought were important. Evolutionary relationships were then based on these characters. Individual bias led to disagreements that could not be resolved objectively.

Computer-assisted numerical approaches permitted systematists to employ a more objective methodology and analyze large quantities of data, gathered from a variety of sources that range from traditional morphology to the most sophisticated molecular techniques. The earliest attempt, *phenetics*, used computers to determine the degree of total similarity between taxa. Unfortunately, this ignored both parallel and convergent evolution.

The methods of cladistics were first formalized in the 1950's and 1960's by Willi Hennig. This approach requires three assumptions to be met: evolution occurs; evolution is monophyletic (that is, lineages derive from a common ancestor); and characteristics passed from generation to generation are either modified or not. Although phylogenetics is concerned with genealogical relationships, the latter cannot be observed; rather, they must be inferred from observable characters (morphological, biochemical, behavioral, and so on) in much the same way as one infers genotypes when constructing a family pedigree. Cladistics is a quantitative method that attempts to recover evolutionary relationships, based on observable characters, and presents the resulting phylogeny in the form of a treelike diagram called a *cladogram*.

When many different organisms are being classified and when many different characters are being analyzed simultaneously, alternative cladograms may result. The most parsimonious tree (the cladogram requiring the fewest evolutionary changes) is generally preferred, because it is assumed that the simplest pathway is the one most likely to reflect the evolutionary history of the plants being examined.

Constructing a Cladogram

The most important decision to make before beginning construction of a cladogram to represent the relationships among a group of plants is the choice of an appropriate *outgroup*. The outgroup cannot belong to the group of plants being analyzed, but it should be closely related. Much of the work of a phylogenetic study is determining an appropriate outgroup to be used for comparisons.

The next step involves construction of a *character matrix*. A character is any feature of a plant. It may be an observable morphological or biochemical feature or an ecological or physiological attribute. Every useful character will have more than one character state. For instance, the character "root type" may have the character states "taproot," "fibrous root," or "adventitious root."

Characters having a common origin are called homologous. Cladistic analysis recognizes two types of homologies: plesiomorphies and apomorphies. *Plesiomorphies* are considered to be the primitive state of a character; that is, the character is unchanged from the ancestral condition. Plesiomorphies are determined by comparison of the character states in the members of the taxa being investigated with the character state in the outgroup. A character state found in both the outgroup and the taxa being examined is considered to be plesiomorphic. Any modification of the character state is considered to be apomorphic; thus, *apomorphies* are derived from plesiomorphies. Apomorphies shared by two or more taxa are called synapomorphies. Identification of *synapomorphies*, assumed to be derived from increasingly recent common ancestors, provides the basis for constructing cladograms.

Qualitative Approach

The first step toward a qualitative approach to constructing a cladogram is to examine the character matrix and list groupings of taxa according to the apomorphic trait for each character. Next, one character to begin the tree is chosen. Any character will do, but it is simplest to begin with a character in which only the outgroup has the plesiomorphic state and all ingroup taxa share the same apomorphy. For instance, a conifer might be the outgroup for classifying flowering trees. The plesiomorphic reproductive structure would be a cone, and the synapomorphy shared by all ingroup members would be flowers. The tree would have the conifer at the base, with a single line extending to the right to a branch point (node) from which all ingroup members diverge. The character state "flower" would be placed on the line between the conifer and the node, indicating that the shared character state, flowers, evolved prior to the divergence of ingroup taxa from one another.

Next, a second character is added to the existing tree. For instance, the conifer and dicot trees would all share the plesiomorphic character of a taproot, but monocot trees, such as palms, would have fibrous roots. The tree should now be extended to the right to form a second node with the character state "fibrous roots" added to the new stem segment and the monocot trees branching off the second node. The monocot taxa diverged from each other after fibrous roots evolved. The dicots do not have fibrous roots, so they are diagramed at the node to the left of "fibrous roots." The tree is continued by the addition of one character at a time until all have been used.

Quantitative Approach

The qualitative approach becomes increasingly difficult as the number of taxa and number of characters are increased. The advantages of the quantitative approach are that the process can be automated and human bias can be minimized. The following example illustrates "by hand" the way computers can be programmed to produce a cladogram. The first step is to code the character matrix to produce a numerical matrix for analysis. Plesiomorphic characters in the data matrix are coded as 0; different apomorphic character states are coded as successive integers, 1, 2, 3, and so on.

The simplest cladogram consists of a Y-shaped diagram representing three taxa, two from the group being studied and a third being the outgroup. The outgroup is placed at the bottom and serves to "root" the tree. The two ingroup species are located at the top of each arm. The point where the two arms diverge is a node and represents the ancestral taxon derived from the outgroup that gave rise to both ingroup taxa—it represents the common ancestor of the ingroup taxa. A numerical algorithm computes what the character states of this ancestral species must have been.

Additional taxa can now be added to the cladogram, one at a time. A series of new trees are constructed in which the new taxon is added between each existing taxon and each existing node. There are three places a fourth taxon could be added to the simple tree: between the root and the node, between the node and the first ingroup taxon, and between the node and the second ingroup taxon. An algorithm computes which of the three possible trees is the most parsimonious, and this tree is used as the basis for adding the fifth taxon (in one of now five possible positions between the four existing taxa and the two nodes). This process is continued until all taxa have been added to the tree and the cladogram is complete.

Marshall D. Sundberg

See also: Coevolution; Evolution: convergent and divergent; Evolution of plants; Molecular systematics; Systematics and taxonomy; Systematics: overview.

Sources for Further Study

Brooks, D., J. Caira, T. Platt, and M. Pritchard. *Principles and Methods of Phylogenetic Systematics: A Cladistics Workbook*. Lawrence: University of Kansas Press, 1984. A "nuts and bolts" introduction to cladistics.

Freeman, Scott, and Jon C. Herron. *Evolutionary Analysis*. Upper Saddle River, N.J.: Prentice Hall, 2001. An introduction to evolution for advanced high school and lower-division undergraduates, with a brief coverage of cladistics.

Futuyma, Douglas. *Evolutionary Biology*. 3d ed. Sunderland, Mass.: Sinauer Associates, 1998. Book for advanced undergraduates and graduate students. Contains a full chapter devoted to modern cladistic analysis.

CLIMATE AND RESOURCES

Category: Environmental issues

Climate is described by the average of weather conditions at a place or in a region, usually recorded as both the mean and the extremes of temperature, precipitation, and other conditions. Resources are the factors and characteristics of the natural environment that people find useful, including climate, land, soil, water, minerals, and wild vegetation. Thus, climate itself is a resource, affecting the character of the plant life and other resources it supports.

The nature and distribution of wild vegetation are to a large degree the products of climate: the temperature, moisture, solar radiation, and other environmental conditions that characterize a region. The major global vegetation types that accompany forest, shrub, grassland, desert, rain forest, tundra, and other biomes reflect climatic controls.

Solar radiation is the basic determinant of climate. The sun's rays are vertical at some time of the year only in the tropics, between the Tropic of Cancer (23.5 degrees north latitude) and the Tropic of Capricorn (23.5 degrees south latitude). These lines determine where the greatest heat supply is found; regions poleward of about 40 degrees north and south latitudes actually have a net loss of reradiation to outer space and depend upon a heat supply from the tropics, which is carried poleward by the general circulation of the atmosphere. The general circulation is the average of wind flow at the surface of the earth and is driven by the surplus of solar radiation in the tropics.

Equatorial Climates

By definition, tropical climates do not experience freezing temperatures, have the least variation in length of day, and consequently experience the least "seasonality" of any latitudes. Seasons in the tropics are characterized more by precipitation contrasts—"dry" and "wet"—than by summer and winter temperatures. The greatest combination of heat and moisture resources on the earth's surface, especially important in creating the conditions under which tropical rain forests flourish, is near the equator.

The depth to which rock and soils are weathered and leached (mineral plant foods dissolved and removed by groundwater flow) is greater near the equator than elsewhere on the earth's surface. Continuous high temperatures work against carbon storage in the soils. Under wild vegetation conditions, where the rain forest canopy protects soils from raindrop impact, erosion rates are not as high as one would expect from the intense rain showers. On sloping land, however, the soils become saturated and flow downslope, often catastrophically in landslides. Where wild vegetation has been removed by human activity, such as farming or development of urban centers, erosion and mass wasting (landslides) are exacerbated during rainy seasons and cause considerable loss of life and property damage.

With increasing distance from the equator, the tropics experience more pronounced seasons, particularly in moisture resources. Precipitation totals decline, and *drought* risk increases. Dry seasons are expected annually because of the shifting of the general circulation of the atmosphere. The timing and extent of this shift determine whether a region experiences drought.

East and South Asia are most affected by shifting atmospheric circulation and the resulting wet and dry seasons. Africa also has pronounced wet and dry seasons. Droughts in this part of the world result in famine: An estimated one million people died in the Sahelian droughts of the late 1960's and 1970's. Thus climate must be defined both in terms of averages and of extremes. Extremes result in hazards that have dire consequences for the inhabitants of the affected region.

The probability of drought increases as precipitation averages decrease. Additionally, most tropical rainfall takes the form of intense thundershowers, which are spatially highly variable. One farm may be drenched by rain while its neighbors con-

PhotoDisc

The eastern sides of midlatitude landmasses are subject to intense summer storms.

tinue to be tormented by drought. In addition to drought risk on the margins of the tropics, a major climatic hazard is the tropical cyclone, also called a hurricane or typhoon. Cyclones rarely affect the equatorial zone but frequent the tropical transition to the subtropics and midlatitudes. Movement of tropical cyclones is easterly in their early and middle stages, following the general circulation known as the trade winds.

The Subtropics

The climates that exist in the *subtropics*, poleward of the tropics, depend on the side of the continent: West sides are *deserts* or subtropical *drylands*; east sides are the *humid subtropics*, a transition zone with cooler temperatures and more risk of frost with greater distance from the equator. The humid subtropics are subject to occasional easterly flow weather systems, including tropical cyclones. While cyclones represent a serious hazard, claiming both lives and property, these easterly systems also deliver moisture and thus reduce the possibility of drought. The generally warm temperatures and moist conditions make these climates some of the most productive for crop growth, exceeding the potential of the tropics.

In the subtropics, leaching of soils and high erosion rates on cleared fields are nearly as great a problem as in the tropics. The west coast drylands, which include all the world's major deserts—Sahara, Atacoma, Kalahari, Australian, and North American—are a consequence of the general circulation of the atmosphere, which in these locations makes the swing from the prevailing westerlies of the middle latitudes to the easterly trade winds. In the process, high atmospheric pressures prevail, and winds are descending or subsiding, and therefore warming—just the opposite of the conditions required for rainfall. Drylands may extend deep into the continents, as in North America and especially in Africa and Asia. The dryness of the Sahara blankets the Middle East and extends northward into Central Asia. Temperatures along the equatorward flank of these five major dryland zones are tropical, and where irrigation water is available, tropical plants may be grown. Most of the drylands are subtropical or midlatitude, and thus they experience frost as well as drought hazard. Weathering and erosion are appreciably less in the drylands, owing to the absence of moisture. Leaching of the soils is virtually absent. Instead, salts in the soils can build up (*salinization*) to levels that are toxic to most plants—another climate-related hazard.

The Midlatitudes

The *midlatitudes* extend from the subtropics to the polar climates of the Arctic and Antarctic. Temperatures follow a transition from warm on the equatorward flank to too cold for agriculture nearer the poles. This is the realm of the westerlies, with extratropical cyclones delivering most of the weather. It is a zone of contrasting conditions, year by year and day by day, ranging from warmer than average to colder than average, from too humid to too dry on the inland dryland border. The hazards of extreme temperature and precipitation often dominate life, as tropical and polar air masses converge to create the cyclones that march from west to east.

Drought risk is most important on the dryland border and results in the world's great grasslands. Summer heat may be a hazard on occasion. Nearly every winter brings storms with freezing rain, high winds, and heavy snowfalls, particularly on the eastern sides of the continents. The eastern sides are also afflicted with intense summer storms, such as the tornadoes of North America (a winter phenomenon in the adjoining humid subtropics) and the tail ends of hurricanes and typhoons, as these storms become caught up in westerly circulation and curve poleward again. The Arctic fringe of the midlatitudes is too cool for significant agriculture but yields the great subarctic forests of Canada, Scandinavia, and Russia.

Neil E. Salisbury

See also: Biomes: definitions and determinants; Biomes: types; Drought; Erosion and erosion control.

Sources for Further Study

Bryson, Reid A., and Thomas J. Murray. *Climates of Hunger: Mankind and the World's Changing Weather*. Madison: University of Wisconsin Press, 1979. Discusses climate change and its effects on food production in various regions.

Grigg, D. B. *The Agricultural Systems of the World: An Evolutionary Approach*. New York: Cambridge University Press, 1988. Climate's greatest impact on resources is in agriculture; a global overview is provided here.

Ladurie, Emmanuel Le Roy. *Times of Feast, Times of Famine: A History of Climate Since the Year 1000*. Translated by Barbara Bray. Rev. ed. New York: Noonday Press, 1988. Presents and explains data on the earth's climate in "recent" history, derived from contemporary evidence of crop yields, harvests, and other reports.

Strahler, Alan, and Arthur Strahler. *Physical Geography*. 2d ed. New York: Wiley, 2001. A popular introductory physical geography textbook containing a readable account of the world's climates.

CLINES

Categories: Ecology; ecosystems; genetics

A cline is one form of geographic variation in which characteristics of a species change gradually through the species' geographic range.

Many plant and animal species have populations that differ in terms of their morphological, physiological, and biochemical characteristics. A *species* is generally defined as a group of organisms that have the potential to interbreed and produce fertile offspring. A *population* is defined as a group of organisms which are actively interbreeding. The following example will clarify the relationship between species and populations and simultaneously introduce geographic variation.

Geographic Variation

The ponderosa pine (*Pinus ponderosa*) occupies a broad geographic range in western North America. Leaves (needles) of ponderosa pines in the Rocky Mountains are bundled into groups of two or three, and cones of these trees are more than 9 centimeters long. In contrast, leaves of ponderosa pines in southern Arizona and northern Mexico are bundled in groups of five, and their cones are less than 9 centimeters long. These differences constitute geographic variation which has developed because reproduction between Rocky Mountain ponderosa pines and Arizona-Mexican ponderosa pines was restricted because of geographic separation. Despite their differences, the two groups belong to the same species because they could produce fertile offspring if their geographic separation were overcome. However, they are members of different populations because they are not currently interbreeding. They are different populations of the same species.

Such geographic variation occurs in many species with broad geographic ranges and is often due to differences in the environmental conditions under which the separate populations exist. The different environments select for different genetic adaptations, resulting in hereditary variation. If such geographic variation occurs gradually over the range of the species, it is *clinal variation*.

Clinal Geographic Variation

In the foregoing example, the geographic variation is too abrupt to be considered clinal. However, ponderosa pines in the Sierra Nevada of California do show clinal geographic variation. The pines at the base of the mountains grow appreciably larger than the pines growing at the highest elevation on the mountains. The change in size is gradual; ponderosa pine trees become progressively smaller as elevation increases.

By taking seeds from trees at several elevations and planting them at the same elevation, scientists showed this size variation to be hereditary. Although all the trees were grown under the same conditions, the largest trees grew from the seeds collected at the base of the mountains, and tree size decreased as the elevation of seed origin increased. The advantages to being small in the relatively harsh environment of the high mountains and tall at the mountain base were important enough to code tree size into the trees' genes. The yarrow (*Achillea lanulosa*) and a number of other plant species show similar clinal variation with elevation in mountains.

In clinal variation, populations are not completely separated from one another, and individuals from adjacent populations do interbreed. However, reproduction between populations is not as common as reproduction between members of the same population. As a result, slight differences between adjacent populations are maintained.

Interestingly, in some clines members of the two extreme populations (the populations at the two ends of the cline) may not be able to interbreed and produce fertile offspring. They are still considered to be members of the same species because they exchange genes through the intermediate populations. The seaside goldenrod (*Solidago sempervirens*) illustrates this. It grows along the Atlantic coast of North America and displays a cline in flowering time. Canadian plants flower in August, plants in the middle Atlantic states flower in September and October, and those in Florida do so in November. These are genetically controlled flowering times, so even if grown together, the plants from Florida and Canada could not interbreed.

However, because Canadian plant flowering times overlap those in the northern United States (which overlap those in the central United States, which overlap those to their south, which overlap those in Florida), there is interbreeding between all adjacent populations and, indirectly, between the Canadian plants and the Florida plants. If the cline were to be subdivided into two or more species, where would the separations be drawn without separating interbreeding organisms into different species? The simplest solution is to consider all members of the cline to be members of the same species.

Local Clinal Variation

Great distances are not always required to establish clines. White clover (*Trifolium repens*), a European native which has been introduced all over the world, affords an example. Some white clover plants release cyanide when parts of the plant are eaten by grazers, such as snails and slugs. Others do not. The cyanide protects the plant from further grazing because it is toxic to the grazers. However, the cyanide-releasing form of clover suffers more frost damage than clover plants that do not release cyanide. Plants protect themselves from the cyanide by sequestering it into cellular compartments. Frost damage occurs when cyanide is released into the plant cells after those compartments are ruptured by ice crystals. Cyanide-storing plants also grow more slowly than forms that do not store cyanide, because some energy that could otherwise be used for growth is required to sequester the cyanide.

Latitudinal and elevational temperature gradients generate clines in the production of cyanide, with greater cyanide production at lower elevations and latitudes. More frequent freezing results in more frequent cyanide damage, and low temperatures result in less grazing, because grazers are not as active. Plants that do not go to the expense of storing cyanide are favored under those conditions. This is a classic geographic cline. However, changes in grazing pressure over much smaller distances also generate clines in cyanide storage. Grazing pressure changes over meters when white clover grows in a garden protected by pesticides and in an adjacent, unprotected field. The result is a cline in cyanide storage by white clover very similar to the geographic clines discussed above but on a scale of meters.

Carl W. Hoagstrom

See also: Adaptations; Coevolution; Evolution of plants; Gene flow; Population genetics; Reproductive isolating mechanisms; Selection; Species and speciation.

Sources for Further Study

Cockburn, Andrew. *An Introduction to Evolutionary Ecology*. Boston: Blackwell Scientific, 1991. Several aspects of clines are covered in the general context of evolutionary ecology.

Endler, John A. *Geographic Variation: Speciation and Clines*. Princeton, N.J.: Princeton University Press, 1977. Old and mathematically involved, but a good discussion of the topic. Understandable without following the math.

Roughgarden, Jonathan. *Theory of Population Genetics and Evolutionary Ecology: An Introduction*. Upper Saddle River, N.J.: Prentice Hall, 1996. Chapter 12 considers several aspects of clines. Many ideas are understandable without following the math.

CLONING OF PLANTS

Categories: Biotechnology; economic botany and plant uses; genetics

Plant cloning is the production of a cell, cell component, or plant that is genetically identical to the unit or individual from which it was derived.

The term "clone" is derived from the Greek word *klon*, meaning a slip or twig. Hence, it is an appropriate choice. Plants have been "cloned" from stem cuttings or whole-plant divisions for many centuries, perhaps dating back as far as the beginnings of agriculture.

Historical Background

In 1838 German scientists Matthias Schleiden and Theodor Schwann presented their *cell theory*, which states, in part, that all life is composed of cells and that all cells arise from preexisting cells. This theory formed the basis for the concept of *totipotency*, which states that since cells must contain all of the genetic information necessary to create an entire, multicellular organism, all of the cells of a multicellular organism retain the potential to recreate, or regenerate, the entire organism. Thus was the basis for plant cell culture research.

The first attempt at culturing isolated plant tissues was by Austrian botanist Gottlieb Haberlandt at the beginning of the twentieth century, but it was unsuccessful. In 1939 Professor R. J. Gautheret and colleagues demonstrated the first successful culture of isolated plant tissues as a continuously dividing callus tissue. The term *callus* is defined as an unorganized mass of dividing cells, such as in a wound response. It was not until 1954, however, that the first whole plant was regenerated, or cloned, from a single adult plant cell by W. H. Muir et al. Thereafter, an increased understanding of plant physiology, especially the role of plant hormones in plant growth and development, contributed to rapid advances in plant cell and tissue culture technologies in the 1970's and 1980's. Many plant species have been successfully cloned from single cells, thus demonstrating and affirming the concept of totipotency.

AP/Wide World Photos

A scientist examines sweet gum plantlets used in an experiment designed to clone and create trees with the best combination of characteristics for use by the forest products industry, which includes fast growth.

Horticulture

By far, the greatest impact of cloning plants *in vitro* (Latin for "in glass," meaning in the laboratory or outside the plant) has been on the horticultural industry. In the 1980's plant tissue culture technologies propagated and produced many millions of plants. Today, many economically important plants are commonly propagated via tissue culture techniques, including vegetable crops (such as the potato), fruit crops (strawberries and dates), floriculture species (orchids, lilies, roses, Boston ferns), and even woody species (pines and grapes).

The advantages of plant cell, tissue, and organ culture technologies include a more rapid production of plants, taking weeks instead of months or years. Much less space is required (square feet instead of field plots). Plants can be produced year-round, and economic, political, and environmental considerations that hamper the propagation of regional or endangered plant species can be reduced. The disadvantages include the high start-up costs for facilities, the skilled labor required, and the need to maintain sterile conditions.

Two other significant considerations must be considered as a result of plant propagation technologies. As illustrated by the Irish Potato Famine of the 1840's, the cultivation of whole fields of genetically identical plants (*monoculture*) leaves the entire crop vulnerable to pest and disease infestations. The second important consideration when generating entire populations of clones, especially using tissue culture technologies, is the potential for introducing genetic abnormalities, which then are present in the entire population of plants produced, a process termed *somaclonal variation*.

Biotechnology

An absolute requirement for genetic engineering of plants is the ability to regenerate an entire plant from a single, genetically transformed cell, thus emphasizing the second major impact of plant cell culture technologies. In 1994 the U.S. Food and Drug Administration (FDA) approved the first genetically modified whole food crop, Calgene's Flavr Savr tomato. This plant was produced using what is termed anti-sense technology. One of the tomato's genes involved in fruit ripening was reversed, thus inactivating it and allowing tomatoes produced from it to have significantly delayed ripening. Although no longer commercially marketed, the Flavr Savr demonstrated the impact of genetic engineering in moving modern agriculture from the Green Revolution into what has been termed the Gene Revolution.

Other examples of agricultural engineering exist today, such as Roundup Ready Soybeans, engineered to resist the herbicide used on weeds where soybeans are grown, and BT Corn, which contains a bacterial gene conveying increased pest resistance. Since 1987, the U.S. Department of Agriculture (USDA) has required field testing of genetically modified crops to demonstrate that their use will not be disruptive to the natural ecosystem. To date, thousands of field trials have been completed or are in progress for genetically modified versions of several crop species, including potatoes, cotton, alfalfa, canola, and cucumbers.

Henry R. Owen

See also: Biotechnology; Cell theory; Endangered species; Genetically modified foods; Green Revolution; Horticulture; Hybridization; Mitosis and meiosis; Monoculture; Plant biotechnology.

Sources for Further Study

Barnum, Susan R. *Biotechnology: An Introduction*. Belmont, Calif.: Wadsworth, 1998. Contains a chapter devoted to plant genetic engineering applications.

Conger, B. V., ed. *Cloning Agricultural Plants via In Vitro Techniques*. Boca Raton, Fla.: CRC Press, 1981. Includes bibliographical references and indexes.

Dodds, John H., and Lorin W. Roberts. *Experiments in Plant Tissue Culture*. 3d ed. New York: Cambridge University Press, 1995. Contains a variety of laboratory exercises, with references.

Kyte, Lydiane, and John Kleyn. *Plants from Test Tubes: An Introduction to Micropropagation*. 3d ed. Portland, Oreg.: Timber Press, 1996. A practical guide to setting up a plant tissue culture laboratory and basic techniques for culturing plants in vitro.

Pierik, R. L. M. *In Vitro Culture of Higher Plants*. Dordrecht, Netherlands: Martinus Nijhoff, 1987. Provides an in-depth historical account of advances in plant tissue culture and descriptions of individual techniques.

COAL

Category: Economic botany and plant uses

Coal is one of the world's most important natural resources based on plant life. Fuel in the form of coal can be any of a variety of combustible sedimentary and metamorphic rocks containing a specified amount of fossilized plant remains.

Coal is a general term encompassing a variety of combustible sedimentary and metamorphic rocks containing altered and fossilized terrestrial plant remains in excess of 50 percent by weight, and more than 70 percent by volume. Categories of coal differ in relative amounts of moisture, volatile matter, fixed carbon, and degree of compaction of the original carbonaceous material. Coal is therefore commonly termed a *fossil fuel*. This key resource is the product of the carbon from ancient plants that have undergone sedimentary and metamorphic transformation over millions of years.

Formation

After dead land-plant matter has accumulated and slowly begun to compact, biochemical decomposition, rising temperature, and rising pressure all contribute to the lengthy process of altering the plant debris into coal. The more common coals are of vascular vegetable origin, formed from the compaction and induration of accumulated remains of plants that once grew in extensive swamp and coastal marsh areas. These deposits are classed as *humic* coals consisting of organic matter that has passed through the *peat*, or earliest coal formation, stage. A variety of humic coals are known.

The swamp water environment within which humic coals form must be deficient in dissolved oxygen, the presence of which would ordinarily cause decay of the plant tissue. Under such near-stagnant conditions plant remains are preserved, while the presence of hydrogen sulfide inhibits organisms that feed on dead vegetation. Analog environments under which coal is currently forming are found within the Atchafalaya swamp of coastal Louisiana and the many peat-producing regions of Ireland. A layer of peat in excess of 2 meters in thickness and covering more than 5,000 square kilometers is present in the Great Dismal swamp of coastal North Carolina and Virginia.

The *sapropelic* class of coal, relatively uncommon in distribution and composed of fossil algae and spores, is formed through partial decomposition of organic matter by organisms within oxygen-deficient lakes and ponds. Sapropelic coals are subdivided into boghead (algae origin) and cannel (spore origin) deposits.

The vegetable origin of coal has been accepted since 1825 and is convincingly evidenced by the identification of more than three thousand freshwater plant species in coal beds of Carboniferous (360 million to 286 million years ago) age. The common association of root structures and even upright stumps with layers of coal indicate that the parent plant material grew and accumulated in place.

Detailed geologic studies of rock sequences that lie immediately above and below coal deposits indicate that most coals were formed in coastal regions affected by long-term sea level cycles characterized by transgressing (advancing) and regressing (retreating) shorelines. Such a sequence of rock deposited during a single advance and retreat of the shoreline, termed a cyclothem, typically contains nonmarine strata separated from overlying marine strata by a single layer of coal. In sections of the Interior coal province, a minimum of fifty cyclothems have been recognized, some of which can be traced across thousands of square kilometers. Such repetition in a rock sequence is most advantageous to the economics of a coal region, creating a situation in which a vertical mine shaft could penetrate scores of layers of coal.

The formation of coal is a long-term geologic process. Coal cannot therefore be considered a renewable resource, even though it is formed from plant matter. Studies have suggested that 1 meter of low-rank coal requires approximately ten thousand years of plant growth, accumulation, biologic reduction, and compaction to develop. Using these time lines, the 3-meter-thick Pittsburgh coal bed, underlying 39,000 square kilometers of Pennsylvania, developed over a period of thirty thousand years, while the 26-meter-thick bed of coal found at Adaville, Wyoming, required approximately a quarter of a million years to develop.

Coal formation favors sites where plant growth is abundant and conditions for organic preservation are favorable. Such climates range from subtropical to cold, with the ideal being classed as temperate. Tropical swamps produce an abundance of plant matter but have very high bacterial activity, resulting in low production of peat. Modern peats are developing in temperate to cold climate regions, such as Canada and Ireland, where abundant precipitation ensures fast plant growth, while relatively low temperatures diminish the effectiveness of decay-promoting bacteria.

The first coal provinces began to form with the evolution of cellulose-rich land plants. One of the earliest known coal deposits, of Upper Devonian age (approximately 365 million years ago), is found on Buren Island, Norway. Between the Devonian period and today, every geologic period is represented by at least some coal somewhere in the world. Certain periods of time, however, are significant coal-forming ages.

During the Carboniferous and Permian periods (360 to 245 million years ago) widespread development of fern and scale tree growth set the stage for the formation of the Appalachian coal province and the coal districts of Great Britain, Russia, and Manchuria. Coal volumes formed during these periods of geologic time constitute approximately 65 percent of present world reserves. The remaining reserves, developed mainly over the past 200 million years, formed in swamps consisting of *angiosperms* (flowering plants). The reserves of the Rocky Mountain province and those of central Europe are representative of these younger coals.

Classification of Coal

With the advent of the Industrial Revolution there was a need for a system of classification defining in detail the various types of coals. Up to the beginning of the nineteenth century, coal was divided into three rudimentary classes, determined by appearance: bright coal, black coal, and brown coal. Through the decades other schemes involving various parameters were introduced.

In 1937 a classification of coal rank using fixed carbon and Btu content was adopted by the American Standards Association. Adaptations of this scheme are still in use, listing the steps of progressive increase in coal rank as lignite (brown coal), subbituminous, bituminous (soft coal), subanthracite, and anthracite (hard coal). Some classification schemes also list peat as the lowest rank of coal. Technically speaking, peat is not a coal; rather, it is a fuel and a precursor to coal.

Coalification is the geologic process whereby plant material is altered into differing ranks of coal by geochemical and diagenetic change. With an increase in rank, chemical changes involve an increase in carbon content accompanied by a decrease in hydrogen and oxygen. Correspondingly, *diagenesis* involves an increase in density and calorific value and a progressive decrease in moisture. At all ranks, impurities include sulfur, silt and clay particles, and silica.

Peat, an unconsolidated accumulation of partly decomposed plant material, has an approximate carbon content of 20 percent. In many classification schemes, peat is listed as the initial stage of coal formation. Moisture content is quite high, at least at the 75-percent level. When dry, peat has an oxygen content of about 30 percent, is flammable, and will freely but inefficiently burn slowly and steadily for months at a low heat-content value of 5,400 Btu's per pound. (The British thermal unit, or Btu, is the

AP/Wide World Photos

A West Virginia truck driver stands on top of approximately 35,000 pounds of coal in the bed of his truck. The coal formed from ancient plants that have undergone sedimentary and metamorphic transformation over millions of years.

quantity of heat required to raise the temperature of one pound of water one degree Fahrenheit.)

Types of Coal

Lignite, or brown coal, is brownish-black in color, banded and jointed, and subject to spontaneous combustion. Carbon content ranges from 25 to 35 percent. With a moisture content around 40 percent, it will readily disintegrate after drying in the open air. Because lignite has a maximum calorific value of 8,300 Btu's, it is classed as a low heating-value coal.

Deeper burial with even higher temperatures and pressures gradually alters lignite to *bituminous coal*, a dense, dusty, brittle, well-jointed, dark brown to black fuel that burns readily with a smoky yellow flame. It is the most abundant form of coal in the United States. Calorific value ranges from 10,500 to 15,500 Btu's per pound, and carbon content varies from 45 to 86 percent. Moisture content is as low as 5 percent, but heating value is high.

The *subbituminous* class of coal is intermediate between lignite and bituminous and has characteristics of both. Little woody matter is visible. It splits parallel to bedding but generally lacks the jointing of bituminous coal. It burns clean but with a relatively low heating value.

Anthracite is jet-black in color, has a high luster, is very hard and dust-free, and breaks with a conchoidal fracture. Carbon content ranges from 86 to 98 percent. It is slow to ignite, burns with a short blue flame without smoke, and, with a calorific value in excess of 14,000 Btu's per pound, is a high heating fuel. U.S. reserves are found mainly in eleven northeastern counties in Pennsylvania. Subanthracite coal has characteristics intermediate between bituminous and anthracite.

Bedded and compacted coal layers are geologically considered to be rocks. Lignite and bituminous ranks are classed as organic sedimentary rocks. Anthracite, formed when bituminous beds of coal are subjected to the folding and regional deformation affiliated with mountain building processes, is listed as a metamorphic rock. Because peat is not consolidated or compacted, it is classed as an organic sediment. *Graphite*, a naturally occurring crystalline form of almost pure carbon, is occasionally associated with anthracite. While it can occur as the result of high-temperature alteration of anthracite, its chemical purity and common association with crystalline rock causes it to be listed as a mineral.

Worldwide Distribution

While coal has been found all over the world, principal mining activity and approximately 95 percent of world reserves lie in the Northern Hemisphere—in Asia, Europe, and North America.

It is estimated that total world coal resources, defined as coal reserves plus other deposits that are not economically recoverable plus inferred future discoveries, are on the order of 10 trillion tons. Of this amount, estimates of world coal reserves, defined as those deposits that have been measured, evaluated, and can be extracted profitably under existing technology and economic conditions, range up to a high of approximately one trillion tons. World reserves can be divided into two categories, with 73 percent composed of anthracite and bituminous coals and 27 percent composed of lignite. Among nations, the United States possesses the greatest amount (approximately 500 billion tons) of total world reserves.

Modern Use

Historically, coal has been industry's fuel of choice. Countries with large coal reserves have risen commercially, while those less endowed with this resource—or lacking it altogether—have turned to agriculture or stagnated in development. Different ranks of coal are employed for different purposes. In the middle of the twentieth century it was common to see separate listings of coking, gas, steam, fuel, and domestic coals. Each had its specific uses. Coal for home use could not yield excessive smoke, while coal for locomotives had to raise steam quickly and not produce too high an ash content. Immediately after World War II fuel coal use, representing 78 percent of annual production, was divided into steam raising, railway transportation, domestic consumption, electric generation, and bunker coal. The remaining 22 percent was employed in the production of pig iron, steel, and gas.

Fifty years later, more than 80 percent of the approximately 1 billion tons of coal produced annually in the United States was used to generate electricity. Industrial consumption of coal, particularly in the production of coke for the steel and iron manufacturing industry, is the second most important use. Other industrial uses of coal are food processing and the manufacture of paper, glass, cement, and stone. Coal produces more energy than any other known fuel, including natural gas, crude oil, nuclear, and renewable fuels.

While expensive to produce, the conversion of intermediate ranks of coal into liquid (coal oil) and gaseous (coal gas) forms of hydrocarbon fuels will become more economically viable, especially during times of increase in the value of crude oil and natural gas reserves. New uses of coal are constantly being explored and tested. Two promising techniques are the mixing of water with powdered coal to make a slurry which can be burned as a liquid fuel and the underground extraction of coal-bed methane (firedamp). Interest in the latter by-product as an accessible and clean-burning fuel is especially high in Appalachian province localities distant from conventional gas resources.

Coal Mining

Coal has been produced by two common methods: underground (deep mining) and surface (strip mining). Underground mining requires digging extensive systems of tunnels and passages within and along the coal layers. These openings are connected to the surface so the coal can be removed. Prior to the development of the gigantic machinery used in open-pit mining, deep mining was the industry norm. This early period was characterized by labor-intensive pick and shovel work in cramped mine passages. Constant dangers to miners included the collapse of ceilings, methane gas explosions, and pneumoconiosis, known as black lung disease.

Today augers and drilling machinery supplement human labor to a large extent. Mine safety and health regulations have greatly reduced a once-high annual death toll. The common method of underground extraction involves initial removal of about 50 percent of the coal, leaving a series of pillars to support the mine roof. As reserves are exhausted, the mine is gradually abandoned after removal of the pillars. Another modern underground mining technique, with a coal removal rate approaching 100 percent, involves the use of an integrated rotary cutting machine and conveyer belt.

Surface mining of coal, accounting for about 61 percent of U.S. production, is a multiple-step process. First the overburden material is removed, allowing exposure of the coal. Coal is then mined using surface machinery ranging from bulldozers to gigantic power shovels. Finally, after removal of all the coal, the overburden is used to fill in the excavated trench, and the area is restored to its natural topography and vegetation. Economics usually determine whether underground or open-pit techniques are preferable in a given situation. Generally, if the ratio of overburden to coal thickness does not exceed twenty to one, surface mining is more profitable.

With increased concern for the environment, and with federal passage of the Coal Mine Health and Safety Act (1969) and the Clean Air Act (1970), the mining of coal in the United States has undergone both geographic and extraction-technology changes. Because the Rocky Mountain province coals, while lower grade than eastern coals, contain lower percentages of sulfur, the center of U.S. production has gradually shifted westward. The burning of high-sulfur coals releases sulfur dioxide into the atmosphere; it is a significant contributor to acid rain.

Western coals are often contained within layers thicker than those found in the East, are shallow in depth, and can be found under large areas—all conditions amenable to surface mining. As a result, the state of Wyoming, with a 1995 production of 265 million tons of low-sulfur coal, became the leading U.S. coal producer. The 1979 version of the U.S. Department of Energy's annual report on the database of minable coal reserves, titled "Demonstrated Reserve Base of Coal in the United States on January 1, 1979," listed reserves at 475 billion tons. At current levels of production it is estimated that the United States has more than 250 years of future production.

Albert B. Dickas

See also: Fossil plants; Peat.

Sources for Further Study

Cobb, James C., and C. Blaine Cecil, eds. *Modern and Ancient Coal-Forming Environments*. Boulder, Colo.: Geological Society of America, 1993. A monograph of nearly 200 pages. Includes bibliographical references, index.

Papp, Alexander, James C. Hower, and Douglas C. Peters, eds. *Atlas of Coal Geology*. Tulsa, Okla.: Association of American Petroleum Geologists, 1998. CD-ROM format requires PC, Pentium, or MMX processor, Windows 95, 16 megabytes of RAM, video color display card. Bibliographical references and index are featured along with the color maps.

Roberts, Laura N. Robinson, and Mark A. Kirschbaum. *Paleogeography of the Late Cretaceous of the Western Interior of Middle North America*. Washington, D.C.: Government Printing Office, 1995. Prepared under the auspices of the U.S. Geological Survey, this resource includes maps and color illustrations showing North American deposits that formed coal. Bibliography, index.

Thomas, Larry. *Handbook of Practical Coal Geology*. New York: Wiley, 1992. The origins, chemistry, and physical properties of coal are discussed. Bibliography, index.

Walker, Simon. *Major Coalfields of the World*. London: IEA Coal Research, 1993. Illustrations, maps, bibliography.

COEVOLUTION

Categories: Ecology; ecosystems; evolution

Coevolution is the interactive evolution of two or more species that results in a mutualistic or antagonistic relationship.

When two or more different species evolve in a way that affects one another's evolution, coevolution is taking place. This interactive type of evolution is characterized by the fact that the participant life-forms are acting as a strong selective pressure upon one another over a period of time. The coevolution of plants and animals, whether animals are considered strictly in their plant-eating role or also as pollinators, is abundantly represented in every terrestrial ecosystem throughout the world where flora has established itself. Moreover, the overall history of some of the multitude of present and past plant and animal relationships is displayed (although fragmentally) in the fossil record found in the earth's crust.

Beginnings

The most common coevolutionary relationships between plants and animals surround plants as a food source. Microscopic, unicellular plants were the earth's first *autotrophs* (organisms that can produce their own organic energy through photosynthesis, that is, from basic chemical ingredients derived from the environment). In conjunction with the appearance of autotrophs, microscopic, unicellular *heterotrophs* (organisms, such as animals, that must derive food from other sources, such as autotrophs) evolved to exploit the autotrophs.

Sometime during the later part of the Mesozoic era, angiosperms, the flowering plants, evolved and replaced most of the previously dominant land plants, such as the gymnosperms and the ferns. New species of herbivores evolved to exploit these new food sources. At some point, probably during the Cretaceous period of the late Mesozoic era, animals became unintentional aids in the angiosperm pollination process. As this coevolution proceeded, the first animal pollinators became more and more indispensable as partners to the plants.

Eventually, highly coevolved plants and animals developed relationships of extreme interdependence, exemplified by the honeybees and their coevolved flowers. This angiosperm-insect relationship is thought to have arisen in the Mesozoic era by way of beetle predation, possibly on early, magnolia-like angiosperms. The fossil record gives some support to this theory. Whatever the exact route along which plant-animal pollination partnerships coevolved, the end result was a number of plant and animal species that gained mutual benefit from the new type of relationship.

Coevolutionary Relationships

Coevolved relationships include an immense number of relationships between plants and animals, and even between plants and other plants. Among these coevolved situations can be found *commensalisms*, in which different species have coevolved to live intimately with one another without injury to any participant, and *symbioses*, in which species have coevolved to literally "live together."

Such intertwined relationships can take the form

of *mutualism*, in which neither partner is harmed and indeed one or both benefit—as in the relationships between fungi and algae in lichens, fungi and roots in mycorrhizae, and ants and acacia trees in a symbiotic mutualism in which the ants protect the acacias from herbivores. In *parasitism*, one partner benefits at the expense of the other; a classic example is the relationship between the mistletoe parasite and the oak tree. Another coevolutionary relationship, *predation*, is restricted primarily to animal-animal relationships (vertebrate carnivores eating other animals, most obviously), although some plants, such as Venus's flytrap, mimic predation in having evolved means of trapping and ingesting insects as a source of food. Some highly evolved fungi, such as the oyster mushroom, have evolved anesthetizing compounds and other means of trapping protozoa, nematodes, and other small animals.

One of the most obvious and complex coevolutionary relationships are the mutualisms that have evolved between plants bearing fleshy fruits and vertebrate animals, which serve to disperse the seeds in these fruits. Over time, plants that produce these fruits have benefited from natural selection because their seeds have enjoyed a high degree of survival and germination: Animals eat the fruits, whose seeds are passed through their digestive system (or regurgitated to feed offspring) unharmed; at times the seeds are even encouraged toward germination as digestion helps break down the seed coat. Furthermore, dispersal through the animals' mobility allows the seeds to enjoy more widely distributed propagation. The coevolutionary process works on the animals as well: Birds and animals that eat the fruits enjoy a higher degree of survival, and so natural selection favors both fleshy-fruit-producing plants and fleshy-fruit-eating animals. Similar selection has favored the coevolution of flowers with colors and smells that attract pollinators such as bees.

Eventually some plant-animal mutualisms became so intertwined that one or both participants reached a point at which they could not exist without the aid of the other. These obligatory mutualisms ultimately involve other types of animal partners besides insects. Vertebrate partners such as birds, reptiles, and mammals became involved in mutualisms with plants. In the southwestern United States, for example, bats and the agave and saguaro cactus have a special coevolutionary relationship: The bats, nectar drinkers and pollen eaters, have evolved specialized feeding structures such as erectile tongues similar to those found among moths and other insects with similar lifestyles. In turn, angiosperms coevolutionarily involved with bats have developed such specializations as bat-attractive scents, flower structures that match the bats' feeding habits and minimize the chance of injuring the animals, and petal openings timed to the nocturnal activity of bats.

Defense Mechanisms

Coevolution is manifested in defense mechanisms as well as attractants: Botanical structures and chemicals (*secondary metabolites*) have evolved to discourage or to prevent the attention of plant

PhotoDisc

As coevolution proceeded over time, animal pollinators and highly coevolved plants developed relationships of extreme interdependence, exemplified by the honeybees and their coevolved flowers.

eaters. These include the development of spines, barbs, thorns, bristles, and hooks on plant leaves, stems, and trunk surfaces. Cacti, hollies, and rose bushes illustrate this form of plant strategy. Some plants produce chemical compounds that are bitter to the taste or poisonous. Plants that contain organic tannins, such as trees and shrubs, can partially inactivate animals' digestive juices and create cumulative toxic effects that have been correlated with cancer. Grasses with a high silica content act to wear down the teeth of plant eaters. Animals have counteradapted to these defensive innovations by evolving a higher degree of resistance to plant toxins or by developing more efficient and tougher teeth with features such as harder enamel surfaces or the capacity of grinding with batteries of teeth.

Frederick M. Surowiec,
updated by Christina J. Moose

Ants and Acacias

One noted example of mutualism can be found in Mexico and Central America, home to the bull's-horn acacia (*Acacia cornigera*). In its native habitat, colonies of stinging ants (*Pseudomyrmex ferruginea*) live inside the acacia's hollow thorns, found at the base of each leaf. There, the ants enjoy plentiful food, such as the tree's carbohydrate-rich nectar and protein-lipid Beltian bodies, nutritive structures located at the tip of each leaflet. The only known function of the Beltian bodies is to provide food for the symbiotic ants, which make their nests only in this species of tree and are utterly dependent on its food sources.

The acacia benefits from this arrangement as well. When an animal of any size brushes against the tree, worker ants swarm out and sting the animal, thus protecting the plant from herbivory. When epiphytic vines or the branches of another plant touch an inhabited acacia tree, the ants cut away the other plant's bark. By doing so, they destroy the invasive plant branch and allow more sunlight to reach the acacia—a crucial factor for plants in densely growing areas of the tropics. Scientists have found that acacias inhabited by ants grow more rapidly than uninhabited acacias, making ants and acacias mutual beneficiaries.

See also: Animal-plant interactions; Biochemical coevolution in angiosperms; Community-ecosystem interactions; Evolution: convergent and divergent; Lichens; Metabolites: primary vs. secondary; Mycorrhizae.

Sources for Further Study

Bakker, Robert T. *The Dinosaur Heresies*. Reprint. Secaucus, N.J.: Citadel, 2001. Although the emphasis of the book is on dinosaur evolution, chapter 9, "When Dinosaurs Invented Flowers," discusses coevolved morphologies and strategies among angiosperm plants and extinct reptiles. A wonderful, thought-provoking book for all reading audiences. Illustrated.

Barth, Friedrich G. *Insects and Flowers: The Biology of a Partnership*. Translated by M. A. Biederman-Thorson. Princeton, N.J.: Princeton University Press, 1991. A delightful study of the coevolution of insects and flowers, highlighting both the complex sensory world of insects and the sometimes bizarre ways in which angiosperms have evolved to take advantage of insect senses in order to assure their own pollination.

Gilbert, Lawrence E., and Peter H. Raven, eds. *Coevolution of Animals and Plants*. Austin: University of Texas Press, 1980. A collection of scientific papers on the subject of evolved plant and animal relationships. The majority of the papers involve analyses of plant and animal biochemical relationships and mechanisms or ecological processes and patterns. Written for professionals in the field, though understandable by the general reader.

Gould, Stephen Jay. *The Panda's Thumb*. Reprint. New York: W. W. Norton, 1992. Chapter 27, "Nature's Odd Couples," focuses on a conspicuous example of a possible coevolution—an inferred mutualistic relationship between the extinct dodo bird and a large tree (*Calvaria major*) that is found on the island of Mauritius. Gould presents the evidence objectively, both for and against the dodo's possible role in facilitating the tree's germination.

Grant, Susan. *Beauty and the Beast: The Coevolution of Plants and Animals*. New York: Charles Scribner's Sons, 1984. This is one of the best books devoted solely to the subject of coevolution written for a general audience. Each of the ten chapters is devoted to a major subtopic within the phenomenon of coevolution, such as specific animal and plant symbioses or antagonisms.

Hughes, Norman F. *Paleobiology of Angiosperm Origins*. Reprint. New York: Cambridge University Press, 1994. The chapter on Cretaceous land fauna sketches out much of what is known about the earliest coevolved relationships between animals and the flowering plants (angiosperms) during the late Mesozoic period. Appropriate for readers who have some familiarity with botany, entomology, and evolutionary theory.

Thompson, John N. *The Coevolutionary Process*. Chicago: University of Chicago Press, 1994. Thompson advances a new conceptual approach to the evolution of species interactions, the geographic mosaic theory of coevolution. Thompson demonstrates how an integrated study of life histories, genetics, and the geographic structure of populations yields a broader understanding of coevolution.

COMMUNITY-ECOSYSTEM INTERACTIONS

Categories: Animal-plant interactions; ecology; ecosystems

Ecosystems are complex organizations of living and nonliving components. They are frequently named for their dominant biotic or physical features (such as marine kelp beds or coniferous forests). Communities are groups of species usually classified according to their most prominent members (such as grassland communities or shrub communities). The interactions between species and their ecosystems have lasting impacts on both.

In an ecological sense, a community consists of all populations residing in a particular area. Examples of communities range in scale from all the trees in a given watershed, all soil microbes on an agricultural plot, or all phytoplankton in a particular harbor to all plants, animals, and microbes in vast areas, such as the Amazon basin or the Chesapeake Bay.

An ecosystem consists of the community of species as well as the environment of a given site. A forest ecosystem would include all living things along with climate, soils, disturbance, and other abiotic factors. An estuarine ecosystem, likewise, would include all the living things present, in addition to climate, currents, salinity, nutrients, and more.

Interactions between species in communities and ecosystems range from mutually beneficial to mutually harmful. One such category of interaction is *mutualism*, which usually involves two species. Both species derive benefit from a mutualism. *Commensalism* is used to describe a situation in which one species benefits without harming the other. If the two species are neither helped nor harmed, a *neutralism* is said to occur, and an *amensalism* happens when one species is harmed while the other remains. During *competition*, both species involved are negatively affected. A number of terms are used to describe a relationship in which one species benefits at another's expense, including herbivory, predation, parasitism, and pathogenicity. The choice of term more often than not depends on the relative sizes of the species involved.

Competition

Plants typically compete for resources, such as light, space, nutrients, or water. One way an individual may outcompete its neighbors is to outgrow them, thus capturing more sunlight for itself (and thus producing more sugars and other organic molecules for itself). Another way is to be more fecund than the neighbors, flooding the surroundings with one's progeny and thereby being more likely to occupy favorable sites for reproduction. For example, in closed forests treefall gaps are quickly filled with growth from the canopy, thus shading the ground and making it more difficult for competing seedlings and saplings to survive.

Plants compete in the root zone as well, as plants with a more extensive root network can acquire more of the water and other inorganic nutrients necessary for growth and reproduction than can their competitors. In semiarid areas, for example, trees often have trouble colonizing grasslands because the extensive root systems of grasses are much more effective in capturing available rainwater.

Sometimes plants resort to chemical "warfare," known as *allelopathy*, in order to inhibit the growth of competitors in the surrounding area. The existence of allelopathy remains a controversial topic, and simpler explanations have been offered for many previously alleged instances of the phenomenon. Allelopathy cannot be rejected outright; however, the controversy most likely proves only that many aspects of nature cannot be pigeonholed into narrow explanations.

Competition involves a cost in resources devoted to outgrowing or outreproducing the neighbors. Because of the cost, closely related, competing species will diverge in their ecological requirements. This principle is known as *competitive exclusion*.

Mutualism, Commensalism, and Parasitism

Many flowering plants could not exist without one of the most important mutualisms of all: *pollination*. In concept, pollination is simple: In exchange for carrying out the physical work of exchanging genetic material (in pollen form) between individual plants (thus enabling sexual reproduction), the carrier is rewarded with nutrients in the form of nectar or other materials. Many types of animals are involved in pollination: insects such as bees, flies, and beetles; birds, particularly the hummingbirds; and mammals such as bats.

Another highly important mutualism is that between plant roots and fungal hyphae, or *mycorrhizae*. Mycorrhizae protect plant roots from pathogenic fungi and bacteria; their most important role, however, is to enhance water and nutrient uptake by the plant. In fact, regeneration of some plants is impossible in the absence of appropriate mycorrhizae. Mycorrhizae benefit, in turn, by receiving nutrients and other materials synthesized by the host plant. There are two types of mycorrhizae: *ectomycorrhizae*, whose hyphae may fill the space between plant roots but do not penetrate the roots themselves; and *vesicular-arbuscular mycorrhizae*, whose hyphae penetrate and develop within root cells.

Few people can envision a swamp in the southeastern United States without thinking of bald cypress trees (*Taxodium distichum*) draped in ethereal nets of Spanish moss (*Tillandsia usneoides*), which is actually not a moss but a flowering plant in the monocot family *Bromeliaceae*. *Tillandsia* is an *epiphyte*, a plant that grows on the stems and branches of a tree. Epiphytism is one of the most common examples of a *commensalism*, in which one organism, for instance the epiphyte, benefits without any demonstrable cost to the other, in this case the host tree. Epiphytes are common in tropical rain forests and include orchids, bromeliads, cacti, and ferns. In temperate regions more primitive plants, such as lichens, are more likely to become epiphytes.

Not all epiphytes are commensal, however. In the tropics, strangler figs, such as *Ficus* or *Clusia*, begin life as epiphytes but send down roots that in time completely encircle and kill the host. Mistletoes, such as *Phoradendron* or *Arceuthobium*, may draw off the photosynthetic production of the host, thus severely depleting its resources.

Herbivory and Pathogenicity

Plants, because of their ability to harvest light energy from the sun to produce the organic nutrients and building blocks necessary for life, are the primary *producers* of most of the earth's ecosystems. Thus, they face an onslaught of macroscopic and microscopic *consumers*. If macroscopic, the consumers are generally regarded as *herbivores* (plant-eating animals); if microscopic, they are *pathogens*. Either way, herbivores and pathogens generally devour the tissues of the host.

Plants have evolved a number of defense mechanisms in response to pressure from herbivores and pathogens. Some responses may be mechanical. For example, trees on an African savanna may evolve greater height to escape grazing pressures from large herbivores, but some large herbivores, specifically giraffes, may evolve to grow to greater heights as well. Plants may encase themselves in nearly indestructible outer coatings or arm themselves with spines in order to discourage grazers.

Other responses may be chemical. Cellulose, one of the important chemical components of plant tissues such as wood, is virtually indigestible—unless the herbivore itself hosts a bacterial symbiont in its stomach that can manage the job of breaking down cellulose. Other chemicals, such as phenols and tannins—the class of compounds that gives tea its brown color—are likewise indigestible, thus discouraging feeding by insects. Plants produce a wide range of toxins, such as alkaloids, which poison or kill herbivores. A number of hallucinogenic drugs are made from plant alkaloids.

Phytoalexins are another group of defensive compounds produced by plants in response to bacterial and fungal pathogens. Substances present in the cell walls of bacteria and fungi are released via the action of plant enzymes and spread throughout the plant. The bacterial and fungal substances function as hormones to stimulate, or elicit, phytoalexin production. Hence, these substances are referred to as *elicitors*. The phytoalexins act as antibiotics, killing the infective agents. Tannins, phenols, and other compounds also serve to defend against pathogen attack.

David M. Lawrence

See also: Allelopathy; Animal-plant interactions; Biomes: definitions and determinants; Biomes: types; Biosphere concept; Community structure and stability; Coevolution; Competition; Ecosystems: overview; Ecosystems: studies; Evolution: convergent and divergent; Metabolites: primary vs. secondary; Mycorrhizae; Nutrient cycling; Pollination; Selection; Succession.

Sources for Further Study

Barbour, Michael G., et al. *Terrestrial Plant Ecology*. 3d ed. Menlo Park, Calif.: Benjamin Cummings, 1999. An overview of terrestrial plant ecology.

Barnes, Burton V., Donald R. Zak, Shirley R. Denton, and Stephen H. Spurr. *Forest Ecology*. 4th ed. New York: John Wiley & Sons, 1998. Discusses many of principles of communities and ecosystems with respect to forests.

Mitch, William J., and James G. Gosselink. *Wetlands*. 3d ed. New York: John Wiley & Sons, 2000. Discusses interactions of species and surroundings with respect to wetland ecosystems.

Morin, Peter J. *Community Ecology*. Oxford, England: Blackwell Science, 1999. A good overview of the topic.

COMMUNITY STRUCTURE AND STABILITY

Categories: Ecology; ecosystems

An ecological community is the assemblage of species found in a given time and place. The species composition of different ecosystems and the ways in which they maintain equilibrium and react to disturbances are manifestations of the community's stability.

The populations that form a community interact through the processes of *competition*, *predation*, *parasitism*, and *mutualism*. The structures of communities are determined, in part, by the nature and strength of these *biotic* factors. *Abiotic* factors (physical factors such as temperature, rainfall, and soil fertility) are the other set of important influences determining community structure. An ecological community together with its physical environment is called an *ecosystem*. No ecosystem can be properly understood without a careful study of the biotic and abiotic factors that shape it.

Energy Flow

The most common way to characterize a community functionally is by describing the flow of energy through it. Based on the dynamics of *energy flow*, organisms can be classified into three groups: those that obtain energy through *photosynthesis* (called *producers*), those that obtain energy by consuming other organisms (*consumers*), and those that decompose dead organisms (*decomposers*). The pathway through which energy travels from producer through one or more consumers and finally to decomposer is called a *food chain*. Each link in a food chain is called a *trophic level*. Interconnected food chains in a community constitute a *food web*.

Very few communities are so simple that they can be readily described by a food web. Most communities are compartmentalized: A given set of producers tends to be consumed by a limited number of consumers, which in turn are preyed upon by a smaller number of *predators*, and so on. Alternatively, consumers may obtain energy by specializing on one part of their prey (for example, some birds may eat only seeds of plants) but utilize a wide range of prey species. *Compartmentalization* is an important feature of community structure; it influences the formation, organization, and persistence of a community.

Dominant and Keystone Species

Some species, called *dominant species*, can exert powerful control over the abundance of other species because of the dominant species' large size, extended life span, or ability to monopolize energy or other resources. Communities are named according to their dominant species: for example, oak-hickory forest, redwood forest, sagebrush desert, and tall-grass prairie. Some species, called *keystone species*, have a disproportionately large effect on community structure. These interact with other members of the community in such a way that loss of the keystone species can lead to the loss of many other species. Keystone species may also be the dominate species, but they may also appear insignificant to the community until they are gone. For example, cordgrass (*Spartina*) is the dominant plant in many tidal estuaries, and it is also a keystone species because so many members of the community depend on it for food and shelter.

The species that make up a community are seldom distributed uniformly across the landscape; rather, some degree of patchiness is characteristic of virtually all species. There has been conflicting evidence as to the nature of this patchiness. Moving across an environmental gradient (for example, from wet to dry conditions or from low to high elevations), there is a corresponding change in species and community composition. Some studies have suggested that changes in species composition usually occur along relatively sharp boundaries and that these boundaries mark the borders between adjacent communities. Other studies have indicated that species tend to respond individually to environmental gradients and that community boundaries are not sharply defined; rather, most communities broadly intergrade into one another, forming what is often called an *ecotone*.

Degrees of Species Interaction

These conflicting results have fueled a continuing debate as to the underlying nature of communities. Some communities seem to behave in a coordinated manner. For example, if a prairie is consumed by fire, it regenerates in a predictable sequence, ultimately returning to the same structure and composition it had before the fire. This process, called *ecological succession*, is to be expected if the species in a community have evolved together with one another. In this case, the community is behaving like an organism, maintaining its structure and function in the face of environmental disturbances and fluctuations (as long as the disturbances and fluctuations are not too extreme). The existence of relatively sharp boundaries between adjacent communities supports this explanation of the nature of the community.

In other communities, it appears that the response to environmental fluctuation or disturbance is determined by the evolved adaptations of the species available. There is no coordinated community response but rather a coincidental assembly of community structure over time. Some sets of species interact together so strongly that they enter a community together, but there is no evidence of an evolved community tendency to resist or accommodate environmental change. In this case, the community is formed primarily of species that happen to share similar environmental requirements.

Competition and Predation

Disagreement as to the underlying nature of communities usually reflects disagreement about the relative importance of the underlying mechanisms that determine community structure. Interspecific competition has long been invoked as the primary agent structuring communities. Competition is certainly important in some communities, but there is insufficient evidence to indicate how widespread and important it is in determining community structure. Much of the difficulty occurs because ecologists must infer the existence of past competition from present patterns in communities. It appears that competition has been important in many vertebrate communities and in communities dominated by sessile organisms, such as plants. It does not appear to have been important in structuring communities of plant-eating insects. Furthermore, the effects of competition typically affect individuals that use identical resources, so that only a small percentage of species in a community may be experiencing significant competition at any time.

The effects of predation on community structure depend on the nature of the predation. *Keystone predators* usually exert their influence by preying on species that are competitively dominant, thus giving less competitive species a chance. Predators that do not specialize on one or a few species may also have a major effect on community structure, if they attack prey in proportion to their abundance. This frequency-dependent predation prevents any prey species from achieving dominance. If a predator is too efficient, it can drive its prey to extinction, which may cause a selective predator to become extinct as well. Predation appears to be most important in determining community structure in environments that are predictable or unchanging.

Disasters and Catastrophes

Chance events can also influence the structure of a community. No environment is completely uniform. Seasonal or longer-term environmental fluctuations affect community structure by limiting opportunities for colonization, by causing direct mortality, or by hindering or exacerbating the effects of competition and predation. Furthermore, all communities experience at least occasional disturbance: unpredictable, seemingly random environmental changes that may be quite severe.

It is useful in this regard to distinguish between regular disturbances and rarer, less frequent catastrophic events. For example, fire occurs so often in tall-grass prairies that most of the plant species have become fire-adapted—they have become efficient at acquiring nutrients left in the ash and at sprouting or germinating quickly after a fire. In contrast, the 1980 eruption of Mount St. Helens, a volcanic peak in Washington State, was so violent and so unexpected that no members of the nearby community were adequately adapted to such conditions.

Natural disturbances occur at a variety of scales. Small-scale disturbances may simply create small openings in a community. In forests, for example, wind, lightning, and fungi cause single mature trees to die and fall, creating gaps that are typically colonized by species requiring such openings. Large disturbances are qualitatively different from small disturbances in that large portions of a community may be destroyed, including some of the ability to recover from the disturbance. For example, following a large, intense forest fire, some tree species may not return for decades or centuries because their seeds were consumed by the fire, and colonizers must travel a long distance.

Early ecologists almost always saw disturbances as destructive and disruptive for communities. Under this assumption, most mathematical models portrayed communities as generally being in some stable state; if a disturbance occurred, the community inevitably returned to the same (or some alternative) equilibrium. It later became clear, however, that natural disturbance is a part of almost all natural communities. Ecologists now recognize that few communities exhibit an equilibrium; instead, communities are dynamic, always responding to the last disturbance.

Long-Term Community Dynamics

The evidence suggests that three conclusions can be drawn about the long-term dynamics of communities. First, it can no longer be assumed that all communities remain at equilibrium until changed by outside forces. Disturbances are so common, at so many different scales and frequencies, that the community must be viewed as an entity that is constantly changing as its constituent species readjust to disturbance and to one another.

Second, communities respond in different ways to disturbance. A community may exhibit *resistance*, not markedly changing when disturbance occurs, until it reaches a threshold and suddenly and rapidly shifts to a new state. Alternatively, a community may exhibit *resilience* by quickly returning to its former state after a disturbance. Resilience may occur over a wide range of conditions and scales of disturbance in a dynamically robust system. On the other hand, a community that exhibits resilience only within a narrow range of conditions is said to be dynamically fragile.

Finally, there is no simple way to predict the stability of a community. At the end of the 1970's, many ecologists predicted that complex communities would be more stable than simple communities. It appeared that stability was conferred by more intricate food webs, greater structural complexity, and greater species richness. On the basis of numerous field studies and theoretical models, many ecologists now conclude that no such relationship exists. Both very complex communities, such as tropical rain forests, and very simple communities, such as Arctic tundra, may be very fragile.

Alan D. Copsey, updated by Bryan Ness

See also: Community-ecosystem interactions; Competition; Ecosystems: overview; Ecosystems: studies; Species and speciation; Trophic levels and ecological niches.

Sources for Further Study

Aber, John D., and Jerry M. Melillo. *Terrestrial Ecosystems*. 2d ed. San Diego: Harcourt, 2001. A wide-ranging, interdisciplinary textbook for beginning students in environmental science. Covers general topics as well as case studies of specific community types.

Begon, Michael, John L. Harper, and Colin R. Townsend. *Ecology: Individuals, Populations, and Communities*. 3d ed. Cambridge, Mass.: Blackwell Science, 1996. Comprehensive college-level text, particularly good at integrating the results of plant ecologists with those of animal ecologists. Part 4 of the book provides a thorough, well-documented discussion of the structure and function of communities; historical and current hypotheses as to the nature of the community; the influence of competition, predation, and disturbance on community structure; and community stability. Excellent bibliography.

Goldammer, J. G., ed. *Tropical Forests in Transition: Ecology of Natural and Anthropogenic Disturbance Processes*. Boston: Springer-Verlag, 1992. Sixteen studies of tropical forest disturbance offer a wide spectrum of approaches to the subject of community structure and stability.

Krebs, Charles J. *Ecology: The Experimental Analysis of Distribution and Abundance*. 5th ed. San Francisco: Benjamin Cummings, 2001. Widely used college-level text, written with excellent clarity and scope. Chapter 25 provides concise, complete coverage of community organization.

Pickett, S. T. A., and P. S. White, eds. *The Ecology of Natural Disturbance and Patch Dynamics*. Orlando, Fla.: Academic Press, 1985. Provocative and wide-ranging, this book is a gold mine of information and examples illustrating current hypotheses about the many kinds of natural disturbance and the ways in which communities respond. Readily accessible to the general reader.

COMPETITION

Categories: Ecology; ecosystems; evolution

The struggle for food, space, and pollinators in order to survive can occur between individuals of different species (interspecific competition) or between individuals of the same species (intraspecific competition).

Competition is a major driving force in *evolution*, the process by which living organisms change over time, with better-adapted species surviving and less well-adapted species becoming extinct. Evolution begins with *mutation*, changes in the nucleotide sequence of a gene or genes, resulting in the production of slightly altered genes which encode slightly different proteins. These altered proteins are the expressed traits of an organism and may give the organism an advantage over its competitors. The organism outcompetes its rivals in the environment, and hence the environment favors

PhotoDisc

Some plant species, including peach trees, release chemicals into the soil that inhibit the growth of other plants that might otherwise compete with them.

the better-adapted, fitter organism, a process called *natural selection*. A mutation may help an organism in one environment but may hurt it in a different environment (for example, an albino squirrel may flourish in snowy regions but may not do as well in warm regions). Mutations are random events whose occurrence can be increased by chemicals called mutagens or by ionizing radiation such as ultraviolet light, X rays, and gamma radiation.

Species Interactions

Natural selection influences the distribution and abundance of organisms from place to place. The possible selection factors include physical factors (temperature and light, for example), chemical factors such as water and salt, and species interactions. According to ecologist Charles Krebs, species interactions include four principal types: *mutualism*, which is the living together of two species that benefit each other (for example, fungi and algae living together as lichens); *commensalism*, which is the living together of two species that results in a distinct benefit (or number of benefits) to one species while the other remains unhurt (for example, plants called *epiphytes* that grow on other plants); *predation*, which is the hunting, killing, and eating of one species by another (examples include insects eating plants or snails eating algae); and *competition*, which is defined as an active struggle for survival among all the species in a given environment.

Competition is related to the acquisition of various resources: food, space, and pollinators. Food is an obvious target of competition. All organisms must have energy in order to conduct the cellular chemical reactions (such as respiration) that keep them alive. Photoautotrophic organisms (plants, algae, cyanobacteria) obtain energy by converting sunlight, carbon dioxide, and water into organic molecules, a process called photosynthesis. Photoautotrophs, also called *primary producers*, compete for light and water. For example, oak and hickory trees in eastern North America grow taller than most pines, thereby shading smaller species and eventually dominating a forest. A few other autotrophs, such as chemoautotrophic bacteria, obtain their energy from inorganic chemical reactions rather than from sunlight.

All other organisms—animals, zooplankton, and fungi—are heterotrophs, also called *consumers;* they must consume other organisms to obtain energy. Heterotrophs include herbivores, carnivores, omnivores, and saprotrophs. Herbivores (plant eaters such as rabbits and cattle) derive their energy from eating plants. Carnivores (meat eaters such as cats and dogs) eat other heterotrophs in order to get their energy needs met. Omnivores (such as humans) eat plants and animals. Saprotrophs (such as fungi and bacteria) decompose dead organisms to meet their energy requirements. Life on earth functions by intricately complex food chains in which organisms consume other organisms in order to obtain energy. Ultimately, almost all energy in organisms comes from the sun.

Types of Competition

Intraspecific competition occurs among individual members of the same population, for example, when sprouts grow from seeds scattered closely together on the ground. Some seedlings will be able to grow faster than others and will inhibit the growth of less vigorous seedlings by overshadowing or overcrowding them.

Interspecific competition involves two or more different species trying to use the same resources. All green plants depend on photosynthesis to derive the energy and carbon they need. Different areas or communities favor different growth characteristics. For plants with high light requirements, a taller-growing plant (or one with more or broader leaves) will have a competitive advantage if its leaves receive more direct sunlight than competitors. If, on the other hand, the species cannot tolerate too much sun, a shorter-growing species that can benefit from sheltering shadows of larger plants nearby will have the competitive advantage over other shade-loving plants.

Competition in nature leads to certain species dominating an environment and evolving to adapt to it, while the outcompeted species move elsewhere or become extinct. Humans are able to utilize the results of competition among species for their own benefit. This principle is useful in agriculture. Certain plant species (such as sunflowers and peach trees) release chemicals into the soil that inhibit the growth of other plants that might otherwise compete with them. Plant competition can be used to fight the growth of weeds and is useful in understanding which plants are compatible.

Competition is also useful in medical microbiological research. Certain fungal species produce and secrete molecules called antibiotics, which kill bacteria. Antibiotics give these fungi an edge in their competition against bacteria. Laboratory production of these antibiotics can treat bacterial diseases in humans.

David Wason Hollar, Jr., updated by Bryan Ness

See also: Allelopathy; Coevolution; Community-ecosystem interactions; Community structure and stability; Ecosystems: overview; Ecosystems: studies; Evolution of plants; Species and speciation.

Sources for Further Study

Barbour, Michael G., ed. *Terrestrial Plant Ecology*. 3d ed. Menlo Park, Calif.: Benjamin/Cummings, 1999. Covers the entire breadth of modern plant ecology, blending classic topics with the results of new research, using as little jargon as possible.

Hartl, Daniel L. *Principles of Population Genetics*. 3d ed. Sunderland, Mass.: Sinauer Associates, 1997. Hartl's work, aimed at graduate-level biologists, is a detailed mathematical approach to species interactions and evolution. Approaches the study of populations from a genetic and statistical viewpoint and provides numerous examples and references.

Keddy, Paul A. *Competition*. 2d ed. New York: Kluwer, 2000. A somewhat controversial yet wide-ranging study of competition in all the natural kingdoms, using many nontechnical examples easily understood by the layperson. Argues that the challenge in ecology is to determine when and where each kind of competition is important in natural systems, focusing on variants of competition such as intensity, asymmetry, and hierarchies.

Krebs, Charles J. *Ecology: The Experimental Analysis of Distribution and Abundance*. 5th ed. San Francisco: Benjamin Cummings, 2001. This advanced textbook is a wonderfully comprehensive survey of ecology. Numerous experiments are cited, with a thorough but understandable discussion of mathematical modeling in ecology. Includes many references.

Raven, Peter H., and George B. Johnson. *Biology.* 5th ed. Boston: WCB/McGraw-Hill, 1999. Beautifully illustrated, clearly written, well-diagramed introductory biology textbook for undergraduates. Chapter 23, "Population Dynamics," is a clear presentation of species interactions, such as competition, that contains numerous examples.

COMPLEMENTATION AND ALLELISM: THE CIS-TRANS TEST

Categories: Genetics; methods and techniques

The cis-trans test is a method of determining whether two mutations are in the same or different genes. Genes are normally found in sets of two copies. If one copy is mutated and the other gene makes a normal product that compensates, the mutation is said to be recessive. If both copies of the gene are mutated and no normal product is made, then the phenotype of the organism will be mutant.

The cis-trans test was developed by Seymour Benzer and his coworkers in the 1950's. They worked with a *bacteriophage* (a virus that attacks only bacteria) called T4, studying one T4 viral gene called *rII*. Mutants of *rII* could be easily identified by the size of the *plaques* they made. A plaque is a clear circle in a continuous lawn of bacteria growing on nutrient agar which results when viruses attack and destroy the bacteria in that area.

Complementation (cis-trans) tests between *mutations* in different bacteriophage, or phage, strains which produce mutant plaques can be performed using a procedure called a *spot test*. Phages of one mutant strain are added to bacterial cells, then spread out on a plate containing nutrients and a solidifying agent (usually agar). Later, a drop of fluid containing phages with a different mutant strain is spotted onto the solid surface. In the area of the drop, some bacterial cells will be attacked by phages of both types. If the mutations complement each other, a normal plaque will result in that area. After looking at hundreds of different mutations, all of which had the same *phenotype* (that is, the expression of the gene) of large plaques, scientists at Benzer's laboratory determined that the *rII* gene was actually made up of two functional groups (which they called cistrons) and that every mutant could be assigned to one or the other of these groups.

Genetic Basis

Genes are located on cellular structures called *chromosomes*. For many years, scientists thought that genes were lined up along the chromosomes like beads on a string, each bead representing a different gene, such as for seed shape in plants. If normal or *wild-type* alleles were represented by black beads, it was thought, a red or white or green bead would represent a changed, or *mutant*, allele. This analogy was popular for many years, and genes, like beads, were thought to be indivisible. The work of scientists in the 1940's and 1950's challenged this theory and led to its eventual demise. The cis-trans test was a key part of this work.

The basis of the cis-trans test is the functional role of the gene. It is now known that each gene is made up of a particular sequence of *nucleotides* in *deoxyribonucleic acid* (DNA), defining the nucleotide sequence of a strand of *ribonucleic acid* (RNA). This RNA may, in turn, be used to define the sequence of amino acids in a protein. A mutant allele is a stretch of DNA in which the nucleotide sequence has been altered, resulting in an altered RNA. The result may be an altered protein, or sometimes no protein at all.

In most higher organisms, chromosomes occur in pairs called homologs. Such pairing means every gene (except those on the sex chromosomes) is normally found in two copies. If one copy is mutated and the other gene makes a normal product that compensates, the mutation is said to be *recessive*. If both copies of the gene are mutated and no normal product is made, then the *phenotype* of the organism (that is, the expression of the gene) will be mutant.

Genetic Crosses

In the cis-trans test, genetic crosses are arranged to yield an organism in which two recessive mutant alleles to be tested are on opposite *homologs*. In other words, one chromosome of a pair will have one mutant allele, and the other chromosome will have the other mutant allele. The alleles are then said to be in a *trans configuration*. If the two alleles are mutations of the same gene, the phenotype of the organism will be a mutant one. If the alleles belong to separate genes, the organism will have a normal, or *wild-type*, phenotype. When alleles produce a normal phenotype, they are said to complement each other. If the mutant alleles are arranged in the *cis configuration*, so that both are on the same chromosome, the resulting phenotype will be normal, whether the alleles are in the same or different genes.

For example, harebells (*Campanula*) normally have blue flowers, the wild-type phenotype, but occasionally mutants have white flowers. If a cis-trans test is done with two white-flowered mutants, in some cases the resulting progeny have blue flowers. This shows that the mutations in these two mutants are in different genes and thus display complementation. It is known that the blue pigment is produced by a biochemical pathway with two steps involving two enzymes that modify a colorless molecule. In other cases when a cis-trans test is done, all the progeny are white-flowered. This occurs when both mutants have mutations in the same gene, knocking out the activity of one of the enzymes, and thus complementation does not occur.

Recombination

Before the structure of DNA was understood, the gene was also defined as the basic unit of *recombination*; recombination within a gene was not thought to occur. In the analogy of genes as beads on a string, recombination was thought to involve the breakage and reunion of the string at positions between beads, never within them. It is now known that, in the process of recombination, pieces of chromosomes break off and rejoin in new combinations, although the rules that this process follows are not fully understood.

Recombination is essential to the cis-trans test. Only through recombination can cis configurations be obtained. The frequency of the occurrence of recombination is related to the closeness of two points on a chromosome. If recombination is desired between two regions that are very close, the chance that crossing over will occur is very low. Recombination can even be used to map mutations within a gene (a process called *fine structure analysis*), but it is a very long process, requiring the analysis of an extremely large number of progeny because the distance separating mutations within a gene will be very small.

Using the cis-trans test and recombination together led to the discovery of a small group of mutations called *pseudoalleles*. When two pseudoalleles are examined using the cis-trans test, an organism in trans configuration will have a mutant phenotype, leading to the conclusion that the alleles belong to the same gene. When recombination is used to try to map these mutations, however, it turns out that they are in separate structural genes. Pseudoalleles, then, are found in separate structural genes that interact in some functions.

Several examples of interaction between pseudoalleles can be found in *Drosophila* (fruit flies). In one such example, the two genes postbithorax (*pbx*) and bithorax (*bx*) are both involved in determining the normal formation of the fly's balancing organs (halteres), a pair of winglike structures that help the fly in its flight. These genes are adjacent on the chromosome but are structurally distinct. The fact that the gene products must interact is revealed by the results of the cis-trans test; when mutants for these genes are placed in trans configuration, the phenotype of the fly is mutant.

Lisa A. Lambert, updated by Bryan Ness

See also: Bacteriophages; DNA in plants; DNA replication; Genetics: Mendelian; Genetics: post-Mendelian; Genetics: mutations.

Sources for Further Study

Benzer, Seymour. "The Fine Structure of the Gene." *Scientific American* 206 (January, 1962): 70-84. The classic review article on the investigation of the *rII* gene of the T4 bacteriophage, written by the scientist who headed the investigations. Designed for readers with some familiarity with viruses and bacteria.

Hawkins, John D. *Gene Structure and Expression*. 3d ed. New York: Cambridge University Press, 1996. A concise guide to gene structure and function, giving an overview of recent advances in the field from both theoretical and practical points of view.

Russell, P. J. *Genetics*. 5th ed. Menlo Park, Calif.: Benjamin Cummings, 1998. A well-written, intermediate-level college text with many touches of humor. Complementation analysis is discussed in the chapter on genetic fine structure and gene function. Includes index, bibliography, and glossary.

Suzuki, D., A. Griffiths, J. Miller, and R. Lewontin. *An Introduction to Genetic Analysis*. 7th ed. New York: W. H. Freeman, 2000. This intermediate-level college text reviews the important concepts in genetics and presents the experimental work that led to the development of important theories. Chapter 12, "The Nature of the Gene," gives an in-depth analysis of Seymour Benzer's viral work. Chapter 15, "The Manipulation of DNA," also explains many of the important techniques of molecular genetics.

Weaver, R. F., and P. W. Hedrick. *Genetics*. 3d ed. New York: McGraw-Hill, 1996. Written for college audiences, the discussion is nevertheless readable by advanced high school students. Unlike many texts, it includes full-color photographs and diagrams. Chapter 13 discusses the cis-trans test using bacterial genes as an example. Seymour Benzer's work with *rII* is covered briefly in the same chapter. Recombination and potential mechanisms of how it works on the molecular level are clearly explained in chapter 7.

COMPOSITAE

Categories: Angiosperms; economic botany and plant uses; *Plantae*; taxonomic groups

In number of species, the family Compositae *or* Asteraceae, *commonly known as the sunflower family, is among the largest families of flowering plants.*

The *Compositae* family consists of more than eleven hundred genera worldwide and possibly as many as twenty-three thousand species. Representatives of the family are found on every continent except Antarctica. Species diversity is high in the southwestern United States and Mexico, in southern Brazil and along the South American Andes Mountain range, along the Mediterranean region, in Southwest and Central Asia, in South Africa, and in Australia.

The plants typically are found in open, sunny habitats, although some species are found in lightly forested areas and at the edges of woodlands. Examples of common genera found in North America are goldenrod (*Solidago*), sunflower (*Helianthus*), daisy (*Chrysanthemum*), *Aster*, and ironweed (*Vernonia*). The predominant life-forms within the family are perennial herbs and shrubs, but some species exist as annuals, vines, lianas (woody vines), and trees. Leaf arrangement is most commonly alternate, although some genera do exhibit opposite leaves. Despite this gross morphological diversity, the feature that unites all the members of the family is the unique form of the compound inflorescence, or flowering head.

Inflorescence

At first glance, the flowering head of many *Compositae*, such as that of the sunflower, resembles a single large flower with what appears to be a set of petals surrounding a cluster of anthers and stamens. Upon closer inspection, this compound inflorescence, or capitulum, is really a collection of highly reduced and modified flowers borne on a flowering stalk (peduncle). The expanded and flattened top of the peduncle is the receptacle and is surrounded by a series of leaflike structures termed bracts that collectively comprise the *involucre*.

Two types of reduced flowers can be present within a capitulum, either ray or disc flowers, or both. The ray flower superficially resembles a petal and occupies the outer circumference of the capitulum. It can be either pistillate or sterile. The corolla is fused at the base, gradually expanding in width and eventually tapering to a tip. At maturity the tip of the ray flower corolla is oriented away from the center of the capitulum. The inner disc flowers can be either perfect (all reproductive parts present) or functionally staminate and possessing a corolla tube consisting of fused petals. At maturity the anthers and the stigmas will protrude from the corolla tubes. At the base of the corolla tube is a set of modified sepals termed the *pappus*. The pappus can consist of fine hairs, scales, or bristles, or it may be absent in some species. The seed is an achene and upon maturity is wind-dispersed with the assistance of the pappus.

DIGITAL STOCK

Dahlia *is one of numerous genera of* Compositae *that are used as ornamental plants.*

Pollination Biology

A variety of mechanisms account for successful pollination in the *Compositae*, wind pollination being a very common mode. "Hay fever," which in humans is an allergic reaction to plant pollen and other allergens, is greatly exacerbated in the late summer and early fall by the wind-pollinated ragweed (*Ambrosia*). Although often falsely accused, the goldenrods (*Solidago*) are not a major cause of hay fever, as they are pollinated by insects, not wind. By far, the majority of the species in the family are pollinated by insects, which are often attracted to the inflorescences by brightly colored ray flowers. The insect-pollinated flowers are typically generalists, attracting a variety of insects rather than a specific insect species. In some tropical taxa, birds or other animals are largely responsible for pollen exchange.

Economic Uses

Genera that contain species used for human consumption include endive and chicory (*Cichorium*), artichoke (*Cynara*), sunflower seeds and oil (*Helianthus*), lettuce (*Lactuca*), and dandelion greens (*Taraxacum*). Pyrethrum is a naturally occurring pesticide obtained from tansy (*Tanacetum*). Numerous genera are used as ornamental plants, such as marigolds (*Calendula*), *Zinnia*, bandana daisy (*Gaillardia*), and *Dahlia*. A number of plants are used for medicinal purposes. Coneflower (*Echinacea*) and fireweed (*Liatris*) are often used as herbal teas. Extracts from species of the *Compositae* are available from many health food retailers and practitioners of folk medicine. A current focus of pharmacological research is in determining those plant compounds responsible for producing beneficial effects in humans. Some species, however, contain compounds that at sufficient levels are quite toxic to animals, such as humans and livestock.

Evolutionary Success

Current opinion is that the family probably originated in South America or in the Pacific region, perhaps as early as the Tertiary period.

Several reasons have been proposed for the biological success of the family. First, the pappus is a very efficient agent for wind-borne dispersal of mature seeds, serving as a lofting device (similar to a parachute) to carry the small seeds on air currents over great distances. Second, the agents of pollination are diverse, ranging from mechanical (wind) to biological (insect or bird). Third, many members of the family exhibit unique chemical compounds (such as sesquiterpene lactones and polyacetylenes) that may discourage large herbivores and insects from foraging on the leaves, thus allowing plants to mature and engage in reproduction. It is the successful production of viable offspring that constitutes the major measure of evolutionary success in the family.

Pat Calie

See also: Angiosperm evolution; Angiosperm life cycle; Angiosperms; Biochemical coevolution in angiosperms; Eudicots; Flower structure; Flower types; Garden plants: flowering; Heliotropism; Inflorescences; North American flora; Plant domestication and breeding; Pollination; Seeds; South American flora.

Sources for Further Study

Bremer, K. *Asteraceae: Cladistics and Classification*. Portland, Oreg.: Timber Press, 1994. A scholarly treatment of the *Compositae*, with discussions of the morphology, anatomy, classification, and evolution of the family. Includes excellent diagrams of inflorescences, tables, index, and extensive references.

Evans, J., ed. *Ultimate Visual Dictionary*. London: DK, 1998. Excellent illustration of the inflorescence of *Helianthus*, with all parts labeled. Includes other illustrations of different flower types for comparison as well.

Judd, W. S., C. S. Campbell, E. A. Kellogg, and P. F. Stevens. *Plant Systematics: A Phylogenetic Approach*. Sunderland, Mass.: Sinauer Associates, 1999. Contains a chapter on the *Compositae*, with a discussion of floral morphology, anatomy, and evolution. Includes illustrations, references, glossary, and index.

Moore, R., W. D. Clark, and D. S. Vodopich. Botany. 2d ed. New York: WCB/McGraw-Hill, 1998. Botany textbook with discussion of angiosperm evolution and the *Compositae*. Includes illustrations, glossary, and index.

Raven, Peter H., Ray F. Evert, and Susan E. Eichhorn. *Biology of Plants*. 6th ed. New York: W. H. Freeman/Worth, 1999. Botany textbook, with illustrations of the *Asteraceae* inflorescence. Includes illustrations, glossary, and index.

COMPOSTING

Categories: Agriculture; gardening; soil

Compost is a mixture of organic ingredients used for fertilizing or enriching land. Composting is the practice of making and using compost.

Composting is a way for gardeners and farmers to enrich and otherwise improve the soil while reducing the flow of household waste to landfills. Essentially the slow, natural decay of dead plants and animals, composting is a natural form of recycling in which living organisms decompose organic matter.

The decay of dead plants and animals starts when microorganisms in soil feed on dead matter, breaking it down into smaller compounds usable by plants. Collectively, the breakdown product is called humus, a crumbly, dark brown, spongy substance. Adding humus to soil increases its fertility. Compost and composting derive from the Old French *composter*, "to manure" or "to dung."

History

The origins of human composting activities are buried in prehistory. Early farmers discovered the benefits of compost, probably from animal manure deposited on or mixed with soil. In North America, American Indians and then Europeans used compost in their gardens. Public accounts of the use of stable manure in composting date back to the eighteenth century. Many New England farmers also found it economical to use fish in their compost heaps.

While living in India from 1905 to 1934, British agronomist Sir Albert Howard developed today's home composting methods. Howard found that the best compost pile consists of three parts plant matter to one part manure. He devised the Indore method of composting, alternating layers of plant debris, manure, and soil to create a pile. Later, during the composting process, he turned the pile or mixed in earthworms.

How Composting Works

Composting is a natural form of recycling that takes from six months to two years to complete. Bacteria are the most efficient *decomposers* of organic matter. Fungi and protozoans later join the process, followed by centipedes, millipedes, beetles, or earthworms. By manipulating the composition and environment of a compost pile, gardeners and farmers can reduce composting time to three to four months. Important factors to consider are the makeup of the pile, the surface area, the volume, the moisture, the aeration, and the temperature of the compost pile.

Yard waste, such as fallen leaves, grass clippings, some weeds, and the remains of garden plants, make excellent compost material. Other good additions to a home compost pile include sawdust, wood ash, and kitchen scraps, including vegetable peelings, egg shells, and coffee grounds. Microorganisms digest organic matter faster when they have more surface area on which to work, so gardeners can speed the composting process by chopping kitchen or garden waste with a shovel or running it through a shredding machine or lawn mower.

The volume of the compost pile is important because a large compost pile insulates itself, holding in the heat of microbial activity, which in turn accelerates decomposition. A properly made heap will reach temperatures of about 140 degrees Fahrenheit in four or five days. Then the pile will settle, a sign that the process is working properly. Piles 3 feet cubed (27 cubic feet) or smaller cannot hold enough heat, while piles 5 feet cubed (125 cubic feet) or larger do not allow enough air to reach the microbes in the center of the pile. These portions are important only if the goal is fast compost. Slower composting requires no exact proportions.

Moisture and air are essential for life. Microbes function best when the compost heap has many air passages and is about as moist as a wrung-out sponge. Microorganisms living in the compost pile use the carbon and nitrogen contained in dead matter for food and energy. While breaking down the carbon and nitrogen molecules in dead plants and animals, they also release nutrients that higher organisms, such as plants, can use.

The ratio of carbon to nitrogen found in kitchen and garden waste varies from 15 to 1 in food waste to 700 to 1 in wood. A carbon-to-nitrogen ratio of 30 to 1 is optimal for microbial decomposers. This balance can be achieved by mixing two parts grass clippings (carbon to nitrogen ratio 19:1) and one part fallen leaves (carbon to nitrogen ratio 60:1). This combination is the backbone of most home composting systems.

Uses and Practice

In the twenty-first century, composting remains an invaluable practice. In landfills, yard and kitchen wastes use up valuable space. These materials make up about 20 to 30 percent of all household waste in the United States. Composting household waste reduces the volume of municipal solid waste and provides a nutrient-rich soil additive. Compost or organic matter added to soil improves soil structure, texture, aeration, and water retention. It improves plant growth by loosening heavy clay soils, allowing better root penetration. It improves the water-holding and nutrient-holding capacity of sandy soils and increases the essential nutrients of all soils. Mixing compost with soil also contributes to erosion control and proper soil pH balance, the amount of acidity or alkalinity present.

Some municipalities collect and compost leaves and other garden waste and then make it available to city residents for little or no charge. Some cities also compost sewage sludge, or human waste, which is high in nitrogen and makes a rich fertilizer. Properly composted sewage sludge that reaches an internal temperature of 140 degrees Fahrenheit

AP/Wide World Photos

Near Vacaville, California, restaurant waste is being transformed by microbial action into compost.

contains no dangerous disease-causing organisms. One possible hazard, however, is that it may contain high levels of toxic heavy metals, including zinc, copper, nickel, or cadmium.

The basic principles of composting used by home gardeners also are used by municipalities composting sewage sludge and garbage, by farmers composting animal and plant waste, and by some industries composting organic waste. Food and fiber industries, for example, compost waste products from canning, meat processing, and dairy and paper processing.

J. Bradshaw-Rouse

See also: Biofertilizers; Organic gardening and farming; Soil; Sustainable agriculture.

Sources for Further Study

Campbell, Stu. *Let It Rot! The Gardener's Guide to Composting*. 3d ed. Pownal, Vt.: Storey Communications, 1998. Contains advice for starting and maintaining a composting system, building bins, and using compost.

Christopher, Thomas. *Compost This Book!* San Francisco: Sierra Club Books, 1994. This complete, nontechnical guide to composting provides step-by-step instructions on how to get started, discusses the history and chemistry of composting, explains why composting has become a vital part of waste disposal, and discusses composting's future.

Minnich, Jerry, and Marjorie Hunt. *Rodale Guide to Composting*. Rev. ed. Emmaus, Pa.: Rodale Press, 1992. A comprehensive treatment of composting.

Ondra, Nancy J. *Soil and Composting: The Complete Guide to Building Healthy, Fertile Soil*. Boston: Houghton Mifflin, 1998. Provides clear and practical information on making compost, taking soil samples, feeding plants using organic methods, dealing with poor drainage, choosing and using mulches effectively, and more.

CONIFERS

Categories: Forests and forestry; gymnosperms; *Plantae*

The conifers, which are woody plants consisting mostly of evergreen trees, make up the phylum Coniferophyta, one of four phyla of gymnosperms that have living representatives. The word "conifer" means cone-bearing. Most conifers bear their reproductive structures in cones.

With 50 genera and 550 of the 700 known gymnosperm species, the phylum *Coniferophyta* includes the bulk of the *gymnosperms*. *Coniferophyta* is also the most widespread and, in terms of numbers of individual trees, the most abundant of the gymnosperm phyla. The abundance and economic and ecological importance of the *Coniferophyta* are out of all proportion to the number of species, which does not begin to compare to the 235,000 species of angiosperms (phylum *Anthophyta*).

Conifers grow over almost the entire world. They are especially abundant in northern temperate and boreal regions. They predominate in the taiga, or northern boreal forest, which covers immense stretches of northern North America and northern Eurasia. They dominate that forest by virtue of large numbers of individuals of only about a dozen species, mainly of spruce, fir, and larch. They show greater diversity in midlatitude, mountain forests of the Northern Hemisphere. Conifers also occur in cool mountain areas of the tropics. In temperate regions of the Southern Hemisphere, they are widespread but less dominant than in the north. Junipers and pines are the most wide-ranging conifers, occurring across the northern continents and into the tropics. In the Southern Hemisphere, the most widely distributed conifer genus is *Podocarpus*.

Due to their abundance and wide range, their typically arboreal habit, and their stem structure, conifers are the only gymnosperms that yield timber on a commercial scale. They are the source of all of the world's softwood timber and nearly half of its total annual lumber supply.

A Long Fossil Record

The oldest known conifer fossils date to the late Carboniferous period of the Paleozoic era, some 300 million years ago. The earliest known angiosperms are less than half this old. Conifers diver-

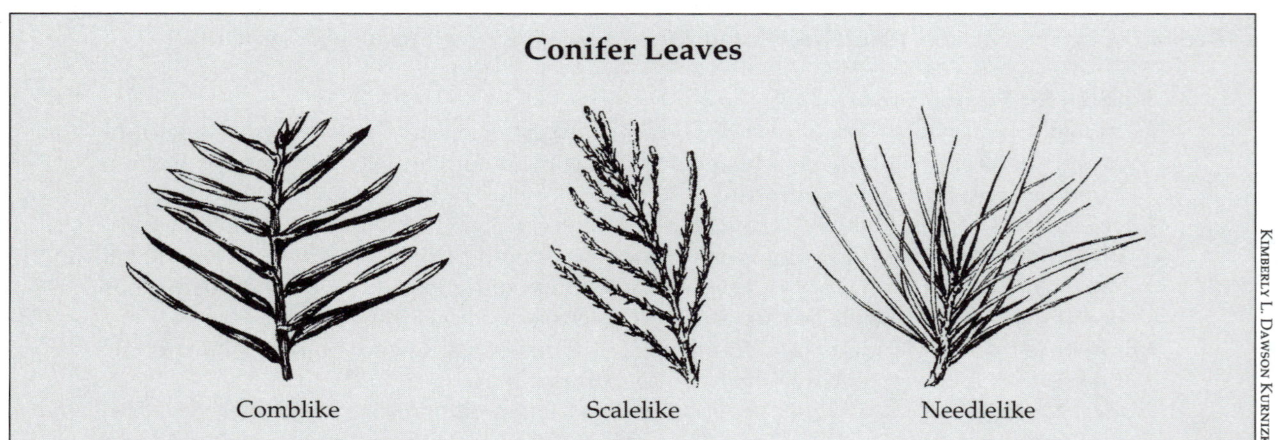

Many conifers are characterized by drought-resistant, needlelike, scalelike, or comblike leaves covered in a waxy outer layer that assists in water retention during cold winters, when water uptake from the frozen soil is limited.

sified greatly during the Paleozoic's relatively dry, cold Permian period (290-245 million years ago), which followed the Carboniferous. The drought-resistant, needlelike leaves characteristic of many modern conifers may have evolved during that dry period.

The modern families of conifers began to differentiate in the Mesozoic era (245 million to 65 million years ago). Conifers were the dominant vegetation during much of that era. They gradually gave way to the angiosperms, which achieved worldwide dominance by about ninety million years ago, during the Cretaceous period of the Mesozoic, and remain dominant today. In the early Tertiary period of the Cenozoic era, which dawned about 66 million years ago, a diverse coniferous flora covered large areas of the Northern Hemisphere.

Classification of Conifers

Class *Pinopsida*
- Order *Pinales*
  - Family *Araucariaceae* (araucarias)
    - Genera: *Agathis, Araucaria, Wollemia*
  - Family *Cephalotaxaceae* (plum yews)
    - Genus: *Gephalotaxus*
  - Family *Cupressaceae* (cypresses)
    - Genera: *Actinostrobus, Austrocedrus, Callitris, Calocedrus, Chamaecyparis, Cupressus, Diselma, Fitzroya, Fokienia, Juniperus, Libocedrus, Microbiota, Neocallitropsis, Pilgerodendron, Tetrachnis, Thuja, Thujopsis, Widdringtonia*
  - Family *Pinaceae* (pines)
    - Genera: *Abies, Cathaya, Cedrus, Keteleeria, Larix, Nothotsuga, Picea, Pinus, Pseudolarix, Pseudotsuga, Tsuga*
  - Family *Phyllocladaceae*
    - Genus: *Phyllocladus*
  - Family *Podocarpaceae* (podocarps)
    - Genera: *Acmopyle, Afrocarpus, Dacrycarpus, Falcatifolium, Halocarpus, Lagarostrobos, Lepidothamnus, Microcachrys, Microstrobos, Nageia, Parasitaxus, Prumnopitys, Retrophyllum, Saxegothaea, Sundacarpus*
  - Family *Taxodiaceae* (redwoods)
    - Genera: *Athrotaxis, Cryptomeria, Cunninghamia, Glyptostrobus, Metasequoia, Sciadopitys, Sequoia, Sequoiadendron, Taiwania, Taxodium*
- Order *Taxales*
  - Family *Taxaceae* (yews)
    - Genera: *Amentotaxus, Austrotaxus, Pseudotaxus, Taxus, Torreya*

Source: Data are from U.S. Department of Agriculture, National Plant Data Center, *The PLANTS Database*, Version 3.1, http://plants.usda.gov. National Plant Data Center, Baton Rouge, LA 70874-4490 USA, and Southern Illinois University at Carbondale, College of Science, http://www.sciencie.siu.edu/landplants/Coniferophyta.

Vegetative Features

Conifers have certain notable vegetative characteristics. For example, many, though not all, of them have needlelike leaves, and such leaves occur only in conifers. Some conifers have scalelike leaves, and a few have bladelike ones.

The stems and roots of conifers are also noteworthy, resembling the woody eudicots and magnoliids of *Anthophyta* in producing a dense mass of *secondary xylem*—that is, wood.

Reproductive Characteristics

In conifers, the ovules, which contain the eggs and eventually ripen into seeds, are not enclosed in ovary tissues within flowers as they are in the angiosperms. Instead, like other gymnosperms, conifers have *naked ovules*. These ovules are borne on scales that are arranged spirally along the axes of cones. Due to a complex evolutionary history, the structure of these cones is compound rather than simple.

In addition to ovulate, or female, cones, conifers produce male cones, which are smaller, shorter-lived, and generally less woody. The axis of a male cone bears reduced leaves, each with two *pollen sacs* on the undersurface. The pollen grains formed within the sacs contain the immature sperm. In most conifers, cones of both genders are present on the same tree. Although gymnosperms other than conifers may produce cones, conifers are distinctive in combining simple pollen cones with compound ovulate cones. The yews (genus *Taxus*) bear their ovules singly, at the tips of branchlets, rather than in cones. The seeds that develop from these ovules are enclosed in a fleshy covering called an *aril*.

In the spring, the male cones of conifers release their pollen to the wind. Some of it reaches female cones, which have their cone scales widely separated to receive it. After pollination, the female cones close their scales. Fertilization usually occurs, within the ovule, about three days to three or four weeks after pollination. In the pines, however, it is delayed for about fifteen months.

In most conifers, the mature seeds, with their contained embryos, are dispersed in the autumn of the year in which fertilization occurs. The most common dispersal agent is the wind, aided in many species by wings on the seeds. Conifer seeds that lack wings generally depend on dispersal agents such as birds or mammals. In some pines, the cones need the heat of a forest fire to open the scales and release the seeds.

Pine Family

Of the families of living conifers, the *Pinaceae*, or pine family, is the largest, containing ten to twelve genera, with about two hundred species worldwide. Most of these are resin-bearing trees.

This family is the most widespread and abundant conifer family in the Northern Hemisphere and very important economically. It supplies a great deal of lumber, pulpwood, and paper products. It is also important ecologically, providing essential cover, nesting habitat, and food for many species of wildlife.

The pine family's range is mainly in Eurasia and North America. Genera that occur all across this expanse are *Pinus* (pines), *Picea* (spruces), *Abies* (firs), and *Larix* (larches). Douglas firs (*Pseudotsuga*) and hemlocks (*Tsuga*) are restricted to North America and Asia. The true cedars (*Cedrus*) occur only in warm-temperate regions of Eurasia and northern Africa. Three genera of the pine family are restricted to China.

Pinus, the oldest genus in the pine family, arose at least 130 million years ago, in the early Cretaceous period. It grows predominantly in the Northern Hemisphere. The pines number more than ninety species, nearly half the total species of the *Pinaceae*. One of the world's oldest known plants is the bristlecone pine (*Pinus longaeva*), which grows in Utah, Nevada, and a small area of western California. Some bristlecone pines are about forty-six hundred years old. In contrast to other conifers, pines produce their adult needles in *fascicles*, or bundles. The number of needles per fascicle depends on the species.

Other Conifer Families

According to a commonly used classification system, there are seven families of conifers in addition to the *Pinaceae*. A notable one is the *Taxodiaceae*, which is today represented by widely scattered species that are the remnants of populations that were much more widespread during the Tertiary period. Bald cypress belongs to the *Taxodiaceae*, as do two redwood species renowned for their size and longevity. The coast redwood (*Sequoia sempervirens*) grows along coastal California and Oregon, and the "big tree" (*Sequoiadendron giganteum*) occurs on the western slope of the Sierra Nevada mountains in California. The coast redwood is generally the taller of the two species, reaching up to 117 meters in height. The big tree, however, generally exceeds the coast redwood in total mass and can live to be several thousand years old.

Another notable member of the *Taxodiacae* is the dawn redwood (*Metasequoia glyptostroboides*). This species was first known to science only from fossils discovered in the early 1940's. Not long afterward, the tree was found growing in China. A more recent living-fossil find belongs to the *Araucariaceae*, a conifer family restricted to the Southern Hemisphere. This tree, the Wollemi pine (*Wollemia nobilis*), had been thought to have gone extinct fifty million years ago. Then, in 1994, a stand was found growing in Australia.

The *Cupressaceae*, or cypress family, includes *Chamaecyparis* (false cypress), *Juniperus* (junipers, often called "cedars" in North America), and *Thuja* (arbor-vitae). These plants are distributed throughout the world. Some are important as timber or ornamentals.

The *Taxaceae* family, which includes the yews, is widely distributed in the Northern Hemisphere. The *Podocarpaceae* is largely restricted to the Southern Hemisphere. The *Sciadopityaceae* and *Cephalotaxaceae* are confined to east Asia.

Jane F. Hill

See also: Asian flora; European flora; Evolution of plants; Forest fires; Gymnosperms; North American flora; Pollination; Seeds; Taiga; Wood.

Sources for Further Study

Krüssmann, Gerd. *Manual of Cultivated Conifers*. Translated by Michael E. Epp, edited by Hans-Dieter Warda. Portland, Oreg.: Timber Press, 1985. A reference work describing the six hundred conifer species known to be in cultivation and many of their varieties. Covers plants' uses and cultivation requirements. Black-and-white photographs and line drawings of the more important plants.

Lanner, Ronald M. *Conifers of California*. Los Olivos, Calif.: Cachuma Press, 1999. Natural history and field guide, combined. Species narratives include detailed identification information and illustrations.

_______. *Made for Each Other: A Symbiosis of Birds and Pines*. New York: Oxford University Press, 1996. A narrative detailing the close relationship between pine seeds that are wingless and the birds that depend on, and disperse, them. Shows how symbiotic organisms can drive each other's evolution.

Pielou, E. C. *The World of Northern Evergreens*. Ithaca, N.Y.: Comstock, 1988. Describes the ten groups of North American conifers, and other species, above 45 degrees north. Identification keys. Written in nontechnical language.

Van Gelderen, D. M. *Conifers: The Illustrated Encyclopedia*. 2 vols. Portland, Oreg.: Timber Press, in cooperation with the Royal Boskoop Horticultural Society, 1996. A reference work on ornamental horticulture. Color photographs.

CORN

Categories: Agriculture; economic botany and plant uses; food

The most important cereal in the Western Hemisphere, corn is used as human food (ranking third in the world), as livestock feed, and for industrial purposes.

Corn (*Zea mays*) is a coarse, annual plant of the grass (*Gramineae*) family. It ranges in height from 3 to 15 feet and has a solid, jointed stalk, and long, narrow leaves. A stalk usually bears one to three cobs, which develop kernels of corn when fertilized. Corn no longer grows in the wild; it requires human help in removing and planting the kernels to ensure reproduction. In the United States and Canada, "corn" is the common name for this cereal, but in Europe, "corn" refers to any of the small-seeded cereals, such as barley, wheat, and rye. "Maize" (or its translation) is the term used for *Zea mays* in Europe and Latin America.

Explorer Christopher Columbus took corn back to Europe with him in 1493, and within one hundred years it had spread through Europe, Asia, and Africa. It is said that a corn crop is being harvested somewhere in the world each month. Corn grows as far north as Canada and Siberia (roughly 58 degrees north latitude) and as far south as Argentina and New Zealand (40 degrees south). Although adaptable to a wide range of conditions, corn does best with at least 20 inches of rainfall (corn is often irrigated in drier regions) and daytime temperatures between 70 and 80 degrees Fahrenheit (about 24 degrees Celsius). Much of the United States meets these criteria, hence its ranking as the top corn-producing country in the world.

Origins and Hybridization

Corn's exact origins remain uncertain, but historical records show the gathering of wild corn, called *teosinte*, in ancient Mexico began around 7000 B.C.E. This corn evolved through unknown means to have tiny, eight-rowed "ears" of corn less than an inch long. Corncobs and plant fragments have been dated to 5200 B.C.E. (and up to a millennium earlier, according to some studies). By 3400 B.C.E., the fossil record shows a marked change in corn, notably increased cob and kernel size, indicating greater domestication. Fully domesticated corn (which could not survive without human help) had replaced the wild and other early types of corn by 700 C.E.

Extensive attempts at *hybridization* began in the late nineteenth century, but the increase in yield was usually a disappointing 10 percent or so. By 1920 researchers had turned to inbreeding hybridization programs. In these, corn is *self-fertilized*, rather than being allowed to *cross-pollinate* naturally. Following a complex sequence of crossing and testing different varieties, the lines with the most desirable traits were put into commercial use, and they often produced 25 to 30 percent gains in yield. Although these early hybrids focused on increasing the yield, researchers later began to look for insect-resistant and disease-resistant qualities as well. One of the hybridizers of the 1920's was Henry A. Wallace, founder of Pioneer Seed Company (the world's largest seed company) and later U.S. vice president under Franklin D. Roosevelt. By the 1950's hybrid corn varieties were in widespread use.

Types and Uses

The types of corn still in use are dent, flint, flour, pop, and sweet. Dent corn, characterized by a "dent" in the top of each kernel, is the most important commercial variety. Flint corn tends to be resistant to the rots and blights known to attack other types; it is also more tolerant of low temperatures and therefore appears at the geographical edge of corn's range. Flour corn is known for its soft kernel, making it easier to grind into flour and thus popular for hand-grinding. A mainstay at American movie theaters and as a snack food, popcorn will, with an optimum moisture content of about 13 percent, explode to as much as thirty times its original volume when heated. Also popular in the United States and eaten fresh, sweet corn is so named because, unlike other types, most of the sugars in the kernel are not converted to starch.

Commercially, corn is used mostly for livestock feed and industrial processing. It is high in energy and low in crude fiber but requires supplements to make a truly good feed. Industrial processing creates a great variety of products found in everyday

Corn, one of the world's three major staple crops, no longer grows in the wild; it requires that humans remove and plant the kernels (seeds) to ensure reproduction.

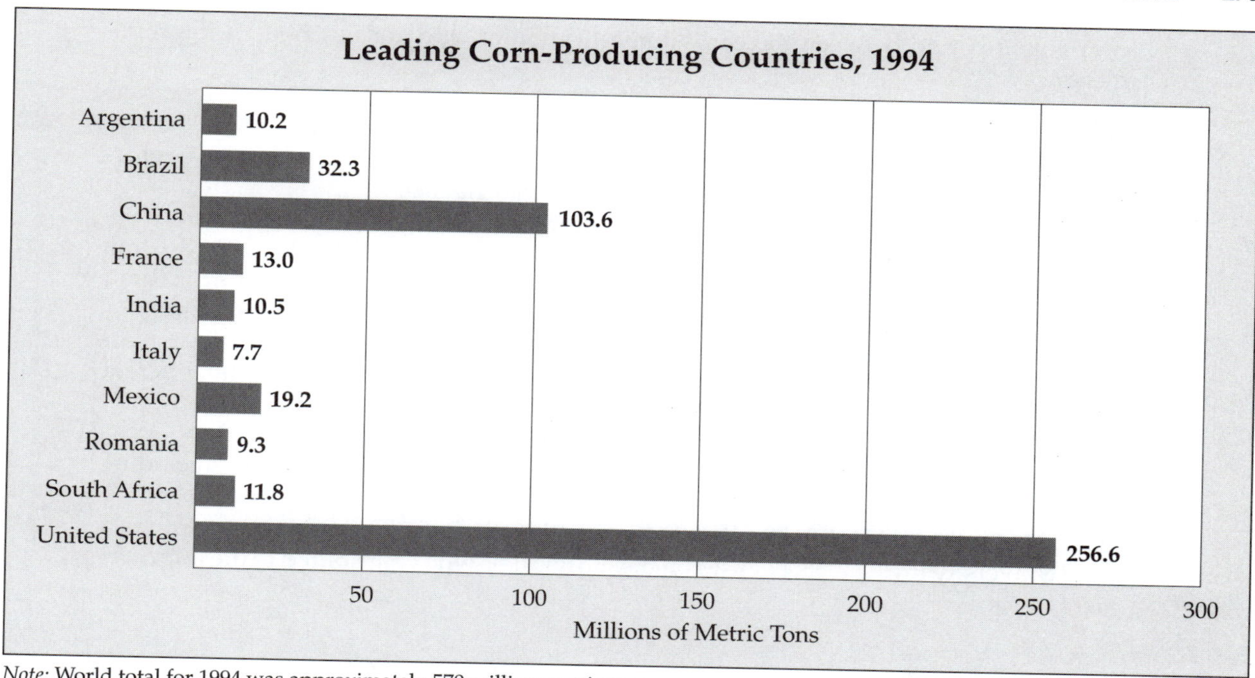

Note: World total for 1994 was approximately 570 million metric tons.
Source: U.S. Department of Commerce, *Statistical Abstract of the United States, 1996*, 1996.

life—underscoring the importance of corn to the world's economy.

Processing takes place in one of three ways: wet milling, dry milling, or fermentation. In wet milling, corn is soaked in a weak sulfurous acid solution, ground to break apart the kernel, and then separated. The resulting by-products are found nearly everywhere. The cornstarch supplies corn syrup (it is sweeter than sugar and less expensive, and billions of dollars' worth is produced for soft drink manufacturers each year), starches used in the textile industry, ingredients for cooking and candymaking, and substances used in adhesives, to name a few. Other by-products provide cooking oil; oil used in mayonnaise, margarine, and salad dressing; soap powders; and livestock feed. Dry milling is a simpler process, involving the separation of the *hull* from the *endosperm* (the food storage organ, which is primarily starch in most corn) and the *germ* (the plant embryo) by repeated grinding and sieving. Dry milling produces hominy, grits, meal, and flour, all of which are used for human consumption. Fermentation of corn changes the starch to sugar, which is then converted by yeast to alcohol. The process eventually results in ethyl alcohol, or ethanol (which is blended with gasoline to reduce carbon monoxide emissions), acetone, and other substances.

Brian J. Nichelson

See also: Agricultural revolution; Agriculture: traditional; Agriculture: world food supplies; Central American agriculture; Grains; Grasses and bamboos; Green Revolution; Hybridization; North American agriculture; South American agriculture.

Sources for Further Study

Bower, Bruce. "Maize Domestication Grows Older in Mexico." *Science News* 158, no. 7 (February 17, 2001): 103. Inhabitants of southern Mexico began to cultivate corn, the major grain crop of prehistoric societies in the Americas, by at least sixty-three hundred years ago, a study has found. This is around eight hundred years earlier than previous estimates.

Hardeman, Nicholas P. *Shucks, Shocks, and Hominy Blocks: Corn as a Way of Life in Pioneer America*. Baton Rouge: Louisiana State University Press, 1981. Addresses the often-overlooked period between the early domestication of corn and its modern improvements.

Johannessen, Sissel, and Christine A. Hastorf, eds. *Corn and Culture in the Prehistoric New World*. Boulder, Colo.: Westview Press, 1994. A collection of twenty-eight essays that examine the relationships between corn and people in the Americas from perspectives of archaeology, anthropology, genetics, linguistics, and biochemistry.

McCann, James. "Maize and Grace: History, Corn, and Africa's New Landscapes, 1500-1999." In *Comparative Studies in Society and History* 43, no. 2 (April, 2001): 246-272. McCann chronicles the historical advance of corn in Africa since its introduction around 1500 and delineates different histories of change through three different areas of its advance.

Mangelsdorf, Paul C. *Corn: Its Origins, Evolution, and Improvement*. Cambridge, Mass., Harvard University Press, 1974. Offers a thorough and balanced discussion of the different theories of corn's origins, with good illustrations.

Sprague, George F., and J. W. Dudley, eds. *Corn and Corn Improvement*. 3d ed. Madison, Wis.: American Society of Agronomy, 1988. Provides articles on all aspects of corn, including history, genetics, cultivation, and uses.

Wallace, Henry A., and William L. Brown. *Corn and Its Early Fathers*. Rev. ed. Ames: Iowa State University Press, 1988. Focuses on the individuals who contributed to the improvement of corn.

CRYPTOMONADS

Categories: Algae; microorganisms; *Protista*; taxonomic groups; water-related life

The phylum Cryptophyta *describes tiny, motile, unicellular organisms with two slightly unequal flagella bearing lateral hairs. Cryptomonads live mainly in marine and freshwater environments.*

Some cryptomonads are alga-like, with blue-green, red, and olive-brown photosynthetic pigments including chlorophylls *a*, *c2*, alpha-carotene, xanthophylls (alloxanthin, crocoxanthin, zeaxanthin, and monadoxanthin), and phycobiliproteins (phycoerythrin and phycocyanin). Cryptomonads are found in a variety of moist places, such as algal blooms in the ocean or in fresh water, and on beaches. Some members are intestinal parasites in animals.

Classification

Historically, botanists and zoologists alike have adopted cryptomonads. Botanically, cryptomonads would be included in kingdom *Plantae*, phylum *Cryptophyta*, class *Cryptophyceae*, order *Cryptomonadales*, and family *Cryptomonadaceae*. Zoologists would place the cryptomonads in kingdom *Animalia*, phylum *Sarcomastigophora*, class *Phytomastigophora*, order *Cryptomonadida*, and family *Cryptomonadidae*. Also, the cryptomonads are currently included in a separate kingdom, *Protoctista* (also known as *Protista*), with phylum, class, order, and family taxa being the same as that for the botanical taxa above.

Synonyms for *Cryptophyta* are *Cryptomonadales*, *Cryptophyceae*, and *Chromophyta*, and they were once placed with the algae. Genera for the cryptomonads include the following: *Chilomonas*, *Chroomonas*, *Cryptomonas*, *Cyathomonas*, *Falcomonas*, *Geminigera*, *Goniomonas*, *Guillardia*, *Hemiselmis*, *Komma*, *Plagiomonas*, *Pyrenomonas*, *Rhodomonas*, *Storeatula*, and *Teleaulax*. *Guillardia theta*, formerly known as *Cryptomonas phi*, has been studied most extensively, and the complete chloroplast genome is known (Genbank accession number AF041468).

Ecology

The cryptomonads are part of the nanoplankton (typically phytoplankton between 2 to 20 micrometers in diameter) and are a relatively small but ecologically and evolutionarily important taxon. Both freshwater and marine representatives are known. Many photosynthetic species likely retain their capacity to eat prey (mixotrophy). Fluctuations in their numbers are correlated with increases in levels of nitrogen in the water in which they live. Some species of cryptomonads form gelatinous colonies. A few cryptomonads have reached the palmelloid or sessile stage of organization, but most are free-living flagellates common in nutrient-rich water. A weakly filamentous member of the cryptomonads is *Bjornbergiella*. Well-known examples of cryptomonads are *Cryptomonas ovata*, *C. similis*, and *Chilomonas paramecium*.

The cryptomonads can form major blooms in Arctic and Antarctic waters as well as in North America's Chesapeake Bay. They can be very important food sources for smaller heterotrophic or mixotrophic plankton, including ciliates and dinoflagellates. Cryptomonads are found in freshwater lakes, ponds, and ditches—especially in colder waters. They are dominant species in many Antarctic lakes, and they are also found in interstitial water on sandy beaches. Reproduction in the cryptomonads is generally asexual in culture, but sexual reproduction has been documented in the *Cryptophyta*.

Characteristics

Cryptomonads are tiny flagellates, 5 to 30 microns (most around 10 to 20 microns). They are flattened dorsiventrally in shape and are asymmetrical, with a *periplast* (a proteinaceous structure that lies inside the plasmalemma and is attached to it). Cryptomonads are mostly algal forms, with an anterior (ventral) groove or pocket and a gullet, which has refractile *ejectosomes* or *trichocysts*. The unequal flagella are inserted nearly parallel to the pocket, not inside the gullet as in the euglenophytes. Mitochondria have flat cristae, while the plastids are secondary with a highly reduced eukaryotic nucleus, the *nucleomorph*.

Although there are a few colorless forms, such as *Chilomonas*, most cryptomonads have a chloroplast. The chloroplast is not contained directly, however, because there is a reduced eukaryote symbiotic within the cell containing a normal prokaryote chloroplast. Usually, there are two chloroplasts, which are secondary plastids. The chloroplast is bound by four membranes (two being chloroplast endoplasmic reticulum, or CER, continuous with the nuclear envelope and homologous to a food vacuole) with a tiny nucleus (nucleomorph) between the middle two membranes. Much like the dinoflagellates (with which the cryptomonads were formerly grouped), it has chlorophylls *a* and *c*; chlorophyll *b* is never present. Thylakoids are paired, and phycobilin pigments are present in the spaces between the thylakoids but not in phycobilisomes, such as in the cyanobacteria and *Rhodophyta*. Food reserves are starchlike, accumulating in the periplastidal space stored between the starch envelope and the chloroplast reticulum. If there is an eyespot, it is inside a plastid not associated with the flagella.

There is a large nucleus at the posterior end. Mitosis is open, and centrioles are not associated with mitosis. Cell division is by furrowing. No histones are associated with the chromosomes. The unique nucleomorph has deoxyribonucleic acid (DNA), is contained within a double membrane, and also has a nucleolus-like region. Molecular data suggest and strongly support the idea that the nucleomorph is a vestigial nucleus from the original endosymbiont, which became the cryptophyte chloroplast. Three chromosomes are associated with the nucleomorph: 240 kilobase pairs (kb), 225 kb, and 195 kb. The bulk of chromosome II (175 kb) is now sequenced with a preponderance of "housekeeping genes" apparently existing for the service of just a few genes encoding plastid proteins. There are parallels between the nucleomorphs of cryptomonads and chlorarachniophytes. The cryptomonad nucleomorph is depauperate in introns, with ribosomal ribonucleic acid (rRNA) genes at the chromosome ends just within the telomeres.

The longer flagellum has two rows of *mastigonemes* (lateral hairs), while the shorter flagellum has a single row. Mastigonemes are two-parted bristles or hairs made up of a rigid, tubular base and, usually, two terminal hairs. These bristles are formed within the endoplasmic reticulum (or nuclear envelope) and are thus transported to the exterior of the cell. The flagella are covered with scales, too.

Trichocysts or *ejectosomes* are in the oral groove, and they are scattered around the cell surface. There is a tightly spooled protein in the trichocysts, which can undergo a very rapid, irreversible conformational change in which it pops out of the cell, pushing the cell backward as a result. The trichocysts are considered a defense mechanism or are perhaps involved in predation.

Evolution

Evolution and phylogeny of cryptomonads have not been well documented using molecular techniques. An 18S ribosomal RNA phylogeny of cryptomonads has been made, however. The nucleus of the endosymbiont (the nucleomorph) does not seem to have a complete complement of genes for photosynthesis, these being relocated to the host nucleus now. Plastid targeting mechanisms of cryptomonads for light-harvesting complex proteins have been studied, though.

The primary plastid is of unknown origin and remains under scrutiny but is probably from a red algal lineage, as the presence of phycobilins is suggestive of the *Rhodophyta* (red algae). Chlorophyll *c* is unknown among the red algae, however. Molecular phylogenetic studies place the nucleomorph close to red algae, and the chloroplast genome map has characteristics suggesting a reduction series from red algae to cryptomonads to heterokonts. This does not imply any descendant relationship among extant groups but rather retention of ancestral character states from common ancestors.

Lateral gene transfer from an ancestral cryptomonad to a dinoflagellate is postulated from sequence analysis of two nuclear-encoded glyceraldehyde-3-phosphate dehydrogenase (GAPDH) genes isolated from the dinoflagellate *Gonyaulax polyhedra*. The plastid sequence forms a monophyletic group with the plastid isoforms of cryptomonads, distinct from all other plastid GAPDHs. This provides the first example of genetic exchange accompanying symbiotic associations between cryptomonads and dinoflagellates, which are common in present-day cells.

F. Christopher Sowers

See also: Algae; Chloroplasts and other plastids; Dinoflagellates; Heterokonts; Photosynthesis; Phytoplankton; Plant science; *Protista*; Red algae; Systematics and taxonomy; Systematics: overview.

Sources for Further Study

Graham, L. E., and L. W. Wilcox. *Algae*. Upper Saddle River, N.J.: Prentice Hall, 1999. This text provides wide coverage of freshwater, marine, and terrestrial algae for ecology students. Origins, technology applications, nutrition, biochemistry, and phylogeny are all discussed. Relationships among algae are covered, including their roles in food webs, harmful blooms, and global biogeochemical cycling.

Margulis, L., J. O. Corliss, M. Melkonian, and D. J. Chapman. *Handbook of Protoctista*. Boston: Jones and Bartlett, 1990. Absolutely requisite work on *Protoctista* groups, done by some of the leaders in phylogenetic thinking.

Raven, Peter H., Ray F. Evert, and Susan E. Eichhorn. *Biology of Plants*. 6th ed. New York: W. H. Freeman/Worth, 1999. This is one of the very best undergraduate texts on botany. Very clear explanations, with diagrams and photographs. Contemporary, well organized, and readable.

CULTURALLY SIGNIFICANT PLANTS

Categories: Economic botany and plant uses; food; medicine and health

Plants are often used as a tool for ceremonial purposes, as artistic media to express indigenous traditions, or as herbal remedies or hallucinogenics to fulfill cultural needs and expectations. Culturally significant plants grow on all continents and are used by all ethnicities.

Historically, humans have appropriated plants for numerous cultural applications. Since prehistoric times, plants have served as symbolic organisms to represent aspects of the life cycle and seasonal changes, to worship, and to make offerings to gods. Plants were incorporated into mythology and legends to show their meanings to various cultures.

Ethnobotany

American Indians are an example of a historic ethnic group which selected specific plant species for cultural uses, such as rituals and ceremonies. These plants have significant roles in the lives of modern American Indians involved in preserving botanical cultural activities by recording information about the horticultural habits of their ancestors, such as knowing where plants grow, when and how they are harvested, how they are prepared and for what functions, and their role in sustaining tribal ethnicity.

Ethnobotanists investigate how cultures identify with and use plants. This scientific field addresses how people throughout time have managed plants, including cultivation methods to improve the quantity and quality of yields and to meet specific needs and requirements. Researchers compare societies' interactions with plants to theorize and determine why people at varying times and places selected certain plants for ceremonies or social uses. The U.S. Department of Agriculture's National Plant Data Center (NPDC) issues culturally significant plant guides which update ethnobotany theories and knowledge.

AP/Wide World Photos

Peyote buttons like the one displayed by this American Indian are used in some religious ceremonies to induce visions.

Religious Uses

Plants, especially trees, have been assigned spiritual roles since prehistory. Humans' daily lives have been influenced by how they perceive and integrate plants into their activities. Christians use plants and flowers, such as the Easter Lily, to represent the history of Jesus Christ. At Palm Sunday ceremonies, palms are the primary Christian botanical symbol for immortality and are used to designate martyrs and divine people and places. Ancient legends describe the Tree of the Cross, which provided the wood for Christ's crucifixion.

Because trees can attain sizable heights and girths, humans have considered them remarkable and worthy of veneration. The strength and longevity of most trees reinforced people's ideas that groves have mythical qualities, and tree cults were formed to protect them. The rarity of trees in many regions, particularly in the Middle East and in African deserts, convinced people that trees were divine and that specific trees could be regarded as sacred.

In Siberia, ancient people were silent when in shaman forests to express respect for gods and spirits they believed dwelled there. Celtic tribal names were derived from tree nomenclature. Many cultures' rituals involved ceremonies with oak trees, because ancient people thought the dead resided inside oaks. Pagan peoples used trees to mark heroes' graves.

Germanic tribes erected trunks as pillars, and in the ninth century the Frankish king Charlemagne had them cut down in his effort to Christianize Europe. Christians destroyed groves where tree cults worshiped and built churches on these sites. The interior decorations of cathedrals often included tree and acorn imagery that reminded worshipers of the groves. Other types of architecture incorporated plants in designs, such as palm fronds and bark on columns and branchlike arches.

Roman legends tell how the Trojan War hero Aeneas and his guide, the Sybil, were permitted entry to the Elysian Fields, a blessed place, by presenting the Golden Bough. Pre-Columbian people living on the American continents gave cacti as sacraments, hoping this would enable them to have communion with the gods. At Zeus's sanctuary in ancient Greece, an oak was believed to be an oracle because its leaves made noises which priestesses interpreted as Zeus's voice.

In Norse mythology, the ash tree Yggdrasil's roots and branches connected the underworld and heaven. One of the most famous culturally significant plants was the Tree of Knowledge of Good and Evil in the biblical Garden of Eden. Medieval miracle plays often featured the Garden of Eden. The Hanging Gardens of Babylon, an ancient wonder of the world, was an artificial, terraced mountain with lush greenery, which archaeologists have been unable to prove existed. Tales about the gardens enticed the imagination in the ancient world.

Trees are significant to many religions' origin. The Bodhdruma, or Bo Tree, is sacred to Buddhists because it is a pipal tree (*Ficus religiosa*, a form of fig tree) under which Siddhartha Gautama sat to receive enlightenment to become Buddha. In modern Thailand, Buddhists sprinkle water on religious statues of Buddha adorned with orchids, and many temples keep orchids in large pottery bowls filled with water to serve this purpose. Tree myths and traditions are also practiced in other parts of Asia. Indians believe that ghosts awaiting reincarnation live in fig trees. Because they are evergreen, pines represent immortality to many cultures.

Ceremony

Many plants are identified with ceremonial activities that commemorate holidays or anniversaries or celebrate a rite of passage in which an individual enters a new phase of life. When celebrating the new year, ancient peoples used plants in symbolic acts, such as drinking cactus juice to appeal to gods to bless them with sufficient rains and large crop yields. Before emancipation, African-American slaves jumped over a broom made from straw as a way to validate publicly their commitment to marriage when such unions were illegal.

In the modern world, vestiges of these ceremonial uses of plants retain some of their symbolic significance. Flowers are often used symbolically in romantic courtship. Plants convey meanings of love, fidelity, and longevity at weddings and are selected in accordance with personal preferences as well as regional and religious customs.

Plants memorialize people. In the United States on Mother's Day, women wear corsages that indicate whether their mother is alive or deceased. Red poppies pay tribute to soldiers' sacrifices on veterans' and memorial holidays. On Decoration (Memorial) Day, many people place flowers on graves as a form of respectful remembrance. Roses are often thrown into oceans near offshore plane crash sites. Flowers form makeshift memorials where people tragically died or at their homes.

At holidays, including Valentine's Day and Christmas, people give symbolic plants with meanings, such as renewal and protection, that have developed from religious and historical customs. Holly was sacred to Romans who feted Saturn, god of agriculture, at Saturnalia festivals during the winter solstice. Early Christians decorated with holly to avoid persecution; later, that plant was incorporated into Christian rituals. People went "wassailing" to pay tribute to apple orchards by anointing their roots with cider to wish for large spring yields.

Uses

Plants are used culturally for nutrition. Ancient peoples from diverse civilizations revered various crop spirits, including the Corn Mother and the Corn Maiden, which appeared in various forms in Europe, the Americas, Asia, and Africa. Often, people believed that spirits of crops resided within specific people in their communities. Human sacrifices and the ceremonial slaying of figures representing agricultural phases were sometimes held in the hope that such offerings would appease gods and assure ample harvests.

Many plants are harvested for medicinal uses. Often, plants are incorporated as drugs in religious rituals. Peyote has been used ceremonially by native peoples since pre-Columbian times for its hallucinogenic properties, which users thought assisted communication with gods and brought on supernatural powers. American Indians are guaranteed legal rights to use otherwise illegal plants for religious purposes.

Plants express cultural values. Plant fibers are used to weave baskets and clothing. Crushed plant parts create dyes which can be used to stain fabric, write, or draw. Plants inspire superstitions: For example, three-leaf clovers are considered lucky because Saint Patrick convinced Druid leaders to convert to Christianity by stating the shamrock was proof of the Trinity. Often, plants have different meanings: Peonies, for example, represent faithfulness to some cultures and shame in others. Literature and works of art—including Aesop's fables, William Shakespeare's plays, the children's story "Jack and the Beanstalk," and Wagnerian operas—have featured plant themes.

Cornfields are shorn into mazes for amusement. In the United States, nostalgia for old-fashioned plants resulted in heirloom seeds becoming popular. Authorities and ethnic groups are attempting to preserve, restore, and reestablish culturally significant plants in natural settings in order to retain traditional use of such plants for ceremonies, food, and craftsmanship and perhaps develop innovative future applications.

Elizabeth D. Schafer

See also: Agricultural revolution; Agriculture: traditional; Corn; Garden plants: flowering; Herbs; Human population growth; Medicinal plants; Plant fibers; Plants with potential; Spices; Textiles and fabrics.

Sources for Further Study

Altman, Nathaniel. *Sacred Trees: Spirituality, Wisdom, and Well-Being*. 1994. Reprint. New York: Sterling, 2000. Describes mythological beliefs about the religious nature of trees. Includes illustrations, bibliographical references, index.

Bernhardt, Peter. *Natural Affairs: A Botanist Looks at the Attachments Between Plants and People*. New York: Villard Books, 1993. Examines human-plant interactions and how they shape folklore and botanical beliefs. Includes color illustrations, bibliographical references, and index.

Frazer, James George. *The Golden Bough: A Study in Magic and Religion*. Abridged ed. New York: Oxford University Press, 1994. A classic work that includes discussion of the complexities of plant superstitions, rituals, and mythology with religious and magical themes based on the author's original 1890 two-volume publication.

Langenheim, Jean H., and Kenneth V. Thimann. *Botany: Plant Biology and Its Relation to Human Affairs*. New York: Wiley, 1982. Explores how people perceive and appropriate plants for social and economic purposes. Includes illustrations, bibliographical references, and indexes.

Lewis, Charles A. *Green Nature/Human Nature: The Meaning of Plants in Our Lives*. Urbana: University of Illinois Press, 1996. Analyzes how humans incorporate plants into their activities and nurture or neglect the environment. Includes illustrations, bibliographical references, and index.

Simoons, Frederick J. *Plants of Life, Plants of Death*. Madison: University of Wisconsin Press, 1998. Explores the relationship between humans and plants and how mythology and folklore have incorporated plant motifs, especially concerning medicinal applications. Includes illustrations, map, bibliographical references, and index.

CYCADS AND PALMS

Categories: Angiosperms; economic botany and plant uses; gymnosperms; paleobotany; *Plantae*

Members of the phylum Cycadophyta, *cycads are descendants of giant, prehistoric seed-bearing, nonflowering plants that thrived when dinosaurs lived. Palms are flowering plants with primitive origins that share characteristics with cycads but are not related to them.*

Cycadophyta is one of the four phyla of *gymnosperms* in the kingdom *Plantae*. At one time, cycads grew on every continent. Fossil cycads have been located in areas where no modern cycads grow, such as Antarctica and Europe, suggesting that those places once had milder temperatures.

Origins and Habitat

The oldest cycad fossils are 245 to 208 million years old, from the Triassic period. The first cycads are believed to have appeared in the Permian period 270 million years ago. Some botanists hypothesize that cycads originated as *progymnosperms* in the Devonian period about 408 million to 360 million years ago. During the Jurassic and Cretaceous periods (65 million to 195 million years ago), cycads flourished on earth, dominating plant life. With the beginning of the Ice Age, cycads gradually diminished in population but did not disappear. Some Paleozoic- and Mesozoic-era cycad genera became extinct.

Modern cycads are classified in 150 to 200 species, 11 genera, and 3 families. The most commonly seen cycad species belong to the genus *Cycas* and exhibit some structural characteristics found in palms, conifers, and ferns. Botanists consider cycads valuable in the study of plant evolution. Modern cycads live primarily in the tropics and other regions with warm temperatures and high humidity and are indigenous to Asia, Africa, Australia, and Latin America. Growing in Florida, *Zamia integrifolia* is the only North American cycad species.

Cycad Structure

Cycads consist of a stem with large, long, narrow leaves similar to palm fronds, gathered at the top in a circular crown. Seeds are located in cones on branch ends. Trunks vary from tall stems as high as 20 meters (66 feet) to short trunks standing about 2 meters (6.5 feet). Some have tuberous stems below the soil surface. A trunk's base and crown are of almost equal thickness. New leaves are tender, hardening as they mature, and their bases form armor on stems. They vary in length from 20 centimeters to 3 meters (8 inches to 10 feet).

Cycads lack growth rings, but their age can be estimated by counting the number of whorls on leaf scars on stems to determine how many annual or biennial leaf productions have occurred. Using this method, botanists have speculated that it is possible, although rare, for some cycads to have life spans of one thousand years.

Seedling cycads have a taproot that is later replaced by secondary roots called *coralloid roots*, and subterranean stems anchor the aboveground stem to the soil. Some roots have bacteria which manufacture helpful amino acids and fix nitrogen. The *Zamia pseudoparasitica* is the only epiphytic cycad species.

Cycad Reproduction

The reproductive cycle can last as long as fourteen months. Female and male cones grow on separate cycads. Pollen with spermatozoids, which are motile sperm, is produced in male cones and fertilizes female cones. Insects (not wind, as was believed until 1993) deliver pollen. Cone-generated heat creates a minty scent appealing to insects. Each cone can produce as much as 200 cubic centimeters (12.2 cubic inches) of pollen annually.

Seeds are scattered when cones fall apart. Only fertilized seeds germinate. Seedlings often require several years before their trunks emerge from the ground. Cycads aid scientists in developing theories about reproduction of extinct seed plants and provide insight on how insect polli-

Rob and Ann Simpson/Photo Agora

Cycads, such as this sago palm, resemble palms outwardly but are gymnosperms.

Classification of Cycads

Class *Cycadopsida*
Order *Cycadales*
Family *Cycadaceae* (cycads)
Genus:
Cycas
Family *Stangeriaceae*
Genera:
Bowenia (natal grass)
Stangeria
Family *Zamiaceae* (sago palm family)
Genera:
Ceratozamia
Chigua
Dioon
Encephalartos
Lepidozamia
Macrozamia
Microcycas
Zamia

Source: Data are from U.S. Department of Agriculture, National Plant Data Center, *The PLANTS Database*, Version 3.1, http://plants.usda.gov. National Plant Data Center, Baton Rouge, LA 70874-4490 USA, and Christopher J. Earle, http://www.botanik.uni-bonn.de/conifers.

nation evolved. Such pollination occurred prior to the existence of angiosperms, or flowering plants (the phylum *Anthophyta*).

Palms

Although cycads and palms can outwardly resemble each other, palms are members of a different phylum and family, the family *Arecaceae* within the phylum *Anthophyta* (angiosperms). Modern palms are represented by approximately twenty-six hundred species. They are the fourth largest group of monocots. (Monocots are one of two large divisions of the angiosperms.) The earliest known palm fossils are eighty million years old and from the Late Cretaceous period. Fossilized palms with pollen indicate that these plants were some of the earliest flowering plant families.

Palms thrived during the Eocene epoch, and a variety of primitive genera grew in Asia, Europe, and North America. Most modern palms live in forests, deserts, swamps, and mountains of tropical regions in Asia, Africa, the South Pacific, the Mediterranean Basin, and the Americas. Some species are confined to specific islands, such as the *Maxburretia gracilis* on the Langkawi Islands. Some palm stands cover hundreds of square miles and consist of millions of plants.

Structurally, palms are similar; all have slender trunks. Most species lack branches and have a leafy cluster on top. Some palms are as tall as 30 meters (98 feet). Seedlings grow a large tip, and most of the bulk forms underground. Upon maturity, trunks maintain the same diameter. Coconut palms have an average diameter of 45 centimeters (18 inches). A network of roots both above and below the surface secures trunks to the ground. Palms are mostly vertical, although some grow horizontally or like vines. Some climbing rattan palms are 182 meters (597 feet) long.

Palm stems consist of masses of densely compacted fibrous strands. Leaves grow from nodes at the tips of stems, causing sharp scars when leaves are shed. Date palms have leaves 3 meters (10 feet) long, and coconut palm leaves average more than 4 meters (13 feet) long. Leaf production offers hints about the age of palms, which average 50 to 100 years, with some *Lodoicea maldivica* attaining 350 years and an Australian *Livistona eastonii* reaching 720 years. Palms first produce seeds approximately eight years after germination and mature at age thirty.

Palms have masses of small flowers, often numbering in the thousands, which are composed of sepals, petals, and male stamens. The female pistil in flowers either forms a soft, berrylike fruit, containing one or more seeds, or a more rigid fruit (a coconut or date), referred to as a drupe, which has a husk and shell encasing one seed. Insects, bats, and wind pollinate palms. Animals eat the fruit, distributing the seeds in their feces.

Uses of Cycads and Palms

Sago is a starch extracted from the stems of palms and cycads. It is used for cooking after the alkaloid is removed. Some leaves and seeds are also edible, but many seeds are poisonous. An amino acid produced by cycads, B-methylamino-L-alanine (BMAA), causes dementia in people who ingest the chemical. Researchers have attempted to discern why cycads manufacture this neurotoxin and do not react like other plants to its presence. The mapping of genes for susceptibility to BMAA may one day help to treat people with neurodegenerative conditions. Cycasin, the carcinogenic compound methylazoxymethanol, a glucoside found in cycad seeds, is also toxic.

Palms yield vegetable oil, waxes, sugars, fats, and saps which are distilled into liquor. Dates and coconuts are popular foods produced by palms. A date palm produces several hundred dates yearly. People consume stems, leaves, nuts, and roots. Rattan palm fibers are used for making furniture, baskets, bags, and other items. Some palm wood is appropriated for veneers, cups, and canes.

Coconut palms can each grow more than one hundred nuts annually. Palms provide essential nutrients for many world populations. Oily coconut liquid and coconut milk are considered delicacies. Leaves from coconut palms make useful thatching material. Coir, the husk's fiber, is twisted into rope. Coconuts have meat that is eaten or processed into *copra* for candles, oil, and other oil-based products. Palm products are crucial in regional and global trade.

Cycads and palms are used ornamentally for landscaping gardens. Some species are suitable for bonsai in interior displays. Because they are often poached to meet collectors' demands, cycads risk extinction. Smugglers locate and ship cycads (such as sago "palms") to clients throughout the world. A limited number of licenses to gather and trade cycads and other protected plants are issued by the Convention on International Trade in Endangered Species (CITES), a 1975 agreement observed by 156 ratifying countries as of 2001 to establish protective legal guidelines to monitor international trade of approximately twenty-five thousand rare plant species and five thousand rare animal species. CITES forbids collection activities that pose risks to wild plant populations.

In Hot Springs, South Dakota, Fossil Cycad National Monument was designated in 1922. In 1957 the site became the first national park service unit to be closed permanently, after collectors removed all of the 120-million-year-old fossils from a prehistoric cycad forest. Cycad conservation efforts in-

Digital Stock

Palms are angiosperms (flowering plants). Date palms like these can produce several hundred dates yearly.

clude guarding habitats and artificially propagating species. Rare specimens such as the South African *Encephalartos woodii* cycad, of which only thirty-eight, all male, specimens existed in 2001, are implanted with microchips to track their shipment to botanical exhibitions.

Nong Nooch Tropical Gardens near Pattaya, Thailand, has one of the world's largest collections of Asian cycad and palm varieties. Botanists saved an undescribed cycad species that grows on limestone cliffs from being destroyed when limestone was gathered for construction material in Bangkok. The gardens hosted the 1998 International Palm Society Biennial and the 2002 Sixth International Symposium on Cycad Biology. In South Africa, the Modjadji Cycad Nature Reserve has both cycads and palms, including the largest stand of the enormous *Encephalartos transvenous* (Modjadji palm).

Elizabeth D. Schafer

See also: Culturally significant plants; Endangered species; Evolution of plants; Fossil plants; Fruit crops; Paleobotany; Plant fibers; Pollination.

Sources for Further Study

Broschat, Timothy K., and Alan W. Meerow. *Ornamental Palm Horticulture*. Foreword by P. B. Tomlinson. Gainesville: University Press of Florida, 2000. Discusses how palms enhance landscapes. Includes illustrations, bibliographical references, and index.

Jones, David L. *Cycads of the World*. Foreword by Dennis W. Stephenson. Washington, D.C.: Smithsonian Institution Press, 1993. Comprehensive account of cycad history, structure, reproduction, and distribution of extinct and modern species. Includes illustrations, biographical references, and index.

_______. *Palms Throughout the World*. Foreword by John Dransfield. Washington, D.C.: Smithsonian Institution Press, 1995. A thorough study of the botanical and cultural role of palms internationally. Includes illustrations, biographical references, and index.

Norstog, Knut J., and Trevor J. Nicholls. *The Biology of the Cycads*. Ithaca, N.Y.: Comstock Publishing Associates, 1997. Provides scholarly theories about cycad pollination, growth, and evolution from Old World to New World genera. Includes illustrations, bibliographical references, and index.

Tomlinson, Philip B. *The Structural Biology of Palms*. New York: Oxford University Press, 1990. Focuses on palm anatomy and development. Includes illustrations, bibliographical references, index.

CYTOPLASM

Categories: Anatomy; cellular biology

The cytoplasm is defined as all of the living matter within the plasma membrane of a cell, except for the nucleus, which is isolated from the cytoplasm by the nuclear envelope.

The cytoplasm, bounded by the plasma membrane, is composed of fluid called the *cytosol* in which floats a large variety of molecules and molecular assemblages, *ribosomes* (responsible for polypeptide synthesis), and a variety of other structures called *organelles* (literally meaning "little organs"). Numerous biochemical processes occur in the cytosol, including protein synthesis (*translation*) and *glycolysis*.

Vacuoles

The *vacuole* is the largest, most noticeable, organelle in plant cells (up to 40 micrometers in length). Vacuoles are membrane-bound sacs that can occupy up to 90 percent of the plant's cell volume. The vacuole serves a wide range of purposes, depending on the cell type. It is filled with water and a variety of salts, sugars, organic acids, pigments, or proteins. The vacuoles of flower petal cells contain blue, red, or pink water-soluble pigments. In other cells, the vacuole is full of toxic compounds that deter insect attack. Vacuoles in seed cells contain storage proteins that are remobilized during seed germination and used for early seedling growth.

Plastids

Plastids are the next largest easily observable organelle (5 to 10 micrometers in diameter). *Plastids* are surrounded by a double membrane and are involved in a variety of biosynthetic reactions in the plant cell. All plastids start as immature *proplastids* in *meristematic* (young and undifferentiated) cells. Then, depending upon which tissue type grows from that meristem, proplastids may develop into one of several different mature plastid types.

The most prominent type of plastid is the *chloroplast*, contained primarily in leaf cells. Chloroplasts are colored by the green pigment *chlorophyll*. They are the site of *photosynthesis* and a number of other biosynthetic reactions. *Chromoplasts* are red, orange, or yellow, as a result of their high levels of *carotenoids*. They are found in the petals of flowers, old leaves, some fruits, and even in some roots, such as the carrot.

Parts of the Plant Cell

Part	*Location in Cell*	*Description/Purpose*
Cell wall	Outer layer	Outer boundary of cell, comprising middle lamella, primary wall, sometimes a secondary wall, and plasmodesmata.
Chloroplasts	Cytoplasm	Two-membrane-bounded organelles where photosynthesis occurs.
Chromatin	Nucleus	Site of the chromosomes (genetic material: DNA, histones).
Chromoplasts	Cytoplasm	Two-membrane-bounded organelles where attractants promoting pollination and seed dispersal are made.
Cytoplasm	Cytoplasm	One of three major parts of the cell, containing plastids and cytosol.
Cytoskeleton	Cytoplasm	Contains microtubules, actin filaments.
Cytosol	Cytoplasm	Viscous fluid in which other parts of the cell (membranes, plastids, and other organelles) are suspended.
Endomembrane system	Cytoplasm	Consists of the endoplasmic reticulum, Golgi complex, and vesicles.
Endoplasmic reticulum	Cytoplasm	Membranes in the cytoplasm that form transport pathways and other compartments. Rough endoplasmic reticulum has ribosomes attached to the cytoplasmic face.
Golgi complex	Cytoplasm	Stacks of membranes located near the nucleus where cell products are modified and prepared for secretion from the cell.
Leucoplasts	Cytoplasm	Two-membrane-bounded organelles that store starch and generate oils.
Microtubules	Cytoskeleton	Hollow cylinders of tubulin protein molecules that form networks in the cytoplasm and the mitotic spindle.

Leucoplasts are a nonspecific group of plastids named for their lack of pigments (the prefix *leuco* means "white"). They include *amyloplasts,* which synthesize and store starch, such as in potatoes, and a variety of other plastids specialized for synthesis of oils, proteins, and other products. Amyloplasts are also found in the cells of the *root cap,* where they are involved in gravity sensing, or *geotropism,* by the root tips. Plastids divide by fission, and there may be ten to one hundred per cell.

Mitochondria

Mitochondria are a medium-sized organelle (1 to 2 micrometers in length) with a double membrane. Mitochondria are the sites of *cellular respiration,* the major chemically derived source of adenosine triphosphate (ATP) in many cells. The outermost of the two membranes controls transport of molecules into and out of the mitochondrion. The inner membrane is the site of the *electron transport chain* and *oxidative phosphorylation,* two components of respiration. The aqueous space bounded by the inner membrane is the site of the *Krebs cycle.* Like chloroplasts, mitochondria divide by fission, and there may be ten to one hundred mitochondria per cell, depending upon cell type.

Microbodies

Small (0.5 to 1.5 micrometers in diameter) organelles called *peroxisomes* are bounded by a single membrane. Some peroxisomes contain the enzymes involved in the recycling and detoxification of the products of *photorespiration*. *Glyoxysomes,* an-

Part	*Location in Cell*	*Description/Purpose*
Middle lamella	Cell wall	Sticky layer between cells, binding them together.
Nuclear envelope	Nucleus	Binds nucleus.
Nucleoli	Nucleus	These substructures are composed of genes that encode RNA.
Nucleoplasm	Nucleus	Fluid-filled matrix contained within the nuclear membrane.
Nucleus	Nucleus	One of three major parts of cell, within cytoplasm.
Oil bodies	Cytoplasm	Organelles that store lipids, primarily triacylglycerols.
Peroxisomes	Cytoplasm	One-membrane-bounded organelle that contains enzymes that produce hydrogen peroxide.
Plasma membrane	Inside cell wall	Outer boundary of cytoplasm.
Plasmodesmata	Cell wall	Strands going through cell walls to move substances between protoplasts of contiguous cells.
Plastids	Cytoplasm	General name for chloroplasts, chromoplasts, leucoplasts: two-membrane-bounded organelles.
Primary wall	Cell wall	Outer wall, site of pit fields, division, metabolism.
Proplastids	Cytoplasm	Precursors to other plastids.
Ribosomes	Cytoplasm, nucleus	Structures composed of RNA, responsible for making proteins.
Secondary wall	Cell wall	Inner wall, rigid, with pits; found only in some cells.
Tonoplast	Cytoplasm	Membrane that surrounds the vacuole in a plant cell; also known as a vacuolar membrane.
Vacuoles	Cytoplasm	Organelle bounded by one membrane (the tonoplast) and filled with fluid; in many plant cells, there is a large central vacuole.

other type of peroxisome, are found in germinating seeds. They contain the enzymes involved in the conversion of fats to sugars during the mobilization of storage lipids.

Endomembrane System

The *endoplasmic reticulum* is a series of connected tubules that traverses the cytosol and is the largest member of the *endomembrane system*. The interior of the tubules is called the *lumen*. *Rough endoplasmic reticulum* (RER) is studded with *ribosomes*, while smooth endoplasmic reticulum (SER) is bare. Proteins synthesized on the ribosomes of the RER are inserted into the lumen during synthesis, where they are delivered to the *Golgi complex* for processing. The SER is typically where lipids are made but is rarely seen in plant cells because plastids perform most of the lipid synthesis in plants.

The Golgi complex is a series of stacked membranes that process and package proteins or polysaccharides for transport within or secretion out of the cell. Individual units are called *Golgi bodies* or *dictyosomes*, while collectively they are referred to as the Golgi complex. The endomembrane system gets its name from the fact that many of its components are connected in various ways.

Ribosomes

Ribosomes are small (25 nanometers in diameter), complex assemblies of proteins and *ribosomal RNA* (rRNA). They have two subunits, one large and one small, each made up of a unique mixture of proteins and rRNA. Ribosomes are the sites of protein synthesis. All proteins are large strings of *amino acids* bonded end-to-end. During protein synthesis, a strand of *messenger RNA* (mRNA) from the nucleus travels to the cytoplasm, and one end binds to a ribosome. The ribosome travels down the mRNA and "reads" the genetic information contained on it. The ribosome then "translates" that genetic information into a protein by stringing amino acids together, one at a time, in accordance with the information on the mRNA. Some ribosomes float freely in the cytosol, and others are bound to the endoplasmic reticulum, which is then called "rough ER" because of its appearance under the electron microscope. Proteins made on free ribosomes are released into the cytosol. Proteins made on ribosomes bound to rough endoplasmic reticulum cross the ER membrane and are released into the lumen of the ER for further processing and transport.

Cytoskeleton

All plant cells are surrounded by a *cell wall*. However, the cytoplasm itself is further structured and organized by components of the *cytoskeleton*. The three major components of the cytoskeleton are *microtubules*, *intermediate filaments*, and *microfilaments*.

Microtubules are hollow tubes 24 nanometers in diameter, made of individual, repeating subunits of the protein *tubulin*. They are involved in positioning and moving the chromosomes during *mitosis* and *meiosis*, in directing the laying down of cellulose strands during cell wall formation, and in determining where a new cell plate will form during cell division (that is, *cytokinesis*). Microtubules do not work alone, and numerous microtubule-associated proteins have been identified.

Intermediate filaments are a broad class of cytoskeletal components, and each type is composed of different proteins. They are all solid rods 8-12 nanometers in diameter. In animal cells (where they have been much more fully studied), intermediate filaments have been shown to provide flexible support to skin, nerve, and muscle cells. Although intermediate filaments have been found in plant cells, their exact roles are not yet fully understood. Much research remains to be done in this area.

Microfilaments are thin filaments (7-8 nanometers in diameter) made up of individual, repeating subunits of the protein *actin*. Microfilaments are involved in moving cellular organelles, such as plastids and mitochondria, around the plant cell and, like microtubules, interact with a variety of other proteins, especially *myosin*. In fact, the actin and myosin interaction is so well documented that many researchers refer to them together by simply calling them actinomyosin.

Robert R. Wise

See also: Angiosperm cells and tissues; Cell cycle; Cell theory; Chloroplasts and other plastids; Cytoskeleton; Cytosol; Endomembrane system and Golgi complex; Endoplasmic reticulum; Eukaryotic cells; Membrane structure; Microbodies; Nucleus; Oil bodies; Peroxisomes; Photosynthesis; Plant cells: molecular level; Plasma membranes; Respiration; Ribosomes.

Sources for Further Study

Campbell, Neil A., and Jane B. Reece. *Biology*. 6th ed. San Francisco: Benjamin Cummings, 2002. An introductory college-level textbook. Chapter 7 covers the cell and cellular constituents. Includes tables, photographs, and colored diagrams.

Hopkins, William G. *Introduction to Plant Physiology*. New York: John Wiley & Sons, 1999. College-level text for upper-level plant physiology course. Cytoplasm and its constituents are covered in chapter 1. Includes tables, photographs, and colored diagrams.

Raven, Peter H., Ray F. Evert, and Susan E. Eichhorn. *Biology of Plants*. 6th ed. New York: W. H. Freeman/Worth, 1999. Textbook for introductory botany course. Chapter 3 covers the cytoplasm and cellular organelles. Includes tables, photographs.

CYTOSKELETON

Categories: Anatomy; cellular biology

The cytoskeleton is a complex network of fibers that supports the interior of a cell. Cross-linked by molecular connectors into systems that support cellular membranes, it holds internal structures, such as the nucleus, in place and controls various kinds of cell movement.

Virtually all eukaryotic cells, including plant cells, have a cytoskeleton. Cytoskeletal systems extend internally from the membrane covering the cell surface to the surface of the membrane system surrounding the cell's nucleus. There are indications that a cytoskeletal support system reinforces the interior of the nucleus as well. The fibers of the cytoskeleton also anchor cells to external structures through linkages that extend through the surface membrane. The cytoskeletal material, rather than being fixed and unchanging, varies in makeup and structure as cells develop, move, grow, and divide.

Structural Elements

The cytoskeleton, depending on the cell type, is assembled from one or more of three major structural fibers: microtubules, microfilaments, and intermediate filaments. *Microtubules* are fine, unbranched hollow tubes with walls built from subunits consisting of the protein tubulin. Microtubules are about 25 nanometers in diameter, have walls about 4 to 5 nanometers thick, and range in length from a few to many micrometers. These structural elements, which may be arranged singly or in networks or parallel bundles, probably provide tensile strength and rigidity to cellular regions containing them. A tubular form combines lightness with strength and elasticity.

Microfilaments, also called actin filaments, are linear, unbranched fibers built up from the protein actin. Microfilaments are solid fibers that are much smaller than microtubules—about 5 to 7 nanometers in diameter, not much thicker than the wall of a microtubule. Microfilaments occur singly, in networks, and in parallel bundles in the cytoskeleton. The consistency of the *cytoplasm* (the living matter of a cell, exclusive of the nucleus), which can vary from highly liquid to solid and gel-like, is regulated by the degree to which microfilaments are cross-linked into networks. Microfilaments are also arranged in parallel bundles that give tensile strength and elasticity to cell regions and extensions. Many cell types contain numerous fingerlike extensions that are reinforced internally by internal parallel bundles of microfilaments.

Both microtubules and microfilaments form the basis for almost all cellular movements. In these motile systems, microtubules and microfilaments are acted upon by motile proteins that are able to convert chemical energy into the mechanical energy of movement. The motile proteins cause the microtubules or microfilaments to slide forcefully, or move cell structures and molecules over the surfaces of the two elements.

Microtubules and microfilaments occur as structural supports of the cytoskeleton of all plant, animal, fungal, and protozoan cells. The third structural element, the *intermediate filament*, is more abundant in animal cells than in plant cells. This type of fiber, called "intermediate" because its dimensions fall between those of microtubules and microfilaments, is about 10 nanometers in diameter. Unlike microtubules and microfilaments, which are each highly uniform in structure and made from a single type of protein, intermediate filaments occur in six different types, each made up of a different protein or group of proteins. Although the proteins making up the various intermediate filaments are different, they are related in both their three-dimensional structures and amino acid sequences.

Intermediate filaments occur in networks and bundles in the cytoplasm. They appear to be much more flexible than either microtubules or microfilaments, so it is considered likely that they form elastic ties holding cell structures in place, much like cellular rubber bands. However, the actual roles of these elements in the cytoskeleton remain uncertain in plant cells.

Assembly-Disassembly Reactions

Both microtubules and microfilaments can be readily converted between assembled and disassembled forms. In the conversion, the protein subunits of microtubules and microfilaments are exchanged rapidly between the fully assembled element and large pools of disassembled subunits in solution in the cytoplasm. Cells can control the balance between assembly and disassembly with high precision. As a result, the protein subunits can be recycled, and cytoskeletal structures containing microtubules and microfilaments can be set up or taken apart as the cell changes its function. As cell division occurs, for example, microtubules and microfilaments forming cytoskeletal structures typical of growing cells are rapidly disassembled and then reassembled into structures that take part in cell division. The assembly-disassembly reactions of microtubules and microfilaments proceed so readily that it is relatively easy to carry them out in a test tube. Microtubules and microfilaments, in fact, were among the first cell structures to be taken apart and put back together experimentally.

Cytoplasmic Streaming and Cell Division

Among the cell activities with which microfilaments are associated is *cytoplasmic streaming*, or cyclosis. The primary function of cytoplasmic streaming, which occurs within all live cells, is unknown. However, moving currents of cytoplasm are thought to facilitate the transport of nutrients, enzymes, and other substances between the cell and its surroundings, and within the cell itself.

A typical plant cell consists of a *cell wall* and its contents, called the *protoplast*. The protoplast consists of the cytoplasm and a nucleus. Within the cytoplasm are *organelles*, membranes, and other structures. Suspended in the cytoplasmic fluid is one or more liquid-filled *vacuoles*, and a vacuole is bounded by a membrane called the *tonoplast*.

In cytoplasmic streaming, the organelles and other substances travel within moving currents in between the microfilaments and the tonoplast. The organelles in the streaming cytoplasm are thought to be indirectly attached to the microfilaments, and this attachment creates a pulling or towing motion, responsible for the movement of cytoplasmic particles.

The microfilaments, in their constantly changing arrays, also facilitate specific activities within the cell, including cell cleavage during *mitosis*. Microfilaments mediate the movement of the cell nucleus before and following cell division. The microtubules, which are longer, move the split chromosomes to the newly forming cells in mitosis, and they play a role in cell plate formation in dividing cells.

In organizing other components of the cell, the cytoskeleton is thus intimately involved in the processes of cell division, growth, and differentiation. The cytoskeleton maintains the cell's overall shape and is responsible for the movement of various organelles within it. In single-celled organisms such as the amoeba, the cytoskeleton is responsible for the locomotion of the cell itself.

Stephen L. Wolfe, updated by Bryan Ness

See also: Cell theory; Cell wall; Cytoplasm; Eukaryotic cells; Mitosis and meiosis; Plant cells: molecular level.

Sources for Further Study

Alberts, Bruce, et al. *Essential Cell Biology: An Introduction to the Molecular Biology of the Cell.* New York: Garland, 1997. Includes many structural and molecular details about the organization of the cytoskeleton and outlines both the supportive and the motile roles of microtubules and microfilaments. The book is clearly written at the college level and contains many informative diagrams and photographs.

Carraway, K. L., and C. A. C. Carraway, eds. *The Cytoskeleton: Signaling and Cell Regulation: A Practical Approach*. 2d ed. New York: Oxford University Press, 2000. Covers current approaches to cytoskeleton experimental research, especially as related to signaling and cell regulation.

Gunning, Brian, and Martin Steer. *Plant Cell Biology, Structure, and Function*. Sudbury, Mass.: Jones and Bartlett, 1996. A popular atlas of plant cell micrographs, with over four hundred micrographs and four pages of full-color plates. Section J illustrates the cytoskeleton.

Hawes, C. R., and Beatrice Satiat-Juenemaitre, eds. *Plant Cell Biology: A Practical Approach*. 2d ed. New York: Oxford University Press, 2001. Covers a wide range of methods for working on living cells, including the application of fluorescent probes, cytometry, expression systems, the use of green fluorescent protein, micromanipulation and electrophysiological techniques. Written for advanced and graduate students and researchers.

Karp, Gerald. *Cell and Molecular Biology*. 3d ed. New York: John Wiley & Sons, 2001. The chapter on the cytoskeleton covers the structural roles of microfilaments, microtubules, and intermediate filaments in the cytoskeleton. Written at the college level.

Menzel, Diedrik, ed. *The Cytoskeleton of the Algae*. Boca Raton, Fla.: CRC Press, 1992. A thorough presentation of the cytoskeleton of the major algal groups. Uses structural, physiological, genetic, and molecular approaches to analyze the possible functions of cytoskeletal components. Intended for graduate students and researchers.

Schliwa, M. *The Cytoskeleton*. New York: Springer-Verlag, 1986. Although written as an introduction at a more technical level, this book contains many sections describing the components of the cytoskeleton, cytoskeletal structures, and the history of developments in this field that can be understood by the general reader. The book has one of the best collections of photographs of cytoskeletal structures assembled in one source.

CYTOSOL

Categories: Anatomy; cellular biology

Within each eukaryotic cell are a number of distinct, membrane-bounded structures, generically called organelles, including the nucleus, mitochondria, the endoplasmic reticulum, and chloroplasts (only found in plants, algae and some protists). Each organelle is a specialized structure that performs a specific function for the cell as a whole. The rest of the cell, excluding the organelles, cell wall, and plasma membranes, is called the cytosol: the fluid mass that surrounds and provides a home for the organelles.

The cytosol is organized around a framework of fibrous molecules and protein filaments that constitute the *cytoskeleton*. Although the cytosol consists mostly of water, it contains many chemicals that control cell metabolism, including signal transmission and reception, cellular respiration, and protein transcription factors. The cytosol makes up more than 40 percent of the plant cell volume and contains thousands of different kinds of molecules that are involved in cellular biosynthesis. Because cytosol has so much material dissolved in it, it has a gelatinous consistency.

Function

The cytosol provides locations in the cell where chemical activities and energy transformations responsible for growth, repair, and reproduction can occur. Through *diffusion* and *active transport*, cytosol collects many essential nutrients from its surroundings, including carbon, hydrogen, oxygen, potassium, nitrogen, phosphorus, and a variety of *micronutrients*. The constant motion of the cytosol provides a mechanism for moving and supplying these vital nutrients by ionic transport to the organelles so that they can perform their specific jobs.

Cytosol also assists with the removal of unwanted waste products from the cell. *Glycolysis*, the initial step in cellular respiration, occurs in the cytosol. In cellular energy transactions, the cytosol helps distribute useful energy and dissipate the associated heat.

Cytosol plays a key role in the vital processes of protein production, sorting, and transportation. All plant proteins are synthesized by *ribosomes* in the cytosol. Cytosol provides the medium that assists in the transporting of *messenger ribonucleic acid* (mRNA) to ribosomes, where they synthesize proteins. The first portion of the amino acid sequence of a protein contains a *signal sequence* that is checked by proteins in the cytosol. Signal sequences are like street addresses that tell where the growing protein is to be transported. Some signal sequences direct a ribosome to the *endoplasmic reticulum* (ER), where the protein sequence is completed, is stored in the ER's *lumen* (the space inside the ER), and is eventually transported in a vesicle to another organelle or is exported through the cytosol out of the cell. Proteins lacking a signal sequence are completed by the ribosome and released into the cytosol.

Alvin K. Benson

See also: Active transport; Cell theory; Chloroplasts and other plastids; Cytoplasm; Cytoskeleton; Endoplasmic reticulum; Eukaryotic cells; Mitochondria; Nutrients; Osmosis, simple diffusion, and facilitated diffusion; Plant cells: molecular level; Proteins and amino acids; Ribosomes.

Sources for Further Study

Gunning, Brian E. S., and Martin W. Steer. *Plant Cell Biology: Structure and Function*. Boston: Jones and Bartlett, 1996. Comprehensive overview of plant cells. Includes illustrations, index.

Salisbury, Frank B., and Cleon W. Ross. *Plant Physiology*. 4th ed. Pacific Grove, Calif.: Brooks/Cole, 1999. A fundamental textbook that covers plant cell biology as well as molecular biology and genetics. Includes illustrations, bibliographical references, and index.

Stern, Kingsley Rowland. *Introductory Plant Biology*. 8th ed. Boston: McGraw-Hill, 1999. Introductory botany text discusses all the basic concepts of plants. Includes color photos, glossary, bibliographical references, and index.

DEFORESTATION

Categories: Forests and forestry; environmental issues

Deforestation is the loss of forestlands through encroachment by agriculture, industrial development, nonsustainable commercial forestry, or other human as well as natural activity.

Concerns about deforestation, particularly in tropical regions, have risen as the role that tropical forests play in moderating global climate has become better understood. Environmental activists decried the apparent accelerating pace of deforestation in the twentieth century because of the potential loss of wildlife and plant habitat and the negative effects on biodiversity. By the 1990's research by mainstream scientists had confirmed that deforestation was indeed occurring on a global scale and that it posed a serious threat to global ecology.

Deforestation as a result of expansion of agricultural lands or nonsustainable timber harvesting has occurred in many regions of the world at different periods in history. The Bible, for example, refers to the cedars of Lebanon. Lebanon, like many of the countries bordering the Mediterranean Sea, was thickly forested several thousand years ago. A growing human population, *overharvesting*, and the introduction of grazing animals such as sheep and goats decimated the forests, which never recovered.

Countries in Latin America, Asia, and Africa have also lost woodlands. While some of this deforestation is caused by a demand for tropical hardwoods for lumber or pulp, the leading cause of deforestation in the twentieth century, as it was several hundred years ago, was the expansion of agriculture. The growing demand by the industrialized world for agricultural products such as beef has led to millions of acres of forestland being bulldozed or burned to create pastures for cattle. Researchers in Central America have watched with dismay as large beef-raising operations have expanded into fragile *ecosystems* in countries such as Costa Rica, Guatemala, and Mexico.

A tragic irony in this expansion of agriculture into tropical rain forests is that the soil underlying the trees is often unsuited for pastureland or raising other crops. Exposed to sunlight, the soil is quickly depleted of nutrients and often hardens. The once-verdant land becomes an arid desert, prone to *erosion*, that may never return to forest. As the soil becomes less fertile, hardy weeds begin to choke out the desirable forage plants, and the cattle ranchers move on to clear a fresh tract.

Slash-and-Burn Agriculture

Beef industry representatives often argue that their ranching practices are simply a form of *slash-and-burn agriculture* and do no permanent harm. It is true that many indigenous peoples in tropical regions have practiced slash-and-burn agriculture for millennia, with only a minimal impact on the environment. These farmers burn shrubs and trees to clear small plots of land.

Anthropological studies have shown that the small plots these peasant farmers clear can usually be measured in square feet, not hectares as cattle ranches are, and are used for five to ten years. As fertility declines, the farmer clears a plot next to the depleted one. The farmer's family or village will gradually rotate through the forest, clearing small plots and using them for a few years, and then shifting to new ground, until they eventually come back to where their ancestors began one hundred or more years before.

As long as the size of the plots cleared by farmers remains small in proportion to the forest overall, slash-and-burn agriculture does not contribute significantly to deforestation. If the population of farmers grows, however, more land must be cleared with each succeeding generation. In many tropical countries, traditional slash-and-burn agriculture can then be as ecologically devastating as the more mechanized cattle ranching operations.

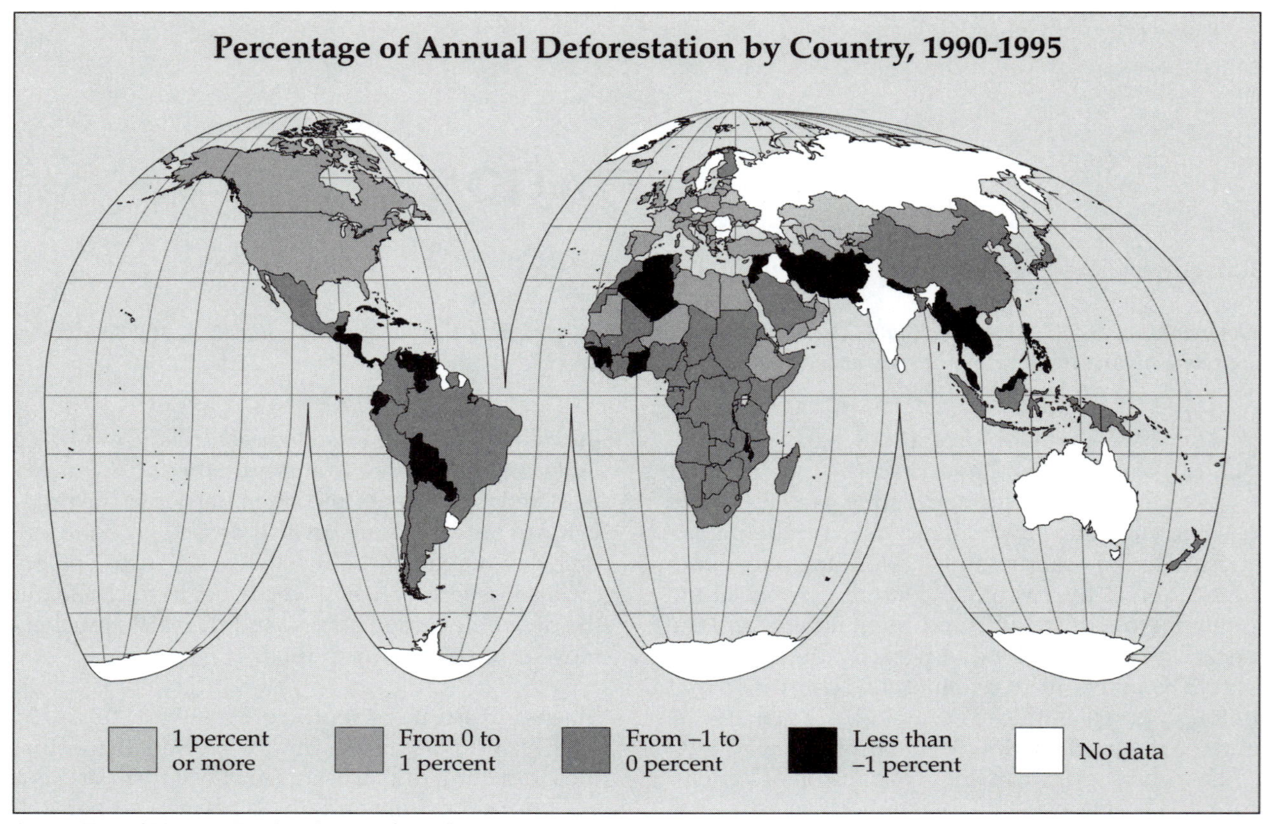

Source: United Nations Food and Agriculture Organization.

Logging

Although logging is not the leading cause of deforestation, it is a significant factor. Tropical forests are rarely clear-cut by loggers, as they typically contain hundreds of different species of trees, many of which have no commercial value. Loggers may select trees for harvesting from each stand. Selective harvesting is a standard practice in sustainable forestry. However, just as loggers engaged in the disreputable practice of *high-grading* across North America in the nineteenth century, so are loggers high-grading in the early twenty-first century in Malaysia, Indonesia, and other nations with tropical forests.

High-grading is a practice in which loggers cut over a tract to remove the most valuable timber while ignoring the damage being done to the residual stand. The assumption is that, having logged over the tract once, the timber company will not be coming back. This practice stopped in North America, not because the timber companies voluntarily recognized the ecological damage they were doing but because they ran out of easily accessible, old-growth timber to cut. Fear of a timber famine caused logging companies to begin forest plantations and to practice sustainable forestry.

While global satellite photos indicate significant deforestation has occurred in tropical areas, enough easily harvested old-growth forest remains in some areas that there is no economic incentive for timber companies to switch to sustainable forestry.

Logging may also contribute to deforestation by making it easier for agriculture to encroach on forestlands. The logging company builds roads for use while harvesting trees. Those roads are then used by farmers and ranchers to move into the logged tracts, where they clear whatever trees the loggers have left.

Environmental Impacts

Despite clear evidence that deforestation is accelerating, the extent of the problem remains debatable. The United Nations Food and Agriculture Organization (FAO), which monitors deforestation worldwide, bases its statistics on measurements taken from satellite images. These data indicate that between 1980 and 1990, at least 159 million hectares (392 acres) of land became deforested. The data also reveal that, in contrast to the intense focus on Latin America by both activists and scientists, the most dramatic loss of forestlands occurred in Asia. The deforestation rate in Latin America was 7.45 percent, while in Asia 11.42 percent of the forests vanished. Environmental activists are particularly concerned about forest losses in Indonesia and Malaysia, two countries where timber companies have been accused of abusing or exploiting native peoples in addition to engaging in environmentally damaging harvesting methods.

Researchers outside the United Nations have challenged the FAO's data. Some scientists claim the numbers are much too high, while others provide convincing evidence that the FAO numbers are too low. Few researchers, however, have tried to claim that deforestation on a global scale is not happening. In the 1990's the reforestation of the Northern Hemisphere, while providing an encouraging example that it is possible to reverse deforestation, was not enough to offset the depletion of forestland in tropical areas. The debate among forestry experts centers on whether deforestation has slowed, and, if so, by how much.

Deforestation affects the environment in a multitude of ways. The most obvious effect is a loss of *bio-*

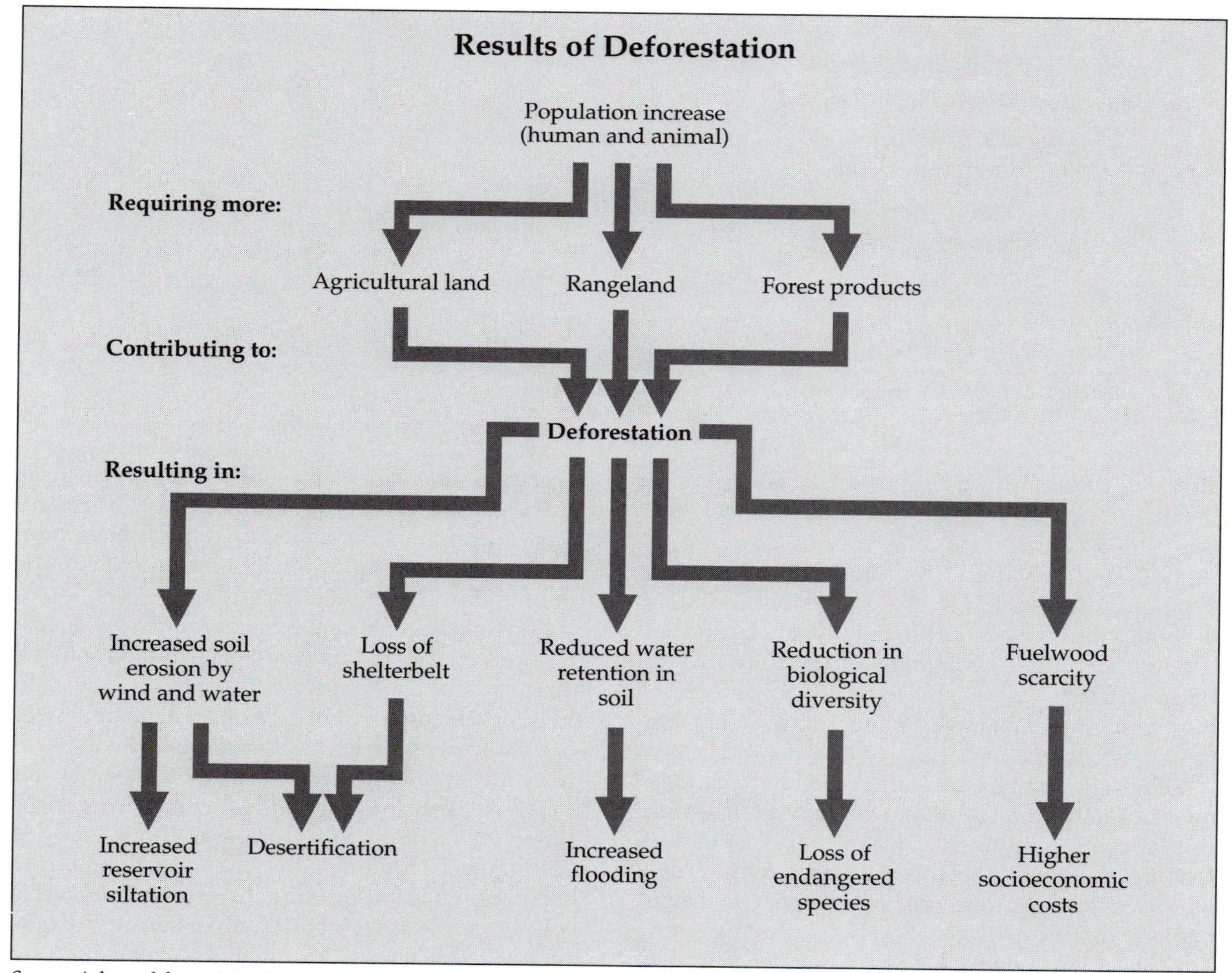

Source: Adapted from A. K. Biswas, "Environmental Concerns in Pakistan, with Special Reference to Water and Forests," in *Environmental Conservation*, 1987.

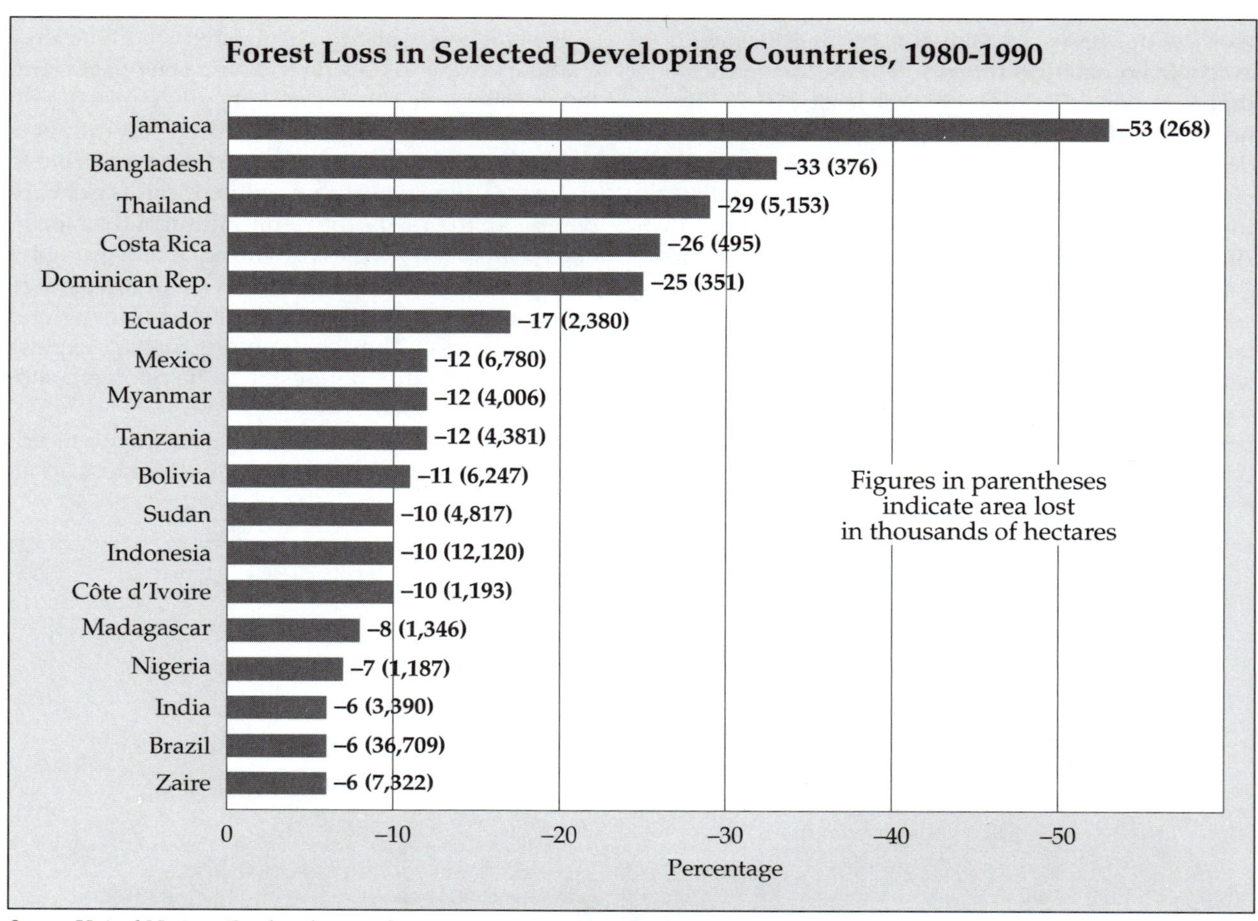

Source: United Nations Food and Agriculture Organization (FAOSTAT Database, 2000).

diversity. When an ecosystem is radically altered through deforestation, the trees are not the only thing to disappear. Wildlife species decrease in number and in variety. As forest habitat shrinks through deforestation, many plants and animals become vulnerable to extinction. Many biologists believe that numerous animals and plants native to tropical forests will become extinct from deforestation before humans have a chance to even catalog their existence.

Other effects of deforestation may be less obvious. Deforestation can lead to increased flooding during rainy seasons. Rainwater that once would have been slowed or absorbed by trees instead runs off denuded hillsides, pushing rivers over their banks and causing devastating floods downstream. The role of forests in regulating water has long been recognized by engineers and foresters. Flood control was, in fact, one of the motivations behind the creation of the federal forest reserves in the United States during the nineteenth century. More recently, disastrous floods in Bangladesh have been blamed on logging tropical hardwoods in the mountains of Nepal and India.

Conversely, trees can also help mitigate against drought. Like all plants, trees release water into the atmosphere through the process of *transpiration*. As the world's forests shrink in total acreage, fewer greenhouse gases such as carbon dioxide will be removed from the atmosphere, less oxygen and water will be released into it, and the world will become a hotter, dryer place. Scientists and policy analysts alike agree that deforestation is a major threat to the environment. The question is whether effective policies can be developed to reverse it or if short-term economic greed will win out over long-term global survival.

Nancy Farm Männikkö

See also: Deserts; Drought; Erosion and erosion control; Forest management; Logging and clear-cutting; Old-growth forests; Rain-forest biomes; Slash-and-burn agriculture; Sustainable forestry.

Sources for Further Study

Bevis, William W. *Borneo Log: The Struggle for Sarawak's Forests*. Seattle: University of Washington Press, 1995. Both disturbing and enlightening as it describes the exploitation of developing nations' resources by industrialized nations.

Colchester, Marcus, and Larry Lohmann, eds. *Struggle for Land and the Fate of the Forests*. Atlantic Highlands, N.J.: Zed Books, 1993. Presents case studies of land and forests from around the world.

Dean, Warren. *With Broadax and Firebrand: The Destruction of the Brazilian Atlantic Forest*. Berkeley: University of California Press, 1997. A solidly researched study of the destruction of one of the world's most endangered forests.

Richards, John F., and Richard P. Tucker, eds. *World Deforestation in the Twentieth Century*. Durham, N.C.: Duke University Press, 1988. One of the few books to discuss the simultaneous reforestation of many industrialized nations and deforestation of developing nations.

Rudel, Thomas K., and Bruce Horowitz. *Tropical Deforestation: Small Farmers and Land Clearing in the Ecuadorian Amazon*. New York: Columbia University Press, 1993. Provides a case study that gives an in-depth analysis of a specific area between 1920 and 1990.

Sponsel, Leslie E., Robert Converse Bailey, and Thomas N. Headland, eds. *Tropical Deforestation: The Human Dimension*. New York: Columbia University Press, 1996. An anthology that shows how deforestation affects human populations.

Vajpeyi, Dhirendrea K., ed. *Deforestation, Environment, and Sustainable Development: A Comparative Analysis*. Westport, Conn.: Praeger, 2001. Contains ten essays analyzing the cause of deforestation in various parts of the world, including the politics involved.

Wilson, Edward O. *The Future of Life*. New York: Alfred A. Knopf, 2001. Harvard biologist Wilson, known for his work on species extinction, speaks out on the staggering impact of human activities on the many species—particularly plant species—worldwide, predicting major extinctions if the present rate continues. He also, however, identifies measures that may be taken to conserve species extinction, given the happy fact that many of them are concentrated in concentrated areas such as the rain forests.

DENDROCHRONOLOGY

Categories: Ecology; forests and forestry; gymnosperms; methods and techniques; paleobotany

Dendrochronology is the science of examining and comparing growth rings in both living and aged woods to draw inferences about past events and environmental conditions.

In forested regions with seasonal climates, trees produce a growth ring to correspond with each growing season. At the beginning of the growing season, when conditions are optimum, the *vascular cambium* produces many files of large *xylem* cells that form wood. As the conditions become less optimal, the size and number of cells produced decreases until growth stops at the end of the growing season. These seasonal differences in size and number of cells produced are usually visible to the unaided eye. The layers produced during rapid early growth appear relatively light-colored because the volume of the large cells is primarily intracellular space. These layers are frequently called *springwood* because in northern temperate regions spring is the beginning of the growing season. Wood formed later, *summerwood*, is darker because the cells are smaller and more tightly compacted. The juxtaposition of dark summerwood of one year with the light springwood of the following year marks a distinct line between growth increments. The width of the ring between one line and the next measures the growth increment for a single growing season. If there is a single growing season per year, as in much of the temperate world, then a tree will produce a single *annual ring* each year.

Tree Rings and Climate

Leonardo da Vinci is credited with counting tree rings in the early 1500's to determine "the nature of past seasons," but it was not until the early 1900's that dendrochronology was established as a science. Andrew Douglass, an astronomer interested in relating sunspot activity to climate patterns on earth, began to record the sequences of wide and narrow rings in the wood of Douglas firs and ponderosa pines in the American Southwest. Originally, trees were cut down in order to examine the ring patterns, but in the 1920's Douglass began to use a Swedish increment borer to remove core samples from living trees. This instrument works like a hollow drill that is screwed into a tree by hand. When the borer reaches the center of the tree it is unscrewed, and the wood core sample inside is withdrawn with the borer. The small hole quickly fills with sap, and the tree is unharmed. Borers range in size from 20 centimeters to 100 centimeters or more in length, so with care, samples can be taken from very large, very old living trees.

Counting backward in the rings is counting backward in time. By correlating the size of a ring with the known regional climate of the year the ring

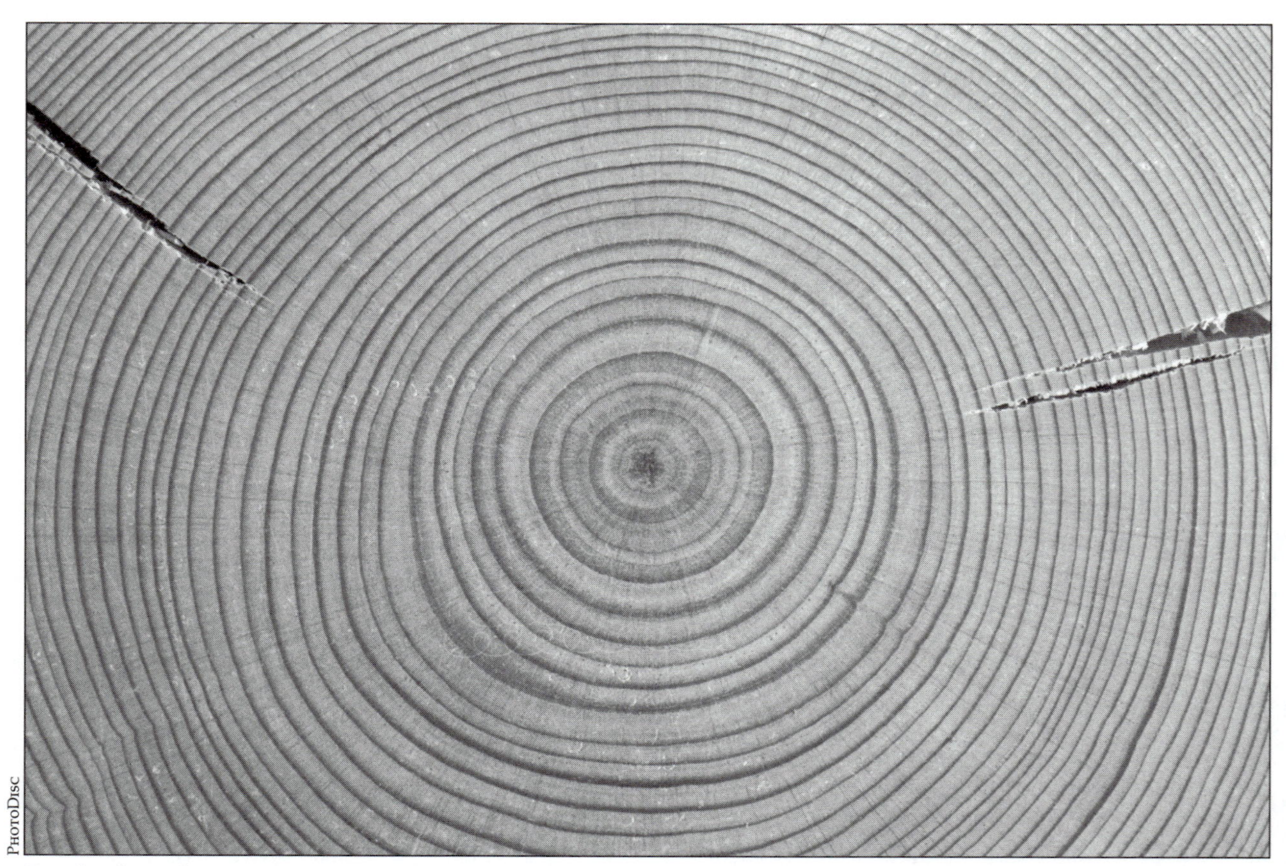

PhotoDisc

Dendrochronology, or tree ring counting, can be used to assess the age of a tree because the width of the ring between one line and the next measures the growth increment for a single growing season.

was produced, a researcher can calibrate a core sample to indicate the surrounding climate during any year of the tree's growth. By extending his work to sequoias in California, Douglass was able to map a chronology extending back three thousand years.

Tree Rings and History

In order to extend his chronologies so far back in time, Douglass devised the method of *cross-dating*. By matching distinctive synchronous ring patterns from living and dead trees of the same species in a region, researchers can extend the pattern further into the past than the lifetime of the younger tree. Archaeologists quickly realized that this was a tool that could help to assign the age of prehistoric sites by determining the age of wood artefacts and construction timbers. In this way archaeologists could calculate the age of pre-Columbian southwestern ruins, such as the cliff dwellings at Mesa Verde, Arizona, by cross-dating living trees with dead trees and the latter with timbers from the sites. In 1937 Douglass established the Laboratory of Tree-Ring Research at the University of Arizona, which continues to be a major center of dendrochronological research.

Fine-Tuning

In the mid-1950's Edmund Schulman confirmed the great age of living bristlecone pines in the Inyo National Forest of the White Mountains of California. In 1957 he discovered the Methuselah Tree, which was more than forty-six hundred years old. The section of forest in which he worked is now known as the Ancient Bristlecone Pine Forest. During the next thirty years, Charles Ferguson extended the bristlecone chronology of this area back 8,686 years. This sequence formed the basis for calibrating the technique of *radiocarbon dating*. In the 1960's, radiocarbon analysis began to be used to determine the age of organic (carbon-based) artefacts from ancient sites. It has the advantage of being applicable to any item made of organic material but the disadvantage of having a built-in uncertainty of 2 percent or more. Tree-ring chronologies provide an absolute date against which radiocarbon analyses of wood samples from a site can be compared.

At about the same time, Valmore LaMarche, a young geologist, began to study root growth of the ancient trees to determine how they could be used to predict the erosional history of a site. By cross-referencing growth ring asymmetry to degree of exposure and slope profiles, he was able to estimate rates of soil erosion and rock weathering, which in turn could be cross-referenced to the climatic conditions predicted by growth rings in the stem. LaMarche and his colleagues, particularly Harold Fritts, continued to "fine-tune" the reading of growth rings to be able to take into account factors such as soil characteristics, frost patterns, and daily, weekly, and monthly patterns.

The Oldest Tree

The Methuselah Tree, mentioned above, is the oldest known living tree. Schulman also cored a forty-seven-hundred-year-old living specimen in the White Mountains, but he did not name it or identify its location. While most of the living specimens older than four thousand years are found in the White Mountains, the oldest living tree was discovered in the Wheeler Peak area of what is now Great Basin National Park in eastern Nevada. This tree, variously known as WPN-114 and the Prometheus Tree, was estimated to be between forty-nine hundred and fifty-one hundred years old when it was cut down in 1964 as part of a research project. The controversy that followed has left many interesting but unanswered questions.

Marshall D. Sundberg

See also: Evolution of plants; Growth and growth control; Petrified wood; Plant tissues; Stems; Wood; Wood and timber.

Sources for Further Study

Cohen, Michael P. *Garden of Bristlecones: Tales of Change in the Great Basin*. Reno: University of Nevada Press, 2000. Includes a good history of the development of dendrochronology.

Cook, E. R., and L. A. Kairiustis, eds. *Methods of Dendrochronology: Applications in the Environmental Sciences*. Boston: Kluwer Academic, 1987. Presents techniques and approaches to the analysis of tree rings, including environmental topics.

Harlow, William M. *Inside Wood: Masterpiece of Nature*. Washington, D.C.: The American Forestry Association, 1970. A well-illustrated introduction to all aspects of wood.

McGraw, Donald J. *Andrew Ellicott Douglass and the Role of the Giant Sequoia in the Development of Dendrochronology*. Lewiston, N.Y.: Edwin Mellen Press, 2001. History examines the development of dendrochronology as a science.

Stokes, Marvin A., and Terah L. Smiley. *An Introduction to Tree-Ring Dating*. Tucson: University of Arizona Press, 1996. One of the most often-cited references in dendrochronology. Well-illustrated, providing good information on the basics of dendrochronology, from mounting core to creating skeleton plots.

DESERTIFICATION

Category: Environmental issues

Desertification is the degradation of arid, semiarid, and dry, subhumid lands as a result of human activities or climatic variations, such as prolonged drought.

Desertification is recognized by scientists and policymakers as a major economic, social, and environmental problem in more than one hundred countries. It impacts about one billion people throughout the world. *Deserts* are climatic regions that receive fewer than 25 centimeters (10 inches) of precipitation per year. They constitute the most widespread of all climates of the world, occupying 25 percent of the earth's land area. Most deserts are surrounded by *semiarid* climates referred to as steppes, which occupy 8 percent of the world's lands. Deserts occur in the interior of continents, on the *leeward* side of mountains, and along the west sides of continents in subtropical regions. All of the world's deserts risk further desertification.

Deserts of the World

The largest deserts are in North Africa, Asia, Australia, and North America. Four thousand to six thousand years ago, these desert areas were less extensive and were occupied by prairie or savanna *grasslands*. Rock paintings found in the Sahara Desert show that humans during this era hunted buffalo and raised cattle on grasslands, where giraffes browsed. The region near the Tigris and Euphrates Rivers in the Middle East was also fertile. In the desert of northwest India, cattle and goats were grazed, and people lived in cities that have long since been abandoned. The deserts in the southwestern region of North America appear to have been wetter, according to the study of tree rings (*dendrochronology*) from that area. Ancient Palestine, which includes the Negev Desert of present-day Israel, was lush and was occupied by three million people.

Scientists use various methods to determine the historical climatic conditions of a region. These methods include studies of the historical distribution of trees and shrubs determined by the deposit patterns in lakes and bogs, patterns of ancient sand dunes, changes in lake levels through time, archaeological records, and tree rings.

The earth's creeping deserts supported approximately 720 million people, or one-sixth of the world's population, in the late 1970's. According to the United Nations, the world's *hyperarid* or *extreme deserts* are the Atacama and Peruvian Deserts (located along the west coast of South America), the Sonoran Desert of North America, the Takla Makan Desert of Central Asia, the Arabian Desert of Saudi Arabia, and the Sahara Desert of North Africa, which is the largest desert in the world. The *arid* zones surround the extreme desert zones, and the semiarid zones surround the arid zones. Areas that surround the semiarid zones have a high risk of becoming desert. By the late 1980's the expanding deserts were claiming about 15 million acres of land per year, or an area approximately the size of the state of West Virginia. The total area threatened by desertification equaled about 37.5 million square kilometers (14.5 million square miles).

Causes of Desertification

Desertification results from a two-prong process: climatic variations and human activities. First,

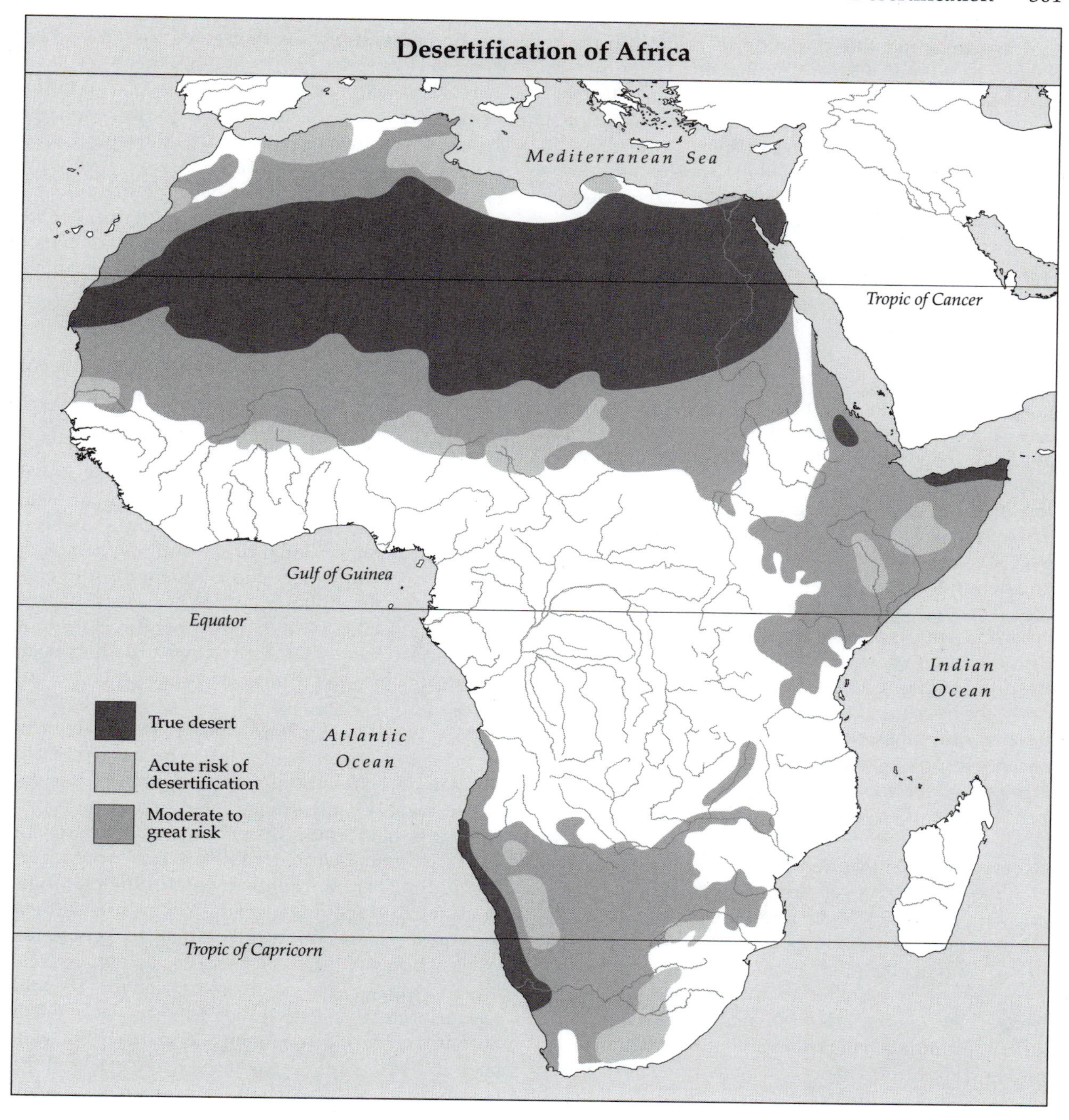

the major deserts of the world are located in areas of high atmospheric pressure, which experience subsiding dry air unfavorable to precipitation. Subtropical deserts have been experiencing prolonged periods of drought since the late 1960's, which causes these areas to be dryer than usual.

The problem of desertification was identified in the late 1960's and early 1970's as a result of severe drought in the Sahel Desert, which extends along the southern margin of the Sahara in West Africa. Rainfall has declined an average of 30 percent in the Sahel. One set of scientific studies of the drought focuses on changes in heat distribution in the ocean. A correlation has been found between sea surface temperatures and the reduction of rainfall in the Sahel. The Atlantic Ocean's higher surface temperatures south of the equator and lower temperatures north of the equator west of Africa are associated with lower precipitation in northern tropical Africa. However, the cause for the change in sea surface temperature patterns has not been determined.

Another set of studies is associated with land-cover changes. Lack of rain causes the ground and soils to get extremely dry. Without vegetative cover to hold it in place, thin soil blows away. As the water table drops from the lack of the natural recharge of the aquifers and the withdrawal of water by desert dwellers, inhabitants are forced to migrate to the grasslands and forests at fringes of the desert. *Overgrazing*, *overcultivation*, *deforestation*, and poor irrigation practices (which can cause salinization of soils) eventually lead to a repetition of the process, and the desert begins to encroach. These causes are influenced by changes in population, climate, and social and economic conditions.

The fundamental cause of desertification, therefore, is human activity. This is especially true when environmental stress occurs because of seasonal dryness, *drought*, or high winds. Many different forms of social, economic, and political pressure cause the overuse of these dry lands. People may be pushed onto unsuitable agricultural land because of land shortages, poverty, and other forces, while farmers overcultivate the fields in the few remaining fertile land areas.

Atmospheric Consequences

A reduction in vegetation cover and soil quality may impact the local climate by causing a rise in temperatures and a reduction in moisture. This can, in turn, impact the area beyond the desert by causing changes in the climate and atmospheric patterns of the region. It is predicted that by the year 2050 substantial changes in vegetation cover in humid and subhumid areas will occur and cause substantial regional climatic changes. Desertification is a global problem because it causes the loss of *biodiversity* as well as the pollution of rivers, lakes, and oceans. As a result of excessive rainfall and flooding in subhumid areas, fields lacking sufficient vegetation may be eroded by runoff.

Greenhouse Effect

Desertification and even the efforts to combat it may be impacting climatic change because of the emission and absorption of greenhouse gases. The decline in vegetation and soil quality can result in the release of carbon, while revegetation can influence the absorption of carbon from the atmosphere. The use of fertilizer to reclaim dry lands may cause an increase in nitrous oxide emissions. Although scientists involved in studies of rising greenhouse gases have not been able to gather evidence conclusive enough to support such theories, evidence of the impact of greenhouse gases on global warming continues to accumulate.

Policy Actions

As a result of the Sahelian drought, which lasted from 1968 to 1973, representatives from various countries met in Nairobi, Kenya, in 1977 for a United Nations conference on desertification. The conference resulted in the Plan of Action to Combat Desertification. The plan listed twenty-eight measures to combat land degradation by national, regional, and international organizations. A lack of adequate funding and commitment by governments caused the plan to fail. When the plan was assessed by the United Nations Environment Programme (UNEP), it found that little had been accomplished and that desertification had increased.

As a result of the 1977 United Nations conference, several countries developed national plans to combat desertification. One example is Kenya, where local organizations have worked with primary schools to plant five thousand to ten thousand seedlings per year. One U.S.-based organization promotes reforestation by providing materials to establish nurseries, training programs, and extension services. Community efforts to combat desertification have been more successful, and UNEP has recognized that such projects have a greater success rate than top-down projects. The Earth Summit, held in Rio de Janeiro, Brazil, in 1992, supported the concept of *sustainable development* at the community level to combat the problem of desertification.

Roberto Garza

See also: Agriculture: modern problems; Climate and resources; Deforestation; Deserts; Drought; Erosion and erosion control; Grazing and overgrazing; Greenhouse effect; Slash-and-burn agriculture; Soil conservation; Soil degradation; Soil salinization; Sustainable forestry.

Sources for Further Study

Bryson, Reid A., and Thomas J. Murray. *Climates of Hunger: Mankind and the World's Changing Weather.* Madison: University of Wisconsin Press, 1977. Provides an overview of climatic change and how it has impacted humankind.

Glantz, Michael H., ed. *Desertification: Environmental Degradation in and Around Arid Lands.* Boulder, Colo.: Westview Press, 1977. A thorough overview of desertification.

Grainger, Alan. *Desertification: How People Make Deserts, Why People Can Stop, and Why They Don't.* London: International Institute for Environment and Development, 1982. A short book on the causes of desertification and how to stop it.

Hulme, Mike, and Mick Kelly. "Exploring the Links Between Desertification and Climate Change." *Environment*, July/August, 1993. Discusses the link between desertification and global warming.

Mainguet, Monique. *Aridity: Droughts and Human Development*. New York: Springer, 1999. At the intersection of environmental science and human biology, this book deals with dry ecosystems—aridity, droughts, and wind—and their influence on soils. Discusses societies affected by these ecosystems and the inventiveness of those living under these conditions.

_______. *Desertification: Natural Background and Human Mismanagement*. 2d ed. New York: Springer-Verlag, 1994. Aims to understand what is commonly called desertification. Each level of technology, excessive or insufficient, creates its own mismanagement. This is reflected in an increase in land degradation and eventually a decrease in soil productivity.

Matthews, Samuel W. "What's Happening to Our Climate." *National Geographic*, November, 1976. Discusses climatic changes for the last 850,000 years.

Postel, Sandra. "Land's End." *Worldwatch*, May/June, 1989. Discusses small-scale efforts to curtail desertification.

DESERTS

Categories: Biomes; ecosystems

Regions characterized by 10 inches or less of precipitation per year as considered deserts. Plants in desert biomes are typically specialized to endure the harsh conditions found there.

Deserts are regions, or biomes, too dry to support grasslands or forest vegetation but with enough moisture to allow specially adapted plants to live. In deserts, hot days alternate with cold nights. Ninety percent of incoming solar radiation reaches the ground during the day, and 90 percent of that is radiated back out into space at night, the result of the absence of clouds, low humidity, and sparse vegetation. The surface of the ground in desert areas is devoid of a continuous layer of plant litter and is usually rocky or sandy. Nutrient cycling in deserts is tight, with phosphorus and nitrogen typically in short supply.

How Deserts Form

Most large landmasses have interior desert regions. Air masses blown inland from coastal areas lose their moisture before reaching the interior. Examples include the Gobi Desert in Mongolia and parts of the Sahara Desert in Africa.

Another factor in the formation of deserts is the *rain-shadow effect*. If moisture-laden air masses bump up against a mountain range, the air mass is deflected upward. As the air mass rises, it cools, and moisture precipitates as rain or snow on the windward side of the mountain range. As the air mass passes over the mountain range, it begins to descend. Because it lost most of its moisture on the windward side, the air mass is dry. As it descends, the air heats up, creating drier conditions on the lee side of the mountain range. Sometimes these differences in moisture are so pronounced that different plant communities grow on the windward and leeward sides.

Latitude can also influence desert formation. Most deserts lie between 15 and 35 degrees north or south latitude. At the equator, the sun's rays hit the earth straight on. Moist equatorial air, warmed by intense heat from the sun, rises. As this air rises, it cools and loses its moisture, which falls as rain; this is why it usually rains every day in the equatorial rain forests. The *Coriolis force* causes the air masses to veer off, to the north in the Northern Hemisphere and to the south in the Southern Hemisphere. The now-dry air begins to descend and warm, reaching the ground between 15 and 35 degrees north and south latitude, creating the belt of deserts circling the globe between these latitudes.

Deserts can also form along coastlines next to cold-water ocean currents, which chill the air above them, decreasing their moisture content. Offshore winds blow the air above cold ocean waters back out to sea. In deserts, rain is infrequent, creating great hardships for the native plants and animals. The main source of moisture for the plants and animals of coastal deserts is fog.

Desert Vegetation

Many typical desert perennial plants, such as members of the *Cactaceae* (the cactus family), have thick, fleshy stems or leaves with heavy cuticles, sunken stomata (pores), and spiny defenses against browsing animals. The spines also trap a layer of air around the plant, retarding moisture loss. Desert plants, many of which photosynthesize using C_4 or CAM (crassulean acid metabolism), live spaced out from other plants. Many desert plants are tall and thin, to minimize the surface area exposed to the strongest light. For example, the entire stem of the Saguaro cactus is exposed to sunlight in the early morning and late afternoon; at noon, only the tops of the stems receive full sun. These traits allow the plants to cope with heat stress and competition for

DIGITAL STOCK

The Gobi Desert, much of which is in Mongolia, is one of the world's largest.

water and avoid damage from herbivores (plant-eating animals).

Where the mixture of heat and water stress is less severe, perennial bushes of the *Chenopodiaceae* (goosefoot family) or *Asteraceae* (sunflower family) form clumps of vegetation surrounded by bare ground. Numerous annuals, called *ephemerals*, can grow prolifically, if only briefly, following rainfall.

Unrelated plant families from different desert areas of the world show similar adaptations to desert conditions. This has resulted from a process called convergent evolution.

Types of Deserts

Deserts are not all alike. Hot deserts are found in lower latitudes, and cold deserts are found at higher latitudes. North America contains four different deserts, which are usually defined by their characteristic vegetation, which ecologists call *indicator species*.

In Mexico's Chihuahuan Desert, lechuguilla (*Agave lechuguilla*) is the indicator species of the Chihuahuan Desert. Fibers from lechuguilla can be made into nets, baskets, mats, ropes, and sandals. Its stems yield a soap substitute, and its pulp has been used as a spot remover. Certain compounds in lechuguilla are poisonous and were once used as arrows and fish poisons. Two of the most common plants in the Chihuahuan Desert are creosote bush (*Larrea divaricata*) and soaptree yucca (*Yucca elata*). Cacti in the Chihuahuan are numerous and diverse, especially the prickly pears and chollas.

The Joshua tree (*Yucca brevifolia*), is the indicator species of the Mojave Desert in Southern California. Nearly one-fourth of all the Mojave Desert plants are *endemics*, including the Joshua tree, Parry saltbush, Mojave sage, and woolly bur sage.

The Great Basin Desert, situated between the Sierra Nevada and the Rocky Mountains, is a cold desert, with fewer plant species than other North American deserts. Great Basin Desert plants are small to medium-size shrubs, usually sagebrushes or saltbushes. The indicator species of the Great Basin is big sagebrush (*Artemisia tridentata*). Other common plants are littleleaf horsebrush and Mormon tea. The major cactus species is the Plains prickly pear.

In the Sonoran Desert in Mexico, California, and Arizona, plants come in more shapes and sizes than in the other North American deserts, especially in the *Cactaceae*. The indicator species of the Sonoran Desert is the saguaro cactus (*Carnegiea gigantecus*).

The Sahara Desert of northern Africa is the world's largest, at 3.5 million square miles. The Northern Hemisphere also contains the Arabian, Indian, and Iranian deserts and the Eurasian deserts: the Takla Makan, Turkestan, and Gobi. Deserts in the Southern Hemisphere include the Australian, Kalahari, Namib, Atacama-Peruvian (the world's driest), and the Patagonian.

Environmental Considerations

Desert ecosystems are subject to disruption by human activities. Urban and suburban sprawl paves over desert land and destroys habitat for plants and animals, some of which are endemic to specific deserts. Farmers and metropolitan-area builders can tap into critical desert water supplies, changing the hydrology of desert regions. Off-road vehicles can destroy plant and animal life and leave tracks that may last for decades.

Carol S. Radford

See also: Biomes: definitions and determinants; Biomes: types; C_4 and CAM photosynthesis; Cacti and succulents; Desertification; Evolution: convergent and divergent; Photosynthesis.

Sources for Further Study

Allaby, Michael. *Deserts*. New York: Facts on File, 2001. Discusses the desert biome in biological, historical, economic, and environmental terms. Covers formation of deserts, climatic events, and adaptations by plants and animals. Includes maps.

Bowers, Janice. *Shrubs and Trees of the Southwest Deserts*. Tucson, Ariz.: Southwest Parks & Monuments Association, 1993. Descriptive guide includes bibliographic references and index.

Larson, Peggy Pickering, and Lane Larson. *The Deserts of the Southwest: A Sierra Club Naturalist's Guide*. 2d ed. San Francisco: Sierra Club Books, 2000. Comprehensive field companion to the plants, animals, geology, topography, climate, and ecology of the American Southwest. Includes line drawings, maps, charts, and diagrams.

DEUTEROMYCETES

Categories: Fungi; taxonomic groups

Deuteromycetes are an artificial group of fungi, of which there exist approximately fifteen thousand species, often referred to as "fungi imperfecti" because their only known reproductive mechanism is asexual.

Deuteromycetes—also known as *Deuteromycota*, *Deuteromycotina*, fungi imperfecti, and mitosporic fungi—are fungi that are unable to produce sexual spores and are therefore placed in their own separate phylum. The deuteromycetes are commonly called fungi imperfecti, that is, "imperfect fungi," a term accepted by many mycologists.

Reproduction

Reproduction in the deuteromycetes occurs in several different forms. Spores, or *conidia*, may be produced directly on the mycelium (the mass of hyphae, or tubular filaments, forming the body of a fungus) or on a structure of specialized mycelial cell called a *condiophore*. Some of these fungi do not produce spores. Nonsporulating fungi are able to propagate themselves by fragmenting the hyphae or by producing a mass of hyphae called a *sclerotium*. Sclerotia can be microscopic in size or as large as several millimeters in diameter.

Conidia can vary in size and shape from small (2-3 microns long) to large (250-300 microns long). Colors can range from clear (hyaline) to a variety of earth tones. Conidia may consist of one to several cells. Shapes of conidia range from simple and oval to elongated and filamentous.

Classification

Classification of the fungi as deuteromycetes is based on the presence of conidia; the kind of conidia according to their shape, color, and size; and whether the conidia are produced in fungal structures called *conidiomata*. Conidiomata may have the shape of a flask made of fungal tissue, called a *pycnidium*; a pin cushion, called a *sporodochium*; or a mass of conidiophores located under either the epidermis or the cuticle of a plant host, called an *aecervulus*.

There are three classes. The *hyphomycetes* contain the fungi that produce conidia and conidiophores on hyphae or groups of hyphae. The *agonomycetes* do not produce conidia. The *coelomycetes* contain the fungi that produce conidia in distinct conidiomata.

Economic and Research Uses

The fungi in the deuteromycetes are extremely important for humanity. Several members of the deuteromycetes are used in industry. Antibiotics, such as penicillin and griseofulvin, are produced by these fungi, especially those of the genus *Penicillium*. These fungi are often found in the soil, and it is believed that they produce antibiotic substances in order to reduce competition with soil bacteria and other fungi.

Enzymes are produced by many of these fungi to enable them to degrade plant residues, from which they obtain nutrients. The enzymes they produce have been used by humans in the manufacture of laundry detergent, paper, and condiments such as soy sauce; the enzymes have also been used in scientific experiments. These enzymes are easily produced under industrial conditions.

Some of the fungi in the genus *Penicillium* are also used in the production of cheeses, including blue cheese and brie. After the cheese is processed and formed into wheels, spores of the fungus are injected into blue cheese, and cheese wheels of brie are dipped into a solution of spores. The cheeses are then allowed to age before entering the market.

Deuteromycetes as Pathogens

Several thousand species of deuteromycetes are pathogenic to plants and plant parts. Many are responsible for the degradation of foods, including decay from rots and molds on grains, vegetables, and fruits. All of the deuteromycetes, like other fungi, are *heterotrophic* (eat or get their food from other organisms) and therefore need to attach to an organic substrate (a living foundation). Food products are excellent substrates for fungi and, within short periods of time, the fungi will consume and destroy these foods. Some fungi produce toxic chemicals that are harmful to those who eat the rotting food. One example is aflatoxin, which is produced by the fungus *Aspergillus flavus*, found on peanuts. A general screening for the fungus can be done using a black light, under which the fungus fluoresces a yellow-green color.

Because all plants and plant parts that serve as food sources for people are affected by deuteromycetes, diseases of plants and animals are one of the more important effects of this fungal group. The fungi use their ability to produce enzymes to enter into growing plant tissue and then destroy the tissue. Annual crop losses caused by fungi in the United States can be measured in billions of dollars. In addition, the small spores of deuteromycetes can affect animals and humans directly. The spores are released into air currents and are blown from place to place. As humans breathe this air, the spores enter the nasal passages and lungs and react with the immune system, creating the allergies.

J. J. Muchovej

See also: Ascomycetes; *Basidiomycetes*; Basidiosporic fungi; Fungi; Lichens; Mitosporic fungi; Mycorrhizae; Yeasts.

Sources for Further Study

Alexopoulos, Constantine J., C. W. Mims, and M. Blackwell. *Introductory Mycology*. 4th ed. New York: John Wiley, 1996. The text details the biology, anatomy, and physiology of fungi. Includes illustrations, bibliographic references, and index.

Deacon, J. W. *Modern Mycology*. 3d ed. Malden, Mass.: Blackwell Science, 1997. An introduction to fungi for botanists and biologists. Emphasizes behavior, physiology, and practical significance of fungi. Contains numerous photographs, line drawings, and diagrams.

Manseth, James D. *Botany: An Introduction to Plant Biology*. 2d ed. Sudbury, Mass.: Jones and Bartlett, 1998. General botany text with section on fungi, their importance, and their biology. Includes illustrations, bibliographic references, and index.

DIATOMS

Categories: Algae; microorganisms; *Protista*; taxonomic groups; water-related life

Diatoms are unicellular microorganisms of the phylum Bacillariophyta *that are abundant in aquatic, semiaquatic, and moist habitats throughout the world, growing as solitary cells, chains of cells, or members of colonies.*

Diatoms, algal organisms of the phylum *Bacillariophyta*, have more than 250 genera and about 100,000 species. A distinctive siliceous cell wall called a *frustule* surrounds each vegetative cell. Diatoms have an extensive fossil record, going back some 100 million years to the Cretaceous period. Deposits of fossil diatoms, known as *diatomite* or *diatomaceous earth*, are mined commercially for use as abrasives and filtering aids. One subterranean marine deposit in Santa Maria, California, is about 900 meters in thickness. More than 270,000 metric tons of diatomaceous earth are quarried annually from a deposit in Lompoc, California. Analysis of fossil diatom assemblages provides important information on past environmental conditions. Although the ancestry of diatoms is obscure, they share similar pigments, food reserves, and plastid structure with the *chrysophytes* and complex, multicellular brown algae.

Classes

Diatoms are frequently separated into two classes based on differences in the symmetry of the cell wall. *Centric diatoms* (class *Centrobacillariophyceae*) may be circular, triangular, or rectangular but typically have surface markings that radiate from a cen-

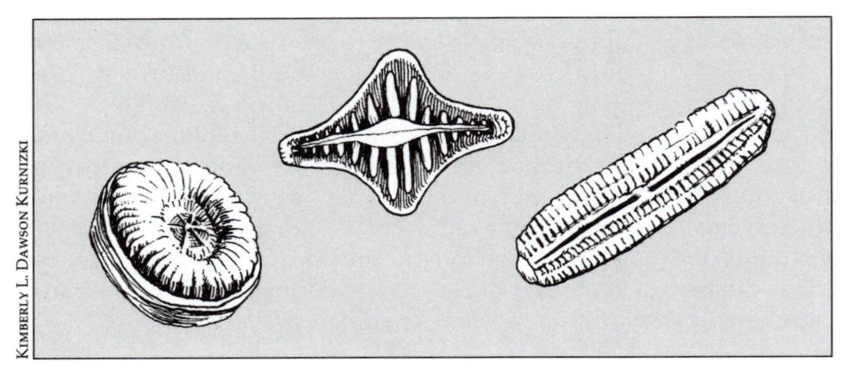

Diatoms are unicellular or colonial algal marine organisms noted for their contribution to phytoplankton, contributing perhaps one-quarter of the world's primary photosynthetic production. They are noted for their frustules: cell walls made of silica and consisting of two halves that overlap each other, with a beautiful variety of ornamentation that makes these tiny organisms appear diverse and jewel-like.

tral area, an arrangement called radial symmetry. *Pennate diatoms* (class *Pennatibacillariophyceae*) are elongated with surface markings at right angles to the long axis, an arrangement called bilateral symmetry.

Cell Wall Components

The hallmark of the diatom is its distinctive and beautifully ornamented, boxlike cell wall, or frustule. The frustule is frequently highly differentiated and is almost always heavily impregnated with silica. An organic layer composed of carbohydrates and amino acids covers the siliceous frustule. The frustule is composed of two halves that fit together like the plates of a petri dish. The larger, overlapping half, the *epitheca*, sits atop the smaller *hypotheca*. The epitheca is composed of a relatively flat upper part (*valve face*) with downturned edges (*valve mantle*), called the *epivalve*, and one or more hooplike girdle bands called the *epicingulum*. Similarly, the hypotheca is composed of a *hypovalve* and a *hypocingulum*. The epicingulum and hypocingulum are collectively known as the *girdle*.

When viewed with the microscope, frustules may be seen in two very different perspectives, depending on the position in which they are lying. If the valve face and outline of the valve are visible, the frustule is said to be in valve view. If the valve mantle and girdle are visible, the frustule is said to be in girdle view.

The varied shapes and beautiful ornamentation of these walls give the diatom cell its intrinsic beauty and have long been used to classify diatoms. Many pennate diatoms possess a *raphe*, a slit along one or both valves, divided by a thickened bridge of silica (a central nodule) and terminated by polar nodules. Raphid pennate diatoms are capable of gliding movement that is caused by the secretion of mucopolysaccharides, derived from vesicles or crystalline bodies, through the raphe. Araphid pennate diatoms lack a raphe and posses a central, unornamented area known as the *sternum* or *pseudoraphe*. No centric diatom has a raphe.

Simple pores (*puncta*), more complex *areolae*, or chambers (*loculi*) are frequently arranged in regularly spaced lines (*striae*) which, in turn, may be strengthened by siliceous ribs (*costae*). Loculi and areolae open externally by a delicate pore plate and internally by a large, round hole. Because the protoplast is completely enclosed by the frustule, the flux of materials across the cell wall primarily occurs through these pores and slits. Additional processes and appendages may extend from the valves of different species.

Electron microscopical studies have revealed two additional openings in the valve: the *fultoportula* (strutted process) and the *rimoportula* (labiate process). The fultoportula consists of a tube that penetrates the wall and is supported internally by two or more buttresses. Fultoportulae are confined to a single order of centric diatoms. The rimoportula, found in centric and pennate diatoms, consists of a tube that opens to the outside by a simple aperture and internally by a longitudinal, liplike slit. Some researchers suggest that the raphe system may have evolved from one or more rimoportulae.

Reproductive Strategies

The most common mode of reproduction in diatoms is asexual by cell division of a diploid vegetative cell to produce two daughter cells. Following mitosis the protoplast expands, pushing apart the valves, and divides by furrowing. Each daughter cell receives the epitheca of the parent and forms a new hypotheca within a *silica deposition vesicle*. The daughter cell that receives the original epitheca remains the same size as the parent. However, the daughter cell that receives the original hypotheca forms a new hypotheca and is usually smaller than the parent cell. Thus, the average cell size of a population of diatoms may become progressively smaller during the growing season. The maximum size of the population is restored during sexual reproduction.

During sexual reproduction, diploid vegetative cells divide by meiosis to form haploid gametes (eggs and sperm in centric diatoms or amoeboid gametes in pennate diatoms). Fusion of the gametes results in a diploid zygote that enlarges to several times its original size by the uptake of water and forms valves. This enlarged zygote (*auxospore*) has a different valve morphology from that of the valves of vegetative cells. Auxospores, which do not serve as resting spores, divide by mitosis to form vegetative cells with the maximum size for the species.

Some centric diatoms form *resting spores* in response to the availability of various nutrients (especially nitrogen), temperature, light intensity, and pH. Resting spores are short cells with thick walls that differ from the walls of vegetative cells. Resting spores are usually formed within the frustule of a vegetative cell. A vegetative cell may give rise to one, two, or four resting spores. Resting spores germinate in light when environmental conditions improve.

Flagella

Flagellated cells are found only in the male gametes (*spermatazoids*) of some centric diatoms. Each spermatozoid bears a single flagellum covered with stiff tubular hairs (*mastigonemes*). Based on ultrastructural studies of two species, it appears that diatoms lack the normal 9 + 2 arrangement of microtubules in the shaft of the flagellum (*axoneme*). The two central microtubules are missing, leaving nine peripheral doublets (9 + 0). Further-

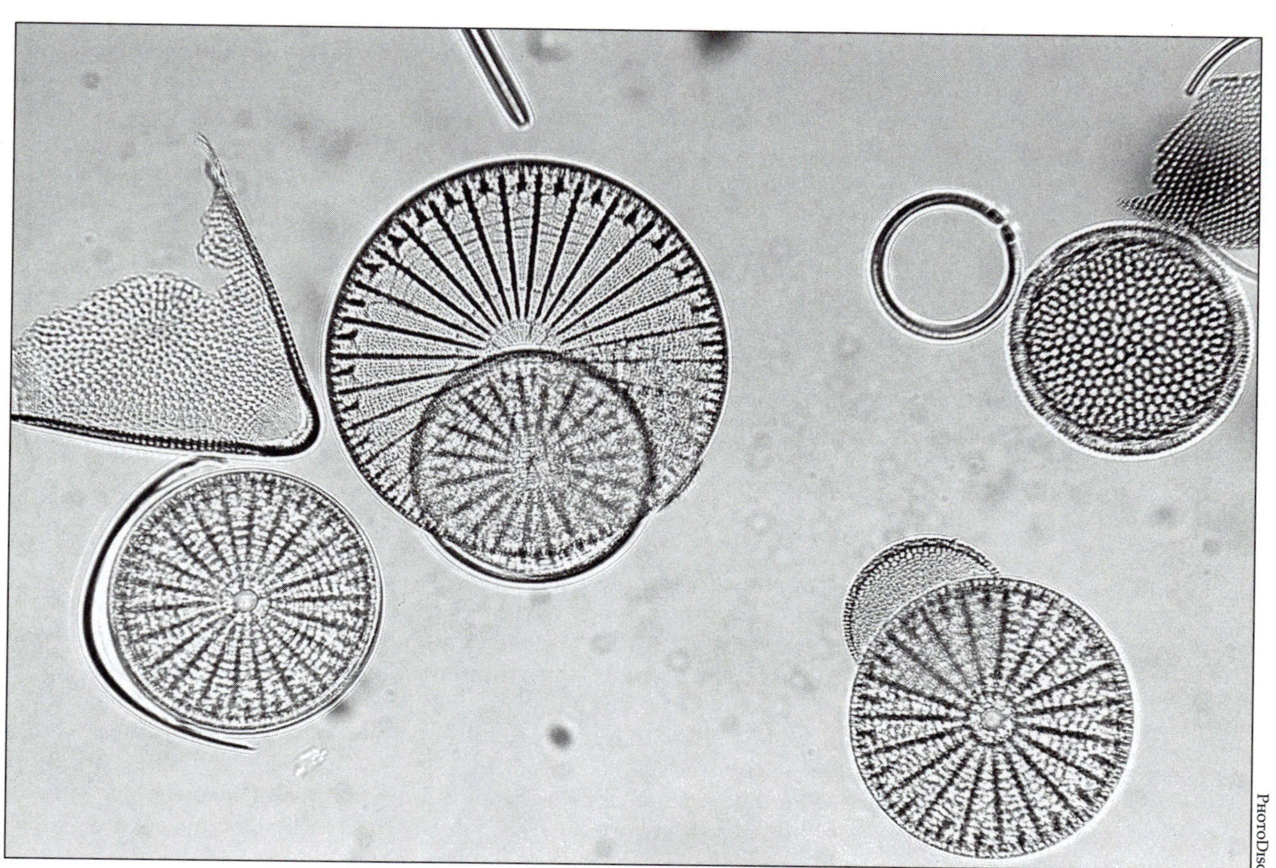

PhotoDisc

Micrograph showing diatoms in fresh water.

more, the basal body of these species consists of nine doublets of microtubules, instead of the normal nine triplets found in other eukaryotes.

Food Reserve

The most important carbohydrate food reserve is the *chrysolaminarin*, a beta-1,3-linked glucan, which is stored in special *vacuoles*. Chrysolaminarin (a water-soluble polysaccharide) is also the most important food reserve in the brown algae and chrysophytes. Diatoms may also accumulate various lipids. The fatty acid composition of these lipids differs somewhat from that found in the green algae and higher plants, notably in the absence of linolenic acid from most species. Lipids may be stored within or outside vacuoles.

Photosynthetic Pigments

All diatoms contain *chlorophylls a* and c_2. Chlorophylls c_1 and c_3 may also be present; however, chlorophyll *b* is never found in diatoms. Although the coloration of diatoms varies, living diatoms are frequently brown due to the presence of the accessory pigment *fucoxanthin*, which masks the green coloration of the chlorophylls. Diatoms also contain other *xanthophylls* (neofucoxanthin, diadinoxanthin, and diatoxanthin) and beta-carotene.

The photosynthetic pigments are stored within membrane-bound organelles called *plastids*. Many pennate diatoms have two large plastids, while centric diatoms generally have a large number of small discoid plastids. Four membranes surround the plastids of diatoms: a double-membrane envelope and a layer of endoplasmic reticulum that is continuous with the nuclear envelope. Each plastid contains more or less parallel lamellae composed of three stacked, flattened sacs (*thylakoids*) and at least one *pyrenoid*, which does not appear to be directly associated with any food reserve product.

Habitat

Diatoms are usually a major component of *benthic* and *planktonic* communities in all but the hottest and saltiest waters. While most diatoms live in water, a few species grow in damp soil and can tolerate extreme drought and heat for some time. Pennate diatoms are common members of the benthos and plankton in marine and freshwater habitats, while centric diatoms are more commonly found in the marine plankton. Some benthic diatoms grow attached to rocks, sand grains, other algae, aquatic plants, and even animals. Other benthic diatoms (usually raphid pennates) live freely on the surface of, or in, the substrate. Planktonic diatoms often produce spring and fall blooms in temperate lakes and oceans and summer blooms at higher altitudes. In coastal waters and lakes, they may produce resting spores to survive between growing seasons. One estimate suggests that 20-25 percent of the total primary production on earth is contributed by marine planktonic diatoms.

Michael C. Amspoker

See also: Algae; Aquatic plants; Brown algae; Chrysophytes; Marine plants; Phytoplankton; *Protista*.

Sources for Further Study

Hoek, C. van den, D. G. Mann, and H. M. Jahns. *Algae: An Introduction to Phycology*. New York: Cambridge University Press, 1995. An in-depth consideration of the biology of diatoms and other algal groups.

Lee, R. E. *Phycology*. 2d ed. New York: Cambridge University Press, 1989. A solid introduction to diatoms and other algae.

Round, F. E. *The Ecology of Algae*. New York: Cambridge University Press, 1981. An outstanding reference for those interested in the ecology of diatoms and other algae.

Round, F. E., R. M. Crawford, and D. G. Mann. *The Diatoms: Biology and Morphology of the Genera*. New York: Cambridge University Press, 1990. A summary of what is known about the cell structure, life cycle, and ecology of diatoms. More than twenty-five hundred scanning electron micrographs plus additional light micrographs and line drawings illustrate many diatom genera.

Sze, P. *A Biology of the Algae*. 3d ed. Boston: WCB/McGraw-Hill, 1998. A well-illustrated introduction to diatoms and other algal groups.

DINOFLAGELLATES

Categories: Algae; microorganisms; *Protista;* taxonomic groups; water-related life

Dinoflagellates, phylum Dinophyta, *are unicellular and colonial algal organisms from the kingdom* Protista *named for the spinning motions that result from the movement of their flagella.*

The two thousand to four thousand species that make up the *Dinophyta* phylum typically have two flagella. Dinokonts (*Dinophyceae*) have one flagellum running in a groove that cuts transversely across the cell and another flagellum, the sulcus, that runs backward in a longitudinal groove and is more or less perpendicular to the transverse one. In desmokonts (*Desmophyceae*), both flagella arise from a point at the front of the cell. The motions of the flagella, which make dinoflagellates spin like a top, help give rise to the name, for *dinos* in Greek means "whirling," and *flagella* means "whip." Single-celled species are the most common, but colonial species exist. The largest dinoflagellate, *Noctiluca,* may grow as large as 2 millimeters in diameter.

In the active phase of their life cycles, dinoflagellates come in two forms, unarmored (naked) or armored (thecate). All species have a complex outer covering, consisting of an outer membrane, flattened vesicles in the middle, and a continuous inner membrane. In thecate forms, however, the vesicles contain plates made of cellulose or some other polysaccharide. The plates may form a structure as simple as a bivalve-type shell; however, in some species they form wings and other appendages that give the beholder the appearance of some fantastic alien spaceship.

Cell Characteristics

Dinoflagellates are not members of the kingdom *Plantae* but rather are protists, and they have both plant and animal characteristics. Some species are autotrophic—in other words, they have their own chloroplasts and can produce their own sugars and organic materials through photosynthesis. Other species are heterotrophic—they have no chloroplasts and typically must prey on or parasitize other organisms or consume organic detritus in order to obtain nourishment. Some of the heterotrophic species, however, can acquire the chloroplasts of prey and become photosynthetic themselves.

The chloroplasts in normally photosynthetic dinoflagellates are unique in that the plastids are enclosed within a triple membrane, rather than the double membrane of chloroplasts of most other organisms, and also in the fact that the chloroplasts in some cases have their own nuclei. These two characteristics, when considered with the large number of nonphotosynthetic dinoflagellates, have led some to argue that dinoflagellate chloroplasts have been secondarily acquired from a eukaryotic endosymbiont. Photosynthetic pigments include chlorophylls *a* and *c*, carotenoids, and xanthophylls.

Photosynthetic dinoflagellates often form symbioses with other aquatic organisms such as sponges, cnidarians (jellyfish, sea anemones, and corals), molluscs (bivalves, gastropods, octopuses, and squids), turbellarians, and tunicates. These symbiotic dinoflagellates, or *zooanthellae,* lack armor. They carry out most of the photosynthesis that fuels the productivity of coral reefs.

Although classified as protists, dinoflagellates have cellular nuclei with characteristics intermediate between those of prokaryotes and those of eukaryotes. As in eukaryotes, the nucleus is surrounded by a nuclear membrane and contains a nucleolus. However, the chromosomes are attached to the nuclear membrane in such a way that chromosomes remain attached to the inner wall of the cell membrane in prokaryotes. Dinoflagellates are also highly unusual in that they have permanently condensed chromosomes—and dinoflagellate deoxyribonucleic acid (DNA) does not form a complex with proteins, as in typical eukaryotic cells. There remains disagreement over whether dinoflagellate characteristics represent some ancient evolutionary lineage or more recent derivation.

Life Cycle

Arguably the most accomplished shape shifters in the living world, dinoflagellates have incredibly complex life cycles. For example, *Pfiesteria* has at least twenty-four distinct stages in its life cycle, one of several reasons why scientists who work with the organism call it the "cell from hell." Dinoflagellate life cycles may include dormant cysts, cells without flagella (including amoeba-like stages), and more typical biflagellated cells.

Dinoflagellates may reproduce sexually or asexually. The cells are generally haploid, except for a zygote produced by the union of two gamete cells during sexual reproduction. The zygote undergoes meiosis shortly after fertilization. Dinoflagellate cells divide asexually in three ways: The parent cell of a naked dinoflagellate simply constricts and pinches off into two daughter cells; some armored types shed the theca prior to or during division; and other armored types split the parental theca, dividing the portions between the daughter cells. Unfavorable environmental conditions may trigger sexual reproduction as well as the formation of dormant cysts. Cysts may be transported large distances by currents, which in large part explains the dispersal of toxic dinoflagellate blooms up and down coastlines.

Pfiesteria exhibits all three forms, with the amoeboid and flagellated stages being toxic to fish. Encysted stages lie dormant in the bottom sediments of estuaries. The active amoeboid and flagellated stages are usually nontoxic. Amoeboid stages, which either inhabit the sediments or are free-swimming, consume bacteria, algae, small animals, or bits of fish tissues. Flagellated stages may ingest prey in a similar fashion but often siphon off the tissue of their prey through a cytoplasmic extension called the peduncle.

In the presence of an environmental trigger—such as substances given off by a school of live fish—amoeboid, flagellated, and encysted cells activate into toxic forms that swim toward the prey. The toxic forms then secrete toxins that immobilize and injure the prey with ulcerated, bleeding sores. *Pfiesteria* then feeds off materials that leak from the sores. Once the fish die, flagellated cells transform into amoeboid forms that feed on the carcass. If conditions suddenly become unfavorable, the active forms encyst and sink to the bottom. The entire cycle can take place in a matter of hours.

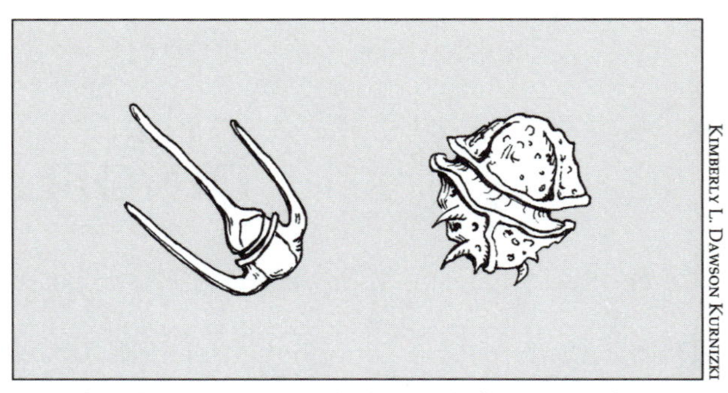

Dinoflagellates are unicellular or biflagellate algal organisms known for their whirling or spinning motion.

Red Tides and Toxins

Dinoflagellates are responsible for most of the *red tides* or *brown tides* that sicken and kill aquatic organisms and humans worldwide. Red tides are known from biblical times; one of the ten plagues reported to have been visited upon Egypt in the Book of Exodus (8:20-21) was most likely a red tide. Red tides were also known in ancient China and among Native Americans in Alaska and the Pacific Northwest.

Typically, the organisms that cause red and brown tides cause no harm until their populations explode or bloom. Adverse effects to other organisms result from oxygen depletion by irritation to skin and other organs, by the blocking of sunlight (in cases where the bloom is visible), or by the production of toxic substances, as in the case of *Pfiesteria*. In general, it is the toxic substances that sicken or kill humans and other vertebrates, such as manatees and birds. There is some controversy over whether human activities in estuarine and coastal waters have caused an increase in the frequency of these *algal blooms*.

In the mid-1990's a spate of horrific fish kills in the estuaries of North Carolina, Virginia, and Delaware began raising alarms up and down the East Coast of the United States. Ghastly lesions appeared on the affected fish, as if they were being eaten alive. People who spent a lot of time on or near the afflicted waters were affected, too, with symptoms ranging from memory loss to skin lesions. The single-celled culprit turned out to be *Pfiesteria piscicida*.

Several toxic syndromes that affect humans are

caused by dinoflagellates: ciguatera fish poisoning, caused by toxins produced by *Gambierdiscus* and other species; paralytic shellfish poisoning, caused by toxins produced by *Alexandrium* and other species; neurotoxic shellfish poisoning, caused by toxins produced by *Gymnodinium breve*; diarrhetic shellfish poisoning, caused by toxins produced by *Dinophysis* species; and *Pfiesteria*-associated syndrome.

Bioluminescence

Dinoflagellates, besides being among the deadlier marine microorganisms, are also among the most beautiful. They are responsible for the *bioluminescence* of the sea that has enchanted mariners for millennia. The light is created through the reaction of oxygen with a substrate, luciferin (which has no relation to luciferins responsible for phosphorescence in other organisms such as lightning beetles), which is catalyzed by an enzyme, luciferase. Bioluminescence in dinoflagellates follows a circadian rhythm in which the maximum occurs at night.

David M. Lawrence

See also: Algae; Animal-plant interactions; Aquatic plants; Bioluminescence; Chloroplasts and other plastids; Chromosomes; Eukaryotic cells; Marine plants; Mitosis and meiosis; Phytoplankton.

Sources for Further Study

Barker, Rodney. *And the Waters Turned to Blood*. New York: Touchstone, 1998. A journalistic account of the havoc caused by and investigation of *Pfiesteria piscicida*.

Graham, Linda E., and Lee W. Wilcox. *Algae*. New York: Prentice-Hall, 2000. An overview of the biology of algae, with emphases on systematics, biogeochemistry, and environmental and economic effects.

Raven, Peter H., Ray F. Evert, and Susan E. Eichhorn. *Biology of Plants*. 6th ed. New York: W. H. Freeman/Worth, 1999. Includes a chapter on protists, including dinoflagellates, with plantlike characteristics.

Ruppert, Edward E., and Robert D. Barnes. *Invertebrate Zoology*. 6th ed. Fort Worth, Tex.: Saunders College Publishing, 1994. A zoological overview of the dinoflagellates.

DISEASES AND DISORDERS

Categories: Diseases and disorders; pests and pest control

The science and study of plant diseases is known as plant pathology, which can be briefly defined as the study of the nature, cause and control of plant disease.

Plant disease is as old as land plants themselves, as shown by the fossil record. Several biblical accounts of plagues have been attributed to plant diseases, and in Roman times cereal rust was so serious that an annual ritual, the Robigalia, was performed to appease the Rust God, Robigo. In the mid-nineteenth century the Irish Potato Famine, a result of *potato late blight* disease, caused the deaths of some 800,000 persons and the emigration of about 1.5 million more, mostly to North America. Similarly, but to a lesser extent, *brown spot* disease of rice caused the Bengal famine of 1943 in India. Plant diseases continue to cost billions of dollars annually worldwide. The combined costs of lost yield, reduced quality, and costs of pesticides and other control measures are inevitably passed on to consumers. No type of plant is free from disease.

Causes and Types

Plant disease results from the continuing action of an irritant that can be either physical or biological. Physical, or abiotic, causes of disease include water stress (either from excess or insufficient

Symptoms of Plant Diseases

Symptom	*Injurious Effects, Sample Diseases*
Conversion of host	Fungal hyphae ramify and totally consume tissue into masses of host tissues, eventually replacing them with pathogen tissue: hard or powdery masses of fungal tissue. Includes rye ergot, cereal smuts.
Leaf mosaics, mottling	Irregular patterns of chlorophyll loss, malformations, localized leaf-cell death, and/or enhanced growth by other leaf cells. Includes tobacco mosaic virus, peach leaf curl disease.
Leaf spots, leaf blights	Death of leaf cells from parasitism and /or toxin release from the pathogen; limited to discrete, necrotic spots, or may coalesce into larger lesions covering most or all the leaf. Includes rose black spot, bean bacterial blight, potato late blight.
Overgrowths, galls	Localized overgrowth of host tissue, typically on stems, trunks, or roots, resulting in fleshy to woody galls. Includes root knot nematode, crown gall disease, pine rust galls.
Root rots	Debilitation or death of roots by soil-borne pathogens, reducing the plant's ability to absorb water and minerals. Includes common root rots of most plants.
Storage rots/molds	Ramification of fleshy storage tissue by fungi or bacteria, accompanied by release of wall-softening enzymes and sometimes by substances toxic to humans and animals. Includes potato soft rot, *Aspergillus* mold of peanut (with aflatoxin formation).
Stunted growth	Reduced activity of various meristematic tissues, resulting in dwarfed, often malformed plants. Includes peanut stunt virus.
Trunk and/or stem blight	Death of outer tissues of host stems; cankers on trunks resulting from parasitic and toxic action of the pathogen. Includes fire blight of pear, chestnut blight.
Wilts	Physical plugging of xylem vessels by the pathogen, with toxins that injure plant vascular tissue and/or formation of ballonlike extensions (tyloses) from adjacent host cells. Includes Dutch elm disease.

quantities), poor nutrition, improper soil acidity, and other environmental factors. Brief, damaging effects such as hail, wind, and lightning are considered injuries, not diseases.

Biological, biotic, causes of disease include bacteria, viruses, fungi, nematodes, and other microorganisms acting as disease-causing agents, or *pathogens*. Such pathogens infect plants, colonize tissues, and extract nutrients by living as *parasites*. Those actions often result in disruptions of normal physiological processes in plants, including photosynthesis, water uptake and movement, nutrient transport, and reproduction. In some instances anatomical abnormalities such as stunting, growth distortions, and gall formation are induced. The resulting physiological and anatomical abnormalities constitute disease *symptoms*.

Typically, and most important to producers and consumers, yield is reduced, both in quantity and in quality. In addition to diseases of growing plants, there are *postharvest diseases*. Those include fruit and vegetable rots and decay of stored grains. They may begin either before or after harvest, but in either case such diseases can continue long after harvest, further reducing the quality and value of food, fiber, and feed products. Some pathogens also produce toxins in plants they infect. Those toxins can be injurious, sometimes even fatal, to humans or animals that eat the contaminated plant materials.

Three interacting components are required for any plant disease to occur: (1) a susceptible host plant, (2) either a biotic pathogen or an abiotic, nonliving, causal agent, and (3) environmental conditions favorable for development of disease. Each of those components may vary in their contribution to overall severity of the resulting disease. The most important influences are the existing level of genetic resistance or susceptibility of the host plant; the existing level of *virulence*, or ability to cause disease, of the pathogen; for abiotic causes, the plant's degree of nutrient deficiency or water deficiency; and the degrees of favorable or unfavorable temperature, humidity, light intensity or other environmental factors that influence the health and vitality of both host and pathogen. A fourth component, a *vector*, may be required for some diseases. A vector is a second organism, most commonly an insect, that transmits the pathogen from a diseased plant to a healthy plant and injects or otherwise introduces the pathogen into the plant while feeding or laying eggs. Most pathogens do not require vectors.

Kinds of Plant Pathogens and Typical Diseases

Most fungi, bacteria, and nematodes are free-living organisms in nature. They generally contribute to the ecosystem and cause no deleterious effects on plants. Only a small percentage of microorganisms are plant-parasitic or pathogenic. The most important plant pathogens include fungi, bacteria and related forms, viruses, viroids, and nematodes; each group contains species that cause serious, economically important diseases. Fungi constitute the largest number of plant pathogens. By their very nature, viruses and viroids function only as parasites within plant or animal cells and cannot exist as free-living organisms. Nematodes are microscopic, wormlike animals that most commonly reside in soil and feed externally or internally on plant roots.

Plant Disease Control and Management

Plant pathology, in contrast to human and veterinary medicine, focuses very little on diseases of individuals but rather on large populations. Exceptions include certain high-value individuals, such as trees of historic or particular aesthetic or economic value. Emphasis is largely on prevention of plant diseases rather than curative therapy. Control of plant disease, in absolute terms, is usually impossible or economically impractical to achieve. The realistic goal is more commonly management of plant disease, with the goal of achieving a level of disease prevention or reduction that is both economically and environmentally sound. Briefly, most plant disease management practices fall within one of five broad categories: exclusion, eradication, resistance, protection or therapy, and adaptation to or modification of cultural conditions or practices.

Exclusion is the practice of keeping pathogens separated from their host plants and can be accomplished in several ways. Quarantines between countries are commonly used, with varying degrees of success, to prevent importation of pathogens into countries where they currently do not exist. Success of quarantines often depends on the extent and physical nature of the separation between the countries and whether the pathogen is wind-, soil-, or seed-borne. A variant of exclusion is *evasion* or *avoidance* of the pathogen: for example, growing plants susceptible to certain bacterial pathogens only in irrigated, semiarid regions unfavorable to the pathogen. The use of pathogen-free, certified seed or pathogen-free nursery stock are other examples of exclusion aimed at preventing or limiting introduction of pathogens into new areas.

Eradication is the practice of killing pathogens either in the environment or in and on the diseased plant. This limits infection, helps prevent further spread to healthy plants in the area, and helps prevent introduction of pathogens by shipment of diseased plant parts into regions where they currently are not present. Eradication practices such as chemical treatment of soil prior to planting, sanitation procedures such as removing infected crop debris, and long-duration crop rotations can all be successful to various degrees. Eradication and destruction of infected, living plants are often practiced for diseases of orchard crops and landscape trees, such as citrus canker and Dutch elm disease.

Resistance, resulting from innate genetic properties of plants or by directed plant breeding, is usually the most cost-effective means of managing plant diseases. Unfortunately, resistant cultivars may succumb to new, more virulent, or more aggressive strains of the pathogen that can arise by natural selection from populations of the pathogen in nature. Several examples also exist of successful application of *genetic modification* or genetic engineering to the development of resistant varieties, by introducing genes from other species, including nonrelated plants, viruses, or bacteria. Successful breeding of plants resistant or more tolerant to plant diseases provides billions of dollars' worth of benefit annually to producers of all types of plants.

Protection is the practice of treating susceptible plants in some manner, often chemically, to prevent infection once a pathogen reaches the plant surface by growth of the pathogen through the soil to seeds or roots, by deposition of windborne fungal spores, or by splash-drop dispersal of bacteria onto aerial plant surfaces. Protective fungicides or bactericides must be on the seed or plant surface prior to arrival of a pathogen to be effective. Protective fungicides do not kill pathogens; rather, they prevent spore germination, host penetration, or other early stages of infection, if spore germination does occur. Protective fungicides and bactericides are widely used throughout plant agriculture. Some recently developed protective fungicides also have limited eradicative activity. Those fungicides not only function as surface protectants but also enter plant tissue and kill pathogens already inside the plant—an example of therapeutic chemical action.

Adaptation to and *modification* of cultural conditions or practices are used to take advantage of conditions that benefit the plant but are also detrimental to development and spread of the pathogen. Examples include planting when soil temperatures are favorable for germination and subsequent growth of seedlings, altering soil pH by addition of lime or sulfur, not overfertilizing, increasing or lowering temperature, lowering humidity, and increasing airflow in greenhouses. The goal of such practices is to foster plant health and survival as well as prevent or reduce plant disease.

Often it is advantageous to use several of these approaches simultaneously. By combining natural host resistance, even if incomplete, with manipulation of environmental conditions unfavorable to the pathogen, by utilizing pathogen-free seed or nursery stock, or other practices, growers can often reduce significantly the amount of chemical pesticides required to manage a disease. Such a combined approach to disease and insect management is known as *integrated pest management* (IPM). Major goals of IPM are to reduce significantly the amount of pesticides going into the ecosystem while limiting plant disease at an economically acceptable level.

Larry J. Littlefield

See also: Acid precipitation; Agriculture: modern problems; Air pollution; Ascomycetes; Bacteria; Biopesticides; Endangered species; Fungi; Genetically modified bacteria; Genetically modified foods; Herbicides; Integrated pest management; Oomycetes; Pesticides; Resistance to plant diseases; Rusts; *Ustomycetes*; Viruses and viroids.

Sources for Further Study

Agrios, George N. *Plant Pathology*. 4th ed. San Diego: Academic Press, 1997. The standard, most comprehensive college-level textbook of plant pathology in the United States. Covers basic concepts of the discipline plus information on all types of plant diseases and their pathogens; well illustrated.

Schumann, Gail L. *Plant Diseases: Their Biology and Social Impact*. St. Paul: APS Press, 1991. Overview of a wide range of plant diseases and pathogens, with emphasis on basic biology and economic, social, and historical impacts of plant diseases.

DNA: HISTORICAL OVERVIEW

Categories: Cellular biology; genetics; history of plant science

The determination of the structure of DNA was preceded by years of research. Its discovery led to an avalanche of significant discoveries in molecular biology. The most important feature of the double-helix model was that it suggested how DNA might produce an exact copy of itself.

Deoxyribonucleic acid, or DNA, a molecule at the core of life itself, has existed for at least a few billion years. However, it was not until 1869 that Friedrich Miescher, a Swiss biochemist, was able to isolate it, though in a highly impure form. He discovered DNA in Tübingen, Germany, when he was a young postdoctoral student doing research on the chemistry of white blood cells. Through a series of chemical operations performed on pus cells, which have particularly large nuclei, he isolated a previously unobserved gelatinous substance. Analysis revealed that the new material contained phosphorus and nitrogen; Miescher named it *nuclein* because of its origin in cell nuclei. He recognized that the new substance's acidity was attributable to phosphoric acid.

Early Research

In the late nineteenth century, Albrecht Kossel, a German physiological chemist, discovered that nuclein contained *purines*, nitrogenous compounds with two rings, and *pyrimidines*, nitrogenous compounds with single rings. Furthermore, he showed that these nucleic acids were made of two different purines, adenine and guanine (A and G), and two different pyrimidines, thymine and cytosine (T and C). He also found a sugar among nucleic acid's decomposition products, but he was unable to identify it. Because nucleic acids were difficult to analyze, it was not until the first decade of the twentieth century that scientists realized that there are two types of *nucleic acid:* the thymus type, which is now known to be rich in DNA, and the yeast type, which is rich in *ribonucleic acid* (RNA).

Phoebus Levene, a Russian-American chemist who studied for a time with Kossel, was able to identify the sugar in yeast nucleic acid as ribose in 1908, but he did not definitively identify deoxyribose in thymus nucleic acid until 1929. He established that thymus nucleic acid contains the bases adenine, guanine, cytosine, and thymine and that yeast nucleic acid contains the bases adenine, guanine, cytosine, and uracil. He also isolated, from various nucleic acids, sugar-base fragments (called *nucleosides*) and base-sugar-phosphate groups (called *nucleotides*). In the 1920's he developed a theory that nucleic acids are linear chains of purines and pyrimidines joined to one another by means of the sugar-phosphate group. He took the simplest route and proposed that nucleic acids are composed of repeated sets of four nucleotides (the *tetranucleotide hypothesis*). Since this monotonous sequence could convey little information, most scientists believed that proteins, with their more than twenty amino acid building blocks, were more likely than nucleic acids to be the purveyors of the cell's genetic messages.

By World War II, scientists knew that DNA was an extremely long molecule, but most continued to believe that proteins were the sole carriers of genetic information and that nucleic acids played, at the most, a facilitating role. That DNA was the true source of genetic information was shown by Oswald T. Avery and his coworkers at the Rockefeller Institute for Medical Research. In a laboratory near Levene's, they discovered that protein-free DNA carried genetic information from a virulent form of *Pneumococcus* bacterial cells to a nonvirulent form. Avery stated that DNA is not merely structurally important but is also a functionally active substance in determining a cell's specific characteristics and biochemical activity. His discovery, which was published in a 1944 paper, greatly surprised those geneticists and biochemists who had long believed that genes, in order to perform their complex tasks, had to be made of proteins. Many of them continued to believe that protein contaminants in Avery's experiments must be the transforming agents, not the DNA itself.

The experiments of Alfred D. Hershey and Martha Chase in 1952 finally convinced most scientists that DNA, not protein, is the genetic material. They experimented with a *bacteriophage* (a virus that infects bacteria). The outer surface of a bacteriophage, or phage, consists of protein, but DNA exists within its head. Phages were known to multiply, in part, by injecting their genetic material into bacterial cells.

Hershey and Chase prepared two populations of phages. They labeled the phage protein in one group with a radioactive isotope. In the other group, it was the phage DNA that they radioactively labeled. The scientists then allowed the phages to attack cells of *Escherichia coli*, a colon bacteria. Afterward, their analysis showed that the phage protein had remained outside the bacteria, but the phage DNA had been injected into the bacteria. This research showed that the phages' genetic material was in the DNA rather than in the protein.

Research into DNA Structure

Only by understanding how covalent bonds link together the atoms of DNA and then by establishing their three-dimensional arrangement in space could scientists learn how this molecule carries genetic information. The British organic chemist Alexander Todd decided to clarify the chemical bonding of the nucleic acids by starting with the simplest units, the nucleosides. By the early 1950's he was able to show that the nucleic acids are linear, rather than branched, polymers and to specify exactly how the sugar ring (ribose or deoxyribose) is bonded to the various bases and to the phosphate group.

During the 1940's, while Todd was working out the detailed bonding of the nucleic acids, Erwin Chargaff, a biochemist at the College of Physicians and Surgeons in New York, was exploring the chemical differences in the base compositions of DNA from different sources. By 1950 his careful analyses of the base compositions of DNA samples from many plants and animals had revealed that the compositions varied widely, but his data also yielded the significant result that the ratios of adenine to thymine, and guanine to cytosine, were always close to one. Chargaff's work disproved the theory that DNA was made of repeating tetranucleotide units, but he did not attach any meaning to the one-to-one base ratios (now known as *Chargaff's rules*) because, along with other scientists, he continued to think of DNA as a single polynucleotide chain.

Linus Pauling, along with many other American chemists, was slow to accept DNA as the genetic material. Beginning in the 1930's, Pauling had done important work on hemoglobin and the structure of proteins. In 1948 he discovered the structure of the protein alpha keratin by using only a pen, a ruler, and a piece of paper: the alpha helix held together in its twisting turns by hydrogen bonds.

Moving Toward a Helical Model

In the early 1950's James Watson and Francis Crick, both of whom had become interested in DNA, were deeply impressed by Pauling's work, not so much because of the helical structure he had discovered ("helices were in the air," according to Crick) but because Pauling had the correct approach to biological problems. He believed that the chemists' knowledge of atomic sizes, bond distances, and bond angles would allow biologists to build accurate models of the three-dimensional structures of the complex molecules in living things. Crick, then in his mid-thirties, was a physicist working on the theory of the X-ray diffraction of proteins. Watson, in his early twenties, was a biologist whose interest in genes had led him to the Cavendish Laboratory of Cambridge University. At first glance, Watson and Crick made an odd team: Crick had expertise in crystallography and Watson in genetics, but neither had much knowledge of chemistry. Nevertheless, they were determined to discover the structure of DNA.

The work that ultimately led to the formulation of DNA's three-dimensional structure was performed by three different groups: Pauling and Robert Corey at the California Institute of Technology; Maurice Wilkins and Rosalind Franklin in John Randall's laboratory at King's College, London; and Watson and Crick at the Cavendish Laboratory. Particularly important were Franklin's X-ray studies of DNA. She showed that DNA existed in two forms, each with a distinctive X-ray picture: The wetter form (called the *B form*) gave an X-shaped pattern, and the slightly drier form (the *A form*) gave a more detailed array of spots. From these X-ray pictures, the workers in Randall's laboratory found that DNA had two periodicities in its B form: a major one of 3.4 angstroms and a secondary one of 34 angstroms. Gradually, their studies also began to suggest that DNA had a helical structure.

Watson and Crick decided to approach DNA's structure using Pauling's model-building approach. It was known that DNA's chemical formula, with its many single bonds that allowed free rotation of various groups, meant that the molecule could assume many different configurations. In their early research, Watson and Crick had to make some assumptions. For example, they assumed that the polynucleotide chain was coiled in a helix, somewhat like a vine winding around a cylinder. They reasoned that the different sizes and shapes of the bases probably led to their irregular sequence along the chain, whereas the orderly sugar-phosphate backbone must be responsible for the regular molecular features predicted from the X-ray photographs. Because of the high density of DNA, Watson and Crick proposed, as their first model, a triple helix. The three polynucleotide strands were arranged with the sugar-phosphate groups on the inside of the molecule and the bases on the outside. Wilkins, Franklin, and others came to Cambridge to see this model, but they thought it was ridiculous. Watson and Crick had understood neither the X-ray data nor the water content of the various forms of DNA.

In late 1952 Pauling, relying on poor X-ray photographs of DNA and faulty information about the density of its two forms, also proposed a triple helix. He had decided that packing the extremely bulky bases in the center of the molecule would be too difficult, and so, like Watson and Crick, he situated them on the periphery of his three-strand model. Before this work, Pauling had speculated, in some of his papers and talks, that the genetic material in living things had to be composed of two complementary strands. In studying the actual density determinations and X-ray photographs, however, he was led to the conclusion that this DNA, the product of laboratory manipulations, must be a triple helix.

Watson and Crick quickly abandoned their triple helix. Their search for another model of DNA was aided by a visit to Cambridge in the summer of 1952 by Erwin Chargaff. After talking with them, he was shocked that neither had precise knowledge of the chemical differences among the four bases. He explained to them his findings about the base ratios, which eventually suggested to them that these bases might be paired in the DNA structure. Watson, in fact, started manipulating paper models of the bases, in an attempt to mate them, but he did not understand which base forms actually existed in DNA. (Chemists had found that these nitrogenous bases could exist, depending on conditions, in two forms, the keto and the enol.)

Fortunately, an American crystallographer and Pauling protégé, Jerry Donohue, had an excellent knowledge of these base forms, which he communicated to Watson, who was still mired in like-with-like schemes. At Donohue's insistence, Watson began matching adenine with thymine and guanine with cytosine. This proved to be the key to the structure, for when these base pairs were joined with hydrogen bonds, their combinations were almost identical in size and shape. Thus, these base pairs could be fit into the interior of the helix, whereas the regular sugar-phosphate backbone at the molecule's exterior could account for its acidity and its interactions with water.

Watson-Crick Model

In the Watson-Crick model, DNA's structure consists of two strands, which may contain several million nucleotide units, that run in opposite directions (because of the asymmetry of the sugar-phosphate linkage). Although the hydrogen bonds that hold the chains together are much weaker than normal chemical bonds, they are so numerous that the union is tight. The two strands are not straight; rather, they wind around an imaginary axis in a helical fashion, as if wrapped around a cylinder. The sugar-phosphate groups form the coil, with their attached bases extending inward toward the axis of the helix.

In this double helix (which represents the B form of DNA), a single turn repeats itself every ten base

California Institute of Technology

Martha Chase, whose experimental work with Alfred D. Hershey in 1952 finally convinced most scientists that DNA, not protein, is the genetic material of all organisms.

pairs. More specifically, the space between one base pair and the next is 3.4 angstroms, and the helix makes one complete turn every 34 angstroms. Scientists later determined that the A form has eleven base pairs per turn.

The double helix is a right-hand helix, which means that an observer looking down the axis of the helix sees each strand, as it goes away from him, moving in a clockwise direction. This helix, which is about 20 angstroms in diameter, has a deep groove on its surface, winding parallel to a much shallower groove. The structure is stabilized by the hydrogen bonding of the complementary base pairs. The purine adenine (A) is always hydrogen-bonded to the pyrimidine thymine (T), while the purine guanine (G) is always hydrogen-bonded to the pyrimidine cytosine (C).

Because of this base-pairing scheme, each chain's sequence determines its partner's sequence. For example, if one strand has the first following sequence, then the other strand would have the second sequence.

TACGCAT

ATGCGTA

The complementary nature of these opposing sequences accounts for Chargaff's rules and, as Watson and Crick noted in their famous 1953 paper in *Nature*, provides a natural explanation for the replication of the molecule. Because of the specific phosphodiester bonding between the sugars, each strand has a top and bottom end, making the two strands antiparallel—that is, positioned relative to each other like two swimmers heading in opposite directions.

In the decades after the discovery of the double helix, scientists found that DNA could exist in structures other than the B-form double helix. For example, the chromosomes of some small viruses have single-stranded DNA molecules. Furthermore, some DNA molecules from living organisms turned out to be circular, which means that the polynucleotide chain formed a closed loop. In fact, the DNA molecules of some viruses interconvert between linear and circular forms, the linear being present inside the virus particle, the circular form existing in the host cell. While studying the conversion of the linear to the circular form of DNA, Jerome Vinograd, a physical chemist, discovered that the axis of the double helix can be twisted to form a superhelix, which can be right- or left-handed.

In the late 1970's and early 1980's, Alexander Rich at the Massachusetts Institute of Technology discovered a left-handed double helix containing about twelve nucleotides per turn. Because the backbones follow a zigzag path, this left-handed structure was called Z-DNA. Instead of the major and minor grooves of B-DNA, only a single groove winds around Z-DNA. Scientists have found the Z form of DNA in nature, specifically in regions of the DNA molecule where G and C bases form an extended, alternating sequence.

Implications of DNA Structure

DNA has been called the master key of life, and the discovery of its structure has been called the most influential event in modern biology. Watson and Crick's discovery of the double helix led to an avalanche of significant discoveries in molecular biology. The most important feature of their model was that it suggested how DNA might produce an exact copy of itself.

Because the model consists of two chains, each the complement of the other, either chain may act as a mold or template on which a complementary chain can be synthesized. As the two chains of DNA unwind and separate, each strand begins to construct a new complement onto itself. When the process is finished, two pairs of double helices exist where there was one before. Thus, the Watson-Crick model helped lead to an explanation, at the molecular level, of how genes duplicate themselves with great fidelity.

The Watson-Crick model also contributed to the solution of the pivotal problem of genetics: how genes control the making of proteins in cells. Though this problem was not solved directly by the double helix, the study of the interactions between DNA and the proteins it indirectly produces revealed that its structure contains, in the sequences of its bases, an intricate code that supervises the construction of the thousands of proteins used by cells. In working out this genetic code, scientists discovered little similarity between DNA's three-nucleotide sequences that code for particular amino acids and the structures of the amino acids themselves. Indeed, the transfer of information from DNA to the cellular mechanisms overseeing protein construction turned out to be quite complex, involving various types of RNA.

Watson and Crick, the originators of the revolution in molecular biology, also played important parts in its evolution. Their work was the catalyst for the development of its "central dogma," which explains how genetic information passes from DNA to the cell's proteins. In the first part of the process, DNA serves as the template for its *self-replication* (duplication). Then, RNA molecules are made on a DNA template (*transcription*). Finally, RNA templates determine all proteins (*translation*). Although DNA can therefore act on itself, the processes of transcription and translation were seen as moving in one direction only: DNA sequences are never made on protein templates, and DNA sequences are never made on RNA templates. In the course of time, some exceptions to these rules have been discovered, but the central dogma remains essentially valid, as does the double helix. Watson and Crick, along with many other scientists, solved, by means of ideas drawn from physics and chemistry, the basic mysteries of the gene.

Robert J. Paradowski

See also: Chromosomes; DNA in plants; DNA replication; Genetic code; Nucleic acids.

Sources for Further Study

Crick, Francis. *What Mad Pursuit: A Personal View of Scientific Discovery.* New York: Basic Books, 1988. This memoir focuses on Crick's experiences before, during, and after the discovery of the double helix. Autobiographical account centers on the fascinating combination of choice and chance that led to the epochal discovery of DNA's structure and how the genetic code was broken.

Echols, Harrison. *Operators and Promoters: The Story of Molecular Biology and Its Creators.* Berkeley: University of California Press, 2001. Intellectual history of the science of molecular biology, based on interviews with scientists and the author's own experiences as a molecular biologist. Offers unusual insight into the process of discovery and its dependence upon the researcher's personality.

Judson, Horace Freeland. *The Eighth Day of Creation: The Makers of Revolution in Biology.* Expanded ed. Plainview, N.Y.: Cold Spring Harbor Laboratory, 1996. Account of how DNA's structure was determined and how the genetic code was worked out. Based on more than one hundred interviews with the scientists involved. Communicates both a sense of wonder about scientific discovery and a clear understanding of many of the scientific ideas behind the revolution in biology.

Olby, Robert. *The Path to the Double Helix.* Reprint. New York: Dover, 1994. The first detailed account of the intellectual and institutional evolution in biology and allied sciences that led to a chemical understanding of the gene. Explores how the concept of the macromolecular was developed, how the nucleic acids came to be seen as the hereditary material, and how the double helix was discovered.

Sayre, Anne. *Rosalind Franklin and DNA.* New York: W. W. Norton, 2000. A biography of Rosalind Franklin, a colleague of James Watson and Francis Crick whose contributions to the discovery of DNA structure was long overlooked, not only by historians but also by Watson and Crick themselves.

Strathern, Paul. *Crick, Watson, and DNA.* New York: Anchor Books, 1999. An account of the discovery of the structure of DNA, amusingly written, which includes a concise primer of modern genetics.

Watson, James D. *The Double Helix: A Personal Account of the Discovery of the Structure of DNA.* Reprint. New York: Simon & Schuster, 1998. A remarkable account of the discovery of the structure of DNA. Based on Watson's memories of the discovery, written down in 1951-1953, while the reality of it was still fresh in his mind. A classic "warts and all" story of the scientific process.

DNA IN PLANTS

Categories: Cellular biology; genetics

DNA is the hereditary or genetic material, present in all cells, that carries information for the structure and function of living things.

In the plant kingdom, DNA, or deoxyribonucleic acid, is contained within the membrane-bound cell structures of the nucleus, mitochondria, and chloroplasts. DNA has several properties that are unique among chemical molecules. It is universal to all living organisms, having the same structure and function in each. It is capable of reproducing itself in a process known as self-replication. This property allows cell division, and thus continuity, growth, and repair. It carries in its structure the genetic code, or set of instructions, for cellular development and maintenance. Finally, it undergoes changes in chemical structure, from both environmental and internal causes, called mutations, which contribute to evolution, diversity, and disease.

Chemical Structure

DNA is a simple molecule, consisting of four nucleotides. Each nucleotide has a five-carbon sugar (deoxyribose), a phosphate, and one of four possible nitrogenous bases: the double-ringed purines of *adenine* (A) and *guanine* (G) and the single-ringed pyrimidines of *thymine* (T) and *cytosine* (C). Most of the properties of DNA relate to the unique bonds that form among the nucleotides: The sugar-phosphate components align themselves linearly, while the nitrogen rings bond perpendicularly. The nitrogen rings further bond in a very specific fashion: A always pairs with T, and G always pairs with C. The DNA molecule thus appears as a ladder, the sides being sugar-phosphate; the rungs, the A-T and G-C pairs.

Further bonding and folding produces a structure shaped like a spiral ladder, known as a *double helix*. This double helix is compactly packaged in ropelike structures known as *chromosomes*, which are visible under a light microscope before and during cell division. During the daily life of the cell, the DNA appears as an indistinguishable dark mass call chromatin (an inclusive term referring to DNA and the proteins that bind to it, located in the nuclei of eukaryotic cells).

The Watson and Crick Model

In the mid-1800's Gregor Mendel, an Austrian monk, postulated that genetic material existed. He discovered the laws of heredity using the pea and other plants in his garden to study the inheritance of such traits as flower color. Nearly seventy years passed before scientists James Watson and Francis Crick, in 1953, proposed the double helix as the most plausible model for each of the unique properties of the molecule. Their model was verified by X-ray diffraction techniques soon afterward.

Several researchers, working at Columbia University and elsewhere in the United States, had led the way prior to Watson and Crick by discovering the chemical composition of this genetic material and the nitrogenous base pairing: The amount of adenine always equaled that of thymine and likewise with guanine and cytosine. The Watson and Crick model also suggested that the two sides, or strands, of DNA run in opposite directions: That is, the phosphate sugar of one side points upward, while the other strand points downward. This property is known as *antiparallel bonding*. The Watson and Crick model could easily explain how DNA replicates during cell division and how genetic information is encoded in its structure.

Self-Replication

Self-replication, which allows the continuity of generations and the growth and repair of individual organisms, occurs during cell division. DNA must be able to produce exact copies of itself. The molecule is uniquely designed for this: A series of enzyme-mediated steps allows the double helix to unwind or unzip, like a zipper, separating the two strands. Next, nucleotides from digested food enter first the cell and then the nucleus. They bind to a corresponding nucleotide: A with T and G with C. The process continues until two new double-stranded molecules of DNA have formed, each a copy of the other and each going into the new cells that resulted from cell division.

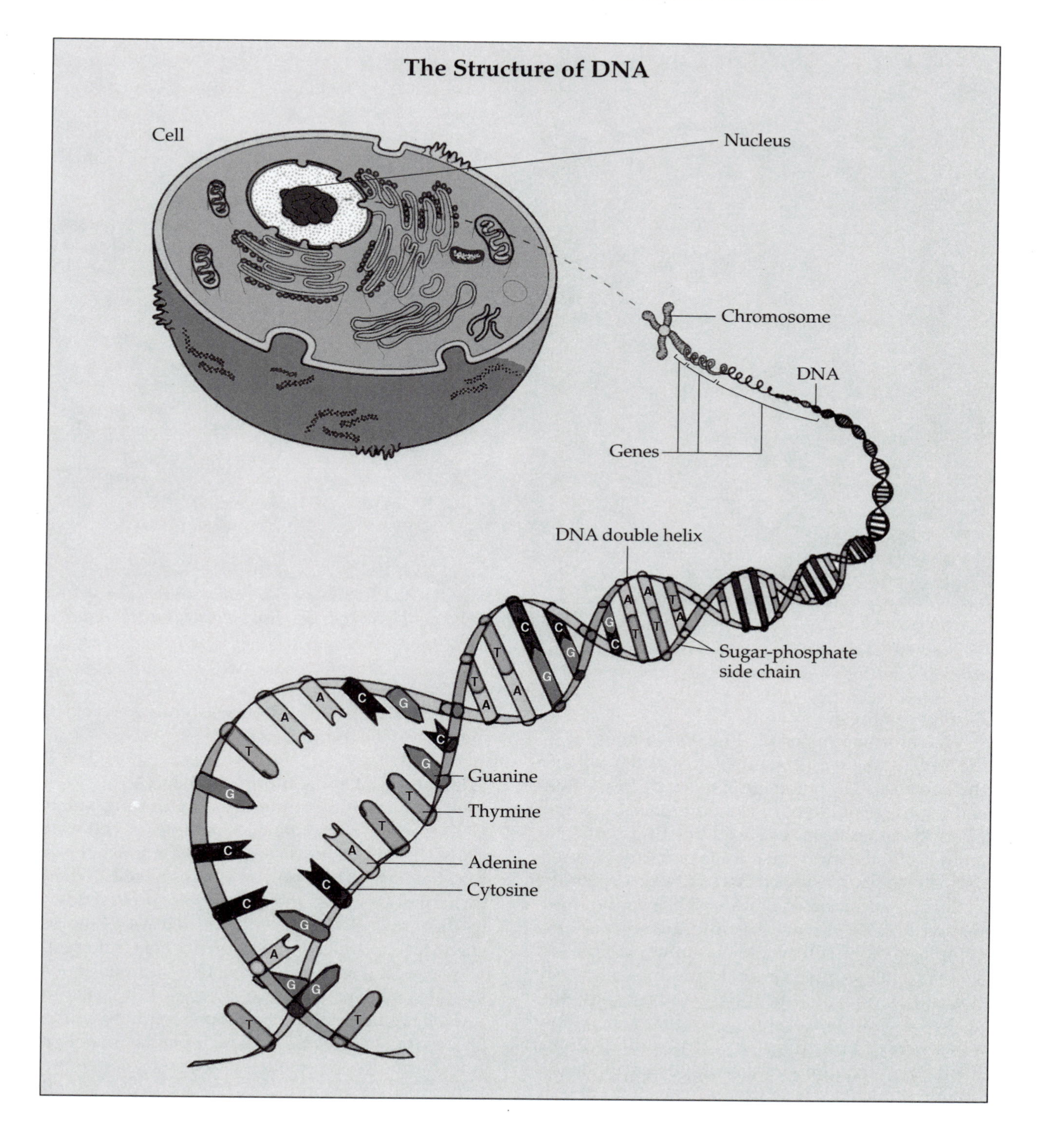

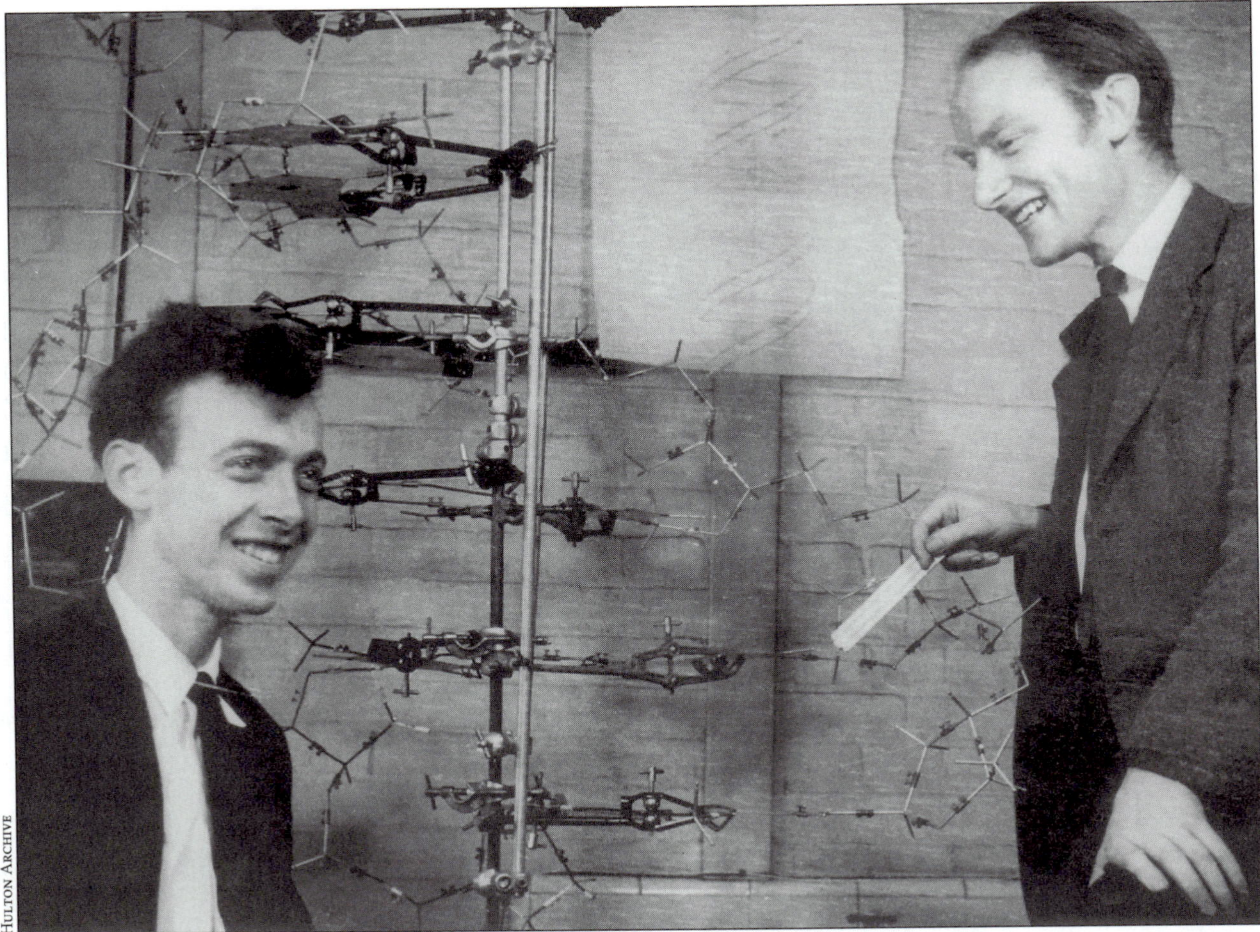

Scientists James Watson, left, and Francis Crick in 1953 proposed the double helix as the most plausible model for each of the unique properties of the DNA molecule.

Protein Synthesis

The information encoded in DNA allows for all the development and maintenance of the cell and the organism. The language of this code lies in a linear reading of adjacent nucleotides on each strand. Every three nucleotides specify or fit a particular amino acid, the individual units of proteins. A second molecule, *ribonucleic acid* (RNA), copies the molecular structure of DNA and brings the information outside the nucleus into the surrounding cytoplasm of the cell, where the amino acids are assembled, in specified order, to produce a protein. Postproduction modifications of these proteins, such as the addition of sugars, fats, or metals, allow a vast array of functional and structural diversity. Plant DNA codes for a variety of substances that are unique to plants. These products sustain not only the plants themselves but also entire ecological niches, as well as humankind.

Mitochondrial and Chloroplastic DNA

A second, independently functioning set of DNA exists in two organelles outside the cell's nucleus, the mitochondria and the chloroplast. It is in the *mitochondria*, the power sources of cells, where carbohydrates, fats, and proteins are broken down to their raw elements with the release of stored chemical bond energy in the form of heat (calories).

The second region in which DNA is housed outside the nucleus is in the *chloroplast*, a structure unique to plant cells. In chloroplasts, photosynthesis occurs, the process by which plants are able to transform carbon dioxide, water, and solar energy to produce sugars and, later, fats and proteins, with the release of oxygen. This critical process undertaken by plants sustains most life on earth.

Both mitochondrial and chloroplastic DNA replicate separately from nuclear DNA during cell division. It is postulated that these organelles once, billions of years ago, may have been independently living organisms that were incorporated into other cells to form the eukaryotic cells that make up nonbacterial life-forms such as fungi, protists, plants, and animals.

Plant Proteins

A large array of proteins that are unique to plants are encoded on plant DNA. A group that has received much attention are the so-called *phytochemicals*, substances with powerful health benefits. Well-studied classes are few, including the flavonoids, phytosterols, carotenoids, indoles, coumarins, organosulfurs, terpenes, saponins, lignans, and isothiocyanates. Each group contains specific proteins that are both antioxidants and anticarcinogens—protecting animal cells from cancer-causing agents. The carotenoids, such as beta-carotene, found in orange and yellow fruits and vegetables, and lycopenes, found in tomatoes, appear to protect animals against heart disease and stroke as well as cancer. The phytosterols, like those found in soybeans, are estrogenlike compounds that mimic female hormones. These appear to protect female organs from cancers and also appear to lower cholesterol.

Plant Hormones

Large segments of plant DNA are devoted to coding for specialized plant hormones. Hormones are substances that are produced by one group of cells, circulate to another site, and affect the DNA of the target cells. In plants, these hormones control cell division, growth, and differentiation.

There are five well-described classes of plant hormones: the auxins, gibberellins, cytokinins, ethylene, and abscisic acid. Among the auxins' functions is allowing phototropism, the property that makes plants bend toward the light. Produced in the roots, auxins travel to stems, making cells elongate on the dark side of plant tissue. Ethylene is a gaseous substance that ripens fruits and causes them to drop from the plant. Abscisic acid contributes to the aging and falling of leaves.

Genetically Modified Plants

Because plants are easy to manipulate, plant DNA is second only to bacterial DNA as a primary experimental subject for bioengineers. The direct modification of DNA by adding or removing a particular segment of genes that code for specific traits is the focus of bioengineering and biotechnology. Because plants provide the major food source for human and livestock populations, *genetically modified foods* have been developed that resist insects, bacteria, viruses, and other pests and decrease the need for external pesticides.

Genetically modified plant crops are designed to enhance a variety of characteristics, from looking and tasting good to growing faster or ripening more slowly to having no seeds. The introduction of genes from other kingdoms, such as the animal kingdom, into plant DNA is allowing scientists to develop future crops that may contain human vaccines, human hormones, and other pharmaceutical products. A tomato was the first federally approved bioengineered food to be sold in the United States. Today, dozens of produce items and livestock feed are in some way genetically modified.

Connie Rizzo

See also: Biotechnology; Chloroplast DNA; Chromatin; Chromosomes; DNA: recombinant technology; DNA replication; Genetic code; Genetically modified bacteria; Genetically modified foods; Genetics: mutations; Hormones; Mitochondrial DNA; Nucleic acids; Nucleus; Plant biotechnology; Proteins and amino acids; RNA.

Sources for Further Study

Jenkins, Morton. *Genetics*. Lincolnwood, Ill.: NTC/Contemporary Publishing, 1999. Along with textbook (below) by Sylvia Mader, this is an outstanding survey of the structure and function of DNA.

Mader, Sylvia. *Biology*. 7th ed. New York: McGraw-Hill, 2001. An introductory college-level biology text with an excellent section on DNA and its structure and function.

Raven, Peter H., Ray F. Evert, and Susan E. Eichhorn. *Biology of Plants*. 6th ed. New York:

W. H. Freeman/Worth, 1999. This well-known introductory college-level text offers good chapters on DNA, genetics, plant heredity, and unique plant products.

Yount, Lisa. *Genetics and Genetic Engineering*. New York: Facts on File, 1999. Excellent chronological explanation of the geneticists who discovered DNA, as well as DNA's structure, function, and future.

DNA: RECOMBINANT TECHNOLOGY

Categories: Biotechnology; cellular biology; economic botany and plant uses; genetics

Recombinant DNA technology makes use of science's understanding of the molecular structure of DNA, the nucleic acid that encodes genetic information, to alter DNA in order to manipulate genetic traits. Such technology has immense implications for agriculture, horticulture, and the generation of medicinal compounds from plants.

Recombinant DNA technology has been essential for understanding DNA sequences. Because of their large, complex genomes, it was difficult to study one gene in eukaryotes, but recombinant DNA technology has allowed the isolation and amplification of specific DNA fragments facilitating the molecular analysis of genes. In addition, the tools of recombinant DNA technology have been used to create genetically modified plants. Such modifications include the introduction of resistance to insects, herbicides, viruses, and bacterial and fungal diseases into plants. Plants have also been made to produce antibodies so that plants can serve as edible vaccines.

DNA Structure

Organisms contain two kinds of nucleic acids: ribonucleic acid, or RNA, and deoxyribonucleic acid, or DNA. DNA is made of a double chain, or helix. The structure of one chain, or strand, is a backbone made up of repetitions of the same basic unit. That unit is a five-carbon sugar molecule called 2′-deoxyribose attached to a phosphate residue. RNA contains a ribose sugar instead. Also attached to the sugar part of the backbone are other molecules called bases. The four bases are adenine (A), guanine (G), cytosine (C), and thymine (T). DNA molecules are double strands that are held together because each base in one strand is paired to (hydrogen-bonds with) a base in the other strand. Adenine always pairs with the base thymine, and guanine always pairs with cytosine. A and T are called complementary bases. Likewise, G and C are complementary bases.

DNA is shaped much like a helical ladder, with the sugar and phosphate backbones being the sides of a ladder and the base pairs that hold the two strands together being the rungs of a ladder. DNA is often represented as a string of letters, with each letter representing a base. The order of A's, T's, G's, and C's (the rungs of the ladder) along a DNA double helix is the sequence of that DNA which contains the genetic information.

Restriction Enzymes

In recombinant DNA technology, scientists are able to use molecules called *restriction enzymes* to make cuts at specific sequences. Some restriction enzymes make cuts straight across the two strands of DNA in the double helix, creating blunt ends. Other restriction enzymes cut the two strands in a staggered pattern, leaving short, specific single strands at the cut sites. These single-stranded regions, called *sticky ends* or cohesive ends, can base-pair (hydrogen-bond) with complementary base sequences from other, similarly cut DNAs. These sticky ends allow joining of DNA from any source cut with restriction enzymes that create the same ends.

Cutting with restriction enzymes creates fragments of DNA with sequence-specific ends that can be spliced into small, self-replicating vehicle, or vector, molecules and introduced into a host cell where the vector molecules with the added DNA

AP/Wide World Photos

Researchers Arun Goyal, left, and Neil Nelson hold tissue samples from a hybrid poplar tree that were made using recombinant DNA technology. Their work led to production of an enzyme which could replace chlorine in the pulp-bleaching process, reducing environmental costs of papermaking.

fragments replicate to produce a large amount of specific DNA for analyses. This process is *recombinant DNA cloning*.

DNA Cloning

One way to clone a specific gene is to clone all the DNA fragments generated from cutting with a restriction enzyme and then screen for the clone containing the desired gene. This method of cloning random DNA segments into a vector is called *shotgunning*. The entire collection of such cloned fragments, which together represent the entire genome of the organism, is called a gene library. Genomic DNA libraries are made by cloning the total genomic DNA of an organism.

Another way to clone a specific gene is to begin with messenger RNA (mRNA) from the organism. (Messenger RNA is a molecule that functions to create complementary copies of DNA strands. At a ribosome the messenger RNA then determine the order of amino acids that are joined to make a protein.) Using *reverse transcriptase* (an enzyme encoded by some RNA viruses that uses RNA as a template for DNA synthesis), scientists can make a DNA copy of the mRNA. The complementary strand is also synthesized to create a double-stranded DNA called cDNA (complementary DNA) that is complementary to the mRNA. These cDNAs are then cloned to create a complementary DNA library. Individual cloned cDNAs can be used to trap the corresponding mRNA on a nitrocellulose filter. At this point, the mRNA can be used in a cell-free protein synthesis system to allow identification of the protein encoded by that cDNA clone. Alternatively, the cDNA can be used to find sequences complementary to it in a genomic library to obtain a clone of the specific gene.

Nucleic Acid Hybridization

The ability to hybridize nucleic acids to find sequences complementary to a particular DNA is another essential tool that offers another way to identify cloned genes. This method is called *nucleic acid hybridization* or Southern blotting (named after E. M. Southern, who developed the method). In this procedure, DNA is cut with restriction endonucleases, and the resulting DNA fragments are separated by size using agarose gel electrophoresis. The DNA in the gel is denatured (made single-stranded) by high pH and transferred to a nitrocellulose filter. The DNAs are immobilized on the nitrocellulose in the same pattern as on the gel (a Southern blot). Aprobe—a specific DNA or RNA—is hybridized to the nitrocellulose. The probe is "labeled" with a radioactive or fluorescent (nonradioactive) tag so it can be detected. The probe is denatured by heat so it is single-stranded and able to anneal (hybridize) with its complementary sequence among the single-stranded DNAs tethered to the nitrocellulose. The probe is then detected to reveal the position of the DNAs that hybridized with the probe.

Polymerase Chain Reactions

Another tool of molecular biology is to use a *polymerase chain reaction* (PCR) to amplify specific segments of DNA in vitro (in the test tube). PCR requires a pair of sequences, called primers, about twenty base pairs long that are complementary to the ends of the region of DNA to be amplified. High temperature is used to denature the double-stranded DNA. At a lower temperature, the primers anneal (base-pair) to their complementary sequences, and a thermal-stable DNA polymerase copies the single-stranded templates. After the replication of the segment between the two primers (one cycle), the newly synthesized double-stranded DNA molecules are denatured by high temperature, the temperature is lowered, primers anneal, and a second cycle of replication occurs. The number of DNA molecules produced doubles with each cycle of replication. As a result, a million copies of a single DNA molecule can be produced in only a few hours, if the appropriate sequences for the two primers are known. PCR is a very sensitive method: Even a single DNA molecule can be amplified. PCR is much faster than recombinant DNA cloning and can produce a large amount of a specific piece of DNA.

Developing ways to determine DNA sequences (the sequences of adenine, thymine, guanine, and cytosine base pairs on the DNA "rungs") has led to the identification of the complete DNA sequences of the genomes of a number of organisms, including the model plant *Arabidopsis* as well as the much more complex human genome.

Susan J. Karcher

See also: Biotechnology; Chromosomes; Cloning of plants; DNA in plants; DNA replication; Electrophoresis; Environmental biotechnology; Gene regulation; Genetic code; Genetically modified bacteria; Genetically modified foods; Genetics: mutations; Genetics: post-Mendelian; Plant biotechnology; RNA.

Sources for Further Study

Hill, Walter E. *Genetic Engineering: A Primer.* The Netherlands: Harwood Academic Publishers, 2000. Provides a concise overview. Illustrations, glossary, appendix, and index.

Kreuzer, Helen, and Adrianne Massey. *Recombinant DNA and Biotechnology: A Guide for Students.* Oxford: Blackwell Science, 2000. Introductory overview. Illustrations and index.

Old, R. W., and S. B. Primrose. *Principles of Gene Manipulation: An Introduction to Genetic Engineering.* Boston: Blackwell Scientific, 1994. Focuses on methods of cloning. Illustrations, appendices, references, and index.

Reece, Jane B., and Neil A. Campbell. *Biology.* 6th ed. Menlo Park, Calif.: Benjamin Cummings, 2002. Introductory textbook. Illustrations, problems sets, glossary, index.

Watson, James, Michael Gilman, Jan Witkowski, and Mark Zoller. *Recombinant DNA.* 2d ed. New York: W. H. Freeman, 1992. Emphasizes applications. Illustrations, reading lists, and index.